CLASSIC CAMARO RESTORATION

Compiled by R.M. Clarke from articles published in the Petersens Automotive Performance Series

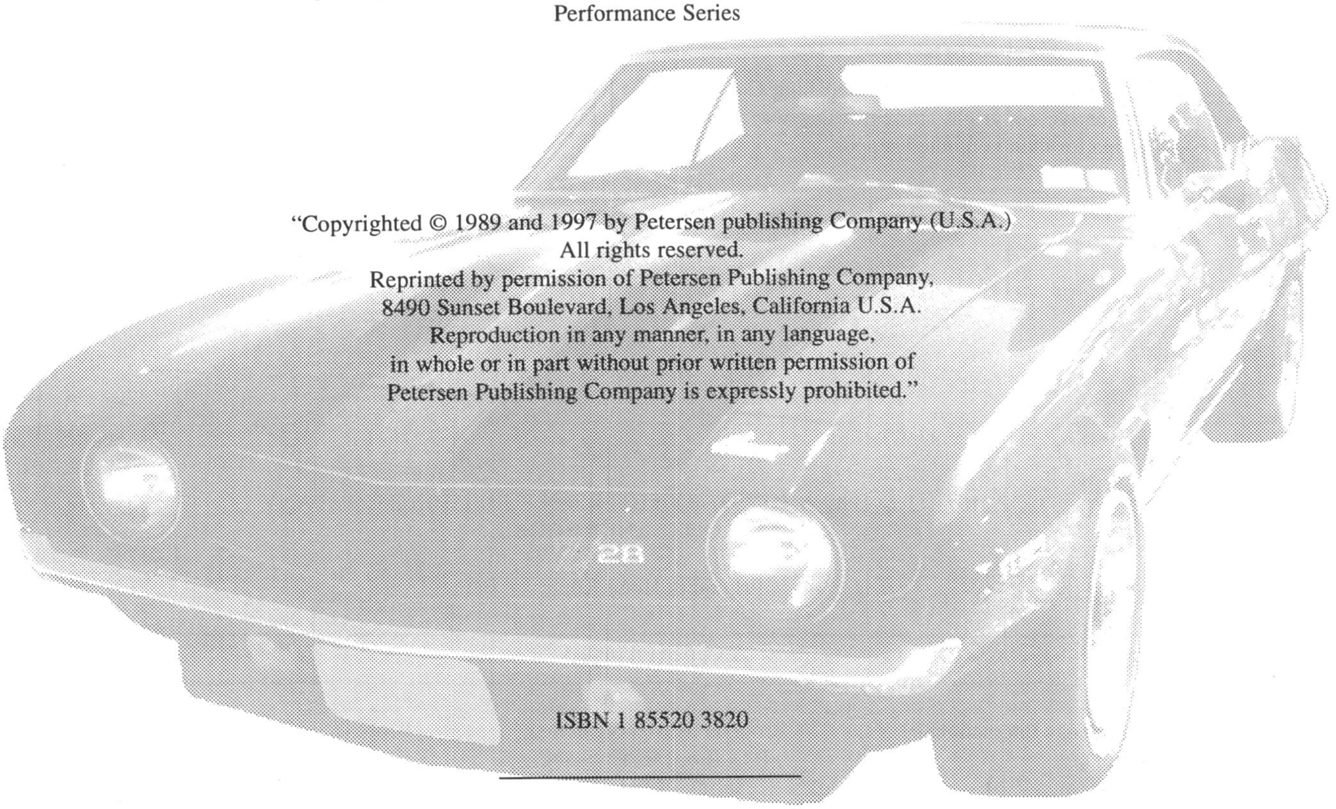

ISBN 1 85520 3820

Published by
Brooklands Books with permission of Petersen Publishing Company
Printed in Hong Kong

A-CAMCR

ACKNOWLEDGEMENTS

For over thirty years Camaros have been bought, driven, raced and enjoyed. The early models have become highly desirable classics and are now rebuilt, restored, beautified and cherished by a new generation of owners. During these three decades the motoring writers and editors of Petersens Automotive Performance magazines have brought us articles, reports and stories on how to maintain, tune, repair and modify these thoroughbred ponycars. This book is the distilled essence of their writings.

Our first book on the Camaro was *Camaro Restoration Tips and Techniques* which enjoyed great success during the early '90's. This book is an enlarged and updated version of our earlier title and contains even more 'How To' procedures and tips for owners of pre '82 Classic Camaros written by those who carried out and perfected these techniques.

Petersens magazines such as *Hot Rod, Car Craft* and *Circle Track* are famous for their easy to follow picture sequences and between these covers will be found over 700 drawings and photographs to help the less adventurous keep their cars in good order, and assist the more daring to tackle major restorations and improvements.

We are indebted to Petersen Publishing Company who for many years have allowed us to reissue their informative copyright stories in this way, and our thanks go in this instance to Lee Kelley and John Dianna for their continued support.

R.M.Clarke

Printed in Hong Kong.

DISTRIBUTED BY

Motorbooks International,
Osceola,
Wisconcin, 54020,
USA.

Brooklands Books Limited,
P.O. Box 146, Cobham,
Surrey, KT11 1LG,
England.

Brooklands Books Limited,
1/81 DarleyStreet,
Mona Vale, N.S.W. 2103,
Australia.

CONTENTS

CAMARO:
From A to Z-28

BY W. R. C. SHEDENHELM

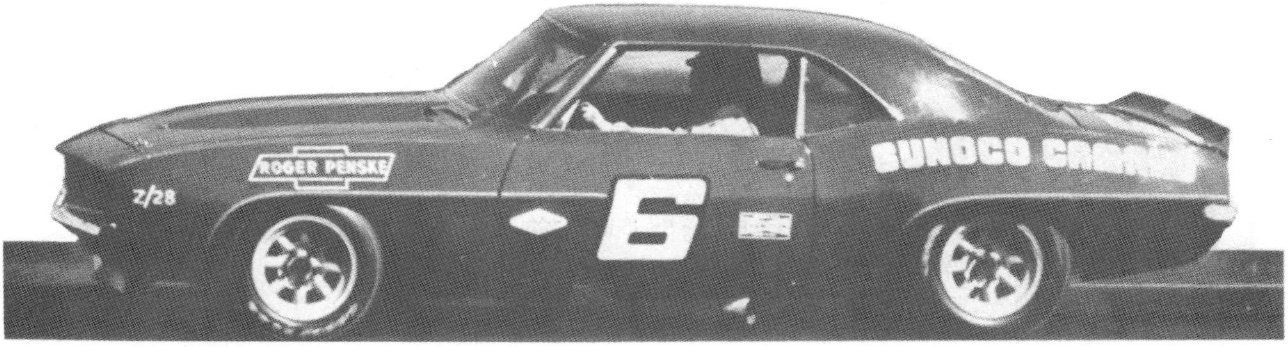

1

The tremendous and unexpected acceptance of Ford's Mustang shook the American automotive experts. With the swiftness of a glacier, Detroit changed its thinking on the youth market: the kids didn't want 2-seater open sports cars, they wanted 2-seater closed sports coupes and sedans. Lincoln-Mercury Division came out with its Cougar, and in 1967 Chevrolet announced its Camaro. The spur was the youth market's interest in the newly formed amateur and professional sedan road racing of the Sports Car Club of America. Suddenly, Detroit iron was running on the same winding road courses as the classy European machinery, and frequently with extremely competitive results. Perhaps the topper of this overworked foreign-vs.-domestic argument came with the running of the 1968 Sebring 24-hour endurance race. The Ford GT's had been wiping out the European cars at LeMans with amusing regularity, but Sebring was the ideal test for the powerful and nimble foreign machinery.

In 1968, Sebring promoter Alec Ulmann ran the Trans-Am sedans in with the Group 6 prototypes and the production sports cars. Everyone snickered. The Camaros and Mustangs would be lucky to finish ahead of the TR4's and MGB's. Twenty-four hours after the started dropped his flag, the overall finishing order was two Porsche Group 6 prototypes and then . . . Lord save us! . . . two SCCA-legal Chevrolet Camaros. The Europeans had many excuses but none of them were very convincing. It wasn't just one Camaro that placed well by some fluke. It was two of the darn things, running fast and true after the toughest 24-hour race in existence!

For the first three years, the Camaro stayed unchanged in looks—for all practical purposes—and was available both as a "sports coupe" sedan and as the seldom-seen convertible. Open cars just aren't selling like they used to; perhaps too many potential buyers now have long hair that would be bothersome in the wind. During the 1967-1969 period the Camaro was sold in three optional upgradings from the bare-bones base car: the Rally Sports (RS), Super Sports (SS), and the SCCA-legal Z-28.

The major differences are both in decorative trim items and in increasingly more stable suspension, plus higher-performance engines in the SS series. The Z-28 came only with the 302-cubic-inch engine, which in stock form cranked out some 50 or 60 more horses than the factory's stated 290.

Racing at the nation's drag strips, and particularly in the SCCA's road races, necessitated a long line of special-but-legal parts and Chevrolet has made them available, if a little difficult for the peppertree mechanic to obtain. There are "heavy-duty parts" (as Chevrolet prefers to call them; Ford would use the more honest term

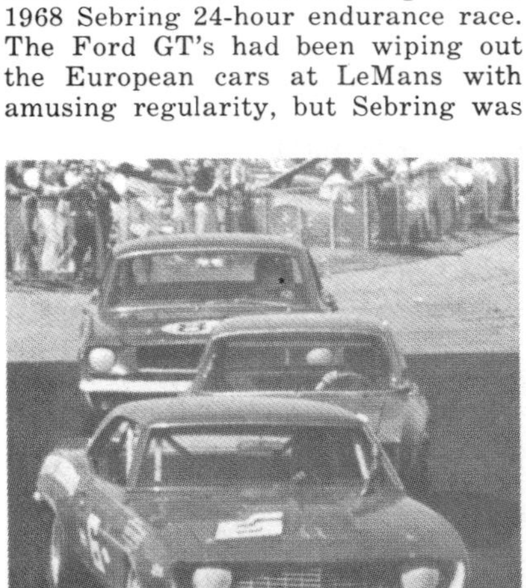
2

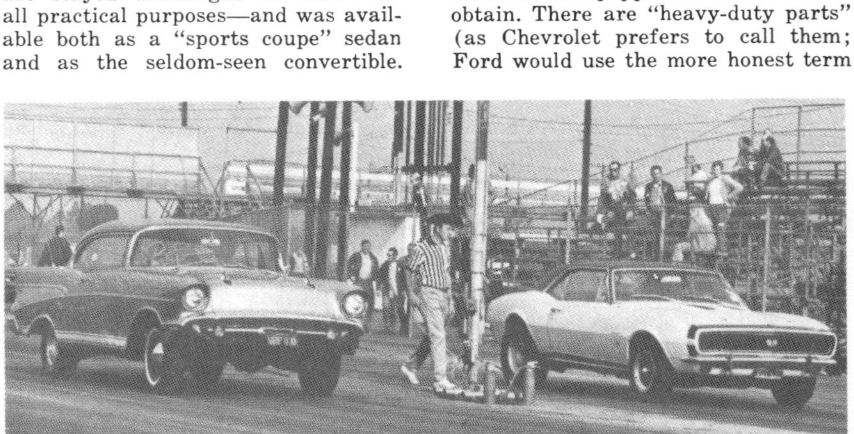
3

1. This is the one that did it. Mark Donohue puts the 1969 Trans-Am title in his hip pocket for Chevy at Sears Point, Calif. course.

2. Donohue again, leading the pack through turn 7 at Riverside in 1969.

3. Camaros are everywhere. Here one faces off with a '57 Chevy at Lions drag strip.

4. In the pits at Orange County International Raceway, there isn't much you can do once you're at the track, but changing carburetor jets sometimes helps.

5. A plug change and a little timing juggling might help the e.t. Do you suppose he is loosening the alternator belt? Every little bit helps.

6. This guy isn't getting much help changing that tire, but maybe they'll let him take the trophy home. Without good rubber, you may as well forget drag racing.

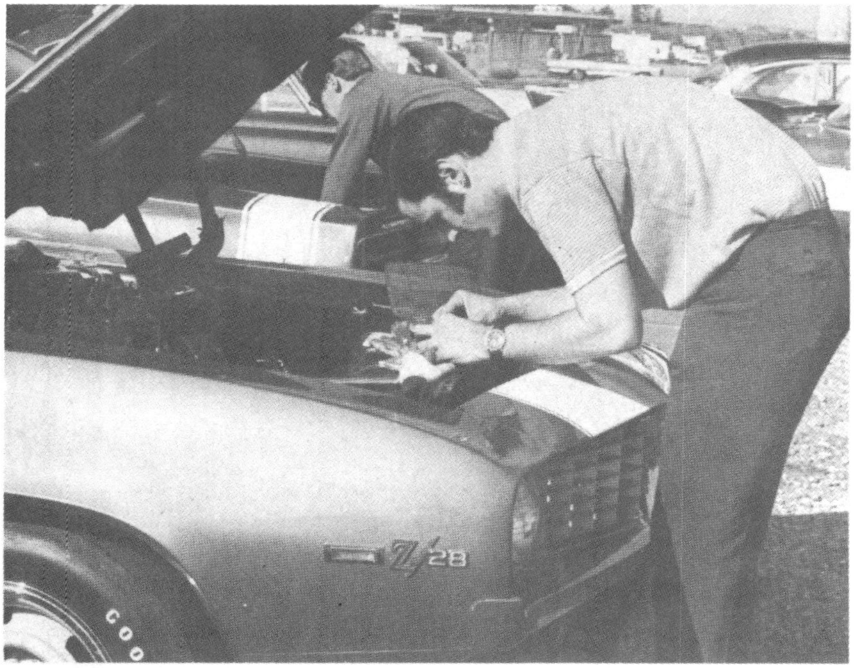

4

"competitive parts") to improve and alter the steering, front springs, sway bars, brakes, rear axle, ring-and-pinion and a few other components. In other words, you don't simply buy a Z-28 Camaro off the showroom floor for $4000 and go out and race it. Not and win anything. This is not to put the Camaro down, mind you. No one with any sense would take *any* car off the showroom floor and race it. However, the Camaro is, particularly in its Z-28 configuration, as close to being a competitive race car as just about anything you can buy on the general market. With the high-revving 302-cubic-inch engine, it feels like a race car from the second you fire it up and take off.

For 1970, Chevrolet has made quite a few (for Detroit) changes in the Camaro, including a 350-inch engine the normal Z-28 option. If you want a 302 in your '70 Camaro it's a do-it-yourself job, using the '69 crank and pistons. It's legal, too, because Trans-Am rules for '70 permit destroking.

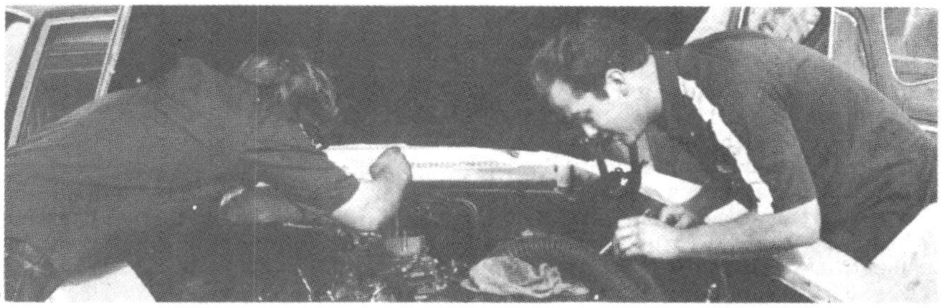

5

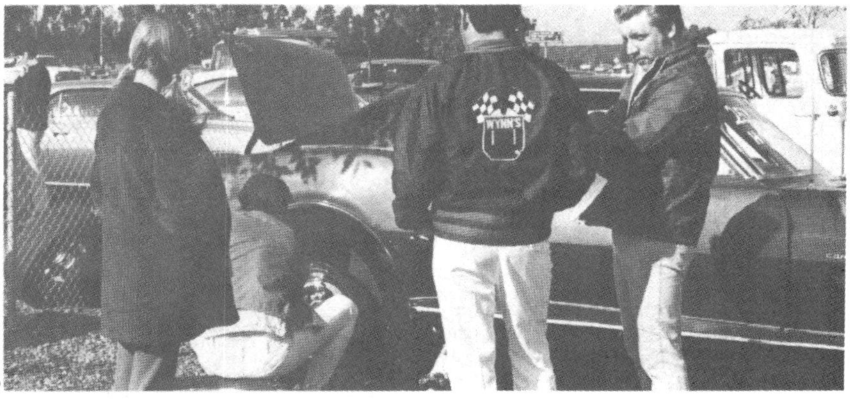

6

FRONT END

The 1970 Camaros use a newly designed front suspension of different dimensions than on the earlier models. Although it is still independent short-and-long-arm (SLA) suspension, the stress at the frame mounting points has been lessened by increasing the span of the upper and lower control arm bushings and the span between the steering knuckle ball joints. With these changes and others on the steering, parts from the 1970 and the earlier Camaros are not interchangeable. As a matter of fact, the whole front subframe is different for 1970, with larger box sections for side rails and a single, larger cross-member up front instead of the two formerly used. The spring seats and crossmember's skidplate are welded to the frame, as was standard racing practice with the earlier models. It was also normal practice, and allowed in the regulations, to reinforce the spring seats and the upper control arm mounting brackets.

SCCA-legal racing practice was to remove all the rubber body mounts and to substitute aluminum or steel spacers to eliminate flexing between the subframe and the unit body.

REAR SUSPENSION

The first Camaros, in 1967, used a single longitudinal leaf spring on each side of the axle tubes, and by the 1968 models this had been changed to a multi-leaf setup, with the shock absorbers biased. Mounting the right shock ahead of the axle with the left directly above the axle helps eliminate rear wheel lift during violent acceleration. The 1970 Camaros use lower-rate rear springs for a softer ride, but there are four heavy-duty spring assemblies available for racing use (plus six front assemblies with varying spring rates). The best combination for road racing has proven to be the 500 lb.-in. front coils with a ⅞, 15/16 or 1-inch sway bar, and the 250 lb.-in. rear leaves. For the courses with high banking, where the downward loadings are excessive, the

CAMARO

Camaro works best with 725 lb.-in. front coils, a ⅞-inch sway bar, and 300 lb.-in. rear leaves.

It should be noted that it is generally agreed that the ideal handling for a race car is to slightly understeer, or plow, with oversteer achievable with the application of power to the rear wheels. Chevy doesn't make any satisfactory racing shock absorbers, so you'll have to buy these from Monroe, Gabriel or Koni.

BRAKES

One of the severe strains that Trans-Am road racing put on all of the pony cars was on their brakes. It was quickly obvious that drum brakes all around just weren't adequate. On the average road course they were having to pull the ton-and-a-half sedans down from well over 100 mph to 20 or 30—time after time, lap after lap, hour after hour. Disc brakes, with their rapid cooling, their self-adjusting features, their long pad life and their resistance to water, were an immediate option for the Camaros, and a necessity on the Z-28's. When Roger Penske found this still wasn't ideal for his Trans-Am race cars, Chevrolet suddenly listed 4-wheel discs as a heavy-duty option. These were not the production discs, which are a little light for serious competition, but the heavier Corvette discs. The front discs fit the normal Camaro knuckles and steering arms, while the rear discs are available only with a complete axle assembly (part #3945131), with 3.73:1 ratio and Positraction.

At this time the 4-wheel discs are not available for the new 1970 Camaro, but will undoubtedly be in the future. The only major change will concern the longer distance between the front knuckles on the 1970 cars.

TIRES AND WHEELS

Standard wheel diameter for all versions of the Camaro except the Z-28 is 14 inches, with the Z-28 at 15. The base inline-6 and V-8 models come with E78x14B Goodyear bias-belted fiberglass weinies, the SS with F70x14B, the Z-28 with F60x15B. The SCCA now limits the over-2-liter sedans' wheel rim width to 8 inches, and you can get mags or aluminum wheels from Chevy or from several other manufacturers to take the wider racing skins. In the early years of the Trans-Am it was not unusual to see 10½- and even 12-inch wheels on the Mustang-Cougar-Camaro cars, which was stretching both the rules and the fender wells immensely.

ENGINES, DRIVETRAINS

In 1967, the first year of the Camaro, the standard engines were the 140-hp, 230-cubic-inch inline-6 and the 210-hp, 327-cubic-inch V-8. Optional engines were a 155-hp, 250-inch inline-6, a 275-hp 327, and a 295-hp 350. The Z-28 engine was a 327 block with a shorter-throw 283 crankshaft that destroked it down to 302 inches to fit into the SCCA's 305-cubic-inch maximum displacement rule.

1

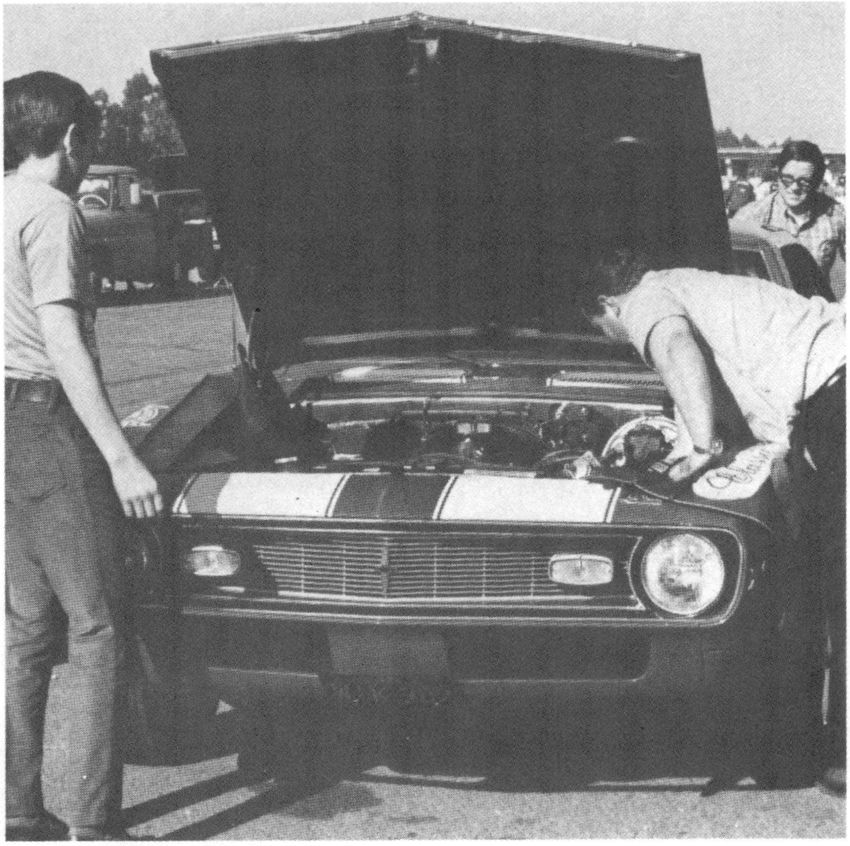

For 1968, the optional 350 and 327 were dropped, with three 396-inch V-8's offered, having 350 and 375 hp with cast iron heads, and 375 hp with aluminum heads. The 302 Z-28 engine was still a special option.

For 1969 the same 230, 250 and 327 were standard engines, with a 255-hp and a 300-hp, 350-inch V-8 and a 325-hp 396-inch V-8 as options, along with the 302 Z-28 Trans-Am engine. The factory 302 was rated at 290 hp at 5800 rpm, with 290 lb.-ft. of torque at 4200 rpm; the Penske-prepared engines in the Camaros that Mark Donohue drove to an impressive series of Trans-Am victories were

2

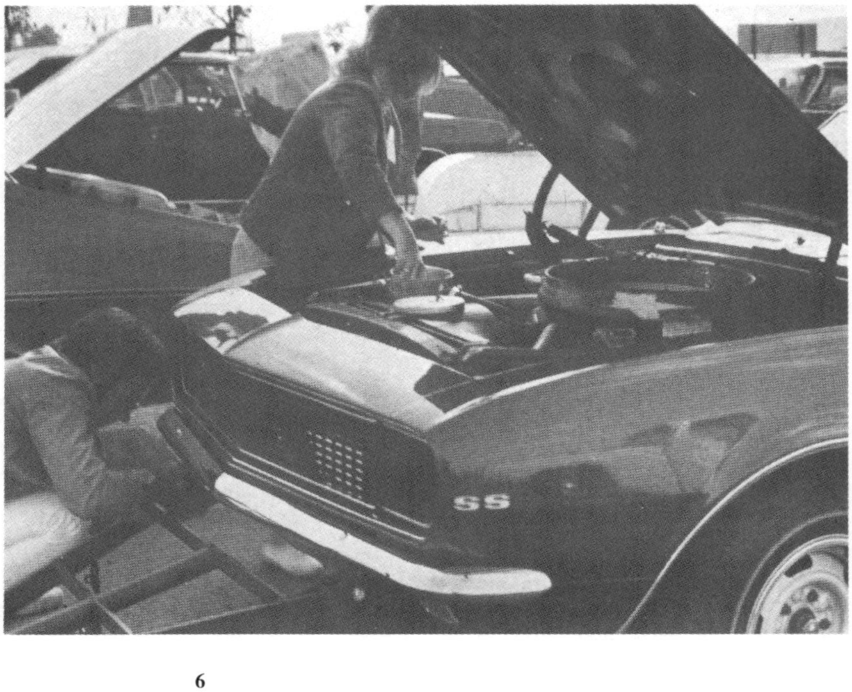

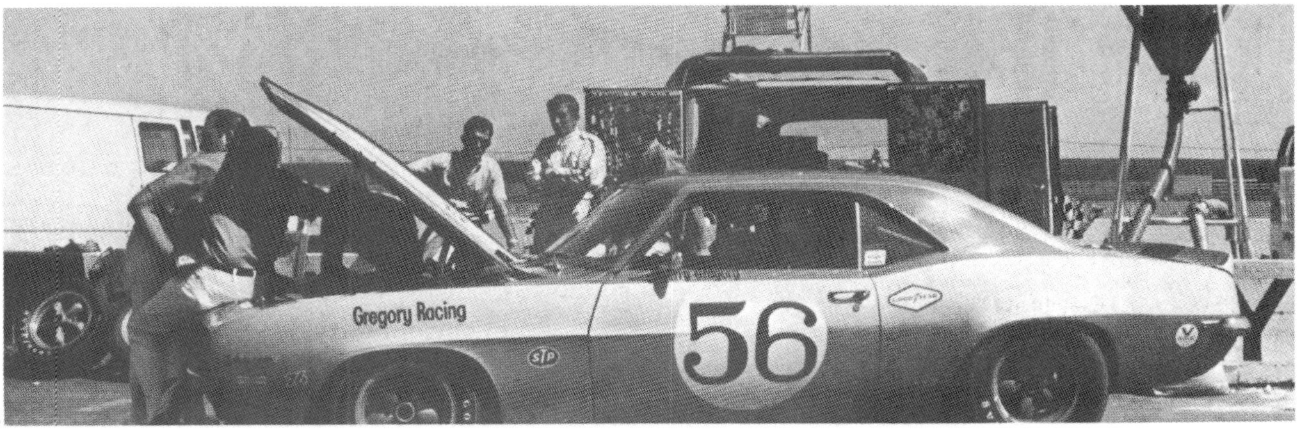

3

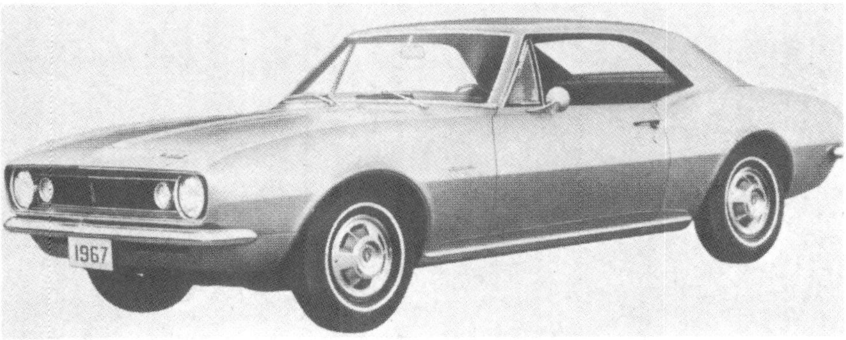

4

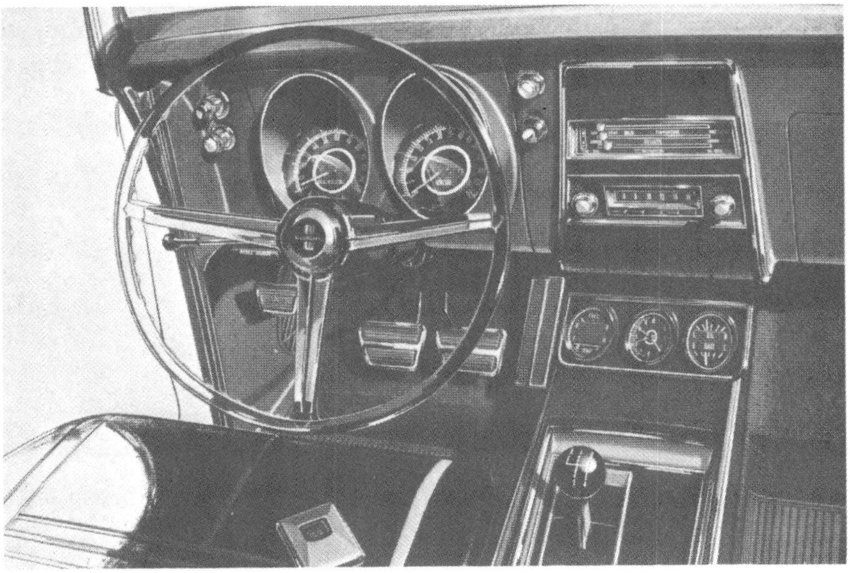

5

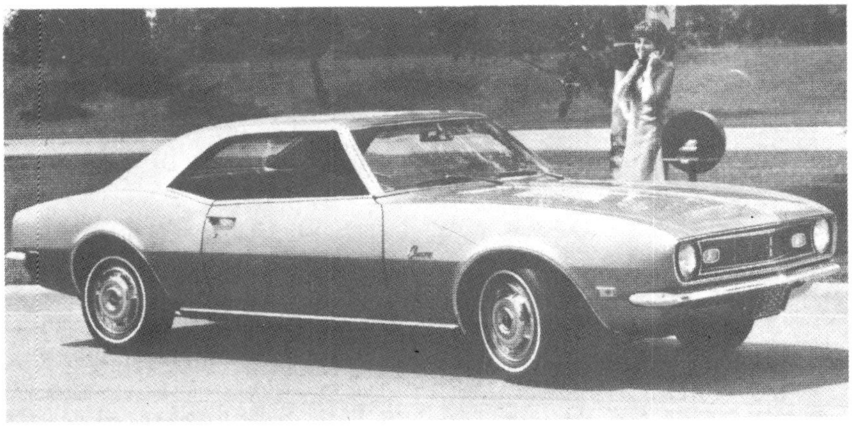

6

getting between 450 and 475 hp. This is unquestionably far from the maximum horsepower of the little 302, as this was with enough durability and reliability to last from two to 24 hours of all-out road racing.

For 1970, with the new overall design for the Camaro, the available engines have been switched around again. This year the 230-inch inline-6 has been dropped, as has the 327 and the Trans-Am 302. The SCCA allows back-dating, so the 302's from last year can be dropped into the 1970 chassis and be completely legal. The 1970 engine selection starts with 250- and 307-cubic-inch 200-hp V-8 regular gas engines as standard or base. Optional V-8's are two 350-inch mills with 250 and 300 hp, using regular and premium gasoline, and a "396," which is actually a 402 and gives 350 hp and 415-lb.-ft. of torque. The Z-28 engine is now an SCCA-illegal 360-hp, 350-cubic-inch high-performance V-8. Thus, for 1970 there is no Camaro that can be bought right off the showroom floor that is SCCA-legal. But then, as we mentioned before, no knowledgeable person ever took a Camaro—or any other pony car—directly from the showroom to the race track.

In looking over the 1970 engines, we'll pass over the regular-gas versions quickly, except to note that they

1. Camaros are for show and go, as evidenced by these scenes at Orange County International Raceway.

2. A cool gal filling a cool can. Better look out for this one, though–it was towed in.

3. Gerry Gregory's car at Sears Point in 1969. Note the tall fueling rig on the right. They are not allowed in 1970.

4. This was the first Camaro, much better looking in our opinion than the original Mustang.

5. The '67 Camaro instrument panel looks a little bit like some Corvettes.

6. In '68 there wasn't much change; the most obvious one was the sidelights for safety.

CAMARO

have cast iron blocks and heads, flat-topped pistons, cast nodular iron crankshafts and hydraulic valve lifters. These are the main features which are decidedly different from those in the higher-performance engine available this year.

Within the three high-performance, premium-gasoline V-8's, both the Turbo-Fire 350 with 300 hp and the Turbo-Jet 396 with 350 hp have 10.25:1

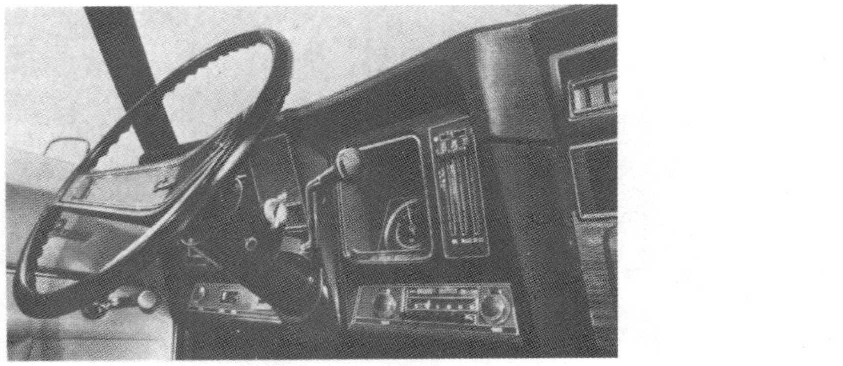

1

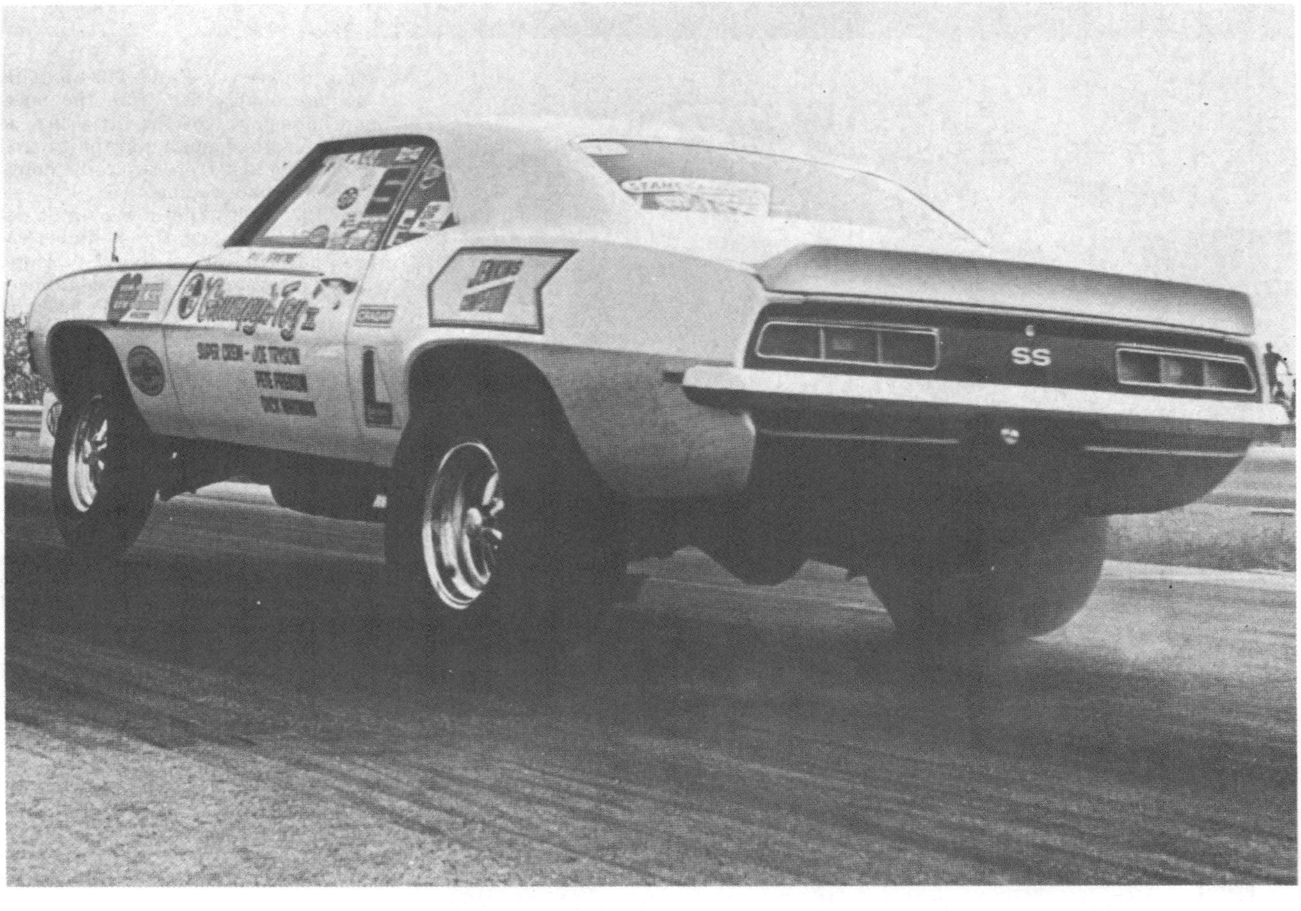

2

compression ratio, with the really hairy 360-hp Turbo-Fire 350—available in the Z-28 Camaro—rated at 11:1. The 396 is not simply a bored or stroked 350; it's the "big block" Chevy, which is completely different.

All of the base and optional engines come with cast iron heads and blocks. The 300/350 comes with flat-topped cast aluminum pistons, as in the lower compression engines. The 350/396 pistons are also cast aluminum, but domed. The Z-28's 360/350 pistons are both domed and made of impact-extruded aluminum. Many of the interior parts are extremely similar throughout the engine variations. All use cast alloy iron compression rings, with the higher-performance engines going to molybdenum inlays on their top compression rings and chromium plating on their lower ones. All piston pins are chromium steel, locked in the connecting rod small end, and all but those in the Z-28 350 are offset .060-inch in the piston in the direction of major thrust. All connecting rods are drop-forged steel, with the two base engines using sintered copper-lead alloy big-end bearings, while the rest are using premium aluminum bearings.

With the crankshafts, there is a major change once one gets into the two highest-performance optional engines—the 360/350 and the 350/396. Both use a forged steel crank in place of the cast nodular iron crank used in the less powerful engines. All the Camaro V-8's have 5-main-bearing cranks, and all use steel-backed replaceable bearings of either copper-lead alloy or premium aluminum, dependent on the engine application.

The valve train for the Camaro engines is similar for all the variations, with a cast iron camshaft and hydraulic lifters (except for the Z-28, which uses solid lifters), and alloy steel valves. All valves use spring dampers inside the valve springs except the 396 engine, which uses double springs on all valves.

Carburetors for all but the Z-28 engine are Rochesters, with a single 2-barrel on the regular-gas V-8's (a single 1-barrel on the straight-6), and with a 4-barrel on the premium gas V-8's. The Z-28 engine comes with a 4-barrel Holley.

The availability of transmission options through the Camaro variations is a little erratic, but in general we can say that the base models can be purchased with a 3-speed manual, a 2-speed Powerglide automatic or a 3-speed Turbo Hydra-Matic automatic. The largest transmission selection is with the Z-28, which can be had with

1. The '69 Camaro instrument panel with its sectionalized treatment.

2. Here's the Grump tying on a good one—again.

3. The Z-28 engine in '69 had the cast rocker covers with fins.

4. Earlier Z-28 engines had chrome rocker covers, and the smog pump set up rather high.

5. This is the 1969 Camaro Spoiler.

6. Here we have the 1970 Spoiler—not much difference, but it's a little bit cleaner design.

7. Like it or not, you'll have to admit it is really different for 1970.

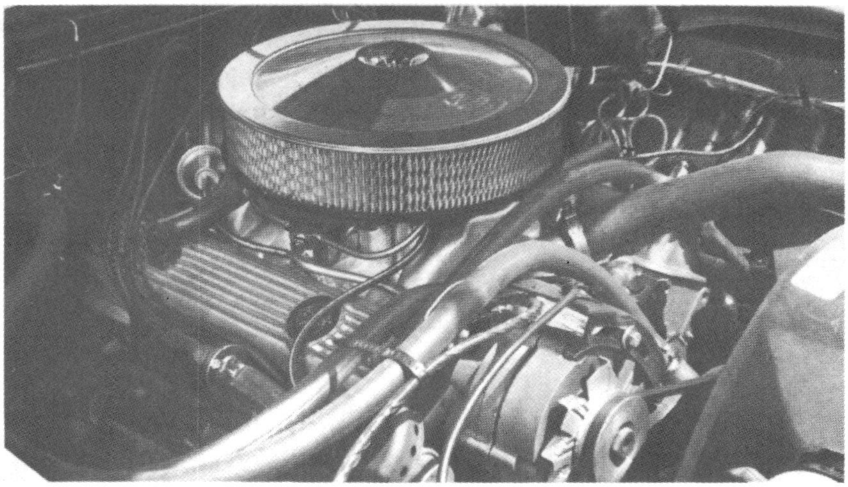
3

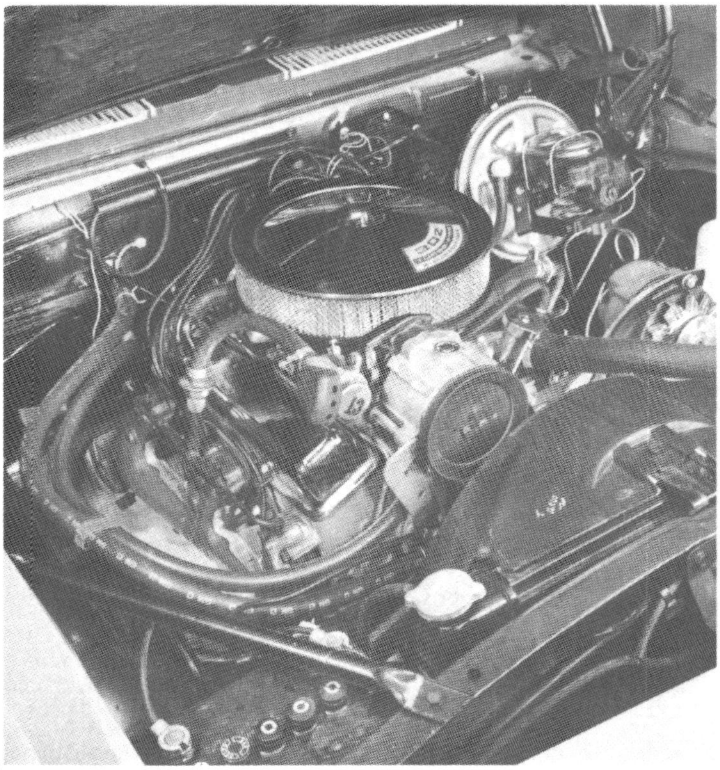
4

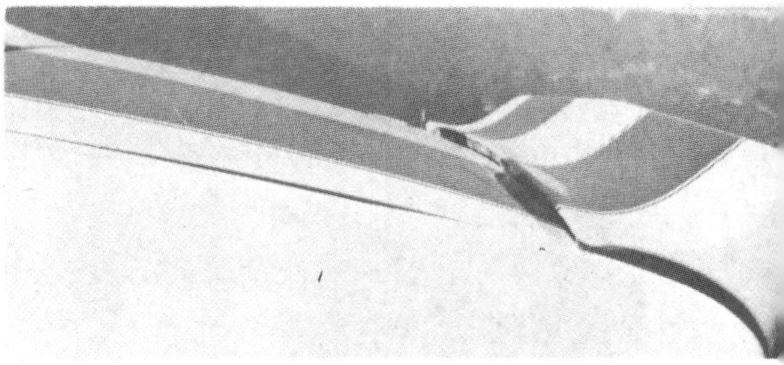
5

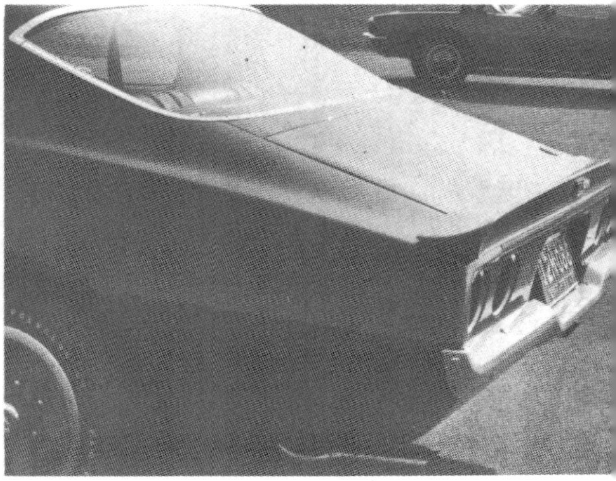
6

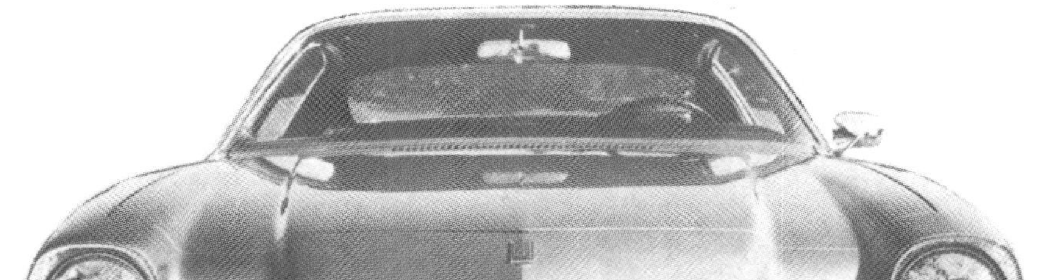

7

the Turbo Hydra-Matic, a 4-speed manual with a 2.52:1 low gear ratio, a close-ratio 4-speed manual with a 2.20:1 low, or a special heavy-duty close-ratio 4-speed manual with a 2.20:1 low. Some of the other high-performance SS types don't list the 2.20:1 transmission as an option, but there is no reason one couldn't be dropped in.

All the 350 and 396 engines use a single dry-disc clutch, with a bent-finger diaphragm clutch spring. The premium-gas versions have an 11-inch woven asbestos facing, the 250/350 a 10.34-inch. From here on back, all Camaros are the same, with a straight propeller shaft, a semi-floating rear axle/differential, with limited slip provided by dual disc clutches. Rear end gears are available in seven ratio combinations—from 2.73 to 4.10—by using one of the six pinion gears with

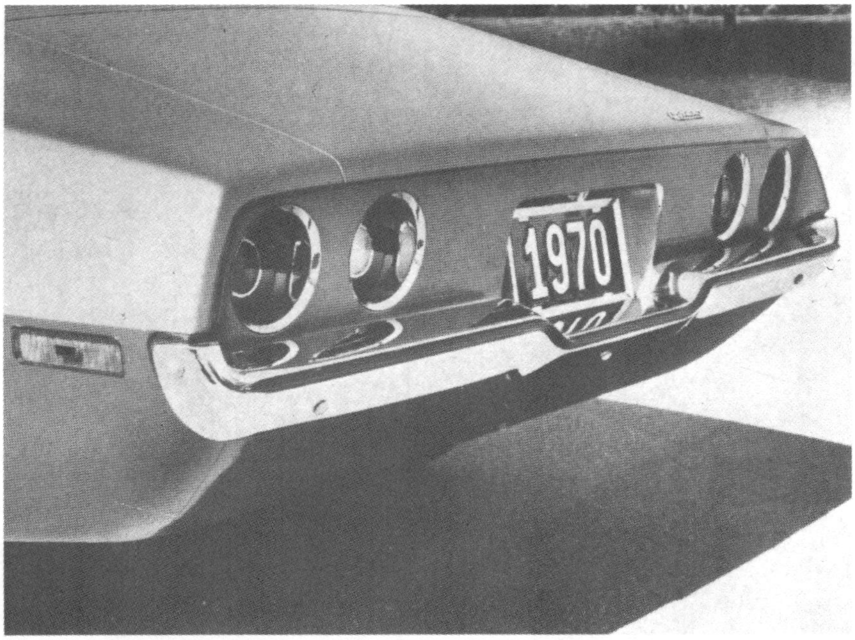

1

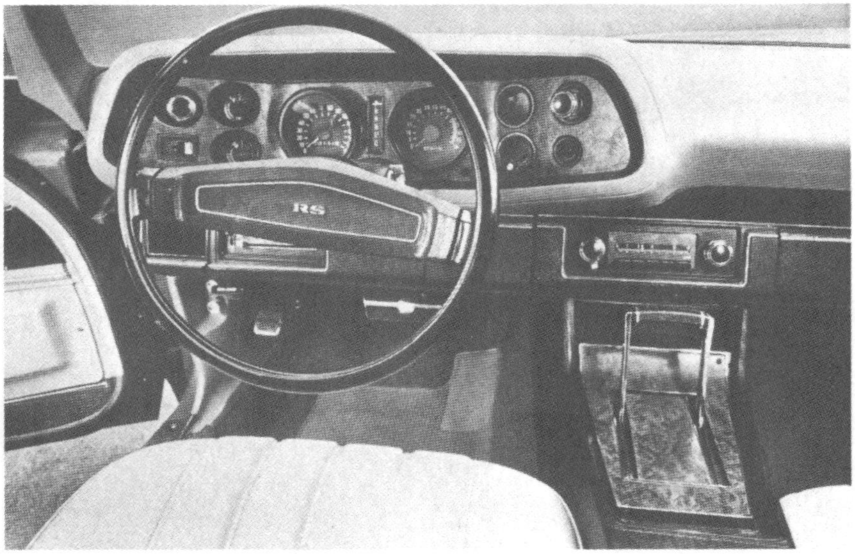

2

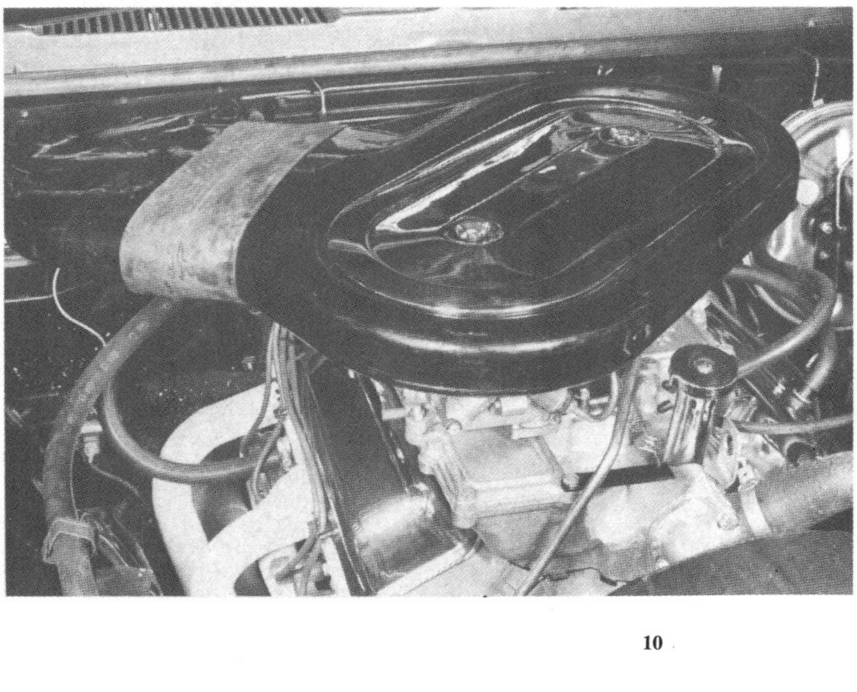

3

CAMARO

from 10 to 15 teeth combined with one of three ring gears having 37-41-43 teeth respectively.

Although the 302-inch Z-28 Trans-Am engine is not an option for 1970, let's see why it's the hottest production racing engine on the American market. The small-block Chevy V-8 had started in 1955 as a 265-inch powerhouse, then had been enlarged to a 283, 307, 350 and finally, 400. People who play around with engines keep track of what parts will interchange, and can come up with a combination of parts to fit almost any need. When the SCCA made its maximum displacement limit 305 cubic inches (5 liters), Chevrolet engineers combined a 327 block with its 4.002 bore, and a 283 crankshaft having a 3.005 throw, and there was the 302-cubic-inch production Z-28 engine—with 3 inches left over for wear.

The factory has always stated that the Z-28 engine produces 290 hp at 5800 rpm, with 290 lb.-ft. of torque at 4200 rpm. Most knowledgeable types consider these figures are quite low, and are designed to make the insurance companies happy; anything in the neighborhood of 300 hp makes the companies frown. One expert tester stated he felt the actual horsepower was closer to 350 at 6200 rpm with the standard 4-barrel Holley. Another estimated 360 with the twin 4-barrels used in racing.

Using the twin quads, special pistons, special cam and extensive "blueprinting," Roger Penske's Camaro team was able to get a reliable 455 hp—and more—by the 1968 racing season. Some have claimed that Penske was getting as high as 475 hp out of Mark Donohue's engines by 1969, but "Roger the Dodger" denies anything over 460. Still, that's a bundle out of a production, small-block Chevy V-8 meant to run in the 5000-7000-rpm range for anywhere from two to 24 hours.

The Z-28 Trans-Am 302 engine, as it comes out of the box, is a short-stroke 90° V-8, with cast iron block and heads and an 11:1 compression ratio. The pistons are impact-extruded aluminum with cast iron alloy compression rings. The top ring is straight-faced with a molybdenum-filled groove; the lower is taper-faced and chromium-plated. Piston pins are chrome steel and lock into the drop-forged steel connecting rod small ends. Premium aluminum big-end bearings mate the rods to the 5-bearing forged steel crankshaft. The factory cam is fairly wild, with a 118° overlap, and would probably be considered a "three-quarter race" grind by most cam makers. It's cast iron alloy, chain driven, and runs in five steel-backed babbitt bearings. Solid lifters are used, and the valves are alloy steel—the same size as those in the 350-inch Stingray engine.

It is obvious from these specifications that there is nothing really unusual or exotic about the Z-28 302 engine. It is simply a well engineered and carefully assembled engine that can take a large power increase without coming unglued.

For normal fun and games on the street, the single 4-barrel Holley carburetor on its special "tuned runner" intake manifold is more than adequate. (It isn't quite cricket to call the manifold a "hi-rise," by the way —that's a Ford term.) For really all-out performance, on the drag strip or the race course, Chevrolet has a legal cross-ram inlet manifold that mounts two 600-cubic-feet-per-minute Holleys (option #3940077) along with the necessary linkage and fuel lines. The carburetors themselves (#3957859) have mechanically operated secondary throttles with separate primary and secondary accelerator pump systems. For racing, the standard jet setup is to use #69 primary and #71 secondary jets. The No. 3 cylinder can be slightly enriched by increasing the right-hand carburetor's primary jets to the larger #71's.

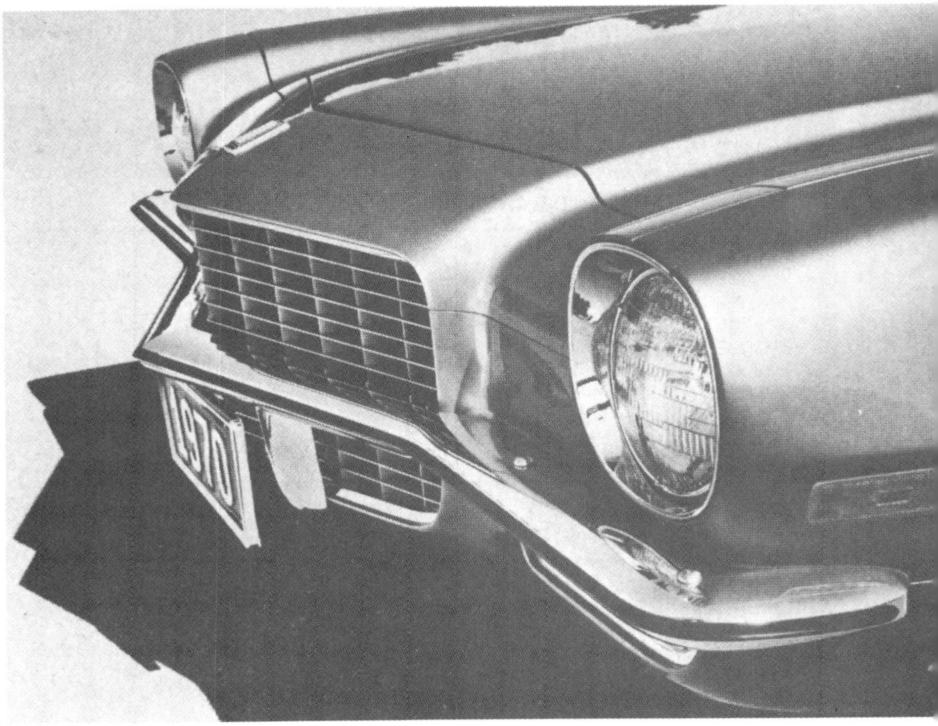

4

For any type of drag or road racing, the fuel line between the gas tank and the carburetor should be at least ⅜-inch inside diameter; ½-inch is better. Although many Camaros use high-capacity electrical fuel pumps, Chevrolet sells a special engine-mounted mechanical pump (#6415325) that is highly reliable. With either type, or a bypass combination of both, a fuel pressure gauge should always be installed to provide constant monitoring of the fuel pressure at the carburetor inlets. If you can't hold at least 4 psi at high revs and with the throttle wide open, the pumps are simply not adequate for maximum power and engine life.

The production 4-quart baffled sump has proven completely adequate for all forms of racing, and can be over-filled by as much as 2 quarts, although a 1-quart overfill is recommended. For sustained high-speed racing you can add a factory oil cooler (#3157804) ahead of and to the side of the water radiator. This connects to the block with an adapter where the normal oil filter mounts, and a remote filter is added to the line returning the cooled oil. The new oil lines should have at least ½-inch inside diameter, and all the oil should circulate through the cooler and filter, with no bypass.

For short-duration drag runs, the standard radiator for the engine's water should be adequate, but for road racing Chevy has a 2.7-inch core radiator (#3016688) that fits the existing mounting points, and therefore satisfies the racing regulations. You can take your choice of a number of 4- and 5-bladed solid and viscous-drive fans that are stock Chevy parts.

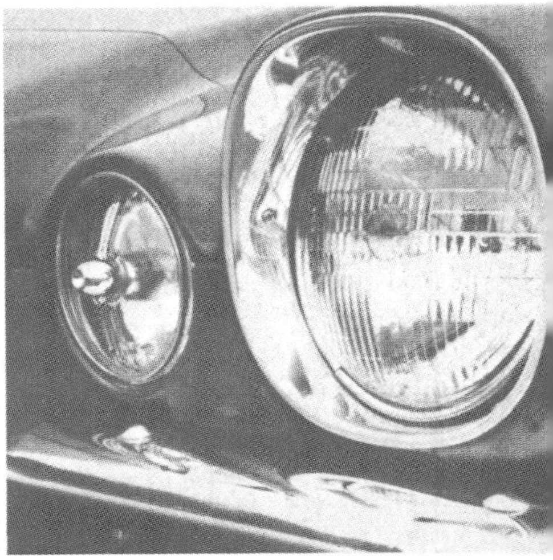
5

SUMMING UP . . .

The Camaro is one of the best of the current pony cars from every aspect. Its handling is rather poor in its base form, but a wide variety of special handling options are available that can give it race car characteristics surpassed by no American production car. The engine selection is wide, particularly if you include the 1967-1969 special order Z-28 Trans-Am engine. The looks of the 1967-1969 sports coupe and convertible and the new 1970 sports coupe are pleasing if not overly original. With prices running from about $3000 in its most base form up through about $4000 as one adds on the various Rally Sport, Super Sport, and Z-28 options and engines, it's a darned good buy. In the Z-28 configuration you've got a real race car—and for less than $4000 all told!

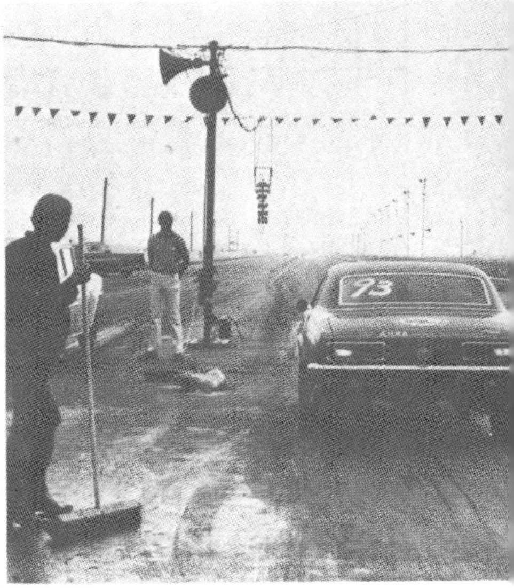

6

1. Rear design is clean, with both license plate and lights in recesses.

2. The Rally Sport for 1970 has this pleasing instrument panel. Note the stirrup shifter.

3. Here's a Hottie. It has the optional headers and dual 4-barrels with the air entry from the cowl.

4. These big headlights on the '70 restyled model are going to take some getting used to.

5. The Rally Sport option includes these little driving lights set in the front sheetmetal facing.

6. "Aren't these guys ever going to quit racing? I've got to sweep up this place, man . . . !"

CAMARO Z-28 HEAVY-DUTY CHASSIS PARTS

PART NO.	DESCRIPTION	QUANTITY	SPECIFICATIONS
STEERING			
3916236	Knuckle—Steering	2	Tuftrided
9748406	Stud & Seal—Upper Cont. Arm Ball	2	Tuftrided
3875067	Stud & Seal—Lower Cont. Arm Ball	2	Production shot-peened
3930030	Socket Assembly—Tie Rod Outer	2	Shot-peened
3930028	Socket Assembly—Tie Rod Inner	2	Shot-peened
3958493	Rod—Steering Relay	1	Shot-peened
3916237	Arm—Steering Knuckle LH	1	Shot-peened
3916238	Arm—Steering Knuckle RH	1	Shot-peened
7806396	Gear Assembly—Steering (RPO N44)	1	20:1 ratio
SPRINGS			
3955727	Spring Assembly—Front	2	400 lb./in. rate
3948989	Spring Assembly—Front	2	507 lb./in. rate
3948984	Spring Assembly—Front	2	561 lb./in. rate
3935784	Spring Assembly—Front	2	615 lb./in. rate
3935785	Spring Assembly—Front	2	723 lb./in. rate
3948988	Spring Assembly—Front	2	777 lb./in. rate
3789664	Spacer—Coil Spring	AR	Spring seat spacer—3/8" thick
3789665	Spacer—Coil Spring	AR	Spring seat spacer—9/16" thick
3948986	Spring Assembly—Rear	2	200 lb./in. rate
3948985	Spring Assembly—Rear	2	250 lb./in. rate
3953673	Spring Assembly—Rear*	2	300 lb./in. rate
3935786	Spring Assembly—Rear	2	350 lb./in. rate
3927507	Spacer—Rear Spring	2	Use with HD springs
3889964	"U" Bolt	4	Use with HD springs
*Use center bolt 3790992, discard shim and spacer			
SWAY BARS			
3962795	Stabilizer Shaft—Front	1	3/4" diameter
3962796	Stabilizer Shaft—Front	1	7/8" diameter
3962797	Stabilizer Shaft—Front	1	15/16" diameter
3961763	Stabilizer Shaft—Front	1	1.0" diameter
3962799	Stabilizer Shaft—Front	1	1-1/16" diameter
3927506	Bushing—Stabilizer Shaft	2	Use with 1-1/16" bar
3927944	Plate—Stabilizer Bracket Reinf.	2	Use with 1-1/16" bar
BRAKES (1967 and 1968 RPO J52)			
5468882	HD replacement Front Brake Pad Unit	1	Corvette HD brake pads
5468886	Caliper Assembly—Front Brake LH	1	Revised for 5468882 shoe
5468887	Caliper Assembly—Front Brake RH	1	Revised for 5468882 shoe
3923535	Pipe Assembly—Front Brake LH	1	
3923536	Pipe Assembly—Front Brake RH	1	
3927510	Hub & Disc Assembly—Front	2	1/2 x 20 wheel bolts
3830635	Shoe & Lining Unit—Rear	1	Metallic shoes rear drums
3927508	Shaft Assembly—Rear Axle	2	Shot-peened 1/2 x 20 wheel bolts
BRAKES (1968 Conversion to Full Disc Brakes)			
5468882	Replacement Front Brake Pad Unit	1	Corvette HD brake pads
5452515	Replacement Rear Brake Pad Unit	1	Corvette HD brake pads
3957992	Front Hub & Disc w/Caliper Unit LH	1	1 1/4" disc w/Corvette calipers
3957993	Front Hub & Disc w/Caliper Unit RH	1	1 1/4" disc w/Corvette calipers
5463775	Caliper Assembly LH	1	HD Corvette caliper
5463776	Caliper Assembly RH	1	HD Corvette caliper
5463751	Master Cylinder Assembly	1	For full disc brakes
3947289	Caliper Mounting Bracket LH	1	For Camaro knuckle
3947920	Caliper Mounting Bracket RH	1	For Camaro knuckle
3945125	Caliper Support Brace	2	
3947283	Bracket—Brake Hose Support LH	1	
3947284	Bracket—Brake Hose Support RH	1	
3947037	Brake Pipe LH	1	Connects brake hose to caliper
3947083	Brake Pipe RH	1	Connects brake hose to caliper
3945118	Hub & Disc Assembly	2	Replacement hub & rotor
3865905	Rear Disc Assembly	2	Replacement rotor
3945189	Spacer Rear Caliper	AR	Use between bracket and axle housing
5469499	Caliper Assembly Rear	2	HD Corvette caliper (use on RH & LH side)
3945187	Bracket—Caliper Mounting	2	Rear axle—attachment LH & RH

REAR AXLE		
Special parts for HD Disc Brake Rear Axle Assembly:		
3945131 Rear Axle Assembly 3.73:1 ratio	1	Complete disc brake axle assembly
3959067 Axle C-Lock	1	.155 thick
3959068 Axle C-Lock	1	.160 thick
3953697 Rear Axle Housing & Spacer Unit	1	Basic axle housing
3945184 Shaft Axle (large diameter)	2	For disc brake assembly
3945176 Positraction Differential Case for 3.07-3.73 ratios (round cross-shaft)	1	HD Positraction—22 plate (1968)
3945177 Positraction Differential Case for 3.91-5.13 ratios (round cross-shaft)	1	HD Positraction—22 plate (1968)
3957937 Gear—Differential	2	Differential side gears
3957938 Pinion—Differential	2	Spider gears
3957939 Spring—Differential Pinion Pin	4	Positraction pressure springs
3957940 Washer—Differential Pinion Thrust	2	Spider gear thrust washers
3957941 Plate—Differential Clutch Pressure	2	Use with springs 3957939
3957942 Unit—Differential Clutch Plate & Disc	1	22-plate Positraction repair kit
7451275 Bearing Assembly—Rear Wheel	2	For 3945184 axle shafts
9422406 Bearing Assembly—Rear Wheel	2	Optional w/7451275
3843648 Seal Assembly—Rear Wheel Bearing	2	For large bearings
3955705 Brake Line—HD Disc Brake Axle	1	Axle pipe assembly—LH
3962780 Brake Line—HD Disc Brake Axle	1	Axle pipe assembly—RH
All other parts standard Camaro:		
3927508 Shaft Axle—Prod. Drum Rear Axle	2	Shot-peened, 1/2 x 20 bolts
RING AND PINION UNITS		
3961406 3.55:1 Ratio Ring & Pinion	1	12 Bolt HD 8 7/8" diameter
3961407 3.07:1 Ratio Ring & Pinion	1	12 Bolt HD
3931564 3.25:1 Ratio Ring & Pinion	1	12 Bolt HD
3961405 3.31:1 Ratio Ring & Pinion	1	12 Bolt HD
3961428 3.42:1 Ratio Ring & Pinion	1	12 Bolt HD
3865997 3.73:1 Ratio Ring & Pinion	1	12 Bolt HD
3931566 3.91:1 Ratio Ring & Pinion	1	12 Bolt HD
3961409 4.10:1 Ratio Ring & Pinion	1	12 Bolt HD
3931567 4.33:1 Ratio Ring & Pinion	1	12 Bolt HD
3961410 4.56:1 Ratio Ring & Pinion	1	12 Bolt HD
3961404 4.88:1 Ratio Ring & Pinion	1	12 Bolt HD
3933095 5.13:1 Ratio Ring & Pinion	1	12 Bolt HD
3961192 4.88:1 Ratio Ring & Pinion	1	Lower hardness for drag racing
3961195 5.13:1 Ratio Ring & Pinion	1	Lower hardness for drag racing
3961435 2.73:1 Ratio Ring & Pinion	1	12 Bolt HD 8 7/8" diameter
MISCELLANEOUS		
3910340 Bolt—HD Rear Axle	AR	7/16-20 x 1.75" long
3851677 Bolt—Rear Wheel	10	7/16 x 20 axle bolts 1 3/8" long
3819780 Bolt—Wheel Hub	AR	1/2-20 x 2 7/8" long bolt
3849110 Bolt—Wheel Hub	AR	1/2-20 wheel bolts 1 3/4" long
3931547 Wheel Assembly (15 x 8" Mag)	AR	.2 + offset
3931548 Seat, Bucket (plastic)	2	Lightweight seat
3916633 Rear Deck Spoiler	1	Production Z-28 (1968)
3949798 Rear Deck Spoiler	1	Production Z-28 (1969)
3943249 Front Valance Air Deflector	1	Large 4" deflector (1968)
3943251 Brace Valance—Outer	2	For large deflector
3943253 Brace Valance—Center	1	For large deflector
3963832 Hood Unit, Ducted Fresh Air	1	1969 fresh air—fiberglass
3963827 Stud, Hood Hold-Down	2	Fiberglass ducted hood
3963828 Pin, Hood Hold-Down	2	Fiberglass ducted hood
3963829 Plate, Hood Hold-Down	2	Fiberglass hood reinforcement
3963830 Cable, Hood Hold-Down	2	Pin retainer
3965713 Rod	1	Fiberglass hood support
3963824 Base Plate	1	Air cleaner base
3963823 Seal	1	Air cleaner base to hood
3963825 Element	1	Air filter
3941146 Cover	1	Air cleaner cover
3016688 Radiator	1	2.7" core
3157804 Radiator	1	Oil cooler
3938689 Front Valance Air Deflector	1	3 3/4" deflector (1969)
5464591 Brake fluid Delco #550	AR	1 gallon can
1051024 Lubricant	AR	Transmission & differential
5467286 Caliper Repair Unit, Front	AR	Seal, boot and O-ring
5467288 Caliper Repair Unit, Rear	AR	Seal, boot and O-ring
5463822 Insulator, Front Piston 1 7/8"	AR	HD Corvette
5452287 Insulator, Rear Piston 1 3/8"	AR	HD Corvette
3949708 Hood Assembly—Steel Fresh Air	1	1969 Camaro ZL-2
NOTE: AR—As Required		

CAMARO RESTORATION Some Camaro owners would rather rebuild an entire engine than repair a door ding. There are F-body fanatics who would drink Powerglide fluid and bathe in positraction oil before they would take on a Camaro paint job. The reasons behind this fear of spraying are varied. Inexperience, a lack of equipment, and confusion about the proper materials are common excuses. There's really no reason, though, to drive around in a primered, dented, and rusty Camaro—not when a "home-blown" paint job can produce professional quality results.

Painting a car has never been easier for do-it-yourself Rembrandts. Paint manufacturers have developed products that can produce outstanding results even when used by novices. New synthetic materials make bodywork a breeze, and the mail order tool companies provide high-quality equipment at rock-bottom prices. Compressors and sprayers can be rented inexpensively for driveway paint jobs; in some cities, fully equipped spray booths are available by the hour. Whether your goal is to restore your car to better-than-new condition, to add some splashy graphics to a race car or street machine, or to repair some serious crash damage, now's the time to enroll in the Camaro color classroom.

Before you pull the trigger on your spray gun, you've got to decide what type of paint job suits your Camaro. If you cover your car's sheetmetal with flawless metallic lacquer, you will forever live in fear of parking lots and careless door openers. Camaros used for daily transportation to school or work are better off with solid colors that can be matched and repaired easily. Show-quality paint jobs also make little sense for race cars, which are always vulnerable to dropped wrenches and sharp belt buckles. On

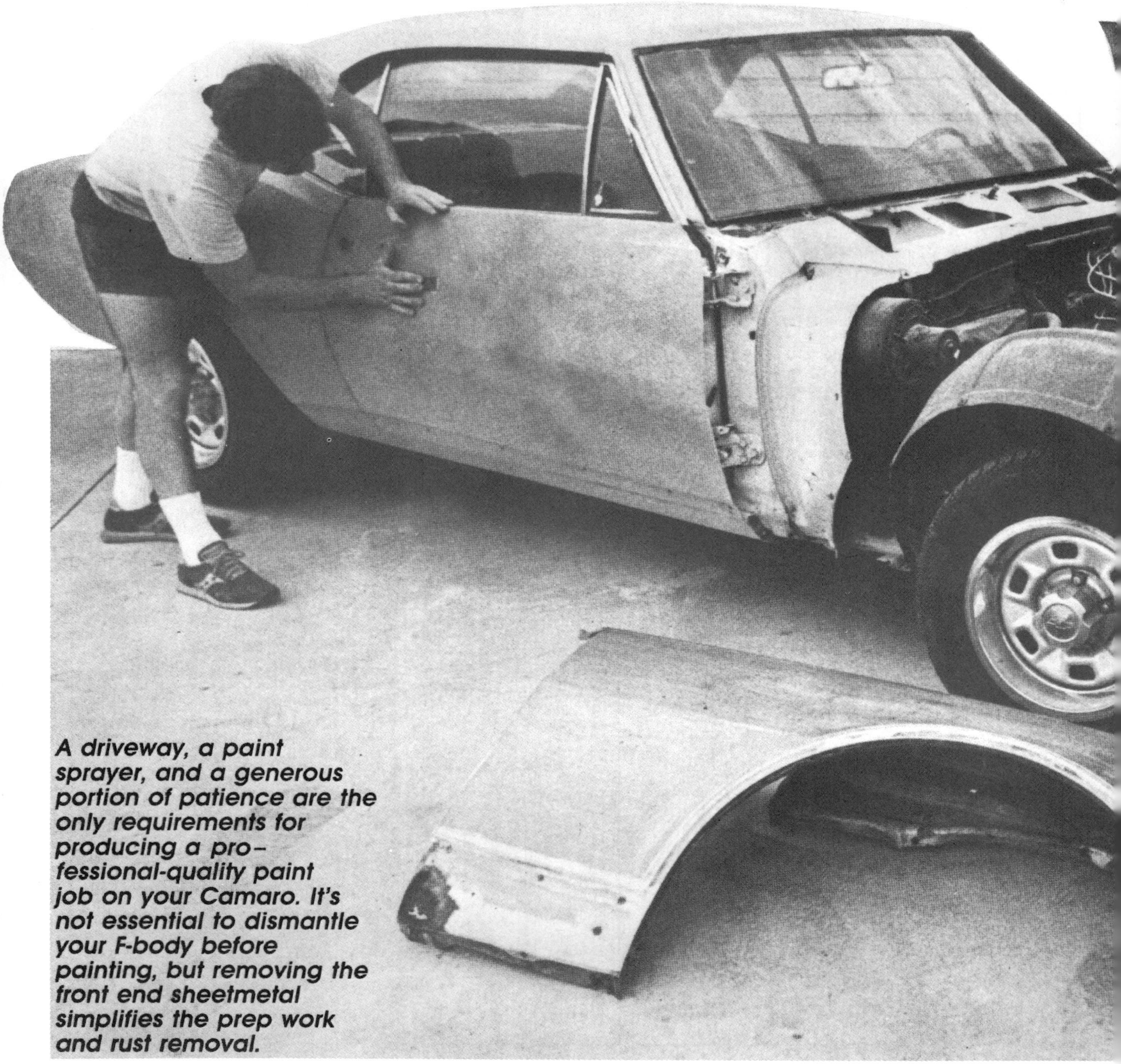

A driveway, a paint sprayer, and a generous portion of patience are the only requirements for producing a professional-quality paint job on your Camaro. It's not essential to dismantle your F-body before painting, but removing the front end sheetmetal simplifies the prep work and rust removal.

the other hand, if you intend your Camaro to be driven only on sunny Sunday afternoons, your car can retain its perfect paint job for years to come.

There are probably as many different ways to paint a car as there are car painters. Every painter has certain products and procedures that work for him. It's not our intention to outline the *only* way to paint a Camaro—rather, our aim is to describe *one* way to refinish an F-body. Jim Stevens, our "guest lecturer" for this Camaro paint course, can turn a rusty hulk into a concours winner. For years, he's turned out some of the smoothest paint in Southern California—often spraying award-winning pigment in the driveway of his Leona Valley home! Jim points out, however, that his recommendations aren't the gospel. Other products and other techniques may produce equal (or even superior) results. Like any craftsman, Stevens is constantly experimenting with new ideas and new technology to improve the quality of his work. As a neophyte Camaro painter, you should be open to trying out various approaches, too. After all, you can always strip off the paint and start over!

PREPARATION

From 50 feet away, it's difficult to see the difference between a $200 "one day" paint job and a $2000 hand-rubbed lacquer masterpiece. Up close, however, it's easy to spot high-quality work. You've probably heard before that preparation is the key to successful painting—and it's true. New paint will spotlight every ripple and imperfection in the sheetmetal, so it's essential to start with a straight body.

Before starting work on your Camaro revival, take an inventory of its present condition. Most Camaros have already suffered the ravages of minor collisions. Rust may be devouring parts of the sheetmetal, and emblems and

CAMARO PAINTING CLASSROOM

A SELF-TAUGHT COURSE IN CAMARO BEAUTIFICATION

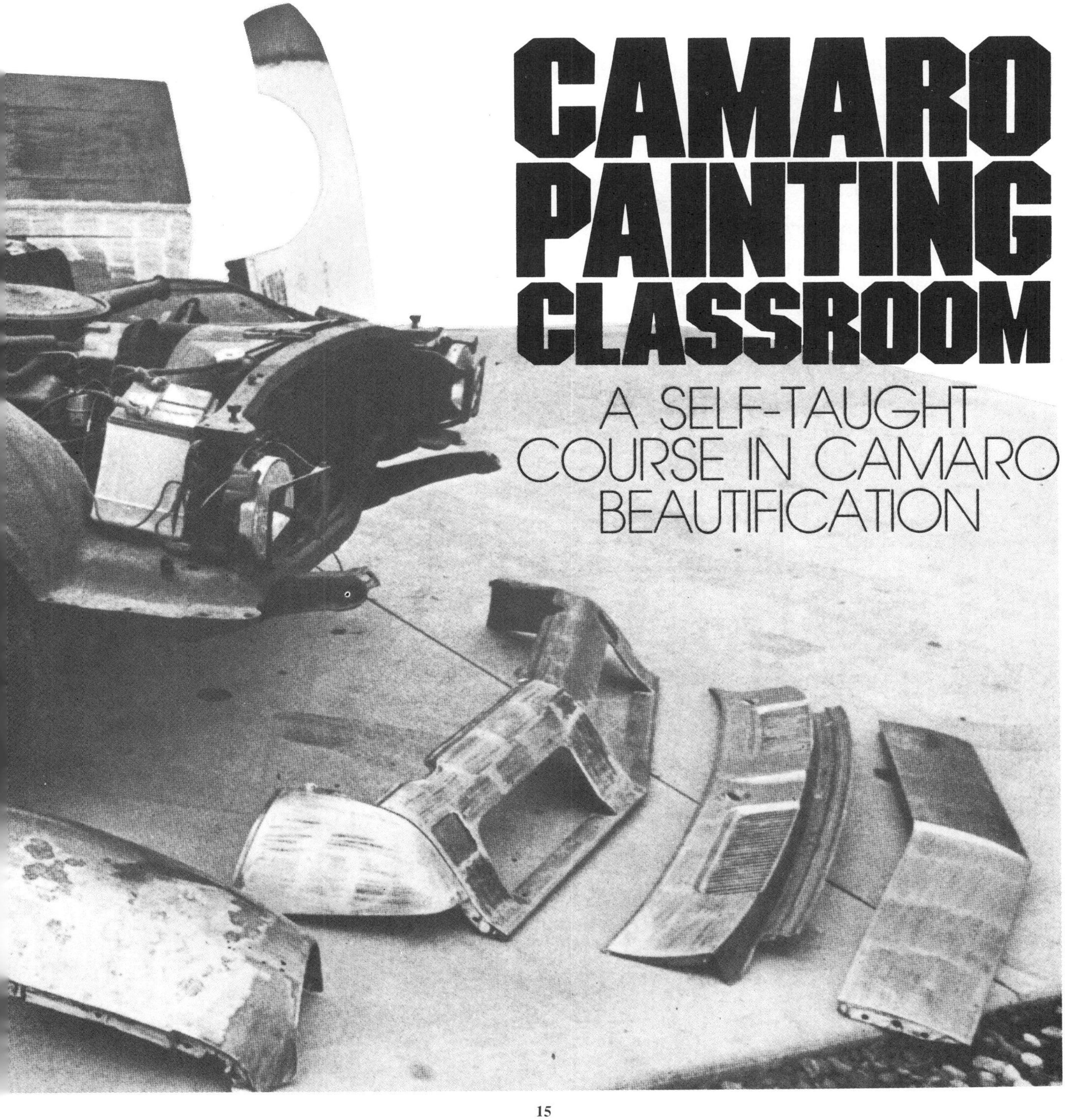

PAINTING

trim pieces may be missing or broken. As you take your Camaro survey, note areas where corrective action is required and make a list of parts to order. You'll have to decide whether it's worthwhile to repair body parts or simply replace them. Major components such as hoods and front fenders are still generally available for most Camaro models—but the supply is running low. It's usually easier to bolt on new sheetmetal than to straighten badly damaged pieces. Of course, rippled quarter panels or creased roofs will require sheetmetal surgery. Several parts suppliers that specialize in Camaro components now offer die stamped "patch panels" which can be used to repair corroded or damaged sheetmetal. These panels are much less expensive than large quarter panel stampings, and are ideal for repairing wheelwells, rear window moldings, and other areas that are particularly susceptible to rust.

You should also note your car's special paint treatments and accessories. If you are restoring your Camaro, take snapshots of any unusual graphics, including body stripes, factory-installed decals, a "bumble bee" nose, or pinstripes. Measure and record the location and spacing of these details so you can duplicate the original paint scheme.

Make sure you add safety equipment to your list of supplies to buy before you begin work. Many of the products you'll be working with are very dangerous. Eye protection, rubber gloves, and a top-quality respirator mask are essential equipment for a do-it-yourself Camaro refinisher.

STRIP SHOW

After you've taken an inventory of your Camaro's condition, it's time to examine its existing paint job. You should plan on stripping the old paint if the finish is checked, crazed, or cracking—these flaws will reappear in the new coat of paint if they are not fixed. If the old finish is peeling or will not "feather-edge" when you sand it, the paint has poor adhesion. Again, it's pointless to put new paint over old pigment that has lost its grip on the underlying sheetmetal. Finally, you should check the thickness of the paint film. Your Camaro may have already been repainted by previous owners. If the paint is too thick, it will be prone to cracking, and it will chip easily. Jim Stevens' rule of thumb is that if a Camaro already has three coats of paint (including its original finish), then a fourth coat is definitely not advised.

Another consideration is the *type* of paint your Camaro is wearing. Early Camaros, for example, were painted with a type of lacquer at the factory. Most body shops use enamel for

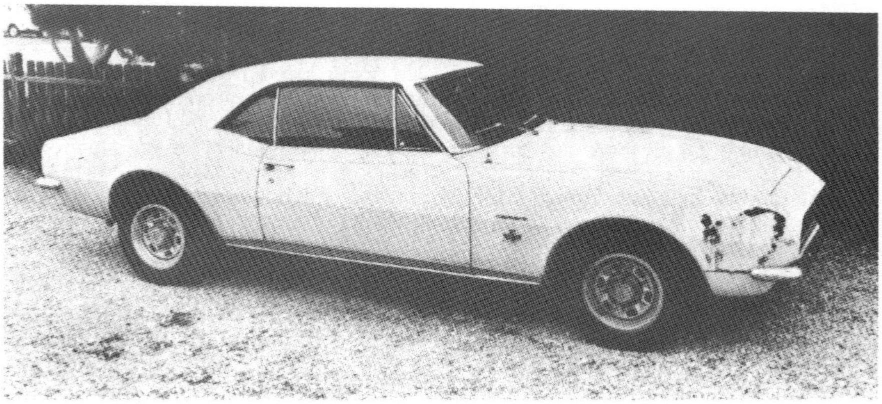

***1** The first step in an F-body beautification program is to take an inventory of its condition. Although basically sound, this '67 Camaro Rally Sport coupe has endured 18 years of surface rust, door dings, and parking lot collisions.*

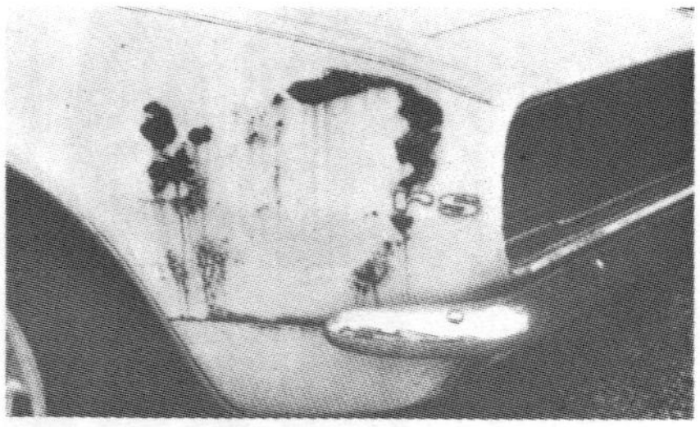

***2** Rust is the bane of all bodymen. Moisture trapped under a layer of paint has corroded this front fender. Overspray on emblems and chrome trim indicates a quick-and-dirty repair job.*

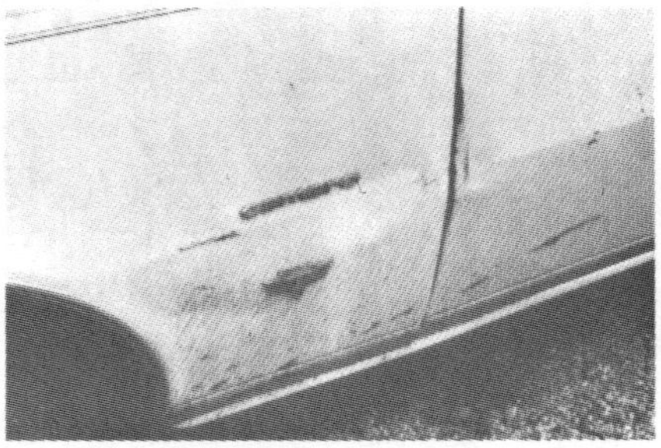

***3** Note missing or broken trim items during your Camaro inventory. Many emblems are still available from Chevrolet; Camaro parts suppliers offer accurate reproductions of rare medallions. You'll have to decide whether to repair or replace dented body panels.*

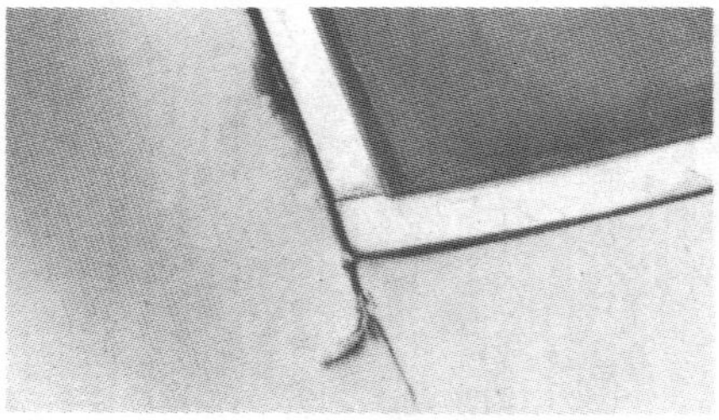

***4** Classic Camaro hardtops are especially prone to rust around their rear windows. Water trapped between the trim and glass eventually corrodes the window channel. Even if there are no external signs of damage, probe the metal with a scribe or penknife to determine if rust has weakened the metal. Beware of rust under vinyl tops! The only permanent solution is to replace the corroded sheetmetal with a patch panel.*

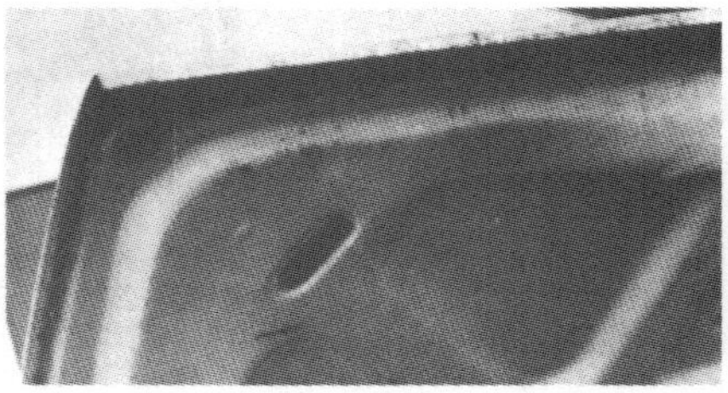

***5** Look for corrosion around body seams. Trunk lids and door bottoms are likely places for rust to gain a toehold. If the damage is not severe, the corrosion can be arrested by removing the rust with a wire brush and spraying the panel with zinc phosphate primer.*

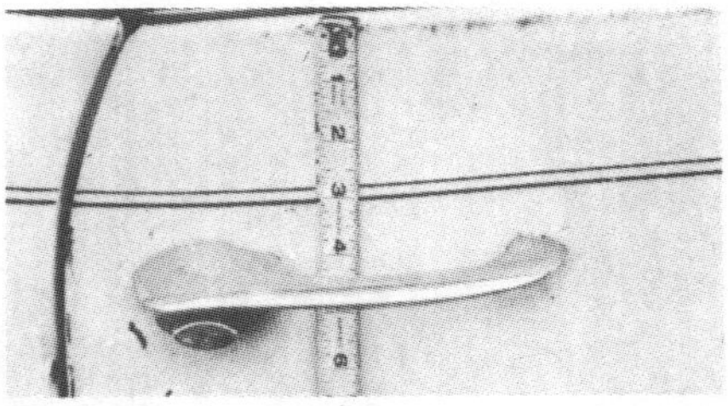

***6** If you plan to restore your Camaro to factory specs, take snapshots and measurements of its special paint graphics (assuming that it hasn't already been repainted). Templates and masks are available from Camaro parts specialists to duplicate original factory stripes.*

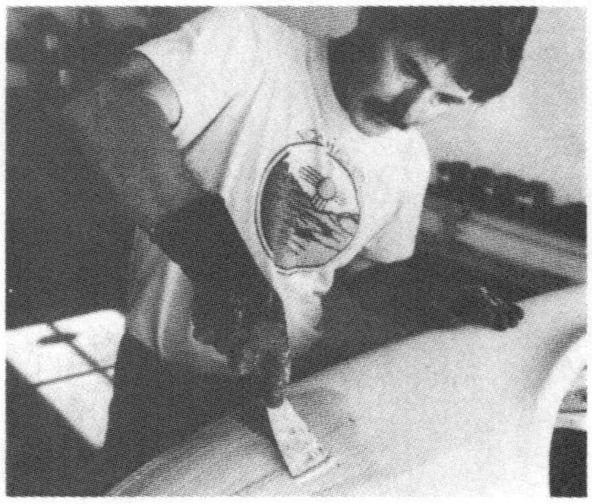

***7** There are several ways to remove old paint, but brush-on stripper is the best method for most backyard bodymen. Apply the stripper, give it time to dissolve a layer of paint, then remove it with a scraper. Wear rubber gloves and eye protection when using paint stripper!*

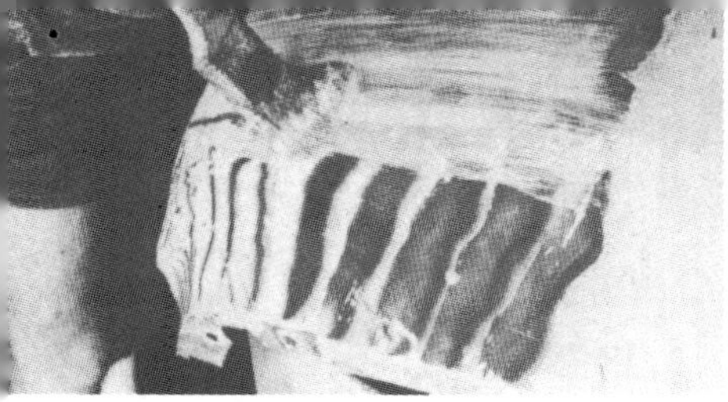

8 *Most strippers form a vapor barrier over the paint surface. For best results, apply the stripper with one brush stroke; don't go back over the area.*

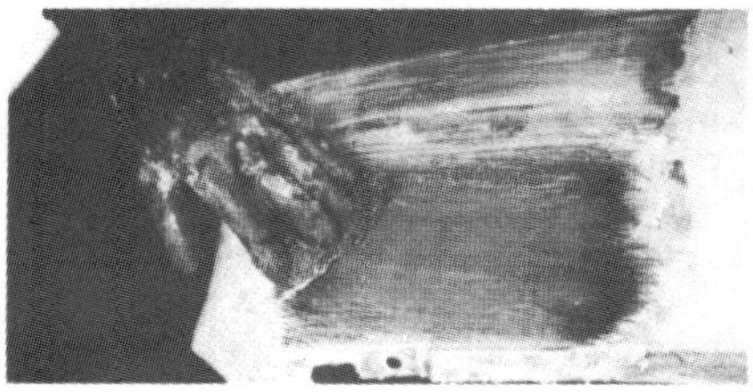

9 *GM primer is difficult to remove. After you have stripped the top coats, scrub off the primer with steel wool. If the factory primer is still in good condition, it can be used as foundation for your Camaro's new paint.*

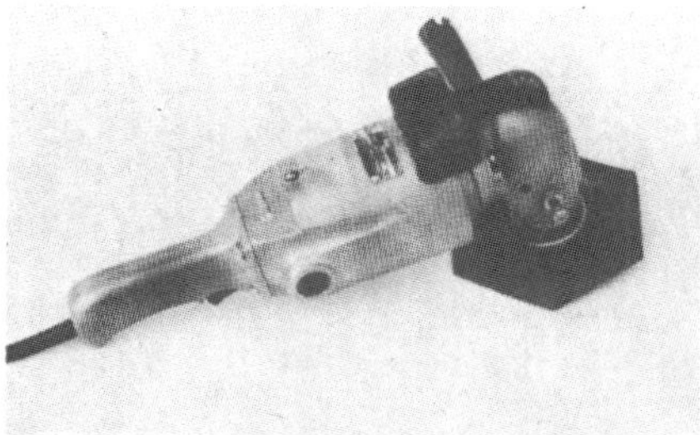

10 *An electric grinder/polisher is an invaluable tool for refinishing F-bodies. You can use it to remove old paint and to polish the new pigment. Remember that a heavy grinder will tire your arms after a few hours of polishing.*

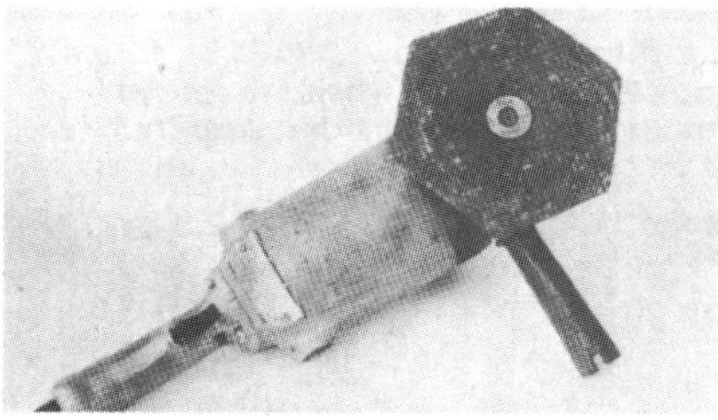

11 *Trimming the grinder pad into a hexagon will make it more "aggressive" when stripping old paint. Be careful not to gouge or score the metal when grinding.*

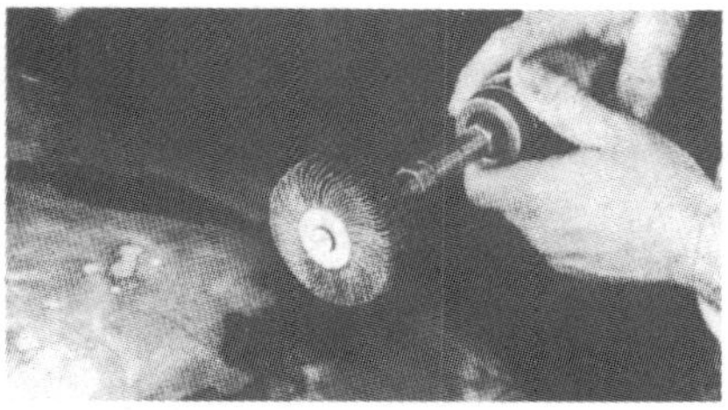

13 *An abrasive "flap wheel" removes surface rust quickly without damaging sound metal. Old body filler and drill holes are evidence of previous repair work.*

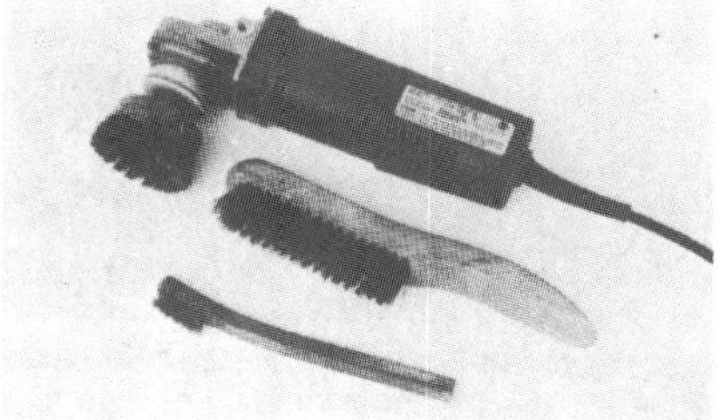

14 *Wire brushes are a backyard bodyman's weapons against rust. Oxidized metal and rust flakes must be removed before applying corrosion treatments.*

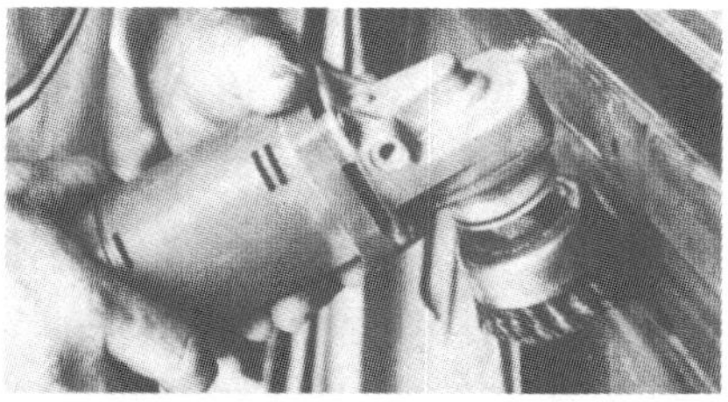

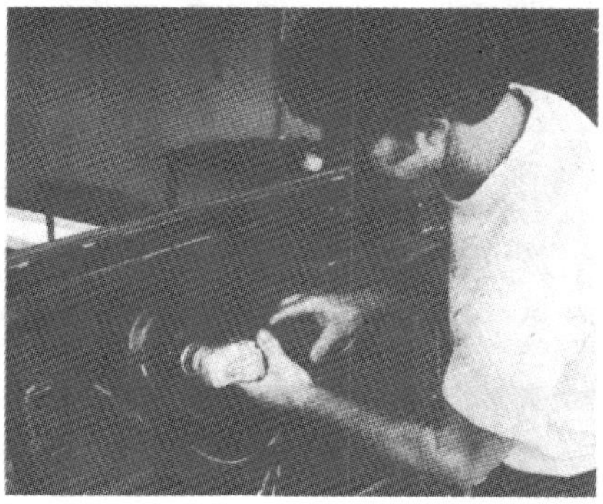

15 *An electric angle grinder and a stiff wire cup brush are very effective when cleaning a Camaro hood. The underhood can be painted satin black for a factory-fresh appearance, or sprayed body color for a custom touch.*

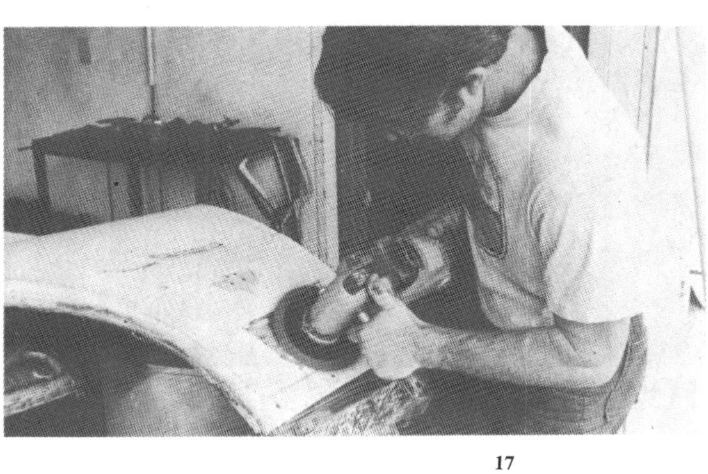

12 *Use an old disc as a backing pad when grinding away rust and paint. The flexible disc conforms to the body contours. Safety goggles are essential!*

inexpensive paint jobs. If you intend to repaint your F-body with lacquer, the enamel coat must be removed. Jim recommends that enamel *never* be sandwiched between two layers of lacquer.

If your Camaro's existing paint gets a passing grade, you can break out the wrenches and begin removing the trim, moldings, emblems, and bumpers. Wash the car thoroughly with detergent and water, then clean the body panels with wax and grease remover (Jim recommends DuPont 3919S or R-M 900 Pre-Kleano). Block sand the body with 180-grit paper and spot prime and fill as necessary. Sand the edges of the doors and hood down to bare metal; old layers of paint build up in these areas, and are very prone to chips and cracks. If you don't plan on priming the entire shell, then it's a good idea to seal the old paint with DuPont 2129S (or equivalent) sealer. The sealer increases the adhesion of the new top coat, and decreases the chances of the new paint irritating the old layers. If the car is especially rough, sand it with 80-grit abrasive and prime the entire body. Your Camaro is excused from the following session on stripping, so you can skip ahead to the section on priming and surfacing.

If your Camaro's present paint is in poor condition, it must be removed. Sandblasting and grinding are two mechanical methods of paint removal; the old finish can also be stripped chemically by dipping the body or applying paint remover. Sandblasting large areas requires a big air compressor, and the work is tedious. Also, it's easy to ripple body panels if the pressure is too high, and it's difficult to clean all the grains of sand out of body seams and crannies. It's best for novice Camaro painters to limit their sandblasting to small areas where there is corrosion damage.

Most large cities have commercial paint-stripping services with tanks large enough to immerse an entire car body. Dipping a Camaro is a big job—you'll have to remove every piece of trim, weatherstripping, and wiring before the shell goes into the tank. If the dipping is done properly, every inch of paint will be removed without damage to the sound metal. Commercial strippers sometimes use phosphoric acid in their solutions which dissolves rust and inhibits corrosion. Before sending a Camaro off to the stripping tank, however, you should do some investigating. Make sure that the formula used by your prospective stripping company is compatible with your Camaro's metal, and ask about neutralizing the solution to prevent future corrosion in inaccessible areas. Remember, too, that dipping a shell removes the paint from places where it can't be replaced. It's relatively easy to repaint the inner panels of an antique car, for example, but it's tough to respray inside a Camaro's rocker panels and under its dashboard.

Grinding off the old paint is an inexpensive and effective technique for do-it-yourself Camaro restorers. All that's required is a grinder (electric or pneumatic), an assortment of 24- and 36-grit discs, and a reverence for the sheetmetal. As Stevens points out, you should leave some metal for the next owner! If your car has previously been repaired or repainted, the panels may already be thin. Most Camaro sheetmetal is relatively light-gauge steel, so use restraint when grinding.

Jim's grinding technique deserves special mention. He replaces the grinder's hard rubber disc backing plate with a worn-out abrasive disc. Then he prepares a fresh grinding disc by trimming it into a hexagon. This allows the disc to conform to the panel contours. Jim continues to trim off the edges of the disc as the abrasive loads up with old pigment. Stevens cautions, however, that this makes the grinder very "aggressive." Exercise caution to prevent gouging or scratching the panels. Lacquer primers can shrink into these underlying scratches six months after you've painted your Camaro, ruining the paint job. Of course, you should also use caution to protect your own body when grinding. Always wear eye protection and a dust mask!

Brush-on paint stripper is probably the best way for a Camaro refinisher to remove old pigment. Stripping a complete F-body will probably take two full days of work, depending on how many coats must be removed. Each application of stripper will usually remove one layer of old paint. Follow the manufacturer's instructions carefully; paint stripper contains some rather nasty compounds. Again, eye protection and rubber gloves are essential! Most strippers perform best when applied with a single brush stroke. Keep the paint saturated with stripper, and work on one area at a time—a fender or door, for example. A putty knife works well to remove the top layers of paint. As you get closer to the sheetmetal, switch to steel wool or plastic scrapers to avoid gouging the sheetmetal.

Various types of paint react differently to paint stripper. Enamel tends to wrinkle up and come off in large sheets after the paint remover is applied; lacquers generally form a gummy solution that must be scraped away. The primer that GM sprays on Camaro sheetmetal is especially difficult to remove. Coarse steel wool and lots of stripper are usually the only way to dislodge this body primer. In fact, if the factory primer is still in good condition, it can be left on the car. This will cut the stripping time in half. Most top coats, in contrast, will dissolve quickly when the paint remover is applied.

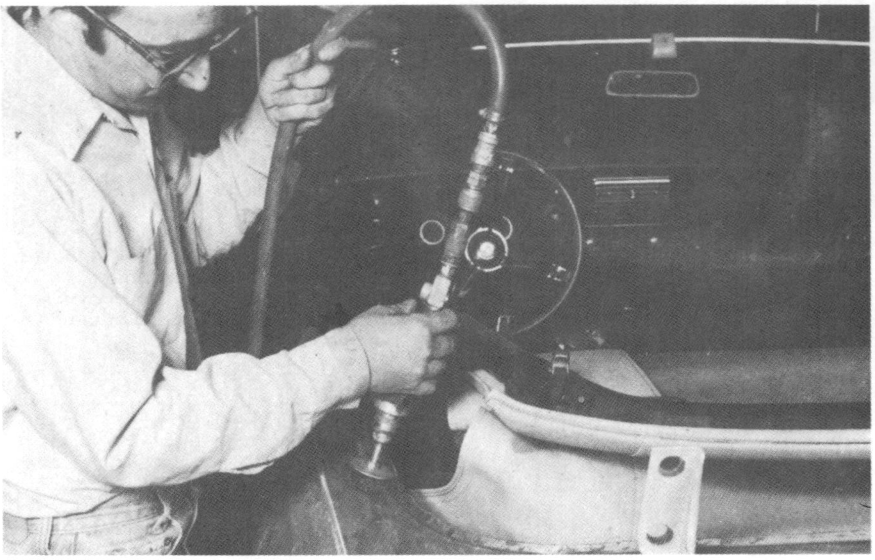

16 *Abrasive Scotchbrite discs quickly remove rust and old adhesive from weatherstrip channels. Pneumatic tools are lighter than electric grinders, but require a high-volume air supply.*

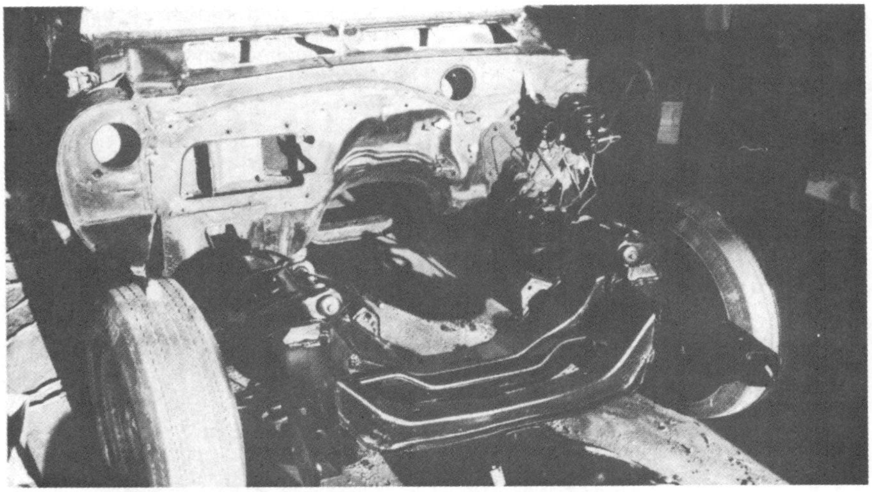

17 *If you want your Camaro to gleam from top to bottom, now is the time to prep and paint the subframe. Removing the powertrain is easy when the fenders are out of the way. Epoxy paint is highly recommended for frames and suspensions because it resists oil, gas, and chipping.*

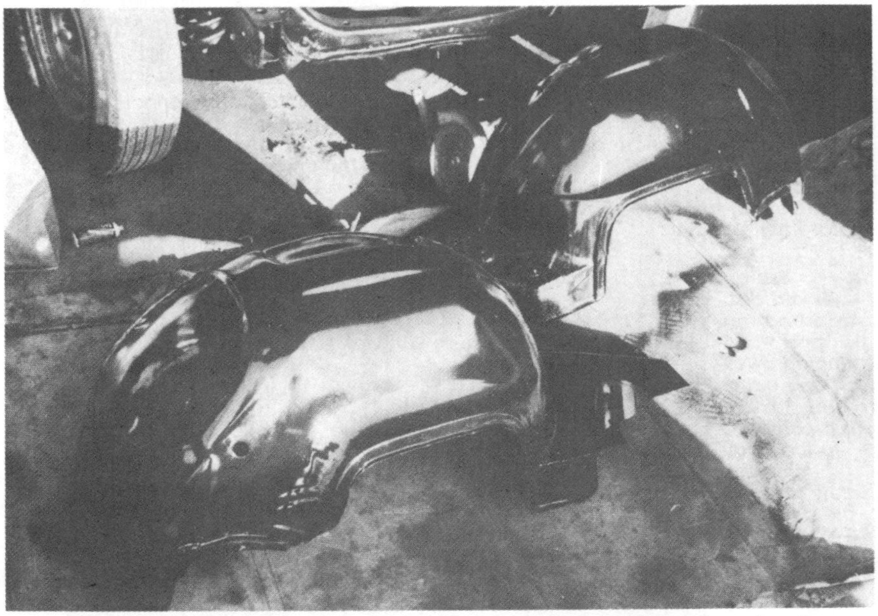

18 *Inner fender panels are also candidates for epoxy paint. Remove rust and oxidized metal under the battery tray before spraying.*

RUST REMOVAL

Corrosion is the curse of every Camaro owner. As singer Neil Young once noted, "Rust never sleeps." Once corrosion begins, it will continue to eat away sound metal until the entire body is oxidized. The only sure cure for Camaro rust infections is to cut out the cancer.

You can easily evaluate the extent of any rust damage after stripping the old paint off your Camaro. Investigate any signs of bubbling or cracking in the sheetmetal; use a penknife or scribe to pick at the rust. If the corrosion is on the *inside* of the panel, your probe will usually penetrate the metal. Body seams are particularly susceptible to rust; examine the fender lips, door panels, and rocker panels carefully.

It's important to distinguish between surface rust (which can be sanded off) and serious corrosion damage. An abrasive "flap wheel" will remove surface rust quickly. For more serious infections, the only solution is to cut out the damaged area and weld in new metal or a formed patch panel. Use minimal heat to prevent warping the surrounding metal, and finish the repair with body filler as described below.

In some areas, it's simply not possible to replace rusty panels. If the corrosion damage is not severe, the rust can be arrested with special metal conditioners and primers. Stevens has obtained good results with R-M 834 ZIP zinc phosphate primer. In fact, he uses the zinc phosphate to prime the entire car after stripping. First the rust scale is mechanically removed with sandpaper, a wire brush, or a sandblaster. Then the bare metal is cleaned with R-M 900 Pre-Kleano wax and grease remover and sprayed with a thin coat of 834 ZIP primer following the manufacturer's directions. This primer should not be built up, and it should not be sanded; when applied correctly, it will be slightly transparent. DuPont metal conditioner is an alternative treatment favored by some painters; ask your paint supplier for recommendations on its use.

BODY FILLER

Plastic body fillers (a.k.a., "bondo") are a bodyman's best friend. Some enthusiasts hold body fillers in low regard; in truth, bondo's bad reputation is not deserved. Most problems with body filler are caused by poor preparation of the underlying metal or improper application.

Moisture trapped under the body filler will foster rust and cause future problems. That's why Stevens favors zinc phosphate primer—it eliminates rust formation under the bondo. It is very important, however, to allow the primer to dry at least 24 hours before applying body filler. This gives time for the solvents in the primer to evaporate completely. It's permissible to put bondo over ZIP primer, but it should never be applied over lacquer primer or

a top coat because the escaping solvents will soften the filler. Always use as little body filler as possible.

Straighten body panels with a hammer and dolly instead of ladling on an inch-thick layer of plastic. Avoid deep fills (over 1/4-inch), and don't use filler to build up edges—it will be fragile. Bondo performs best when it is skimmed. Stevens points out that it is better to have 1/32-inch of filler than 10 coats of primer/surfacer.

When mixing two-part body filler, always use as little catalyst as possible. The final mix should be light pink; if the bondo is salmon colored, the percentage of catalyst is too high.

PRIMER/SURFACER

Priming and surfacing your Camaro's body panels properly can make the difference between an average paint job and a knockout. It's the job of the primer/surfacer to produce an arrow-straight foundation for the color coats.

The problem with conventional lacquer and enamel primers is the solvents they contain. When these primers are applied thick enough to fill the surface irregularities, the solvents trapped in the primer may take weeks to escape. If the primer is sanded before these solvents evaporate, the paint will shrink into the irregularities. The result is that grinder marks, sanding scratches, and other roughness may not show up until after the top coat is applied.

One solution to this problem is to use polyester resin primer such as Morton's Eliminator 100. This primer does not rely on solvent evaporation to cure; instead, it cures through catalytic action. This type of primer/surfacer can also be applied thicker than conventional primers without inviting cracking problems. Jim advises that polyester resin primer should not be applied over fresh paint that may still have solvent left in it; the primer will seal-in the solvent and prevent the underlying paint from curing. Polyester primer can, however, be sprayed over a thin coat of zinc phospate primer that has dried for 24 hours.

Jim usually applies two thick coats of Morton's Eliminator 100, then sprays a third coat over any rough spots. After the primer/surfacer has cured, he sprays a "guide coat." This is a light, fog coat of contrasting color made by mixing one part red oxide primer with nine parts of reducer. Jim sprays this guide coat at low pressure (35 psi). Its purpose is to reveal the high and low spots when the body is block sanded.

The longer the sanding block you use, the straighter your Camaro's sheetmetal will be. A flexible plane sander is good for roughing work because it conforms to the metal's contours. A common paint mixing stick

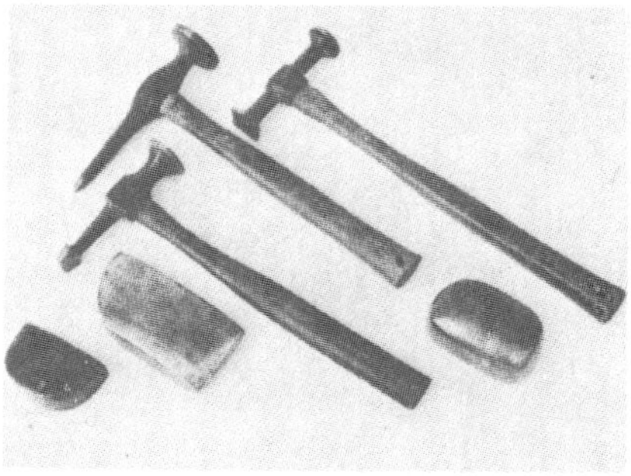

19 ***Don't try to straighten your Camaro's sheetmetal with a claw hammer. Body hammers, picks, and dollies are the right tools for the job.***

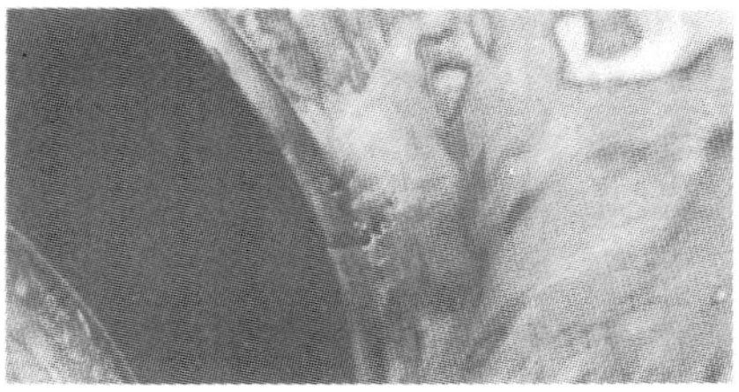

20 ***Bubbling paint is often a sign of rust in the inner panels. Water trapped between body seams fosters corrosion. Use a pick to determine the extent of the damage.***

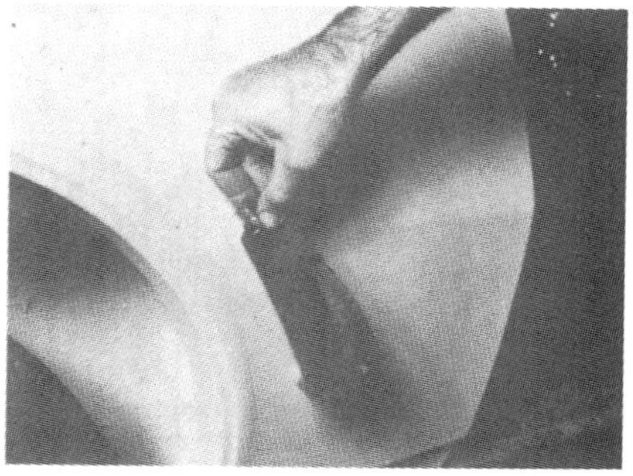

21 ***The only sure cure for rust is to cut out the cancer. You can purchase replacement panels to repair large areas, or form small patches from sheet steel. Use minimum heat when welding or brazing patches to prevent distorting the adjacent sheetmetal.***

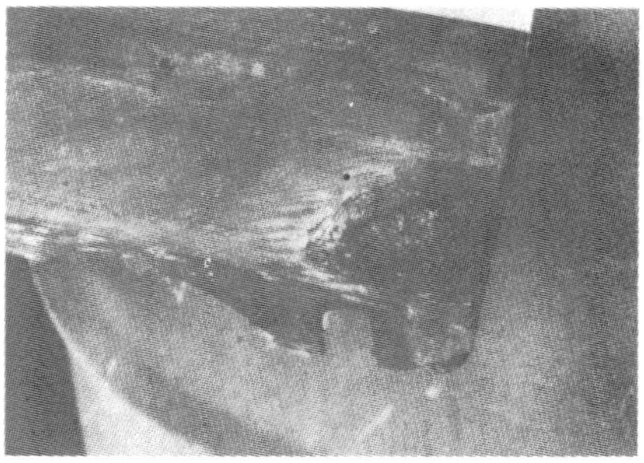

22 ***Check the bottom edges of the front fenders carefully for rust. Mud and leaves are likely to collect there, trapping moisture that leads to corrosion.***

PAINTING

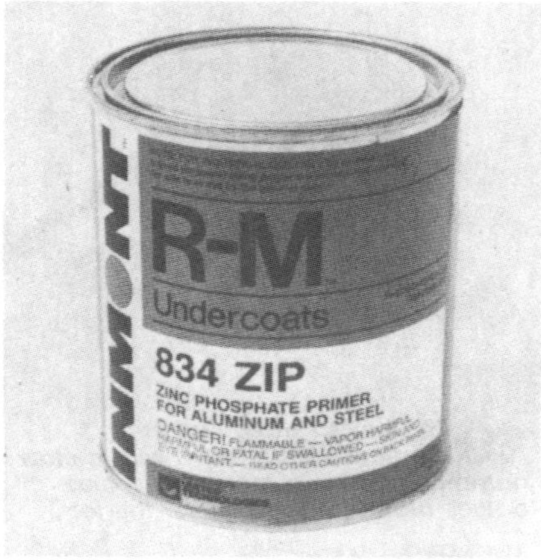

23 *Zinc phosphate primer will prevent further damage to areas already attacked by corrosion. Spray two thin coats; primer should be slightly transparent after application. Read and follow all directions on label! Do not build up zinc phosphate primer, and do not sand it.*

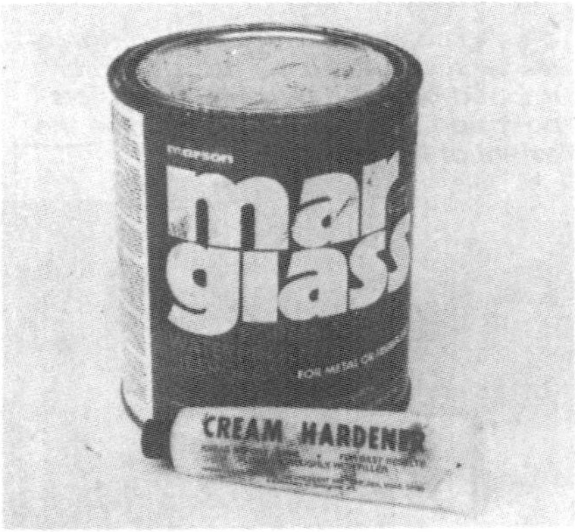

24 *Corrosion damage can be patched with a high-strength filler such as Marglass, a mixture of polyester and fibers.*

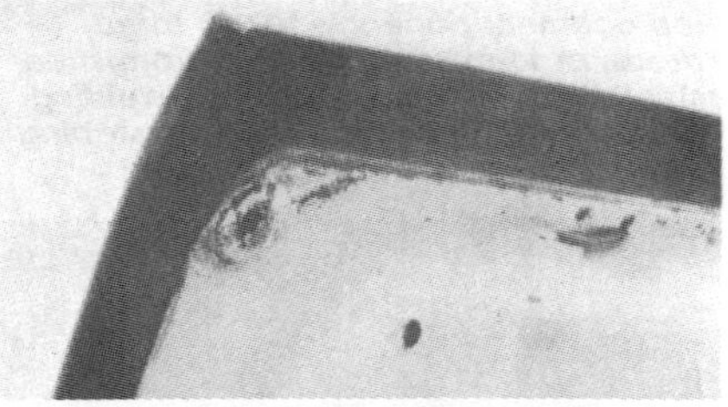

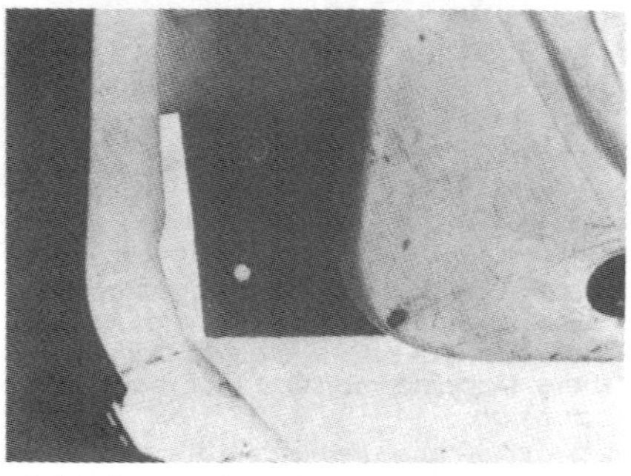

25 *To repair rust around trunk lids and door edges, first grind or sandblast the oxidized metal. Then coat the surfaces with zinc primer and patch with Marglass. Then caulk the seam to seal out water.*

also works well. Jim has found that rubber sanding blocks usually aren't long enough to produce straight body panels.

Stevens selects 180-grit sandpaper for the first pass over the guide coat. He wets the panels with a hose, bucket, or sponge to prevent the guide coat from clogging the abrasive. The goal is to sand off all of the red oxide primer; low spots that do not sand out should be filled with plastic body filler. Fortunately, this filler is perfectly compatible with the polyester primer, and can be applied over it. Jim then seals the filler and any bare metal spots with a light coat of DuPont Prepar-a-Kote enamel primer; he does not apply more polyester primer at this point.

After the entire area has been blocked and filled, Jim applies another light guide coat and then block sands it again with 320-grit wet-or-dry paper. This is the final straightness check; the guide coat should sand off evenly across the entire surface. If Jim intends to use lacquer for the final coat, he makes one more quick pass with 360-grit paper.

MASKING

Don't shortchange your Camaro by using cheap masking tape. Buy good quality tape that adheres well and leaves sharp edges for the paint to follow. You can use old newspapers to mask off windows and wheelwells, but professional masking paper is less porous and has fewer loose fibers that might contaminate the top coat. Stick the masking tape to the paper's edge on a workbench, then put the paper and tape on the car. You'll find this technique is easier than trying to hold the paper in position while you apply the tape.

When masking off chrome trim, leave a small gap between the tape and the body panel. If the tape is right against the metal, the paint will bridge to the masking and leave a ridge when it is removed later. Let the paint cover the underside of the chrome trim slightly. Even if the paint was completely cleaned off, the color would still be reflected under the chrome.

After masking the entire car, Stevens wipes down the body with DuPont 3812 synthetic reducer and a clean, damp cloth. Then he wipes it again with a dry cloth, and finishes up with a tack rag.

CHOOSING A TOP COAT

Today a Camaro enthusiast has several types of paint to choose among. The color, the car's intended use, and the skill of the painter all influence the decision on which kind of paint is best for a particular car. Acrylic lacquers are easy for a novice to spray, especially in metallic colors.

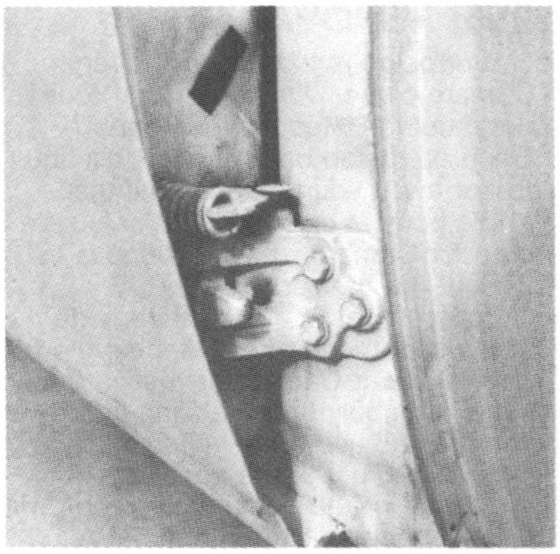

26 *Door hinges are easily accessible after front fenders are removed. Loosen bolts and adjust door position to produce consistent body seams.*

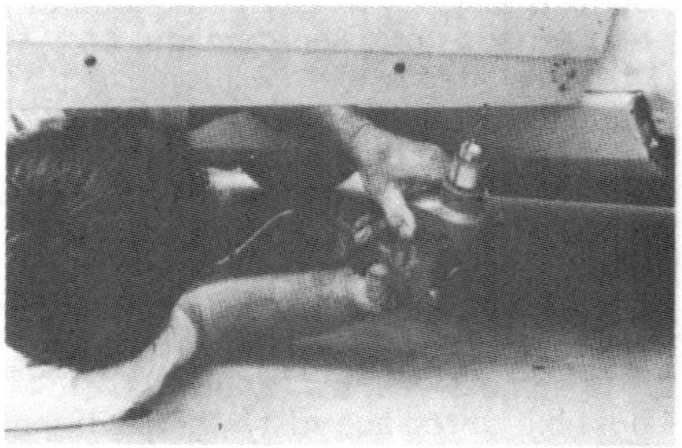

27 *Clean out drain holes in bottom of door to prevent future rust damage. Additional holes are good insurance against corrosion.*

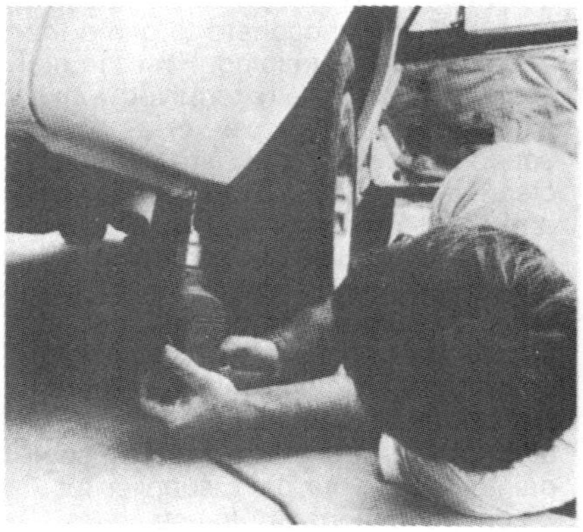

28 *Rear fenders are also water traps if trunk weatherstrip leaks. Drill several small holes to drain water.*

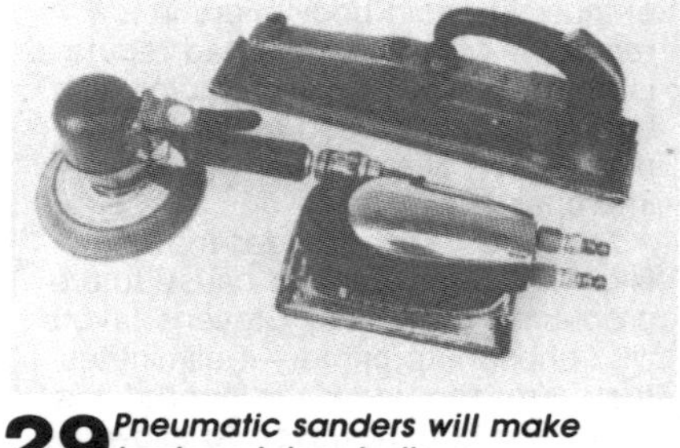

29 *Pneumatic sanders will make bodywork less tedious. Straight-line sander smoothes flat panels; orbital sander is adept at grinding body filler. Jitterbug sander is useful for feathering bondo. Sanders are used to contour and smooth body filler—never for finish sanding.*

Acrylic lacquer dries quickly; a car can be painted in a driveway on a still morning with acrylic lacquer. Acrylic lacquer's fast drying time makes it less susceptible to damage from stray bugs and floating dust particles. Despite these advantages, Stevens still prefers acrylic enamels for solid colors. In Jim's view, acrylic enamel offers better coverage than acrylic lacquer, and does a better job of filling imperfections. He's also found that enamel is more durable and less apt to chip. Finally, acrylic enamel can equal the beauty and shine of lacquer when it is color sanded and polished.

There are many excellent paint products on the market. Jim's personal favorites are DuPont Centari acrylic enamel for solid colors, and DuPont or R-M acrylic lacquer with Ditzler Delclear DAU 75 clear coat for metallics.

SPRAYING THE COLOR

It's essential to read and follow the manufacturer's instructions when applying any paint. *Always wear a respirator-type (charcoal filter) mask when spraying paint.* Breathing the fumes from conventional enamels and lacquers is bad enough, but modern acrylics can be poisonous. Protect your health with proper safety precautions so you can enjoy driving your refinished F-body!

Stevens applies the top coats as wet as possible without causing runs. This allows the paint to flow out smoothly. If the paint goes on too dry, it will orange-peel or become rough. Jim feels it's better to apply the paint on the wet side because it's easier to repair a run than to fix orange-peel.

The air temperature and humidity affect the paint's behavior. Jim mixes his paint according to the manufacturer's recommendations, then sprays an old trash can or fender to test the paint's performance. If the paint does not flow out properly, he adjusts the spray gun pressure and the percentage of thinner to achieve the desired consistency. The paint can be applied heavily to the car's horizontal surfaces (hood, roof, and trunk) without fear of runs; these panels should be especially smooth because they are more visible to onlookers.

Stevens uses four to five quarts of Centari acrylic enamel when painting a Camaro with a solid color. He thins the enamel with 8093S reducer until it flows well. Jim sprays on a heavy first coat, then follows up with a light second and third coat. Metallic colors require a slightly different technique. Jim generally uses about three quarts of acrylic lacquer color, thinned 1.5:1. The trick to getting good results with metallic colors is to spray a series of light coats, being careful to distribute the metallic particles evenly across the panel surfaces. If metallic paint goes on too "wet," the aluminum powder it contains settles and causes a blotchy appearance. After the acrylic lacquer

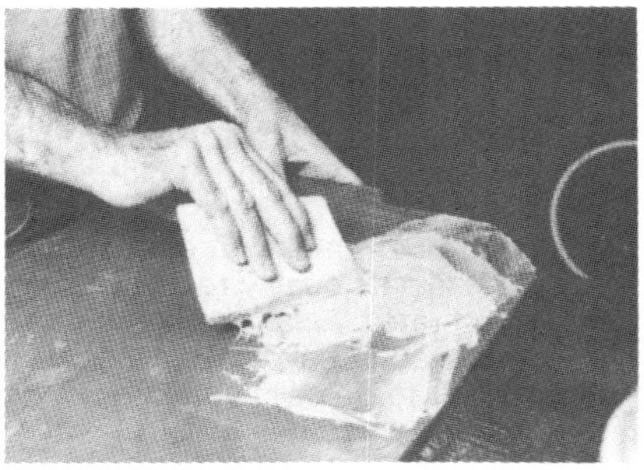

***30** Mix bondo and catalyst thoroughly to obtain uniform pink color.*

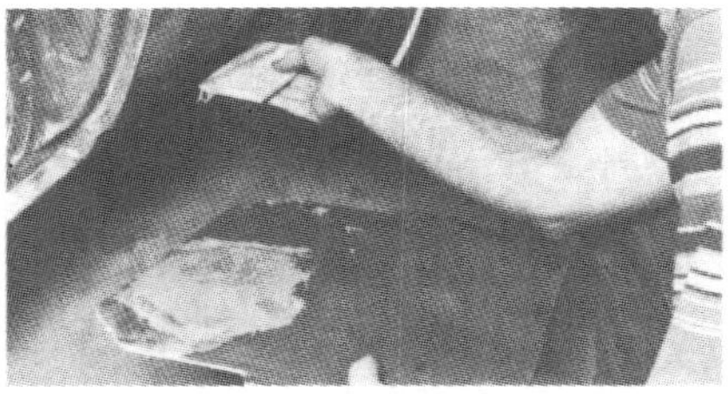

***31** Apply body filler with flexible rubber squeegee. Avoid using bondo on deep fills (over ¼-inch) and on edges of body.*

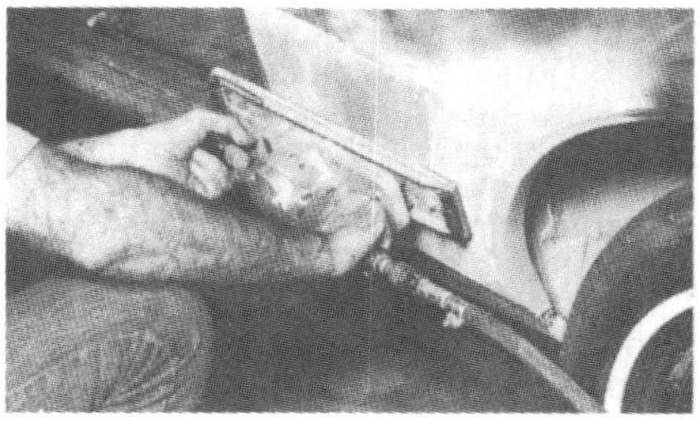

***32** Straight-line sander highlights low spots in body filler. Add more bondo and sand again until surface is smooth. Avoid body creases when using pneumatic sander—it's easy to grind through the metal.*

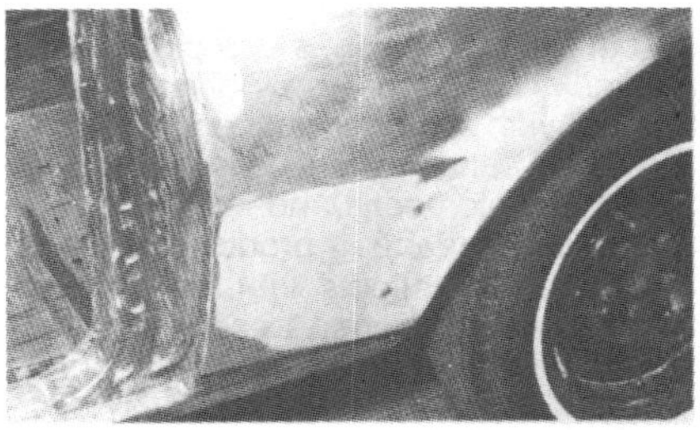

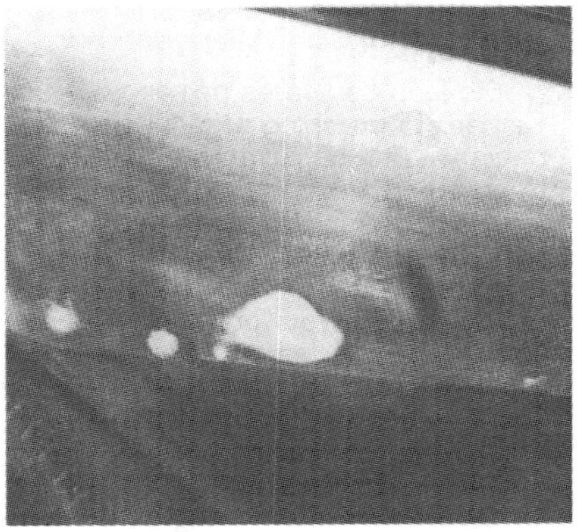

***33** Feather edges of body filler, then prime to prevent bondo from absorbing moisture.*

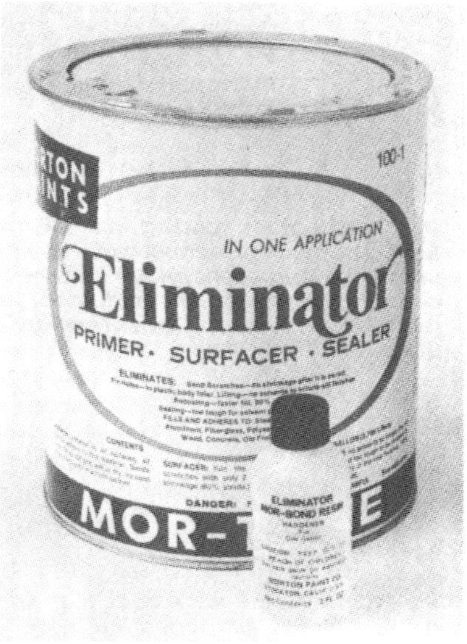

***34** Morton's Eliminator 100 polyester primer cures by catalytic action, not evaporation like conventional primers. This primer fills irregularities quickly, and won't shrink into cracks and sanding marks as it dries.*

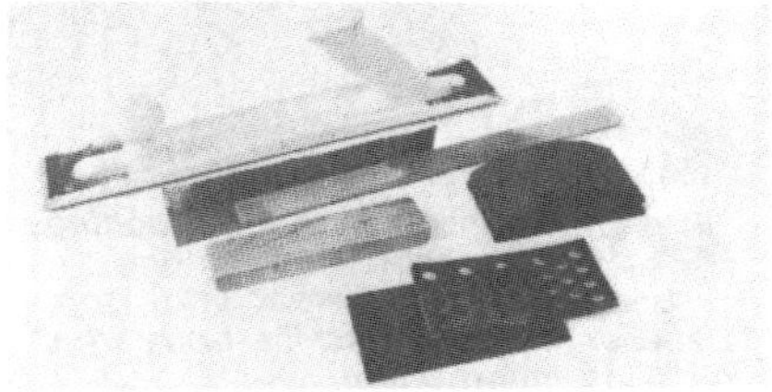

***35** Careful block sanding is the key to a perfect Camaro body. A flexible wooden paint mixing stick is ideal for sanding flat body panels; sponge rubber pads work well around curves and in corners. Rubber sanding blocks are too short for block sanding large panels.*

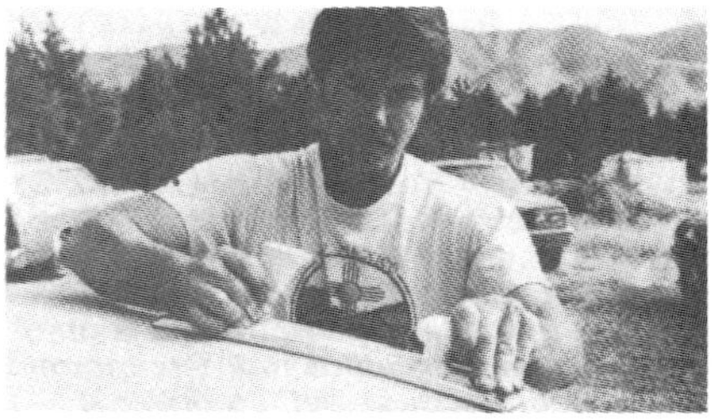

***36** Use a plastic plane sander with 80-grit paper to straighten large panels; flexible tool conforms to body contours.*

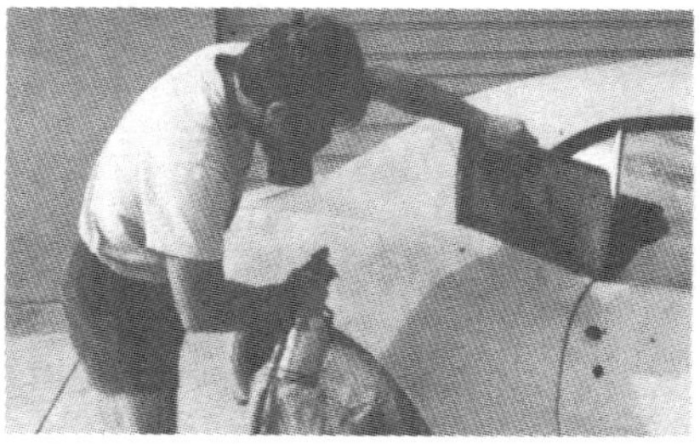

***37** Spray thin guide coat with contrasting primer to check body straightness. Reduce primer 200 percent and spray at low pressure. Always wear respirator mask when painting!*

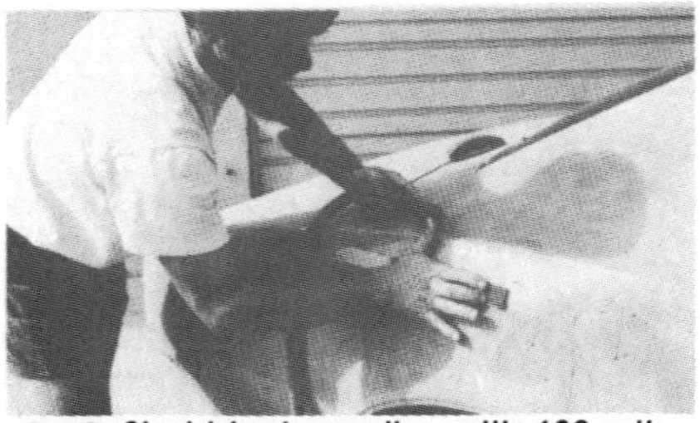

38 *Start block sanding with 180-grit paper. Keep panel wet with hose or sponge to prevent primer from clogging grit. Sand until guide coat is removed; fill low spots by skimming panel with body filler.*

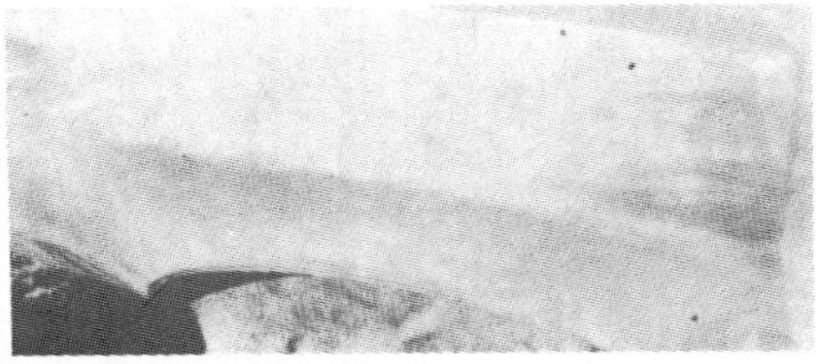

39 *Spray final guide coat, then sand with 320-grit paper. The guide coat will sand off uniformly if body is straight. Go over primer with 360-grit paper if you plan to spray lacquer.*

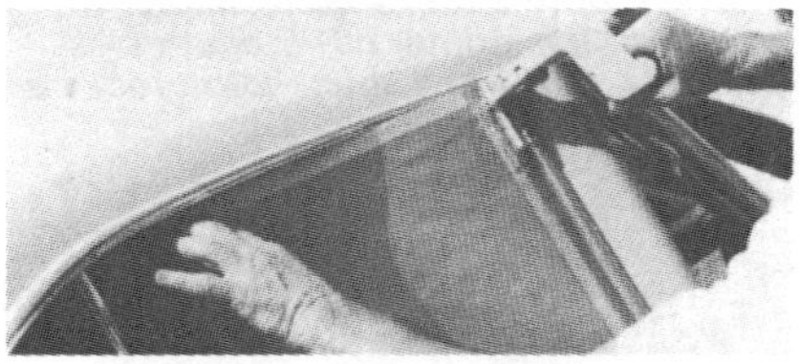

40 *Inexpensive masking machines are available at paint and building supplies. Leave a small gap between the tape and the body when masking chrome trim.*

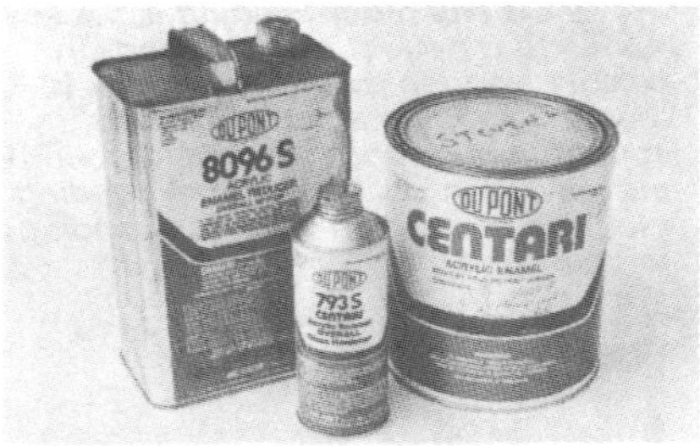

41 *Stevens recommends acrylic enamel for solid color paint jobs. He's found that it rivals the beauty of lacquer after color sanding, and has better durability and adhesion. Enamel is more tolerant of flaws in the preparation work than lacquer, but it requires a spray booth for best results.*

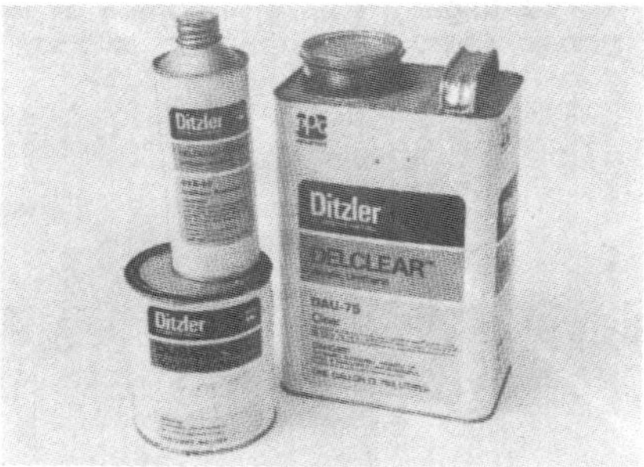

42 *Acrylic lacquer is preferred for metallic colors. Lacquer dries quickly; it can be sprayed in a driveway on a cool, still morning. Metallic paint must be covered with a clear top coat before color sanding.*

has dried overnight, Jim finishes the painting with a gallon of Delclear acrylic urethane. The purpose of this clear top coat is to protect the metallic powder when the finish is color sanded.

A paint booth makes it easy to turn out professional-quality results, but many first-rate Camaro paint jobs have been sprayed in driveways. Hose down the asphalt to deter dust, and spray in the morning when it is still. The car should be in the shade to keep the panels cool and to make it easy for your eyes to adjust to the light. If you are spraying lacquer, the temperature should be less than 75 degrees. Set your alarm clock and get up early on painting day!

COLOR SANDING

Color sanding is the salvation of first-time Camaro painters. Color sanding makes it possible for anyone to do top-quality paint work. It covers up mistakes and shortcomings in a novice's spraying technique. For color sanding to work its magic, however, the body underneath the paint must be *straight.* If your Camaro's sheetmetal is rough or bumpy, you'll sand right through the high spots.

Allow two weeks for the top coat to dry thoroughly before color sanding. Use 3M ultra-fine color sanding paper (approximately 1000-grit) to remove all traces of orange-peel and dust spots. Jim wraps the paper around a soft rubber flex-pad when color sanding; a small 1x2-inch piece of wood is a perfect sanding block for knocking down bug and dust specks. Sand in a straight line; the marks will be less visible than sanding in circles. Wet the sandpaper with a hose, sponge, or wet towel, and be careful not to trap pieces of grit under the abrasive. Sand the top coat until the finish is completely dull. Check your progress by drying the panel with a rubber squeegee.

Color sanding removes the microscopic peaks and irregularities in the paint surface. It produces a uniform surface that reflects light like a mirror. It may seem strange to sand your Camaro's fresh paint until it becomes dull, but have faith—you're really doing your F-body a favor.

POLISHING

Stevens compares painting a Camaro to polishing a gemstone. You begin with rough abrasives, then switch to finer and finer grits until you produce a flawless finish. Polishing compound puts the final luster on your four-wheeled diamond.

You can cut your polishing time by going back over each panel with a well-used piece of ultra-fine sandpaper. This will give the finish a little gloss. You'll need an electric polisher that turns at approximately 1700 rpm and an assortment of polishing bonnets for this phase. Apply 3M Heavy-Duty Rubbing Compound to a two-foot-square area, then spread the

PAINTING

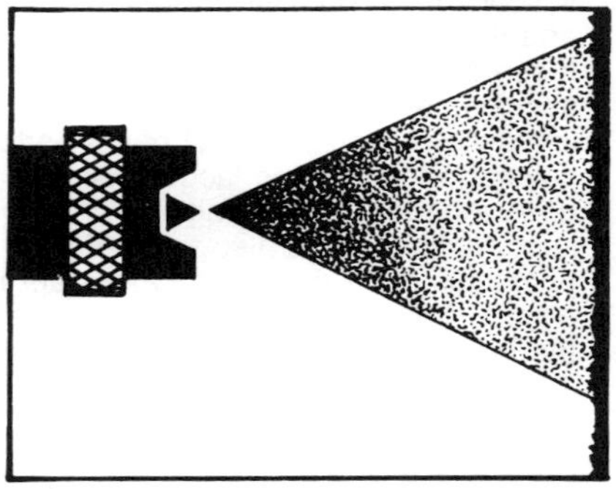

WRONG: Thin coat; rough; dry; no lustre.

REASONS: Fluid control valve is not open enough; gun is too far away; paint is too thin; too much air; stroke is too fast; not enough overlap.

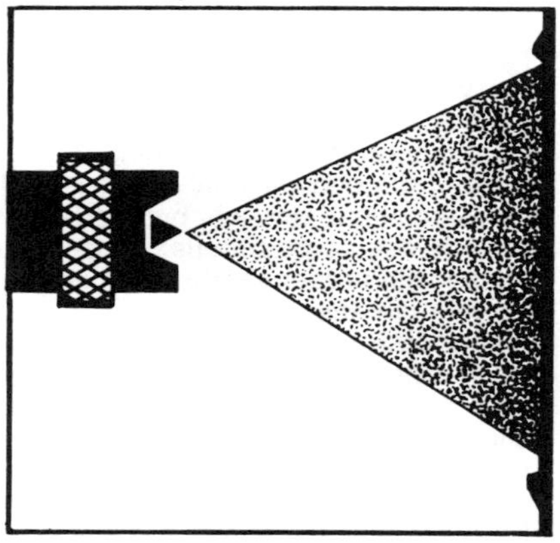

WRONG: Heavy coat with sags, ripples, or orange-peel.

REASONS: Dirty air nozzle; gun is too close; paint is too thin or too thick; low air pressure; stroke is too slow; too much overlap.

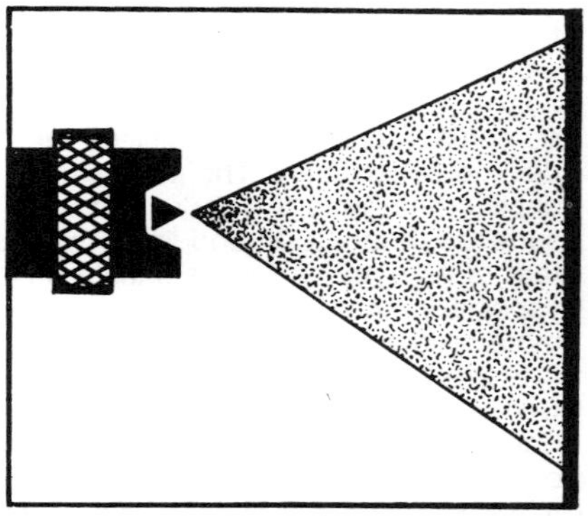

RIGHT: Medium coat; good flow-out with hardly any orange-peel, and no sags.

REASONS: Gun is clean and properly adjusted; gun distance is okay; paint is properly thinned; right amount of air; stroke is okay; overlap is 50 percent; correct amount of reducer or thinner used.

43 *Practice your spraying technique before painting your carefully prepared Camaro. Experience is the best instructor, but you can diagnose common spraying problems by comparing your results with this chart.*

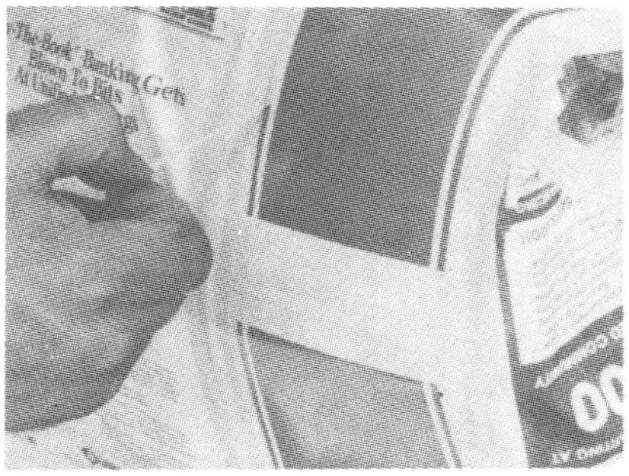

44 *Add special trim accents and stripes after applying the top coat. High-quality masking tape will produce sharp edges. Templates and stencils are available from Camaro parts suppliers to duplicate factory paint schemes.*

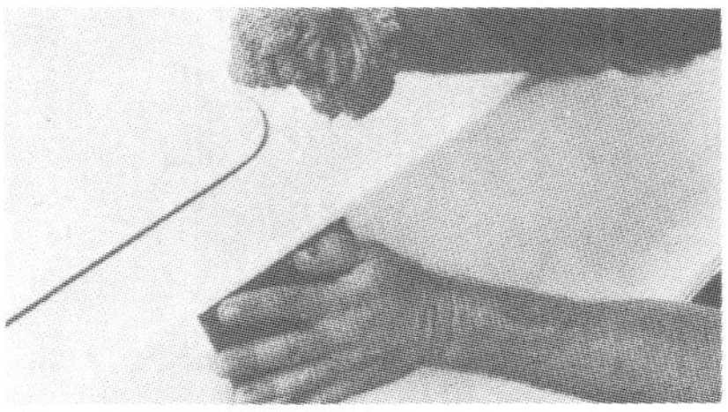

45 *Color sanding is the secret of a mirror-smooth paint job. Sanding and polishing the paint surface removes high spots and corrects spraying mistakes. Sand by hand with ultra-fine paper and a flexible backing pad. Wet the panel with a sponge or towel to prevent clogging the sandpaper.*

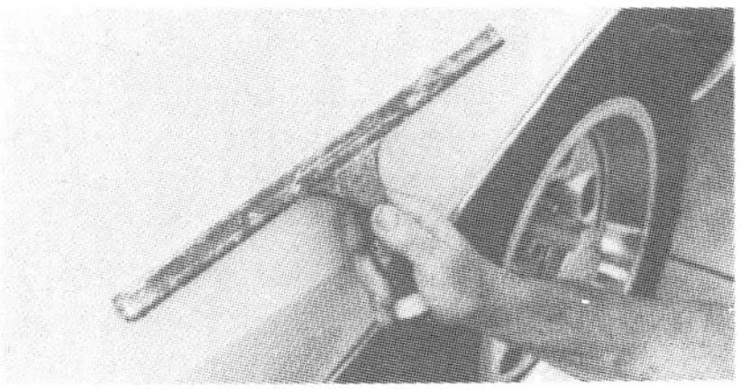

46 *Dry the panel with a soft rubber squeegee to check your progress. Sand until the entire surface is dull.*

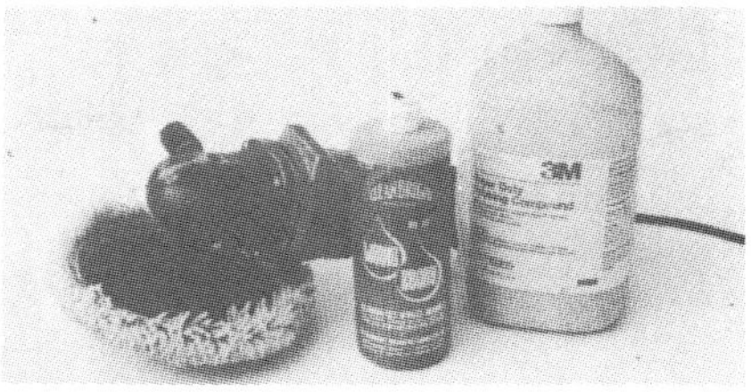

47 *Polishing the smooth top coat restores the paint's shine. Start with super-duty rubbing compound, then switch to Liquid Ebony for a dazzling finish.*

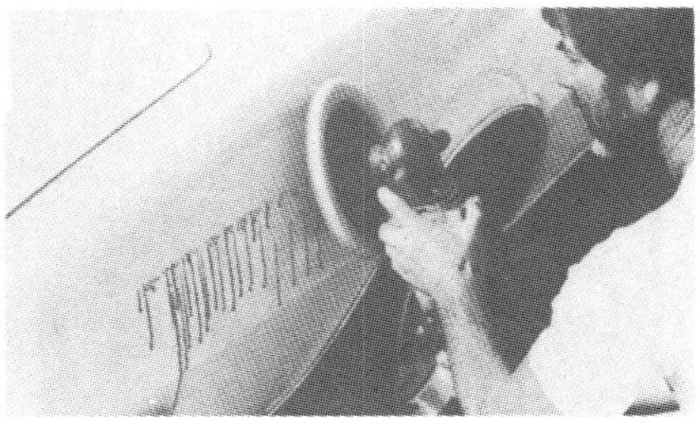

48 *Tilt the polishing bonnet as you finish each section. Avoid polishing the body creases and edges when polishing; you might cut through the color coat.*

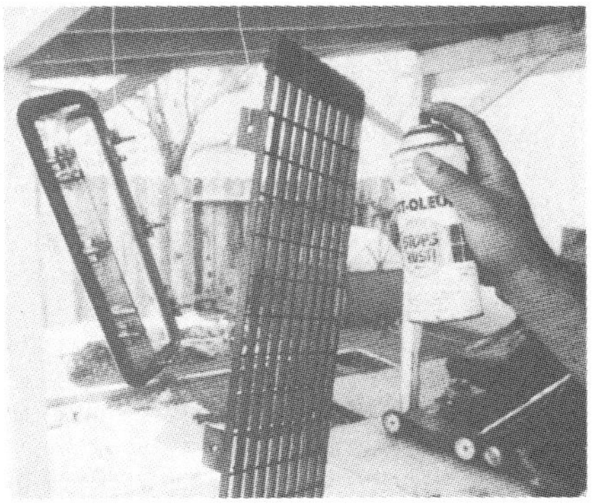

49 *Spray trim pieces with a high-quality aerosol paint or touch-up gun.*

compound with the polishing bonnet. This compound cuts the paint surface when it is wet, then it begins to polish as it dries. Go over each panel twice with the rubbing compound to remove all the color sanding scratches. Clean the polishing bonnet occasionally by running a screwdriver tip across the pad to remove used abrasive. As you finish each area, tilt the bonnet so you are polishing with its dry outer edge. Always avoid the body's peaks and ridges when sanding or polishing; you may cut through the color coat. Some painters protect these body creases with masking tape during the polishing phase.

The final step is to polish the finish to a high gloss with Liquid Ebony. The Liquid Ebony has a mild abrasive which removes the swirl marks left by the rubbing compound. Put a new pad on your polisher and squirt some Liquid Ebony over a two-foot-square area. Spread the liquid with the polishing bonnet. As the compound dries and the paint starts shining, reduce the pressure on the polisher. Wipe the surface with a dry rag, and stand back. Your eyes will behold the smoothest, shiniest Camaro in the state!

Painting your own car has immediate rewards. You save money—lots of it. The job is done right—and if it isn't, you have only yourself to blame. Best of all, the results are there for everyone to see. You might build the fastest small-blocks in the county, or the strongest 4-speeds in the city, but few will ever appreciate your efforts. An outstanding paint job, in contrast, is a visible testament to your Camaro skills. Just keep that vision of a perfect paint job firmly in mind when you're in the middle of your third weekend of block sanding. Always remind yourself, "It's gonna look great"—because it will.

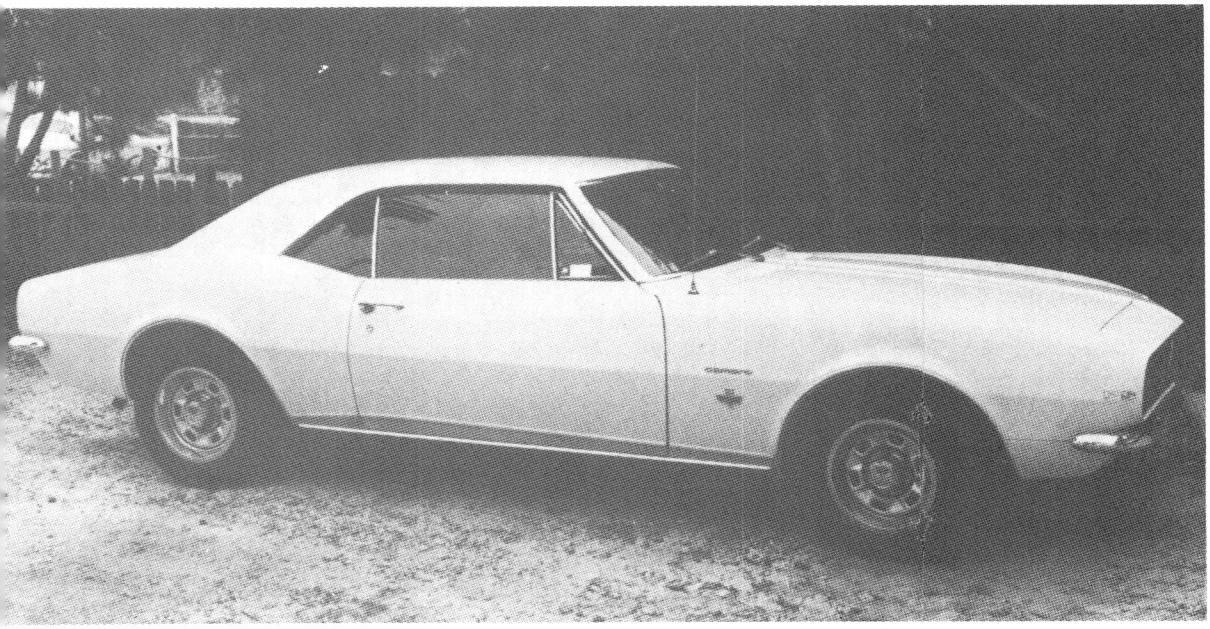

50 *Yes, this is the same RS (Rusty Sport) pictured at the beginning of this section. Four weeks of work and a fresh coat of Ermine White enamel are responsible for the Camaro's startling comeback!*

Camaro-Dynamics

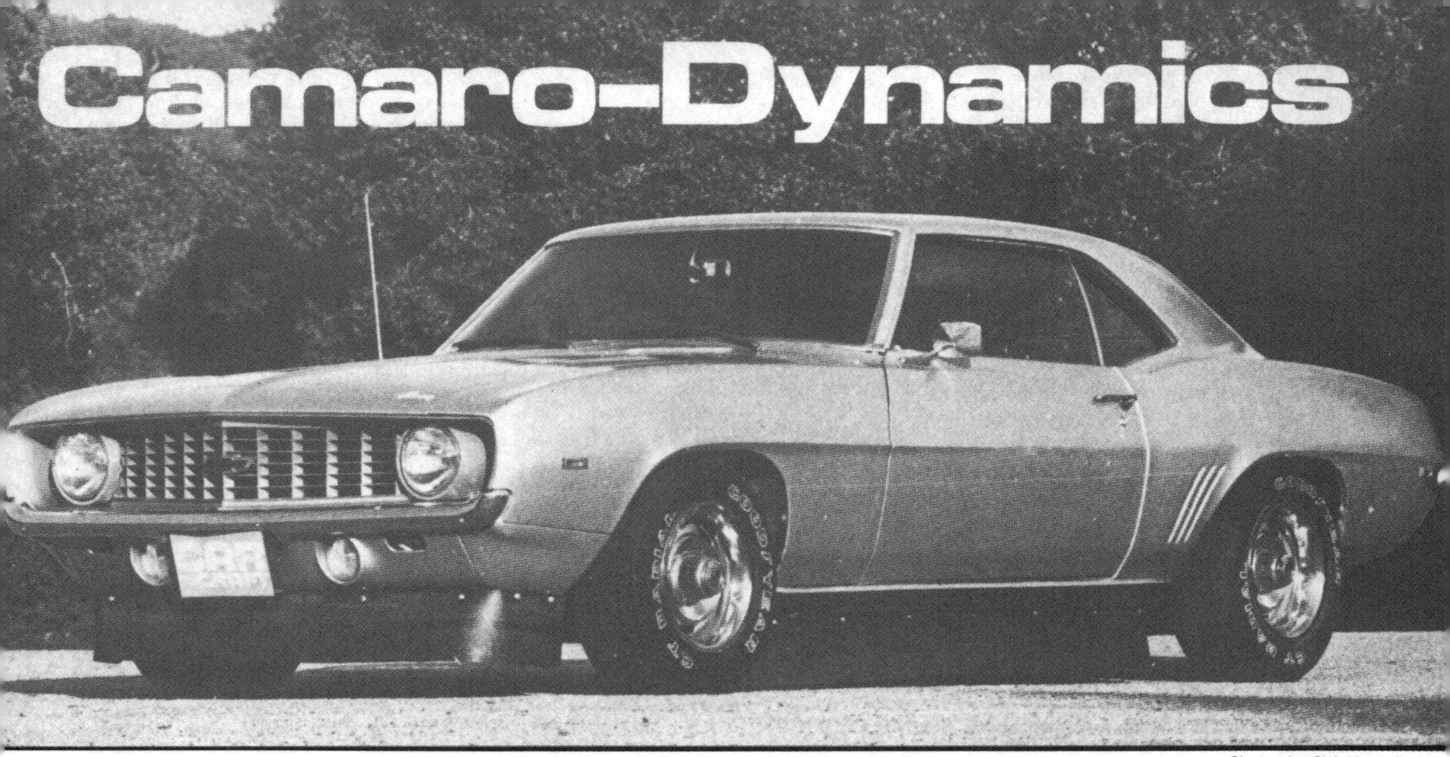

Photos by Rick Voegelin

An Investigation Into Applied Aerodynamics

By Rick Voegelin

The winds of change have reached hurricane force in the realm of the automobile. Years ago, Bob Dylan sang that you didn't need a weatherman to know which way the wind was blowing—but the automakers will quickly point out that you do need a wind tunnel, an army of engineers and enough instrumentation to launch a space shuttle to tell you how the wind is behaving around an automobile. The auto manufacturers' sudden concern for aerodynamics was prompted by the realization that reducing drag increases fuel economy. Racers have long exploited the ironclad relationship between aerodynamic drag and horsepower. Building a slippery body is like finding "free" power, whether the vehicle in question is a Formula One racer or a Funny Car. Until recently, though, aerodynamic effects were regarded as inconsequential at speeds under 100 mph. Even performance-savvy engineers scoffed at the notion that paying attention to airflow was a worthwhile pastime even in a car as fast as a Pro Stock. Nowadays these same engineers are reshaping trunk lids and redesigning door handles to help squeeze the last bit of mileage out of 55 mph econoboxes.

Building a low-drag body is an exercise that begins on a drawing board. A form's wind resistance is a product of its size, frontal area and shape. If a low drag coefficient is the primary concern, you end up with teardrop-shaped cars which would look right at home on the Bonneville Salt Flats. For most car enthusiasts, though, a streamliner is not a very practical solution to the problems of everyday transportation. In other words, you've got to work with what you've got.

What we had was an early model Camaro. Early Camaros may be a product of Sixties' engineering, but they are the quintessential race cars/street machines of the Eighties. There are enough of them on the highways and dragstrips to make an aerodynamic investigation worthwhile. And the techniques we applied can be adapted to any freeway flier. It's all a matter of applying current thinking to 13-year-old sheetmetal.

What we lacked was a wind tunnel, an army of engineers and sophisticated instrumentation. For our "real world" evaluation, all that was required was a smooth, level road, a willing copilot, and a collection of odds and ends that looked like the leftovers from a scavenger hunt. For an investment of $20 in supplies and an afternoon's time, we came up with data that can potentially make race cars run faster and street cars run farther on a gallon of gas. The answers to aerodynamic questions, my friend, are blowing in the wind.

THE $3 MANOMETER

Aerodynamic investigations require a measuring device to detect changes. Fortunately, most of the answers to your airflow questions can be answered by a simple U-tube manometer. Professional manometers are expensive, delicate and astoundingly accurate; our homebuilt version gives away a little precision, but it's dirt cheap and reliable as a rock.

A U-tube manometer measures pressure differences by moving a column of liquid—in this instance, water with a little red food coloring added. Both ends of the U-tube are open to the atmosphere. When one end of the tube is in a high pressure area (such as the leading edge of a car at speed), the pressure causes the column of water to rise. By noting the water's movement and comparing readings taken at various points on the body, it's possible to come up with an accurate picture of the air movement over the automobile. And once you know where the air is going, you should have a good notion of where hood scoops and cold air induction pickups belong.

A U-tube manometer can be constructed from a length of clear plastic tubing and a piece of cardboard. We refined the concept somewhat by adding a plywood back, a graph paper scale, and a 29¢ plastic ruler. The clear plastic tubing is attached to the plywood with metal clips used to hold wiring in place. To fill your homemade manometer, submerge one open tubing end in colored water, then

Car enthusiasts will spend weeks nitpicking a new engine assembly or massaging a trick suspension setup, but won't devote an hour to figuring out the potential gains from applied aerodynamics. The advent of the pocket calculator means that anyone who can punch buttons in sequence can come up with the right answers to aerodynamic questions.

The force required to overcome aerodynamic drag can be calculated as follows:

$$\text{force}=(.0025)\ (Cd)\ (A)\ (V)^2$$

where A=frontal area in square feet
V=vehicle velocity in mph
Cd=drag coefficient

A bit later we'll tell you how to come up with values for the variables. For now, though, we'll put the equation into terms more familiar to car enthusiasts—namely, horsepower required to overcome aerodynamic drag.

$$\text{horsepower}=(.0027)\ (\text{force})\ (V)$$

Without getting bogged down in numbers, these two equations say a lot about what it takes to go fast. Doubling the car's speed requires four times the force to overcome air resistance and eight times the horsepower! That's why reducing drag is just like stumbling over stacks of free horsepower.

So you're convinced, and want to take advantage of your new expertise in aerodynamics. The first thing to do is determine your car's total drag from the following formula, using the numbers you get from your coastdown test:

$$D=.0456W\ \frac{\Delta V}{\Delta T}$$

where D=resistance (in lbs.)
W=vehicle weight (in lbs.)
T=time interval (in seconds)
V=change in speed (in mph)

Now suppose you want to know how much power it will take to make your car run 120 mph. After you've come up with a value for "D" in the equation above, you can calculate the horsepower requirement from the following:

$$hp=\frac{D\times V}{375}$$

Determining a car's drag coefficient is a fairly straightforward procedure based on the numbers coming out of repeated coastdown tests. Space limitations prevent us from running through the whole drill; you can use .5 as a representative figure for most popular ponycars. Finding a car's frontal area is a snap; measure from door handle to door handle, and from the ground to the roof. Multiply the two dimensions and convert your answer to square feet.

Perhaps you might like to know if your new hood scoop is adequate to meet the demands of the engine. There's a way to figure that one out, too, using the readings you get from your homemade manometer. The air velocity in feet per second is equal to 66 times the square root of the manometer water level. Multiply this velocity by the area of the scoop opening (in square feet) and you've got the flow rate (in cubic feet per second) for the vehicle speed you tested at.

Isn't science wonderful?

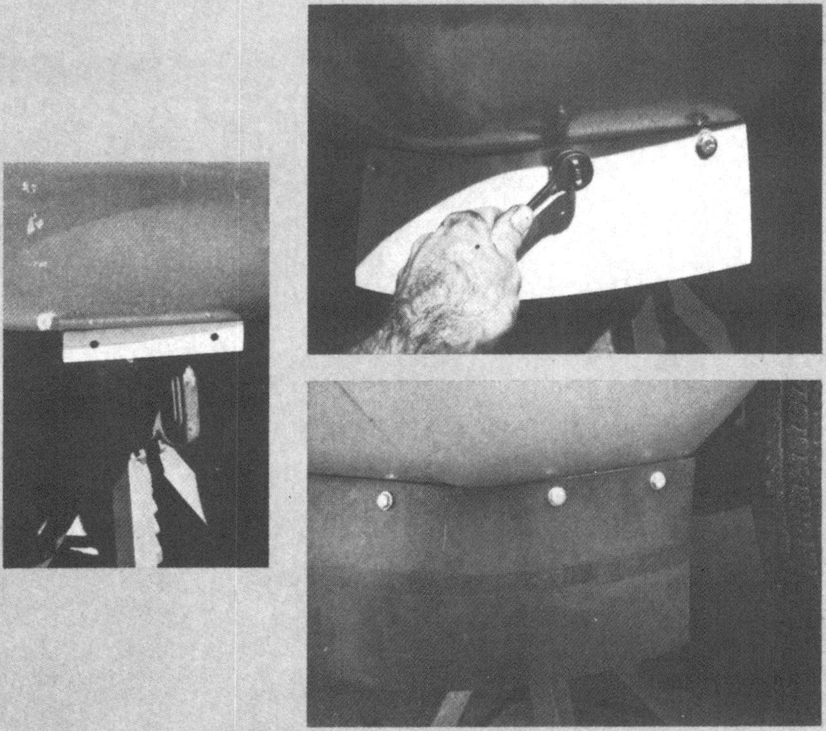

Building a ground-scraping front spoiler was a one-hour project. Factory 4-inch spoiler provides backing for 7-inch wide flexible neoprene. Six-inch length of 1x1 aluminum angle is attached to valence panel in front of tire. Then a 4-inch wide strip of flexible fiberglass is screwed to aluminum angle and OEM spoiler to form curve. Final step is to attach neoprene to backing with ¼-inch bolts and self-locking nuts.

draw a small amount of water into the "U" by applying suction to the other end.

Mount your manometer in the test vehicle so that it is level and easy to read. For best results, the manometer should be perpendicular to the direction of travel; mounting it parallel to the car's motion makes the manometer water react to changes in speed. (Come to think of it, this might be the world's most inexpensive accelerometer!) To use the manometer, tape the pressure end to the area you wish to test. Accelerate to a reasonably rapid speed (mayhaps just the other side of legality) and let the manometer reading stabilize. Then perform the same test, traveling in the opposite direction to negate the effect of wind. Average your two readings, note the result, and then move the probe to another position. Remember that you're seeking relative differences, so strive to keep conditions constant throughout your test. Always take your readings while the car is moving at the same speed, always make two runs in opposite directions, and keep the windows barely open and the vents closed to minimize pressure changes inside the car.

MEASURING AERODYNAMIC DRAG: THE COAST-DOWN TECHNIQUE

If you have a level road, an accurate speedometer and a wristwatch, you can measure an automobile's aerodynamic drag with acceptable accuracy. There are a number of methods which can be used to determine a vehicle's drag coefficient (see sidebar), but most enthusiasts are simply interested in whether a change helped or hurt their car's ability to cut the wind. The coast-down technique is actually preferred over wind tunnel tests by many engineers because it takes into account the interaction between the moving car and the stationary roadway.

Do your coast-down tests on a

Camaro-Dynamics

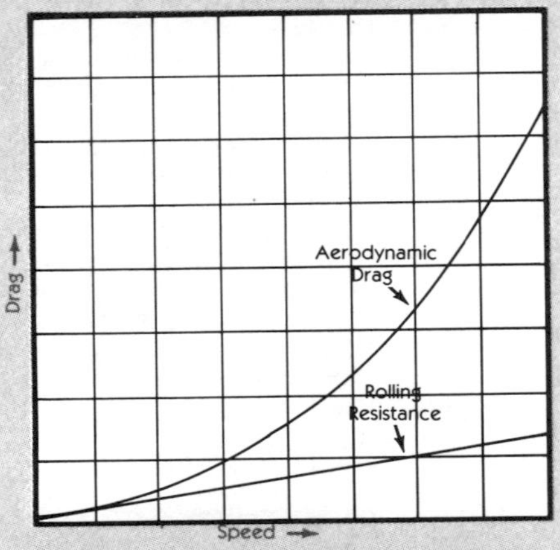

Test	Coastdown†	Calculated Drag ††	Comments
1. 7-inch flexible spoiler	23	.42	
2. Factory 4-inch spoiler	24	.39	Least drag of any configuration tested
3. No spoiler	23	.42	Same as 7-inch spoiler
4. No spoiler, windows open	20	.52	Drag increased 23 percent
5. Reinforced 7-inch spoiler, windows closed	23	.42	No improvement over Test No. 1

† Intervals in Seconds (Two-way average) †† Coefficient (Lower number = less aerodynamic drag)

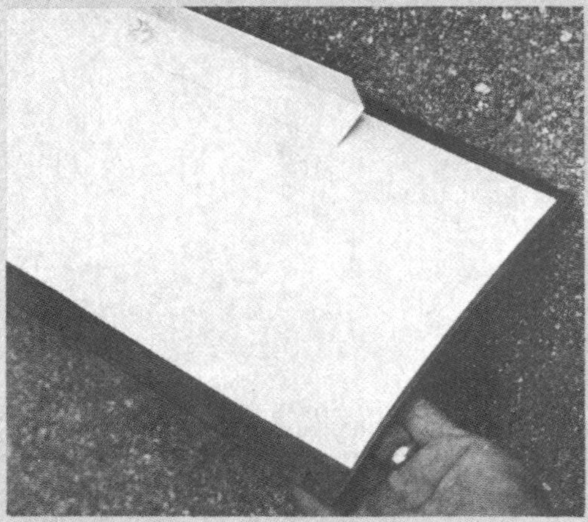

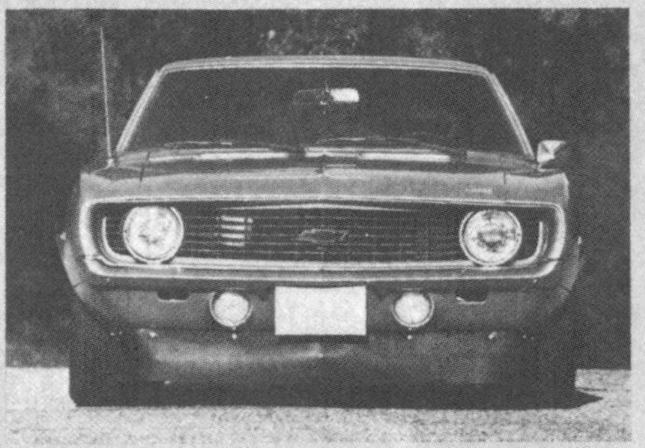

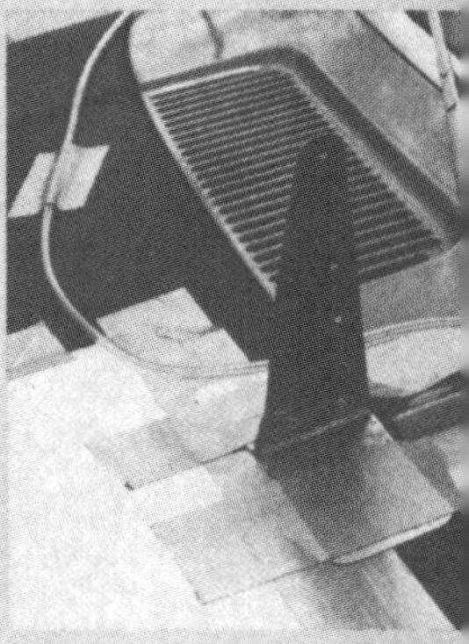

Flexible front spoiler effectively "hides" aerodynamically dirty undercarriage, yet resists damage from curbs and parking lot berms. Our tests indicate that increased frontal area produced by deep spoiler actually increased aerodynamic drag.

Side skirts are similar in construction to tall front spoiler. Aluminum angle is screwed to lower rocker panel. Flexible fiberglass backing provides support for neoprene skirt. ¼-inch bolts spaced on 6-inch centers fasten fiberglass and neoprene to aluminum.

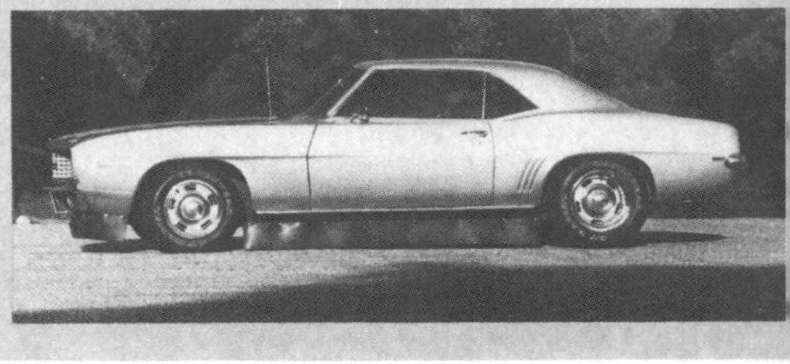

Formula One is in the eye of the beholder. Side skirts on "sucker" Camaro are more cosmetic than functional at highway speeds—but they do collect inquiring looks and double takes. Ground effects principle increases cornering speeds by producing low pressure area under car.

level stretch of road with uniform pavement, on a calm day with a minimum of traffic. You can choose from several methods. The first is a speed/distance function. Select two landmarks a reasonable distance apart (the greater the distance, the more accurate your data will be). Telephone poles, road signs and mile markers are all good. Accelerate to a predetermined speed, the faster the better. At the first mark, shift the car into neutral and let it coast. Note the car's speed at the instant it passes the second mark. Then turn around and repeat the test from the opposite direction, starting at the same speed, to minimize the effects of road grade and wind. Average the two final speeds, and you have your test "baseline." Changes which reduce aerodynamic drag will result in higher speeds at the end of your coast-down; changes which increase drag will produce slower speeds.

An alternate method is based on time. With the car traveling at a constant speed, shift into neutral. Then measure how long it takes for the car to slow to a predetermined speed—from 70 mph, say, to 40 mph. Perform the same test over the same road going the opposite direction to establish the baseline. Changes you make which reduce the vehicle's drag will increase the time it takes for the car to coast down; alterations which increase drag will cause the car to slow down more quickly.

TURBULENCE AND YARN TUFTS

Air moving over an automobile leaves the marks of its passing. Raindrops scurrying across the windshield and the streaks of dirt that remain after a storm are the signatures of aerodynamic forces. A less messy means of recording air flow is the yarn tuft technique. This consists of taping 3-inch long pieces of yarn across the car body at regular intervals and photographing the car at speed. Areas with turbulent flow are indicated by yarn segments which are lined up at odd angles, separated from the body, or blowing forward. Often one aerodynamic fix may cause

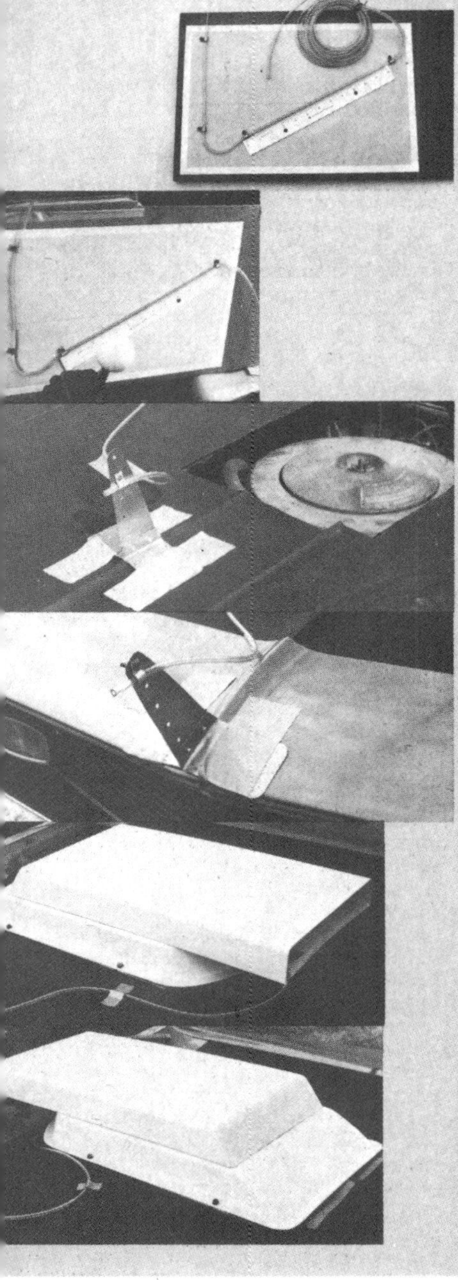

The do-it-yourself hood scoop test: First you need a manometer, which can be fabricated from a ruler, 15 feet of ¼-inch clear plastic tubing, a piece of plywood, four wire clips and colored water. Our test manometer was mounted level on trans tunnel, then taped to dashboard. First series of tests was intended to reveal best location for hood scoop inlet. Middle of hood and leading edge of grille both proved good, providing the probe was several inches above body. Positioning probe at base of windshield produced lower pressure readings. In second part of test, probe was placed inside dummy scoop. Forward-facing scoop proved superior to rear-entry scoop with inlet at base of windshield.

Test No.	Probe Position	Average Manometer Reading " H_2O	Comments
1	Front edge of hood valence panel, flush w/body	2	
2	Front edge of hood valence panel, 4 inches high	2¼	Increasing height increased pressure
3	7 inches from front, 1 inch high	2¾	Probe at hood/ valence junction
4	7 inches from front, 4 inches high	2¾	
5	24 inches from front, 1 inch high	2	Probe in "normal" scoop location over carburetor
6	24 inches from front, 4 inches high	2¾	Increasing height increased pressure
7	24 inches from front, 5 inches high	2¾	Raising probe did not affect reading
8	Base of windshield	1	Lowest reading of test
9	Ventilator plenum at base of windshield	1	No improvement over No. 8
10	Grille just above front bumper	2⅞	Highest reading of test
11	Dummy hood scoop, 5 inches high, 12 inches from front	2½	
12	Dummy scoop, 5 inches high, 24 inches from front	2¼	Scoop centered over carb
13	Scoop mounted backward, inlet at base of windshield	¾	Windshield entry ineffective

problems in another area—installing a boundary layer hood scoop, for example, may create turbulence behind the scoop which can be remedied by extending the base to the windshield.

SPOILERS, SKIRTS AND THE "SUCKER" CAMARO

In aerodynamic terms, the "dirtiest" area of any car is its undercarriage. The underside of a street car is a chamber of aerodynamic horrors, with suspension parts, drivetrain components and floor pan contours all adding to drag. Air passing underneath a vehicle also affects handling; as the air separates at the front of the car, it produces a low pressure area as it rushes over the hood and a high pressure area as it "stacks up" under the front end. The result is lift, which unloads the front suspension and reduces traction needed for high-speed handling. Thus reducing airflow underneath the car is one of those rare modifications which can improve both performance and fuel economy.

In general, a vertical dam completely shielding the front tires and having minimal ground clearance is the most effective design. Unfortunately, life on the streets is tough on spoilers; steep driveways, potholes and parking lot berms will reduce most really effective spoilers to rubble in short order. The factory Camaro chin spoilers are fairly resistant to roadside encounters, but they lack sufficient depth to be aerodynamically effective.

Road racers have developed a solution to the problem of building a ground-hugging spoiler that will survive. They extend their fiberglass or aluminum spoilers to pavement level with flexible rubber skirts. The rubber has to be rigid enough to resist deflection from the wind, yet forgiving enough to bounce back from collisions with the asphalt.

You can build your own flexible spoiler from neoprene strips which are sold at most industrial hardware stores. The material we found measures 7 inches wide by ⅛-inch

Continued on next page

thick, and costs $1.40 a foot. The factory spoiler was left in place to provide a stiff backing for the neoprene; a dozen bolts, flat washers and self-locking nuts attach the flexible spoiler extension.

Research has shown that the area forward of the windshield accounts for approximately 65 percent of a vehicle's aerodynamic drag. Clearly, then, anything that helps "clean up" airflow over the grille, bumper and headlights is an asset. Even though air is prevented from getting under the car by an effective front spoiler, it can still find its way under the car through the grille and radiator opening. Blocking off the entire grille is effective aerodynamically, but costly for engine cooling. You can, however, restrict the size of the grille opening with plexiglass or sheet metal. Engine cooling can actually be improved, because air will flow from the high-pressure area in front of the bumper to the very low-pressure area under the car through the only path open—the radiator core. Similarly, plexiglass headlight covers are worthwhile aerodynamic improvements. And "hide-away" headlights should always be hidden for the best aerodynamic effect.

A novel aerodynamic device which has recently become popular with the road race set are side skirts—"aprons" which prevent air from getting under the side of the car just as a spoiler prevents its entrance in front. Jim Hall's Chapparal ground effects car used Lexan skirts to seal the underside while engine-driven fans evacuated air from underneath it. The car literally sucked itself onto the pavement like a giant vacuum cleaner. Hall's breakthrough was banned, but the seeds of his idea have blossomed in current Formula One technology. The current crop of ground effects cars utilize the internal shape of their side pods to generate downforce. In effect, the entire car is an upside-down wing with an ungodly attraction for the track surface.

Well, a Formula One Lotus may be out of our reach, but the image is not. And so, armed with tin snips, neoprene and unconscionable audacity, we set out to duplicate Grand Prix technology on a shoestring. Enter the "Sucker Camaro," complete with full side skirts. The skirts were attached with aluminum angle along the ridge where the rocker panels join the floor pan. Plastic backing provides rigidity, and screws with self-locking nuts attach the neoprene skirts. Now the considerable flaw in this entire grand plan is, of course, that a Camaro body is not an airfoil. But the skirts do keep air from getting to the undercarriage as it spills around the front tires, and that counts for something. And the looks of uncomprehending motorists are worth the effort. (A note from our attorney: If the local constable objects to your aerodynamic experiments, tell him that the side skirts are simply long mudflaps.)

AN ATTITUDE CASE

Since a car's entire sheetmetal area is an aerodynamic device (whether good or bad), it stands to reason that modifications which effect the body as a whole have significant impact on drag. Vehicle attitude (or rake) is an important factor in road performance. Experiments with a variety of cars have shown that a nose-down attitude of 1 to 1½ degrees provides the best combination of low drag and acceptable downforce. If the car is raked any more steeply, the drag figures go out of sight, although downforce continues to increase at a moderate rate. Thus cars with their trunk lids in the sky are not only evil handling machines, but they're aerodynamic disasters as well.

It's important that the car have the proper attitude at speed. It may have the optimum 1-degree rake sitting in your garage, but on the road the suspension may be loading or unloading enough to alter the car's angle of attack. Drag racing cars are especially susceptible to nose-up attitudes, since drag racing shock absorbers have very little jounce (upward) control. After the initial launch, the shocks have to let the nose settle back down quickly, or else the high-pressure air under the front bumper will keep the front end up throughout the entire run. The best way to diagnose this problem is to photograph the car at the finish line.

Altering vehicle attitude for testing can be accomplished easily with simple equipment. Rear air shocks can be inflated to change the car's rake, or helper springs added to increase the angle of attack. Final adjustments should be made by changing the car's springs, however. This may require trimming the car's front coils or re-arching the rear springs. The idea is to have the car as low as possible and still maintain the correct attitude with acceptable ground clearance.

HOOD SCOOP STRATEGY

Most drag racers' introduction to aerodynamics is courtesy of hood scoop science. Hood scoops have gone through as many fashion changes as hair styles: Rear-facing lumps, ramp-style scoops and "boundary layer" or snorkel scoops have all been popular during the last decade. Our own hood scoop investigation was prompted by two disturbing observations. Our race car-in-residence would run the exact same e.t. and speed regardless of whether its scoop inlet was open or taped closed. Second, a towel stuffed in the scoop opening would invariably be pulled out rather than pushed in while the car was driven home on its transporter.

Most current hood scoop technology is derived from Pro Stock data. We theorized that an early Camaro, with its blunt front end, would exhibit significantly different airflow than its shovel-nosed successors. Our hood scoop investigation had two distinct parts: First, we needed to know what the air was doing, and then we had to locate the scoop where it would take advantage of airflow patterns. (The importance of hood scoop design becomes apparent when you consider that a one percent horsepower increase can be expected for each 10-degree drop in intake air temperature, and two percent for a 150 mph inlet velocity.)

Since repeated runs to test airflow can be tough on a race car with a short engine life expectancy, we used a street car set up to mimic the racer's angle of attack. A quiet country road was the test site, and all runs were made twice in opposite directions to minimize the effects of wind and the road's crown. The first portion of the test focused on mapping the air pressure across the front end; our homemade manometer's probe was taped in various positions from the leading edge of the hood to the base of the windshield. The second half was intended to reveal the pressure which the carburetor would "see" by placing the probe inside a sealed dummy hood scoop.

Going in, we anticipated that the front edge of the hood and the base of the windshield would be the areas of highest pressure. This proved to be dead wrong. In fact, the windshield base and the ventilator plenum beneath it produced the worst numbers of any area we tested. The best figures came when we taped the probe to the grille just above the front bumper—although it would be difficult to duct air from this location to the carburetor. It soon became apparent that the probe's height above the hood was more important than its location. In other words, it seems more important to have a scoop as high as possible than to worry about its position. (There are limits, of course. A tall scoop would probably have more drag than a petite one, which might offset some of the gains.)

This was confirmed by the second segment of the test, using the dummy scoop. Manometer readings were consistent as we moved the scoop across the hood surface. As a check on our earlier results, we tried turning the scoop around so that its inlet was at the base of the windshield; again, pressure readings dropped significantly. Our conclusions? A low, rear-entry scoop seems to be the worst combination, and a high, boundary layer design the best configuration.

AERO CAMARO

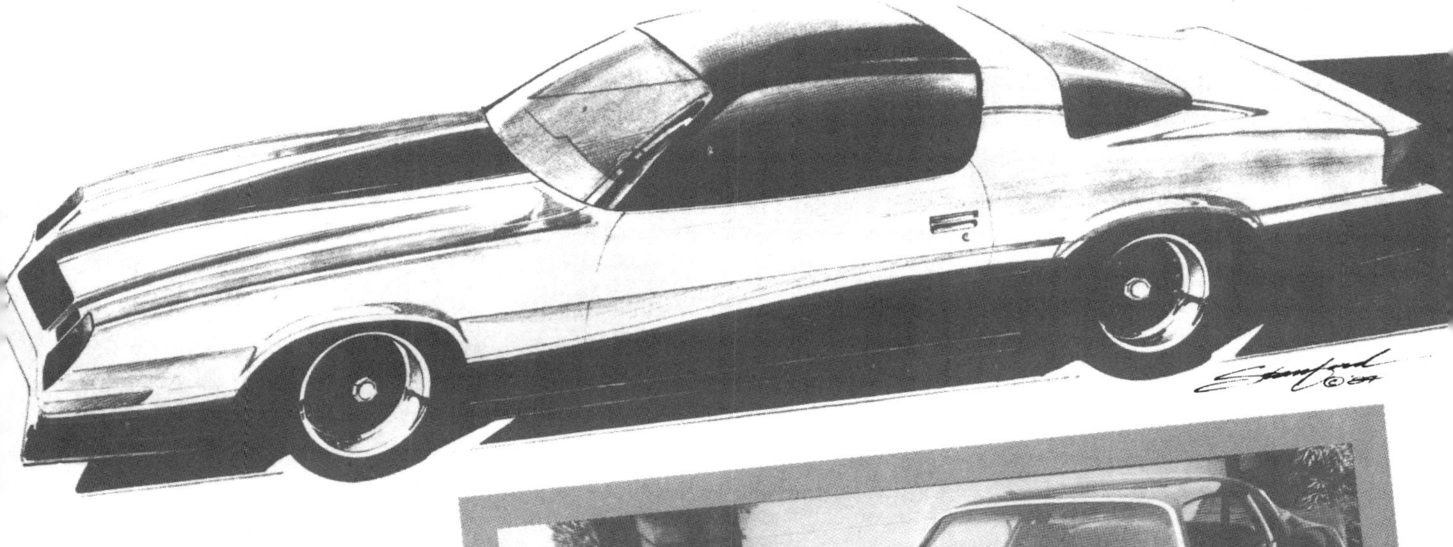

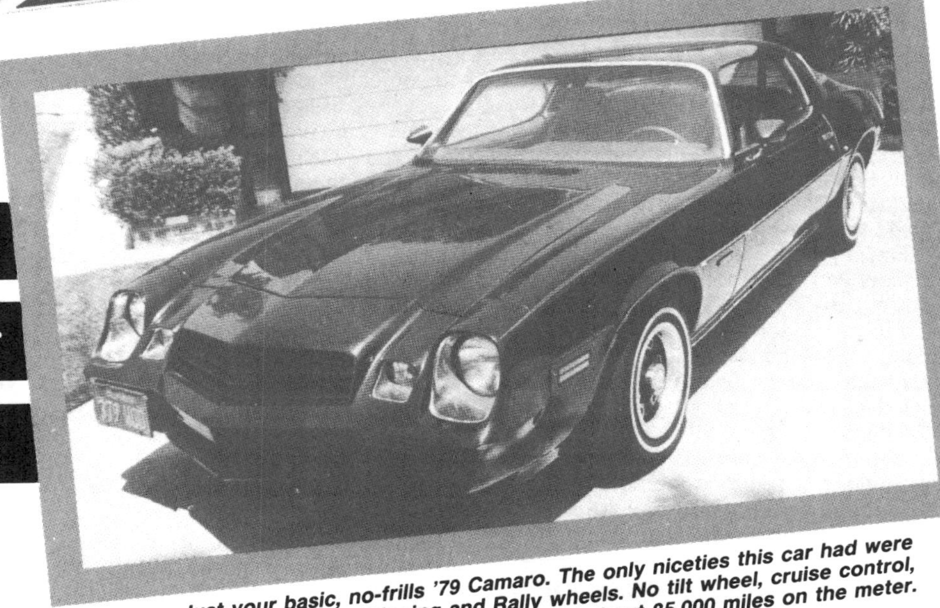

How To Build a Contemporary Street Machine

Part One

Just your basic, no-frills '79 Camaro. The only niceties this car had were air conditioning and Rally wheels. No tilt wheel, cruise control, or power windows, and about 85,000 miles on the meter.

Text & Photography: Kevin Boales

This month we begin a new series intended to illustrate the way to design and construct your own street machine, from buying the raw car all the way through to painting and finishing the project. We're going to use as many photos as possible in order to make the story easier to understand and, hopefully, more fun to read.

The project was started in response to reader letters, comments from people we talk with every day, and verbal tomatoes we've dodged in conversations with event participants. Basically, the comments are about building a car the average person can afford—relatively simple, fast in the HOT ROD tradition, and nice to look at.

The approach you'll see here is a complete, frame-to-roof rebuild. This car is going to make the rounds on the show circuit, so there's an abundance of chrome goodies that could have easily been cadmium plated or even painted. Street machines should express what *you* see in the car you drive.

One of the other considerations in a construction like this is to use parts that are readily available, either from catalogs or easy to purchase through the mail, which we have accomplished with the help of Mr. Gasket Company, and their subsidiaries: Lakewood, Interpart, Cragar, Pro Trac, and American Exhaust. In addition to the Mr. Gasket family, we selected components from A&A Fiberglass, Herb Adams VSE, and Chevrolet.

The base car for the project is a '79 Camaro, chosen because there are so many in existence, and there will be thousands of readers who can literally duplicate the steps shown here. Basically, the process could be applied to any car, especially the tear-down and finishing sections; and upholstery, chassis parts, and engines are available from the suppliers we listed to fit anything you now drive.

By the way, this car is to be duplicated cosmetically and replicas will be given away at car shows around the country. Group Promotions, Inc., an organization that arranges shows, will award the cars during selected events throughout the following year.

Relax, study up, and watch as we take you on a tour through the Mr. Gasket Aero Camaro . . . .

(continued)

A. Buying the Raw Material . . . Beauty or the Basket?

More than a quarter-million '79 Camaros were built during the production runs that year. The car used for the project shown here was selected specifically for the following reasons: it was priced right and was *missing* the correct options. In order to keep the construction simple, the car we wanted wouldn't have cruise control, power seats, power windows, or a vinyl top since they would be removed and would leave holes to be filled or (in the case of cruise control) a steering column to be replaced. The Aero also had an automatic trans instead of a stick.

Finding the car of choice can take a while if you're too picky; don't turn down a reasonable Camaro because the driver's door pull is ripped off the door . . . they almost all look that way after a few years, and the parts are still available (and affordable) to repair the handle.

Your mechanical skills also enter the picture in selecting the right deal. If you're a rocket at bodywork, but don't get off on engine stuff, you might want to look for a wrecked body with a fresh drivetrain; even a pair of clips and a fresh engine and trans might do in a pinch. If neither bodywork nor wrenching are a problem, your car will be the one with the least mileage for what you can spend.

If asking all your friends and groveling a little don't produce any automotive raw material, hit the papers and grab a phone; the search is on. Actually, the papers are a reasonably good place to find out what the "going rate" will be for the kind of car you want. In the case of General Motors cars, and some Chrysler, Ford, and AMC models, front ends and rear bumper and valance treatments are all that distinguish one model year exterior from another; if you're into the project, you can find the nose you like in a "severely used and abused" yard (we don't call 'em junkyards any more) and install it on your car as the body takes shape.

Interiors are a little tougher, but only because new or excellent pieces are harder to find and usually more expensive than most sheetmetal stuff. All the major paint manufacturers, including R-M, Ditzler, and DuPont, make interior colors in finishes designed for vinyl, ABS plastic, and urethanes you'll encounter inside the car. They can be used to freshen the interior you have now, or to color-match pieces you find on dealer shelves or in *those* yards.

Suspension pieces are usually the last consideration when you're buying a used car, but take a quick look under the car to see that nothing has been mangled in a serious accident. Giveaways include: bent spring shackles, poor fit between all or some of the body panels, doors that won't close properly, leaky windows, etc. Some of these problems are easily repaired, but the chassis should be straight if nothing else is.

Mileage is another indicator of how many suspension problems you might run into, but it's really variable; some cars with more than the century mark on the odometer will have survived better than others with half the mileage. Drive the car and feel for loose or noisy suspension components. Most of the front end pieces are reasonably priced, but their installation can be expensive if you have it done professionally. We'll be showing you some front end basics, but if you have any doubts, save your bucks and get a pro to do it.

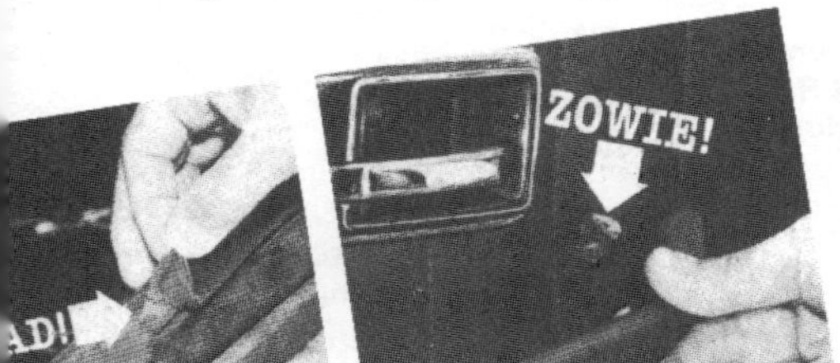

Typical problem areas that will sometimes sour a buyer (but not you, if you're smart) included the usual body dents, dead rubber, broken door pulls, little holes here and there, and upholstery that looks older than dirt. . . .

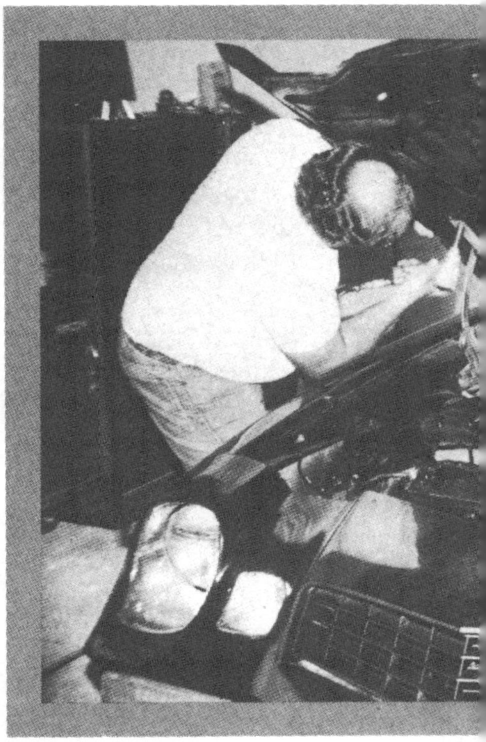

B. Teardown . . . Dirty Laundry

Once you have the right car, the next step is to find a place to work for the next few weeks. With the help of one or two friends, disassembling the average GM car will take about three days with common tools. An impact wrench will speed things up considerably, especially in the areas of the front and rear suspension.

The simplest way to make this a painless operation is to remove the entire front sheetmetal/nose assembly as a unit. It's located with about a dozen bolts, and once the harnesses and plumbing are disconnected, lifts right off the subframe. With the clip removed, you can easily get to the engine and suspension parts.

As you can see in the photos, the Aero Camaro was *entirely* disassembled . . . down to the basic body

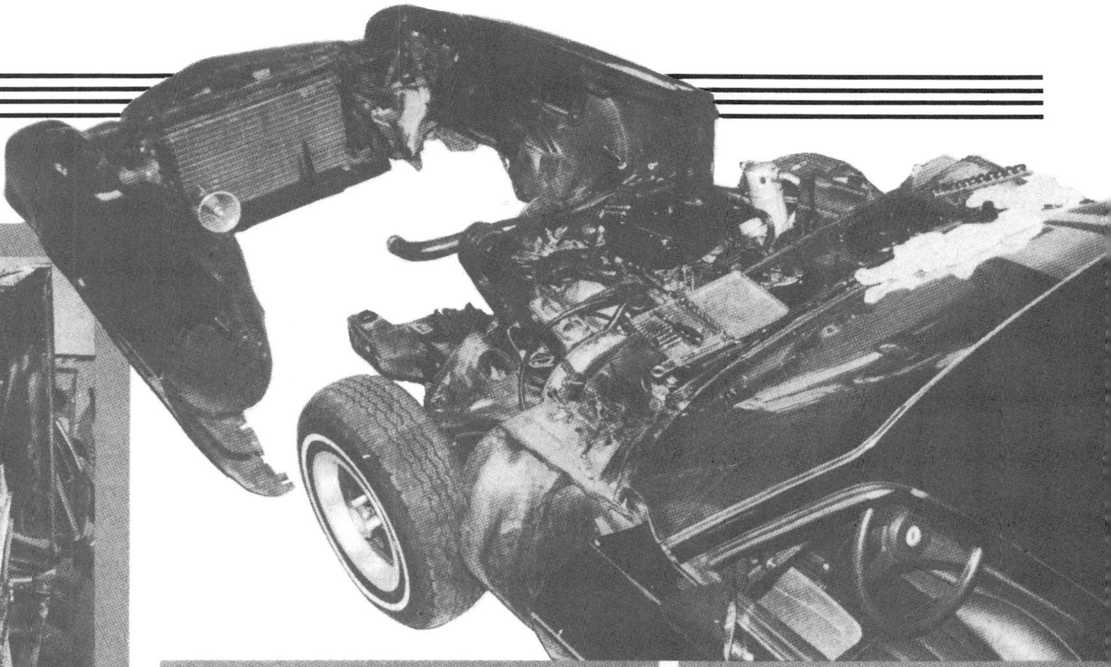

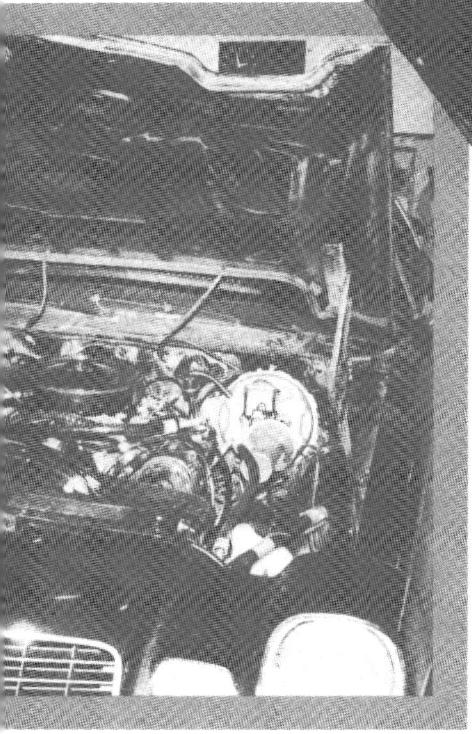

First thing to go, of course, is the hood. This one will be replaced with a 'glass hood later.

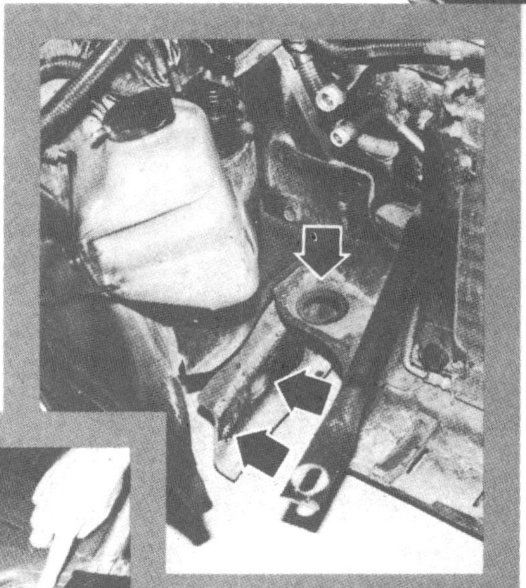

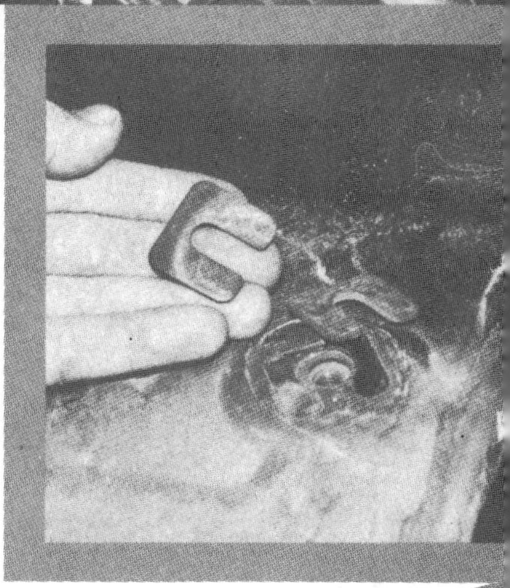

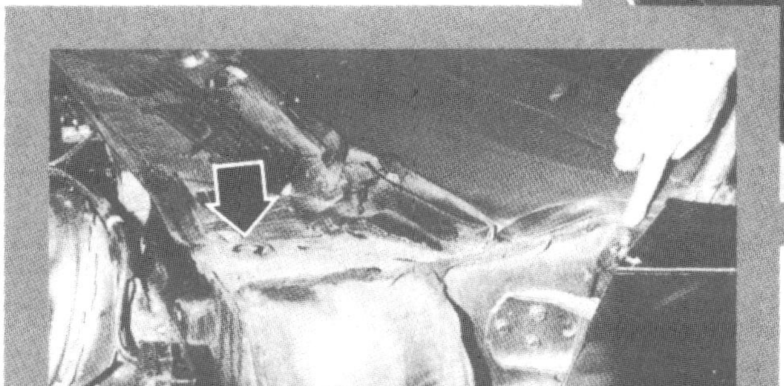

The entire front sheetmetal assembly can be removed by taking out about a dozen bolts; after that it's mostly downhill since now everything is in view. Save the body shims because they will be needed to realign the body panels. The location of the hardware on this car is shown with arrows in these photos. Yours should be similar, regardless of the car.

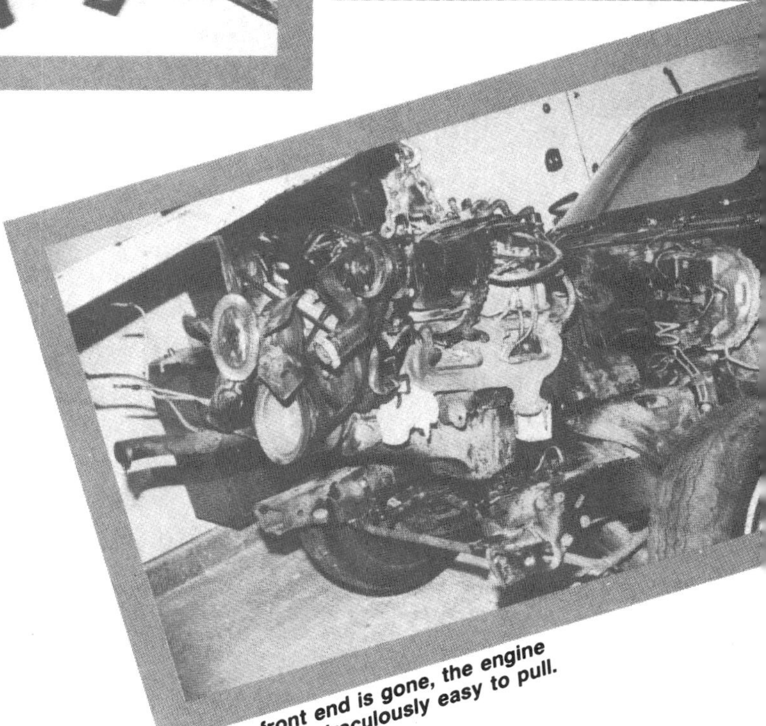

Once the front end is gone, the engine and trans are miraculously easy to pull.

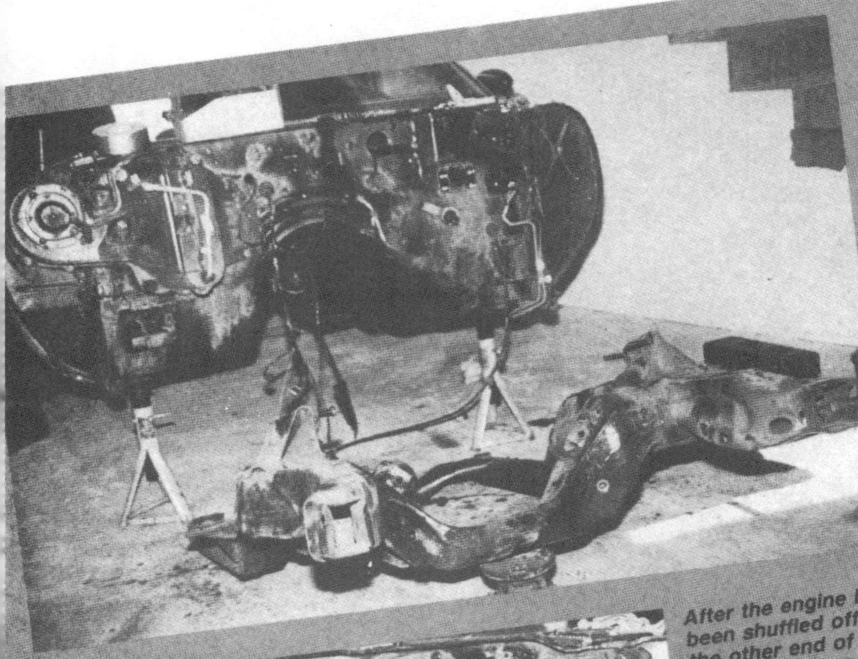

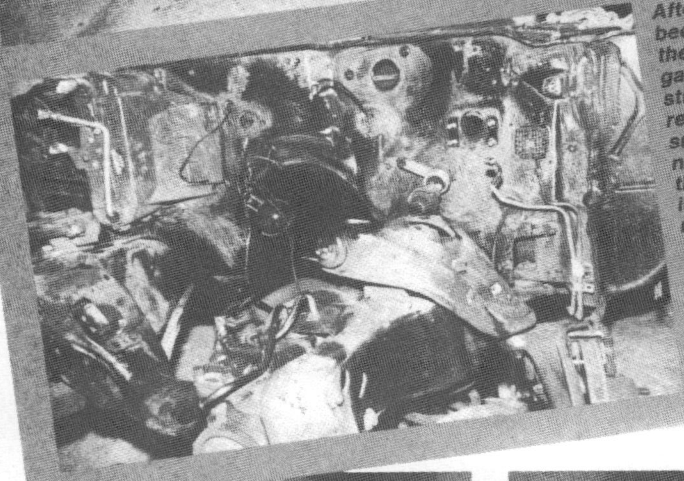

After the engine has been shuffled off to the other end of the garage, you can start stripping the firewall, removing the front suspension, and, finally, attending to the subframe itself. It is vitally important to retain all the stock front suspension hardware, replacing any that appears damaged or corroded with OEM components.

shell. After the front clip was gone, the engine and trans were removed, followed by the front suspension components and, finally, the front subframe.

Working our way back, the exhaust system was "hot wrenched" out of the way, making removal of the fuel tank, rear axle, and springs a lot easier. All the small brackets and hardware were shuffled off to the plater, the underside of the car was scrubbed (it had been steam cleaned entirely while it was still a driver, making for a more pleasant time for any of the people working on the car) with a 3M scuff pad and Ditzler DTR-601 enamel reducer.

The interior was last, and the hot tip for removing one is to get some "Zip-lock" bags for the screws and stay organized about where they go. One way to do this is to tape the bags in areas under the seats or "behind" the upholstered panels while they're out being freshened up.

Next month we'll start seeing the parts as they came together from all over the country. They were plated, painted, made from scratch (some of the upholstery), and bought off the shelf. Are you having fun yet? **HR**

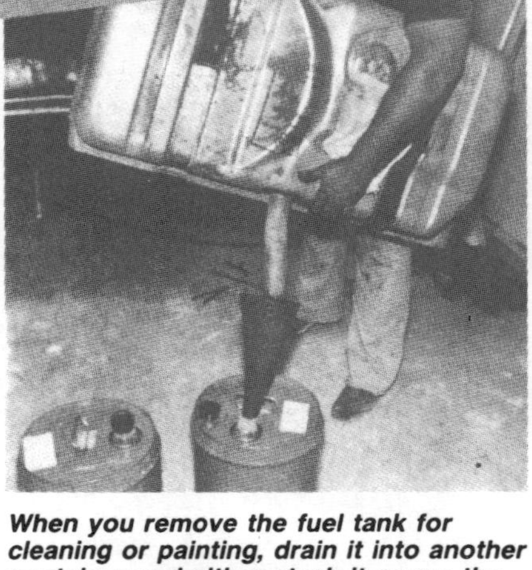

When you remove the fuel tank for cleaning or painting, drain it into another container and either stash it or run the fuel in your Weedeater; don't clean parts with it. These cans were full of Chevron (Standard Oil) solvent originally, and came in handy for storing the gas siphoned from the car.

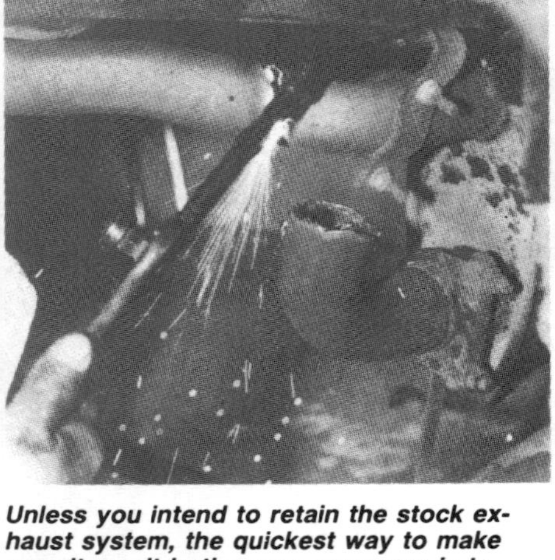

Unless you intend to retain the stock exhaust system, the quickest way to make sure it won't bother you any more is to hot wrench it. In the East, Mother Nature may have already started this process for you.

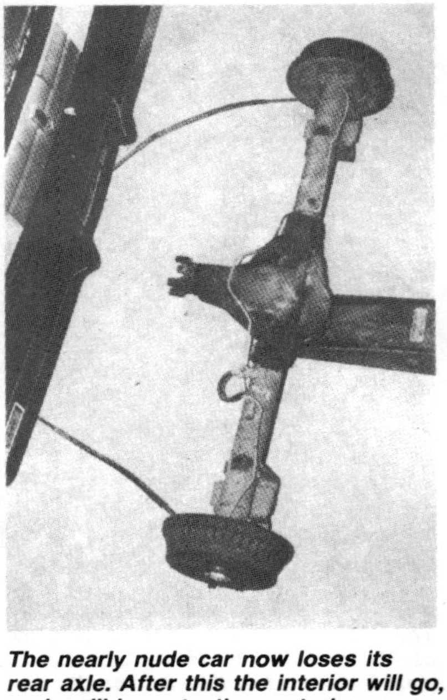

The nearly nude car now loses its rear axle. After this the interior will go, and we'll be onto the next phase: refinishing and rebuilding.

AERO CAMARO

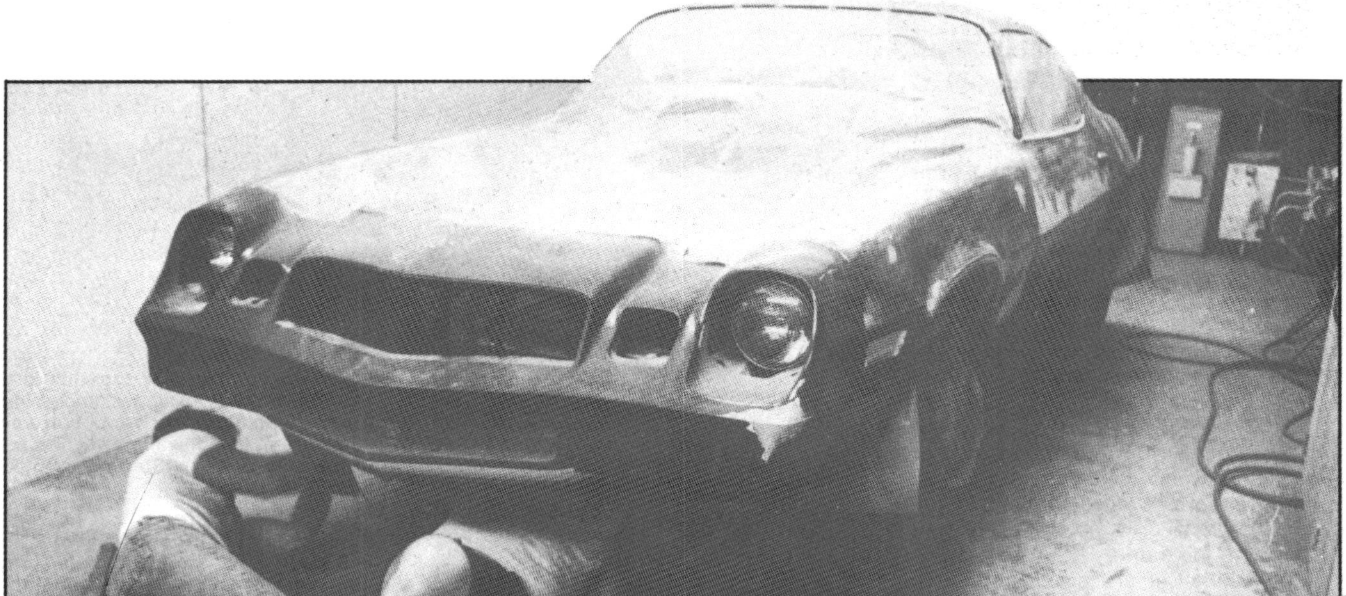

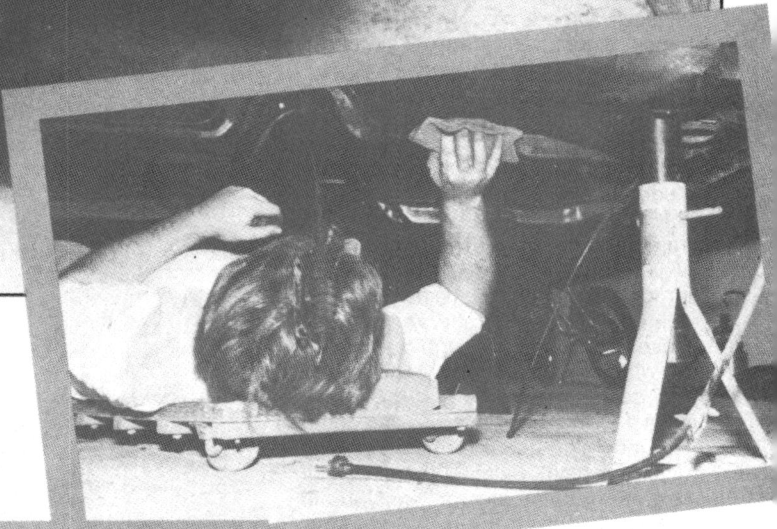

This is the fun part if you're into S&M. Rick Leonard, whose work you'll see all through this project, is shown scrubbing the underside of the Camaro in preparation for the Delstar acrylic enamel. He's washing the car with Ditzler DTR 601 reducer and a 3M Scotch Brite (green) pad.

Part Two

How to Build a Contemporary Street Machine

Text & Photography: Kevin Boales

Last month, we shopped for a car and promptly rendered it *parked* by removing nearly everything that could be wrenched or torched from it. This month, we're beginning to reverse the process by painting, plating, welding or upholstering the individual components.

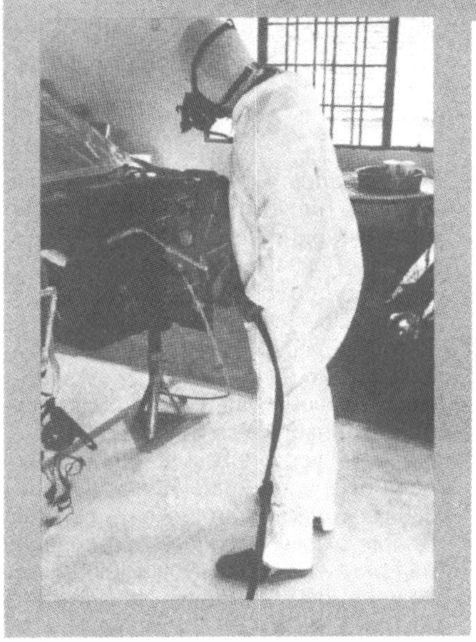

One small step for man . . . when shooting catalyzed enamels, a respirator in good condition is an absolute necessity. Rick is wearing the space suit for when he climbs under the car; the overspray and reducer mist gets a little rough on skin and clothes.

Aero Camaro

A. Chassis Detailing

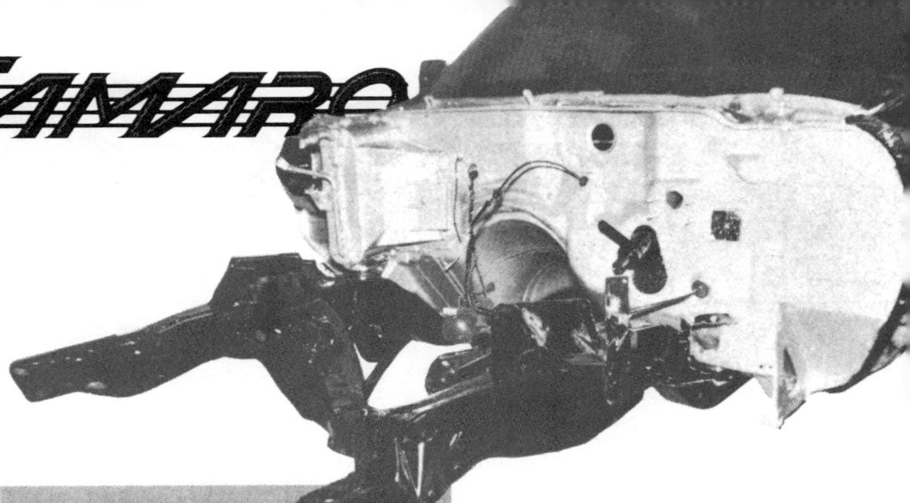

It's finally coming back together in this shot. With the firewall and underside of the body painted, the front subframe is shown reinstalled, waiting for the return of the control arms. Starting to look like a car again, isn't it?

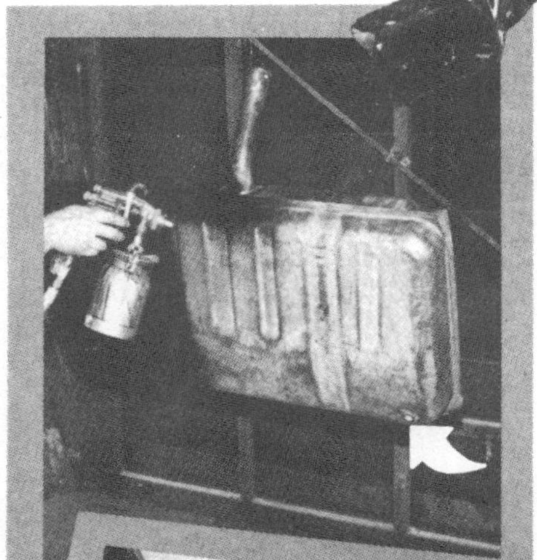

Along with the suspension components and frame rails, the gas tank is sprayed with black Delstar after stripping and installation of new outlet fitting (arrow).

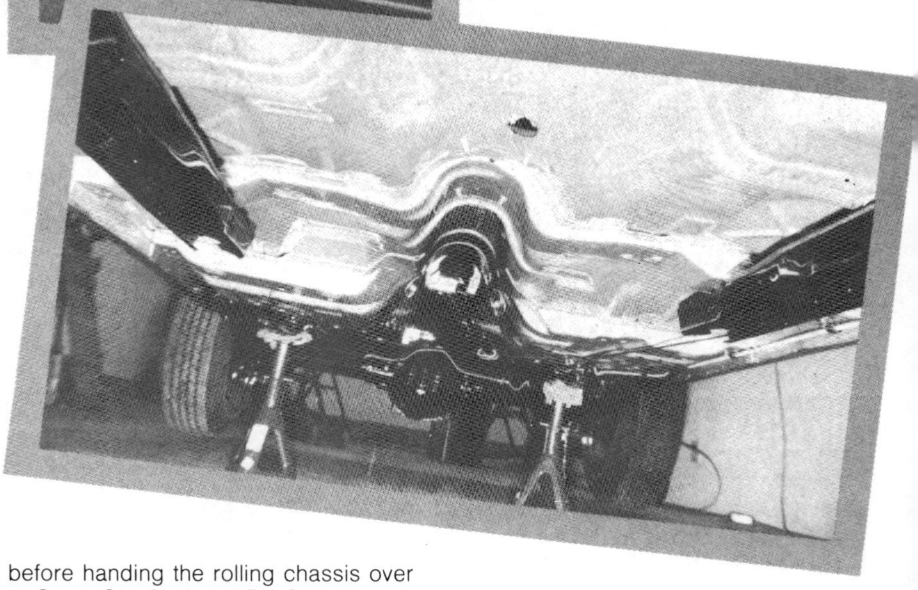

Freshly sprayed in silver Delstar, the underside of the car is lookin' good at this point.

As you recall, the front subframe was one of the first casualties. The upper and lower control arms were shipped off to Herb Adams VSE for rework; Herb makes his own steel and nylon bushing kits for these cars. The idea behind this step is to make the car handle more predictably by eliminating suspension geometry changes under load. The system works for his Trans Ams, and this isn't the first Camaro he's put the whammy on, either. For the rear of the car, Herb supplied new, reworked forward spring perches that have higher locating points for the front spring eye, lowering the car slightly and attaching the spring a little more solidly to the frame.

Since the Camaro is a unibody-style chassis, some areas of concern for suspension tuners are the front subframe attachment points and firewall/front sheet metal integrity. Herb also supplied pieces to strengthen the entire front end of the car by adding struts from the firewall to the subframe. Other details include aluminum "donuts" to replace the soft rubber insulators between the subframe and the body. The result is a slight reduction in ride height and a major increase in chassis stiffness. The ride quality doesn't suffer much from these alterations, although that depends on whether you like driving pillows or *real cars*.

Last month we wound up scouring the underside of the Camaro in preparation for paint. Painting the underside of a car is a real stunt unless you manage to raise the body high enough so you can work comfortably. With so many of the major heavyweight components removed from the Camaro, the remaining exoskeleton is light enough to be supported with wood blocks or straps from an overhead frame.

We masked off the rear frame rails and wheel wells, which will be painted black. Then, following normal preparation for paint, the underside of the car was sprayed with two coats of silver Ditzler Delstar Acrylic Enamel, the same color as the exterior with the exception of the pearl clear coat.

On the outside of the body, the only changes made were the mirrors and the front spoiler, which came from GM. Bodywork was minimal on the car, since it had been selected with a *primo* body being a must. Side moldings were removed, the spoiler was added, and all the panels were carefully re-aligned before handing the rolling chassis over to Steve Stanford and Bill Carter, who were given the task of designing and applying the exterior paint.

All the chrome you see on these pages either was applied by Carter Chrome Plating in Burbank, California, or came from Mr. Gasket Co. As we said last month, the pieces could have been cadmium plated, iridited, or even painted if you prefer that look. If you paint over some materials, specifically stainless (some trim pieces), aluminum, flexible plastic color-matched noses and bumpers or chrome, you should first take the time to prepare the surface so the paint will stick. Ditzler, DuPont and R-M make special primers and additives for these situations; chrome and flexible body panels are the trickiest.

Powder coating is another treatment used on the Aero Camaro. This coating is a tough, electrostatically-applied material that is baked on. The finish is usually very consistent, although for small batches you might have to settle for a "stock" color from the powder coater's catalog. They always have black, red, yellow, white, and gray. This is the right stuff to use if the part is metal and may

Photo: Rick Vogelin

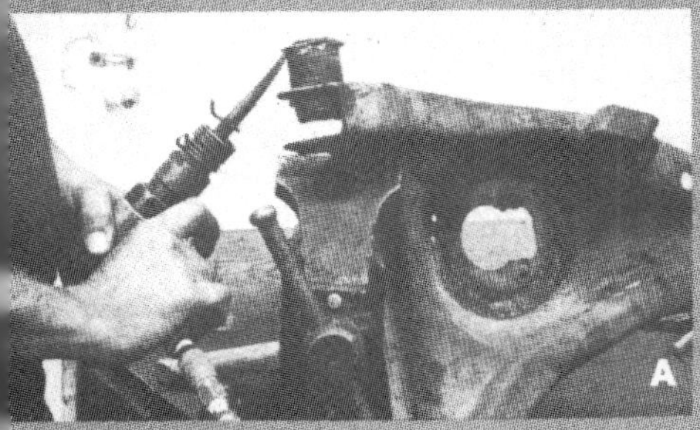

A

be exposed to less than perfect conditions, like bumpers, for instance. For maximum adhesion, the parts should be stripped of any existing coating or plating. This holds especially true for chrome. Most chrome platers will etch the chrome from steel bumpers for a reasonable fee.

If you can't get a powder coating to match the color on your car, you can color-match the bumpers or other parts, after having the chrome stripped, by using an epoxy primer (like Ditzler's DP-40 system) and a topcoat of acrylic enamel, acrylic urethane, or polyurethane. If you strip the chrome, it's a good idea to paint the inside surface of the bumper to prevent corrosion; cheap insurance, since the protective chrome isn't there any more.

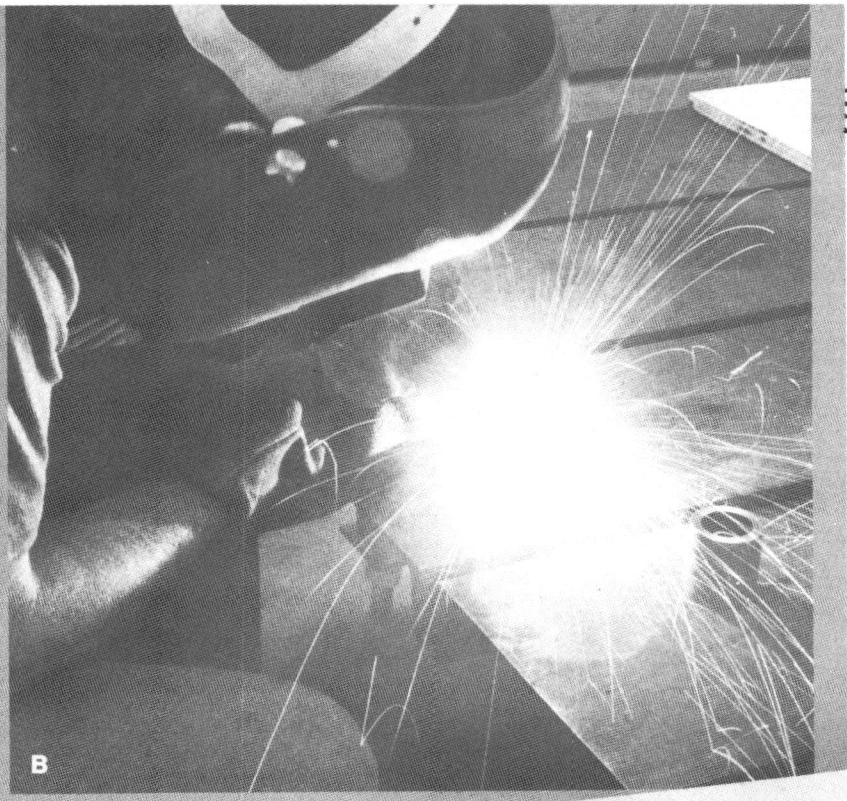

B

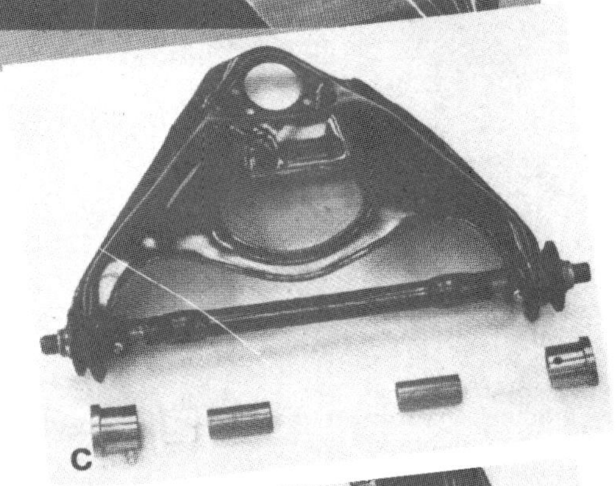

C

First thing to do to the well-used control arms is to blow out the stock bushings (a). Afterward, the bores are enlarged to accept Herb Adams' steel and nylon versions, which are then fitted to a welded-in tube (b). Photos (c) and (d) show the bushings and control arms after modifications are complete. Photo (e) shows the aluminum replacements for the stock, eraser insulators used between the front subframe and body.

B. Suspension Build-Up

Once the control arms came back rejuvenated from Herb's crew, they were reinstalled on the subframe and the brakes were rebuilt, using semi-metallic pads in the front and organic shoes in the rear. In the normal process of rebuilding brakes, the important things to remember are cleanliness and organization; mixing the bearings on the front or the brake return springs on the rear can really wreck your day later. The rotors were marked left and right and the bearings were kept with their respective rotors to avoid problems.

Another change made while the car was dismembered was the addition of a new outlet for the fuel tank. Rather than the standard center top position, a new fitting was welded into the bottom right hand corner of the tank, where it will be routed to the fuel pump. The fuel level sender and float assembly will be re-

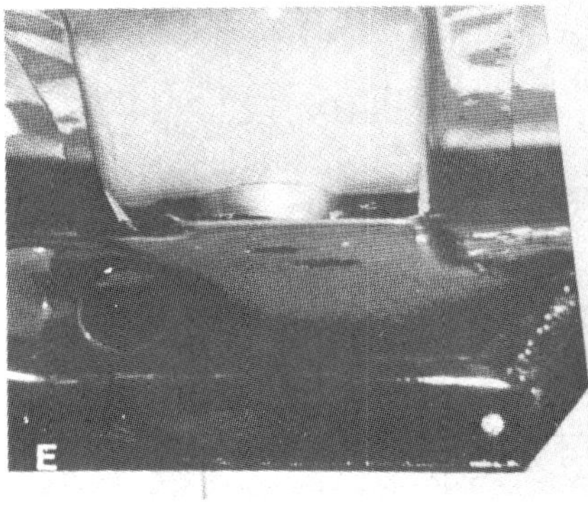

E

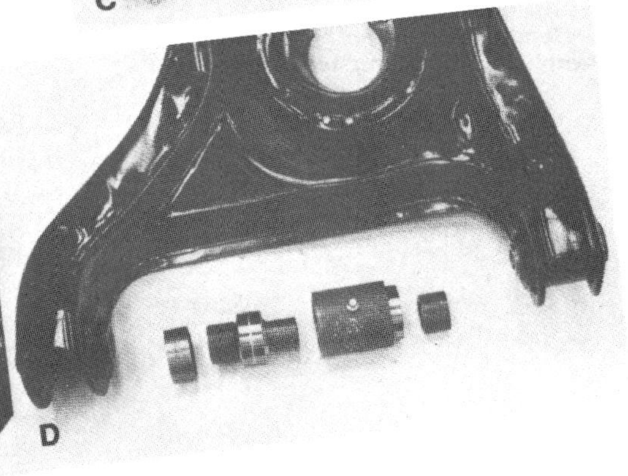

D

tained in the stock position. Fuel tank venting will be accomplished with a vented cap.

New instruments are being installed to wake up the appearance of the stock dash. Auto Meter supplied the water temp and oil pressure gauges, ammeter, and tachometer used in the car. The stock panel can be reworked to accept the small faced gauges, which replace the idiot lights. In addition, a 165 mph speedometer was fabricated using the stock works and a new face from Corvette Clinic, Inc.

What happens next month? Engine, transmission, and rear axle rebuilds. It'll be the hard core, nuts and bolts HOT ROD style thrash, too, because this car will be making its first appearance at this year's SEMA (Specialty Equipment Market Association) Show this fall. No time to lose . . . after that we still have to cover the paint work and detailing. Uncle! **HR**

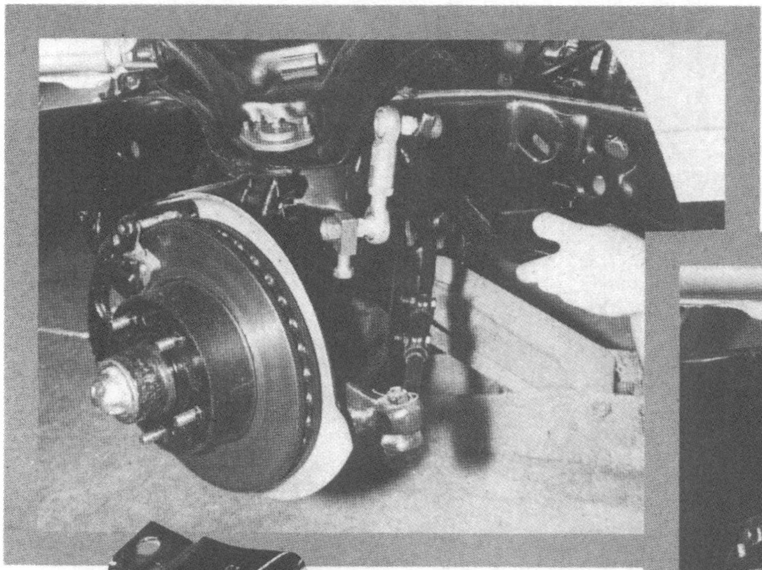

Mounting the VSE swaybars is very easy because they bolt in place of the stock swaybar. End links with Heim joints give the bar quick response.

Aftermarket shock mounting is simply a case of R&R (remove and replace). The correct shock/spring/swaybar relationship is critical.

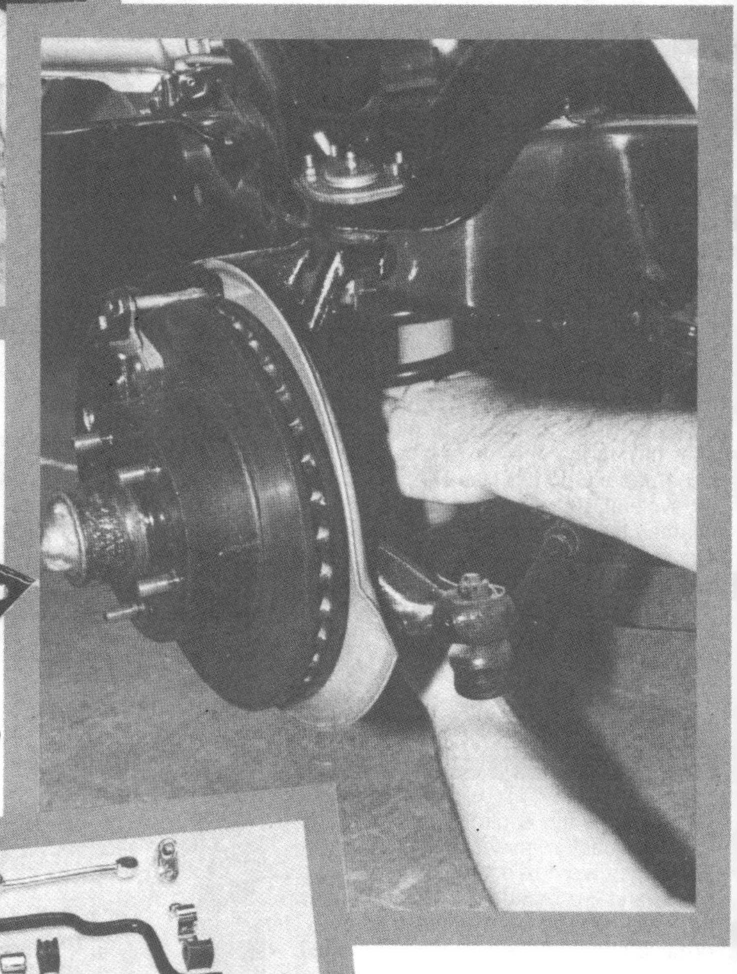

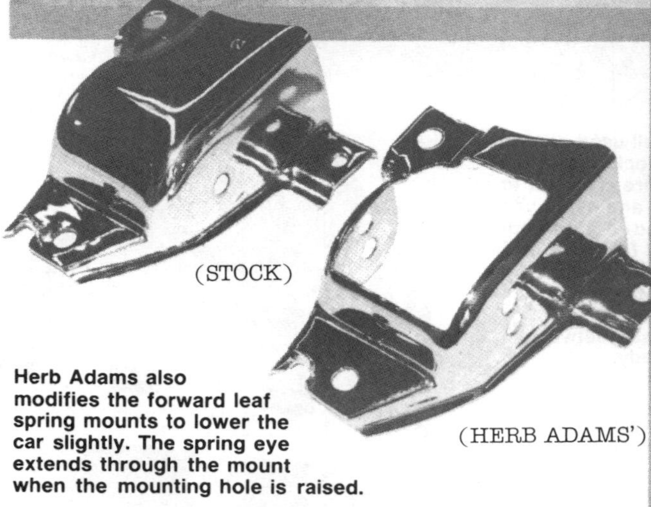

Herb Adams also modifies the forward leaf spring mounts to lower the car slightly. The spring eye extends through the mount when the mounting hole is raised.

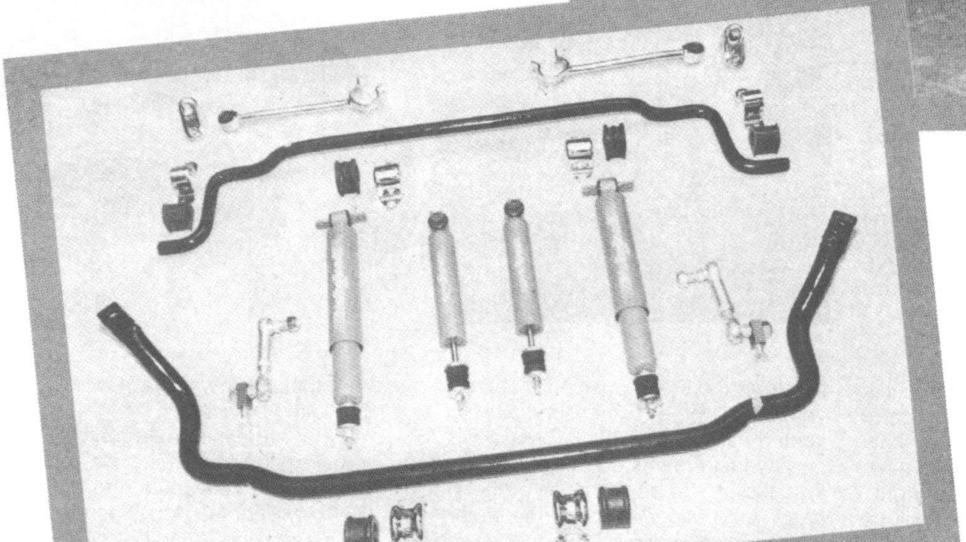

Camaro's handling will be greatly improved by Herb Adams' VSE sway bars and Mulholland shocks. Mounting hardware was plated to match the rest of the undercarriage.

Aero Camaro

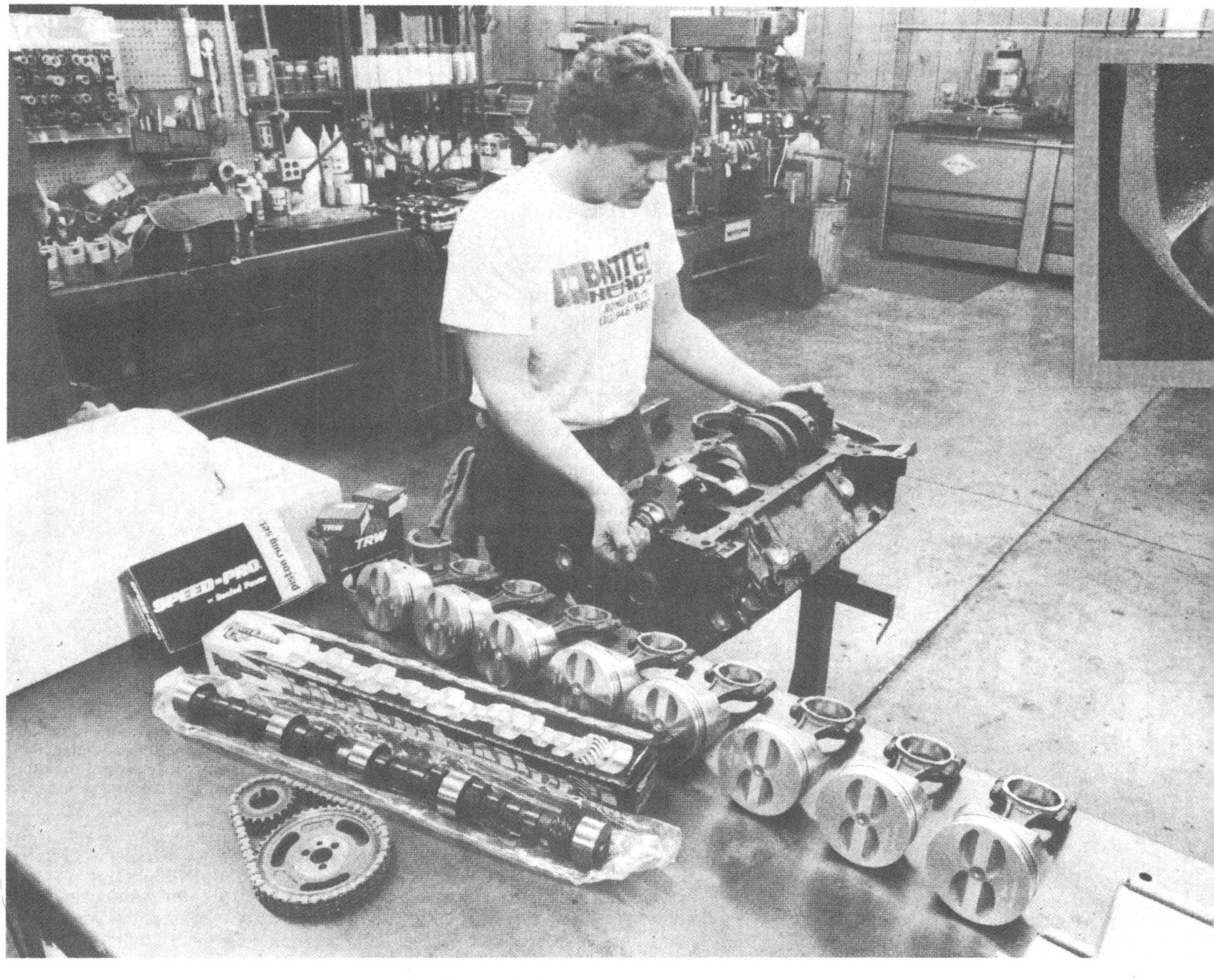

Part 3: The Drivetrain

How To Build a Contemporary Street Machine

By Kevin Boales

Last month we promised to show you some coverage of the drivetrain being built for the Mr. Gasket Camaro. In keeping with the theme of this car, the Performance People at Mr. Gasket wanted to keep things as simple as possible, using either parts available from catalogs, over the counter at GM dealerships, or at local speed shops.

The engine is a good place to start. In addition to a thriving race engine business, John Lingenfelter Racing (1557 Winchester Rd., Decatur, IN 46733, 219/724-2552) builds some fine engines for sale via mail order. The choice for the Camaro is a combination of 350 and 400 parts to make a 383, using a 350 block, 400 crank with 350-sized mains, 400 rods, and 440x heads with 2.02-inch intakes and 1.60-inch exhaust valves. The block is prepared by Lingenfelter in similar fashion to all their racing mills. After align honing, the block is decked, bored .030-inch, and honed using plates. The 5.565-inch rods are equipped with TRW flat top forgings, and when combined with the x-heads' chamber volume, produce a compression ratio of 8.2:1.

The cylinder heads are pocket ported, equipped with screw-in studs and pushrod guides, and port matched to the intake manifold. TRW valves are operated by Competition Cams' 1.52:1 rocker arms and TRW pushrods, all leading down to a Competition Cams 280 hydraulic grind camshaft.

The dyno results? A healthy, streetable 404 horsepower at 5200 rpm, with a torque peak of 414 lb.-ft. at 4500 rpm.

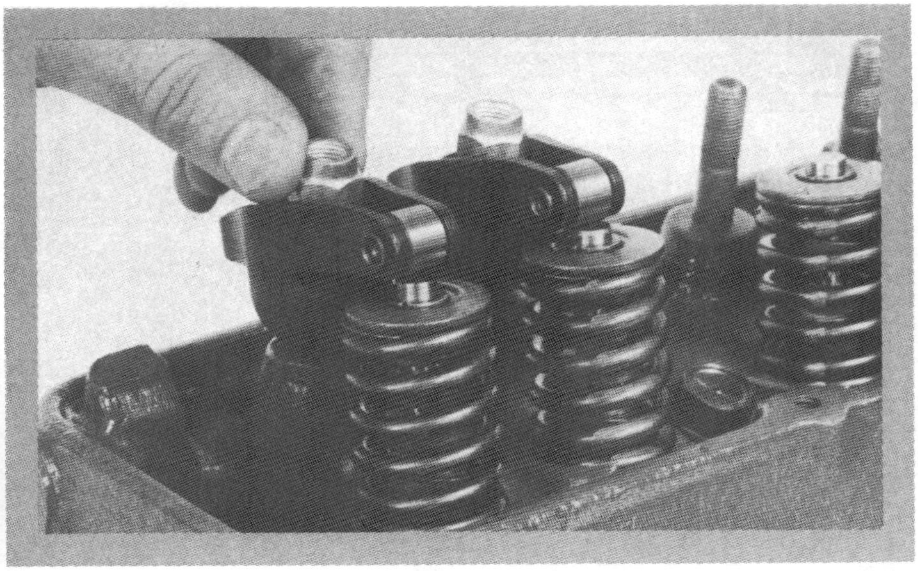

Competition Cams' new Magnum Roller Tip rocker arms (1.52:1 ratio) are used with their potent 280 Magnum camshaft installed at a 103.5 intake centerline. The combination also includes Comp Cams' lifters, pushrods, springs retainers and timing set with Mr. Gasket guideplates installed after the cylinder heads are machined for clearance.

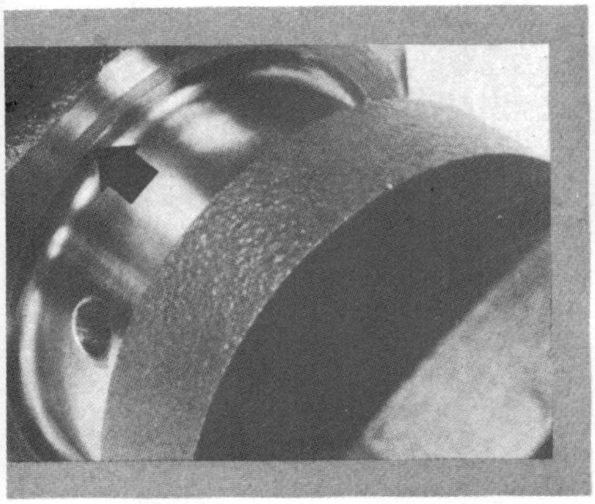

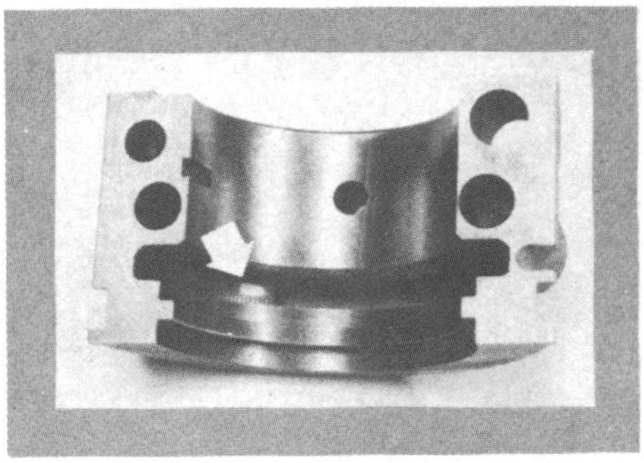

Main journals on the 400-inch small-block crankshaft are turned down 0.100-inch; allowing installation in the 350 main journal saddles. The diameter of the oil slinger flange on the rear main is also reduced slightly to prevent a potential interference problem caused by installing the 400 crank into the smaller 350 rear main cap as shown in the inset photo.

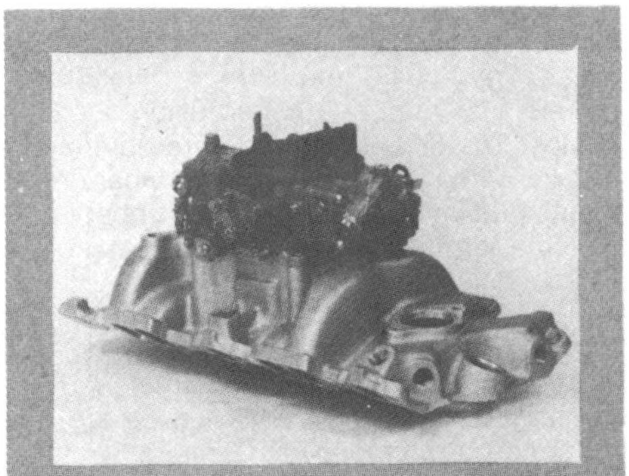

Mr. Gasket's #101 intake set seals the aluminum bow-tie intake manifold to the cylinder heads, and a Holley #4779 750-cfm double-pumper carburetor tops off the package.

Moving rearward, we find a slightly massaged Turbo 350, assembled by Darrell Young (1130 N. Hollywood Way, Burbank, CA 91505, 818/846-2564) using GM internal parts and a Mr. Gasket Tough Shift-3(tm) kit to reprogram the valve body. The kit allows you three optional programs, from slightly firm shifts to neck-snapping drag racing performance. The TH 350 trans is a proven performer, although reputed to be less tolerant of serious drag racing chores than the early TH 400's. Darrell equips all of his racing transmissions with early, steel components where later trans assemblies have plastic or other less durable materials. According to him, GM has continuously "upgraded" their automatics to make them less expensive by substituting materials wherever possible, based on estimated length of service; the 400's originally were heavily overbuilt, lasting a quarter million miles under normal conditions. When GM was through, they lasted about a hundred thousand under the same conditions. There is planned obsolescence in this world.

To understand how trans kits work, you first need to know how the trans operates. All current automatics are basically planetary transmissions, with hydraulically operated clutches engaging or releasing to direct the power flow to each particular gearset. The transmission fluid, placed under pressure by a pump, is directed to each of the clutches by the valve body, which is the brain for the whole transmission.

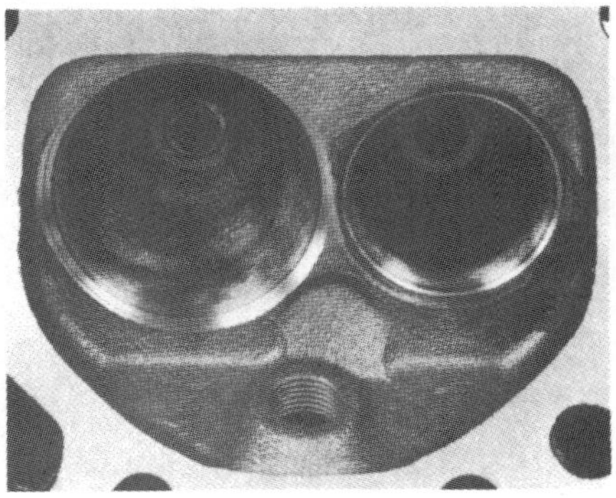

Open chamber 440X casting cylinder heads are reworked with minor pocket porting as shown here. Heads also receive the full rebuild treatment including surface grinding, bronze-wall guides, and intake gasket port matching.

Inside the valve body is a plate, sandwiched between the valve body and the portion of the trans case containing the passages to the various clutches and a band (in the 350's case). The plate can be considered a "jet plate," since it is drilled with orifices for the control of fluid flow. By enlarging strategic holes, flow can be increased to the 1-2 and 2-3 shift points, making them happen faster and harder. In addition, trans kits generally include a recalibrated spring that replaces a pressure regulator spring in the trans, raising the operating pressure in the trans over the stock point. Most automatics normally operate at about 60-120 psi; with most trans kits, the pressure is boosted to 100-200 psi.

Winding things up in the rear is the axle, which was an open carrier to begin with. For the Aero Camaro, Hooper's Rear End Exchange was contacted to build a posi carrier. The carrier used was located in a boneyard, but it turns out GM uses two different carriers, each having a range of gear ratios it will accept. Since most posi carriers made

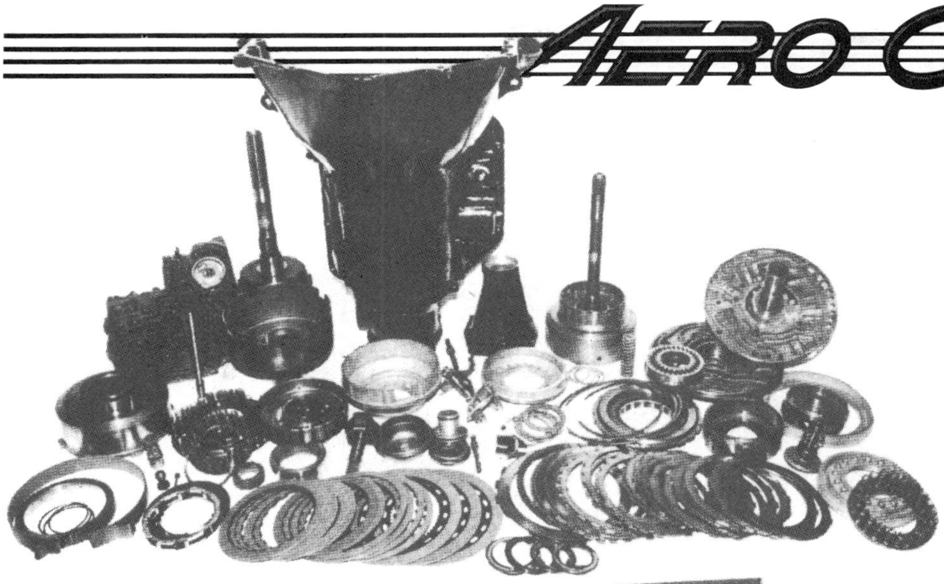

This is your worst nightmare if you figure you're going to do it at home without the right tools. Darrell Young has assembled more TH 350's and 400's than most people will see in a lifetime. For the Mr. Gasket Aero Camaro, the plan was a high-performance rebuild and installation of a Mr. Gasket Tough Shift 3 (right) shift kit and Mr. Gasket chrome oil pan.

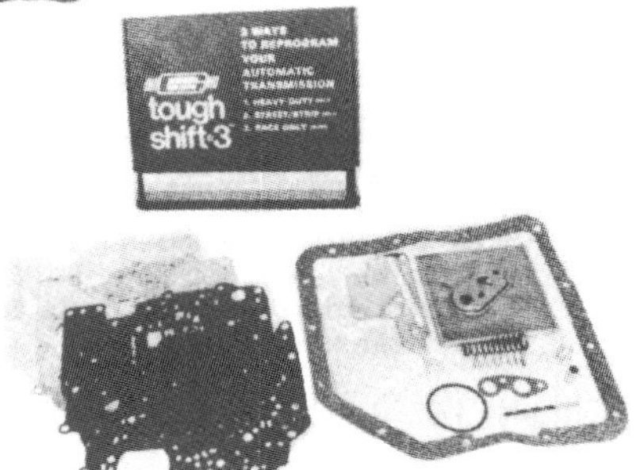

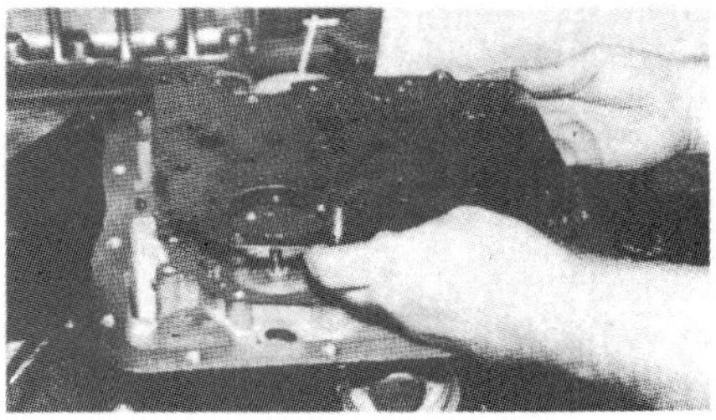

The Mr. Gasket Tough Shift 3 kit comes with two different valve body metering plates; the plates have certain holes enlarged over the stock size to allow higher flow to the clutches for harder shifts. Think of the metering plate as a giant "jet plate," designed to control fluid flow through the valve body.

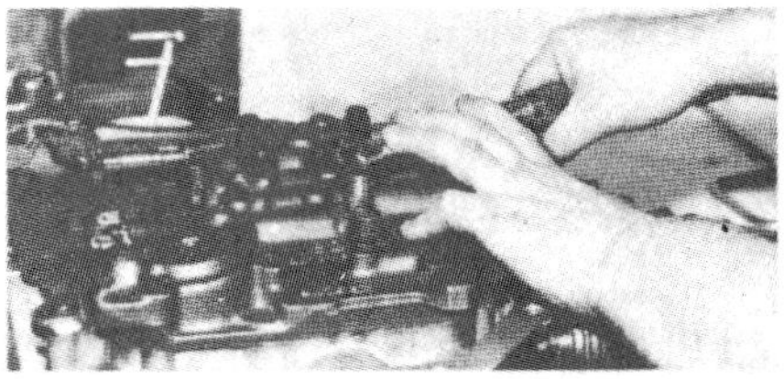

Carefully retorquing the valve body hardware can prevent leaks that may happen internally due to the increased operating pressure in the trans. Torque values vary, and are usually listed in service manuals for your particular trans.

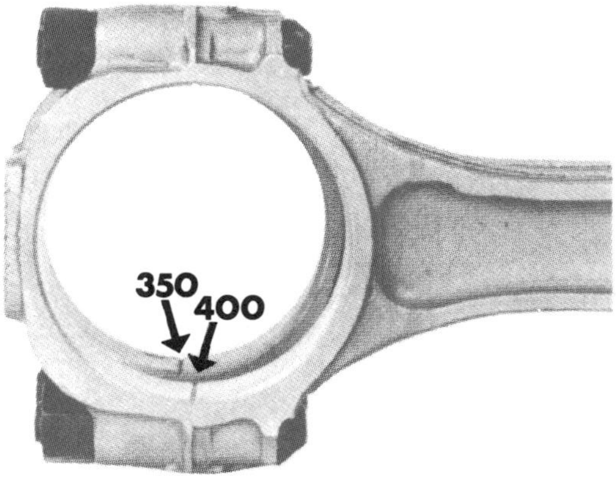

All 350 and 400-cubic-inch engines share a common compression height, hence the piston pin is in the same location on both engines. To use the 350 piston with the 400 crank, you have to use the 400 connecting rod which is .135-inch shorter than a standard 5.7-inch 350 rod. The difference in rod length is illustrated by the rod cap parting line (arrow). The 400 rod is in the foreground.

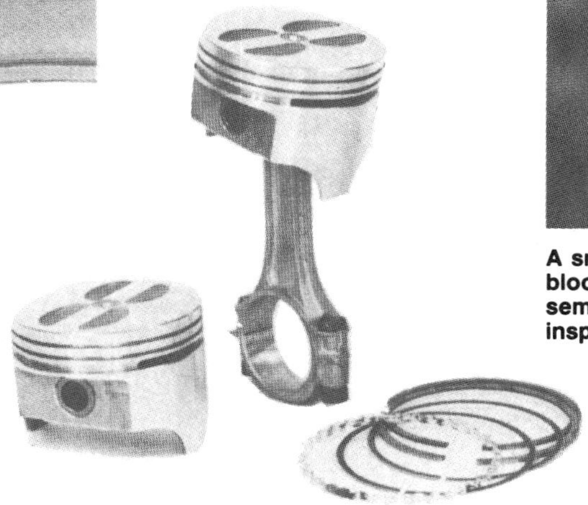

TRW forged pistons (PN L-2256F) are equipped with Sealed Power rings and swing from reworked 400 small block connecting rods. These pistons are listed for late-Sixties applications in 4-barrel 350s. In this application the compression ratio is only 8.2:1 due to the reduced rod length and subsequent lower deck height.

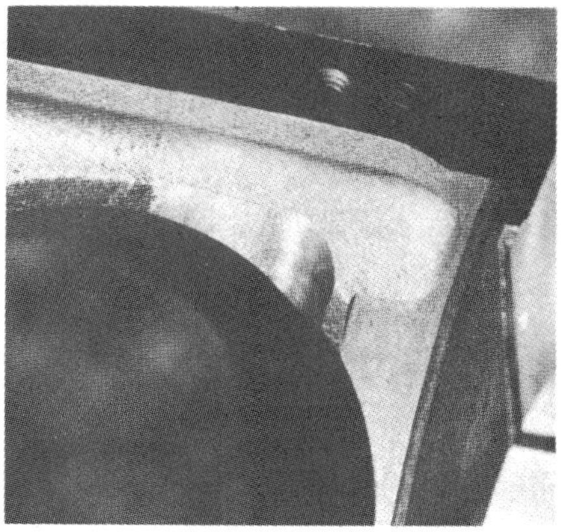

A small relief is cut in the bottom of the block to clear the 400 rod and crank assembly. A mistake here can allow you to inspect the water jacket the hard way.

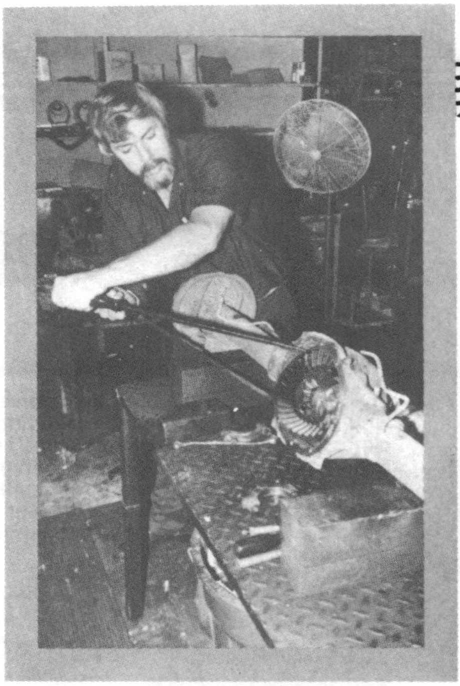

Since the carrier is fairly snug in the case, after removing the retainers on either side, you'll need to assist it with a little friendly persuasion. The name of this game is "heavy-duty parts require heavy-duty tools."

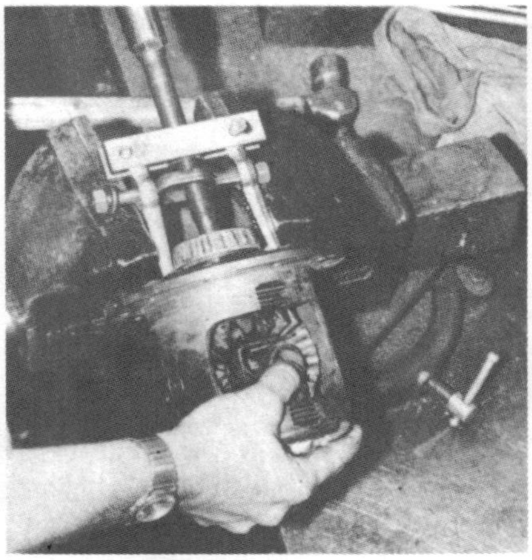

New bearings will replace these tired veterans. Note how thin the flange on this carrier is; on carriers built to handle 3.73's from the factory, the flange is about twice as thick. Although some builders shy away from these carriers, thinking the use of a spacer will weaken the assembly, Hooper's hasn't found that to be true if the ring gear bolts are properly installed and torqued.

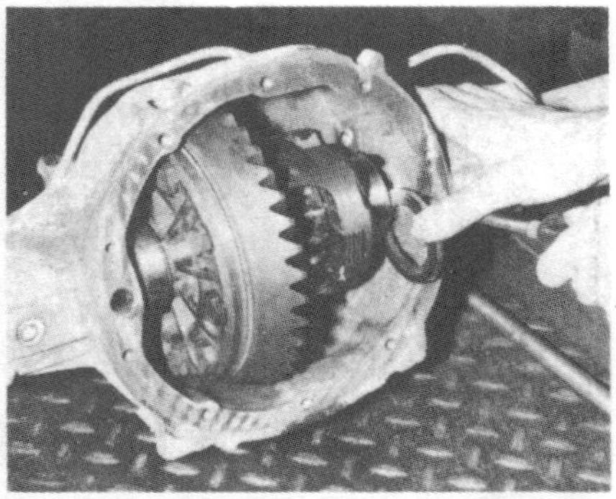

Side load and position are determined by adding or subtracting spacer thickness; at Hooper's they stock dozens of different sizes.

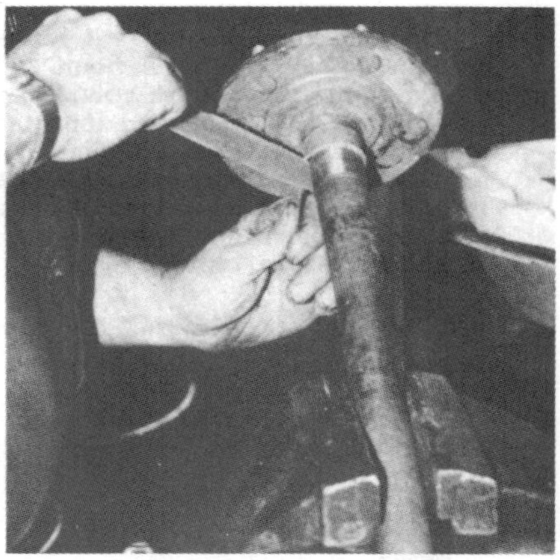

Before reinstalling the axles, any scoring should be removed by polishing with emery cloth.

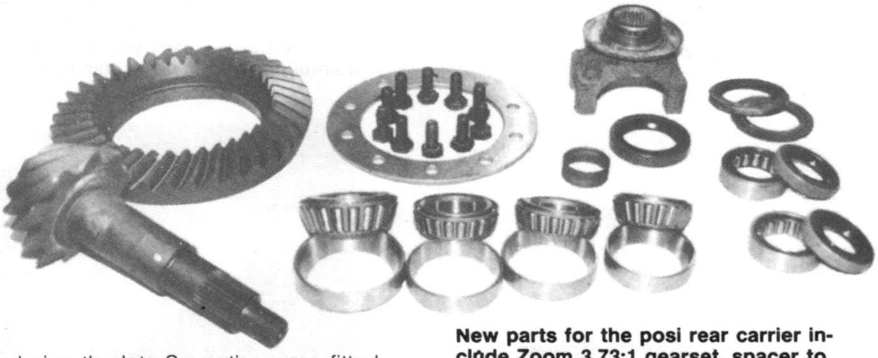

New parts for the posi rear carrier inclûde Zoom 3.73:1 gearset, spacer to allow the gears to be used on the thin flange carrier, and all new bearings and seals.

during the late-Seventies were fitted with relatively "tall" gears (in the two's), a different carrier was made having a thinner mounting flange for the ring gear. Lower ratios typically have thinner ring gears and require a thicker mounting flange on the carrier for correct pinion alignment. Zoom came to the rescue with a 3.73:1 gearset and a spacer to make up the difference in the mounting flange.

In the 10-bolt series carriers, the carriers fitted with 2.41:1 or higher (2.38:1, etc. . . .) have thin flanges, while carriers that accommodate 2.56:1 gears or lower have the thick flange. By the way, this applies only to the 8½-inch corporate carriers used in the second-generation Camaro; GM makes several other gear carriers, each accommodating their own variety of ratios.

Next month? Paint, interior, tunes, and a quick road test to make sure everything works as well as the Performance People intended. We're betting on a strong finish. **HR**

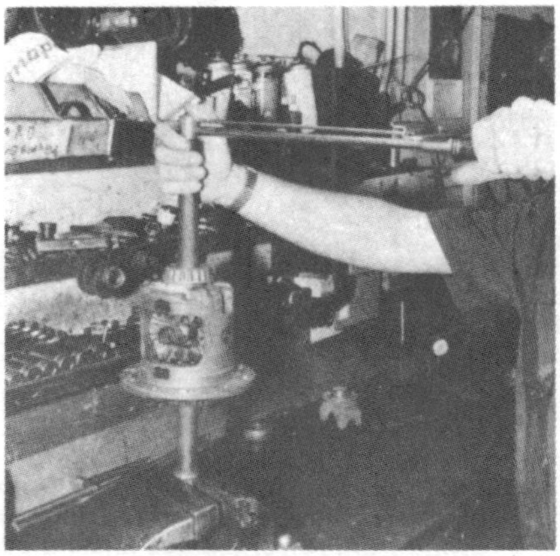

The resistance of the axle is bench checked by looking for a specific torque value. The check is performed before and after the axle is rebuilt; with fresh clutch plates, the torque on this axle should be about 90 lb.-ft.

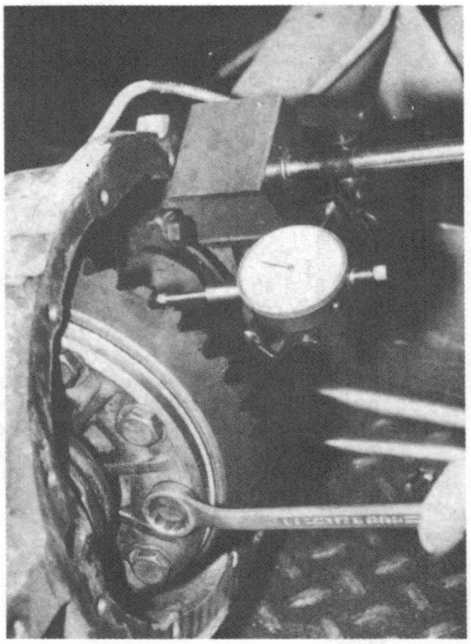

After applying Prussian Blue to the ring gear teeth, a check is made to determine whether or not the gear should be repositioned side-to-side. At this point, the pinion is installed, and may require removal to correct its position also.

Putting the Finishing Touches to the Mr. Gasket Street Machine Buildup

After several months of dismantling, refinishing, plating, and plumbing, the Mr. Gasket Aero Camaro is a rolling, running reality. Loaded with a John Lingenfelter 383 small-block, Darrell Young TH 350, and a GM posi rearend handled by Hooper's, the drivetrain is definitely up to the task of a little street action. This final chapter in the Aero Camaro story deals with the interior treatment, preparation of the body for paint, plumbing of the engine compartment, and all the other touches that make the car appear detailed and properly dressed as a street machine.

The Camaro was purchased complete with a rear deck lid spoiler, a la GM. In front, a stock GM Z28 spoiler was added to complete the look, and a hood scoop from A&A Fiberglass was bonded to the hood. It's not a functional scoop, although it could be made to draw air from the cowl area by removing a portion of the hood's sheetmetal and building a new air cleaner baffle. The bracing under the Camaro hood is strategically placed so you'd only need to cut one layer of steel rather than several.

There is a brief photo sequence on how to bond a scoop to the hood (contained in this portion of the Aero Camaro story), and next month we'll elaborate on the many steps involved in doing the job correctly. The techniques used for the scoop could of course be applied to bonding on fender flares, rear spoilers, air dams, and side panels.

How To Build a Contemporary Street Machine

By Kevin Boales

Part 4:

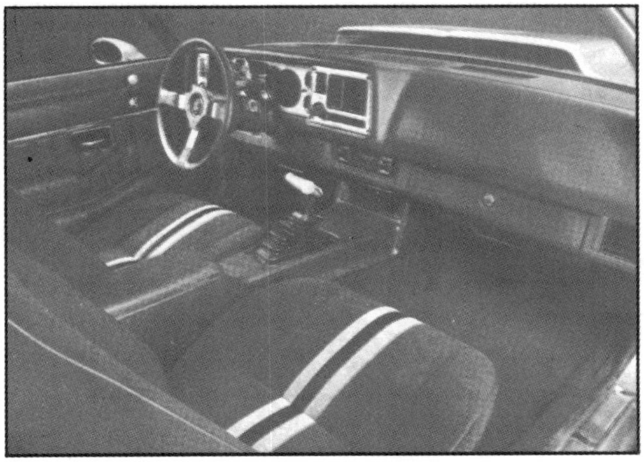

Inside, threads by Denny Nish, tunes by Pioneer, gauges by Auto Meter, and a Nardi steering wheel complete the creature comforts.

Under the hood, Earl's plumbing looks right at home with all those Mr. Gasket accessories (RIGHT). **Air conditioning lines and heater hoses look far better than stock plumbing.**

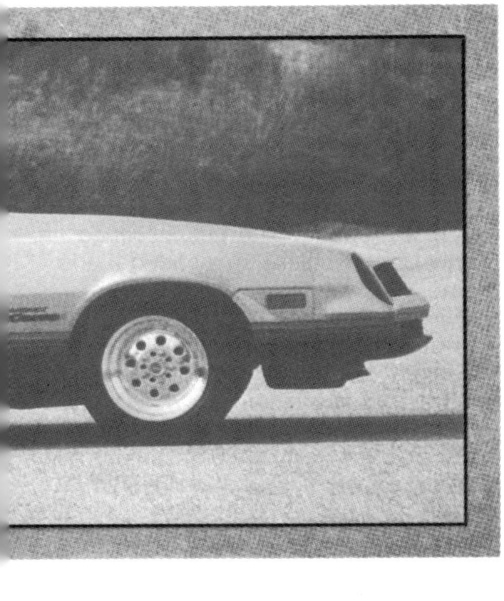

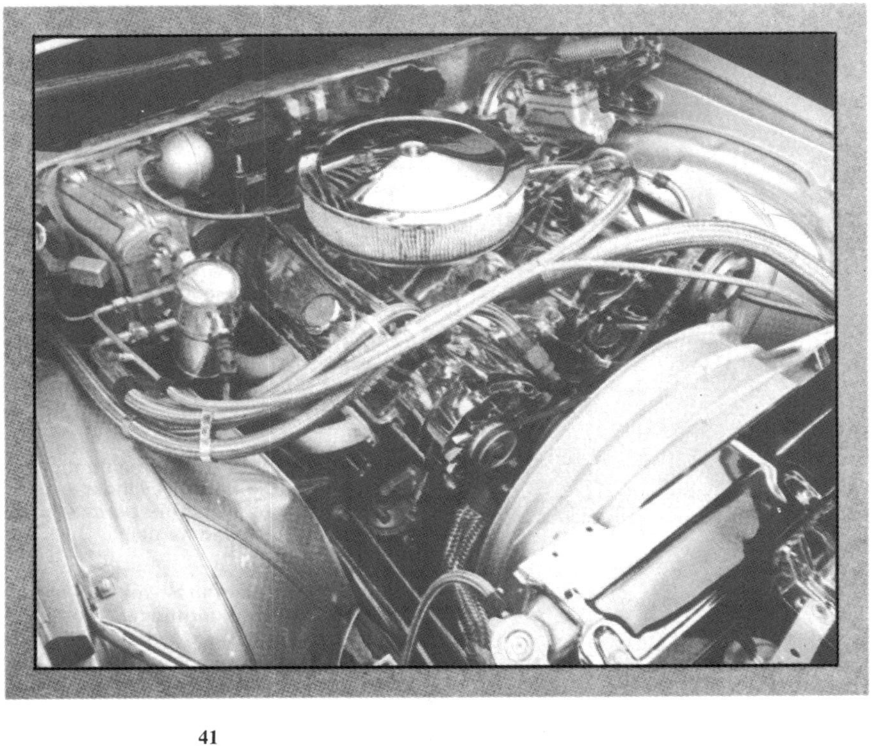

Aero Camaro

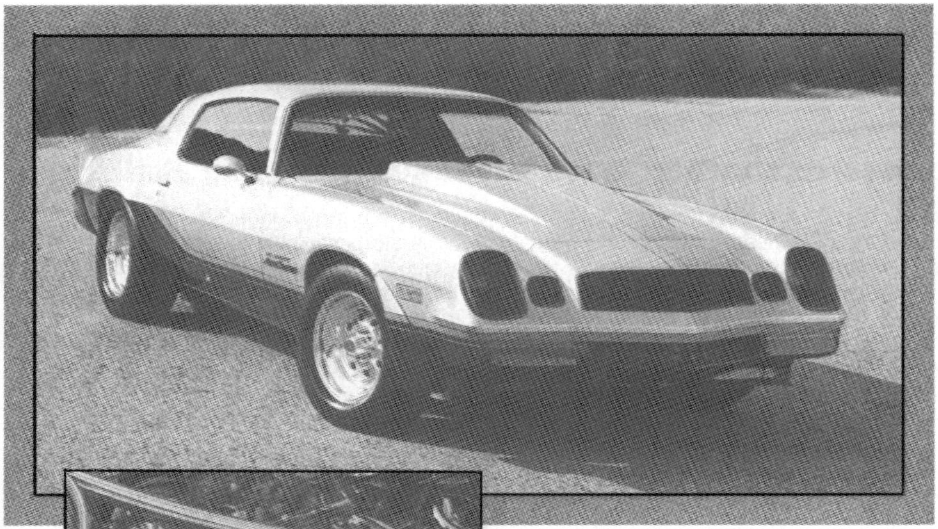

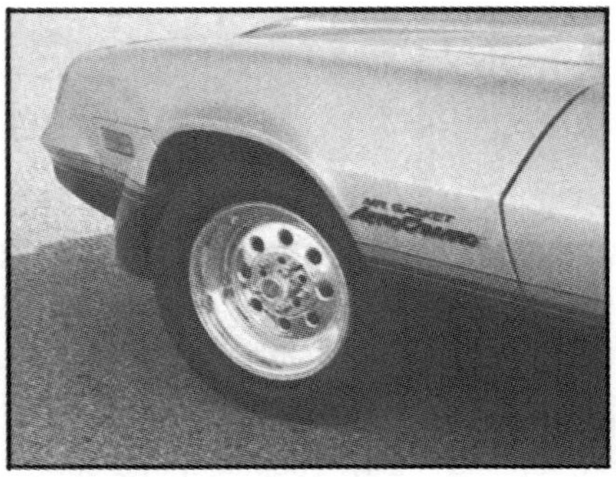

Headlight covers from the A&A catalog and a stock Z28 front spoiler highlight the front-end of the Camaro. Hood scoop is also from A&A.

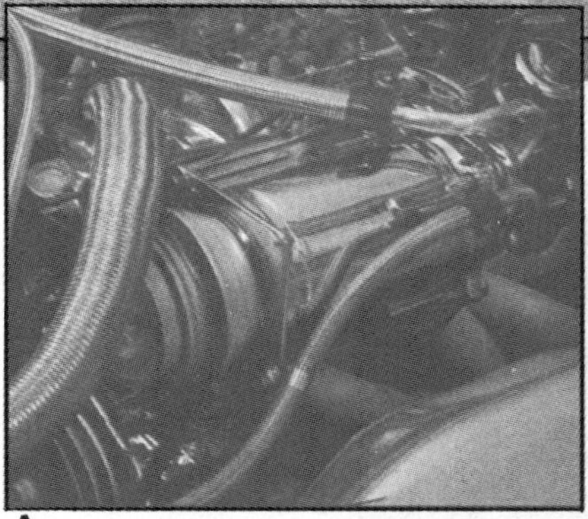

A

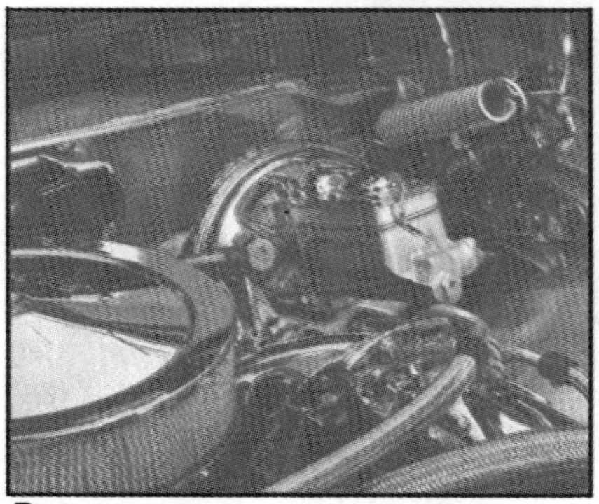

B

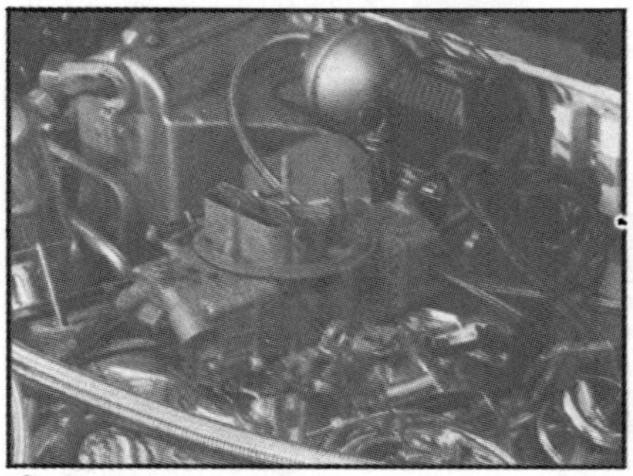

C

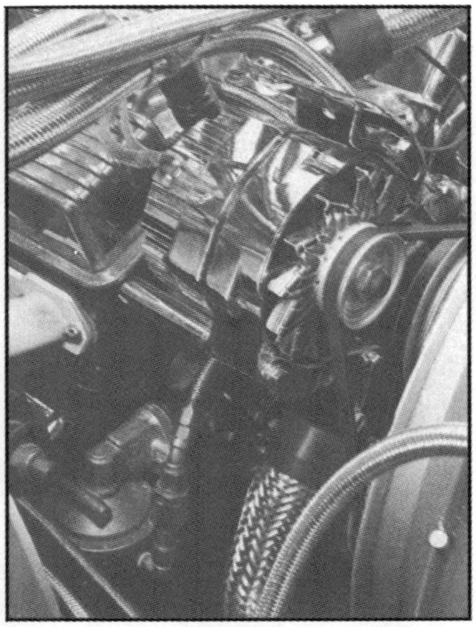

Total Discount Auto Parts disassembled the alternator for chroming and, for accent, painted the rotor assembly red. Note the Earl's plumbing for the Carter high-flow mechanical fuel pump.

(A) **To chrome the A/C compressor, it must be disassembled. The task was given to Lucas Radiator & Refrigeration in Van Nuys, California. The chrome you see everywhere on the car was applied by Carter Chrome Plating of Burbank, California.**

(B) **After grinding off the casting mold parting lines, the master cylinder exterior was tuned up by filling the pores with plastic filler, followed by paint.**

(C) **In order to use the Holley and early Z28 manifold, an older-model throttle cable with a ball end was substituted, and a special bracket was fabricated for the thermostat housing-to-alternator brace.**

Because the car was so straight to begin with, very little bodywork was necessary. We've run bodywork features in the past, though. If you'd like to review them, check out the feature on the HOT ROD Magazine Flashback '57's sheetmetal rework in the March '84 HRM, or the one in the August '83 issue.

With the Aero Camaro, most of the work done on the body was in preparation for paint, involving a lot of priming and blocking. After Rick Leonard roughed out the body and attached the spoiler and hood scoop the car was taken to Bill Carter's Pro Paint, where the blocking began. After a silver base coat, the black panel and stripes were laid on according to the graphic plan developed by Steve Stanford for the project. A urethane clear seals the finish from the air in Los Angeles.

Carter likes to use a hint of pearl in his silver paintwork, and for the Aero Camaro, a platinum pearl was selected. Pearls should be used sparingly over dark, solid colors because they tend to appear as metallic flecks in the paint rather than as the desired pearlescence. Over lighter colors, an excess of pearl will give the car a dirty look, but over silver you can get away with quite a bit. They're used typically in the clear coat rather than in the paint. Since solid colors are opaque, the pearlescence would be lost unless suspended in the clear.

When we left you last month, the car had progressed to the point where the drivetrain was completed and installed, but was missing plumbing, wiring, and so forth. A quick trip to Earl's Supply cured that problem. A combination of Earl's products, including their Auto-Fit Hose Ends, Econo-Braid Hose, adapter fittings, and Tu-Braid nylon sheathing is

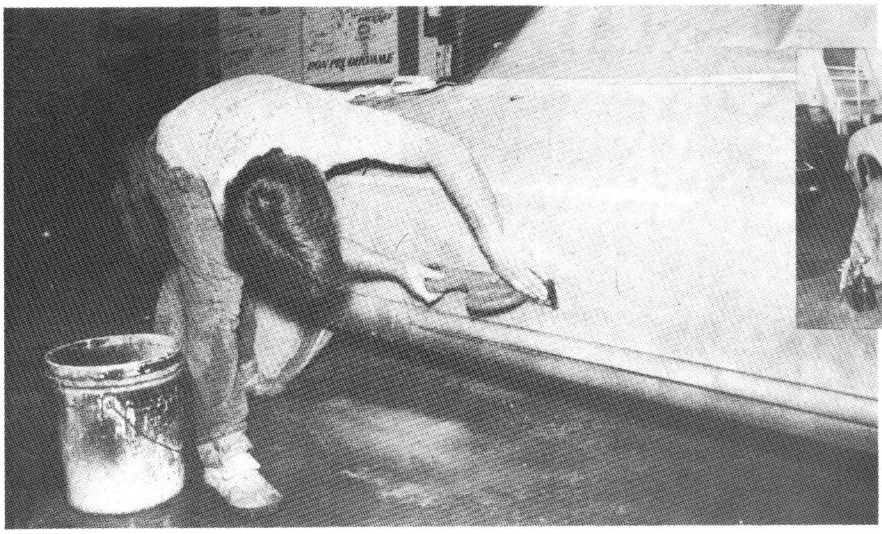

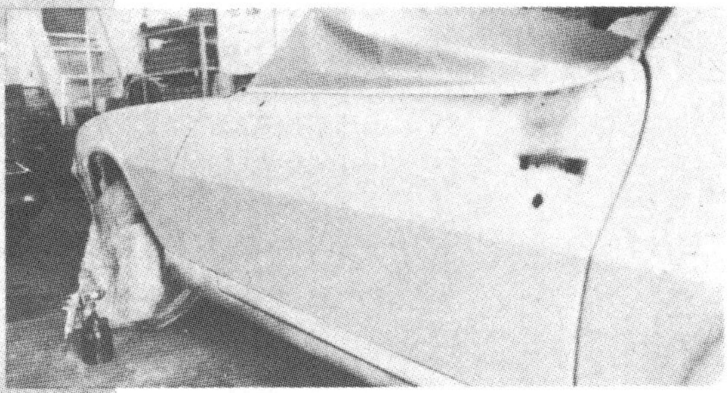

At this point, the car is looking very straight. The basic rule for this type of work is "k-e-e-e-e-e-p sanding!"

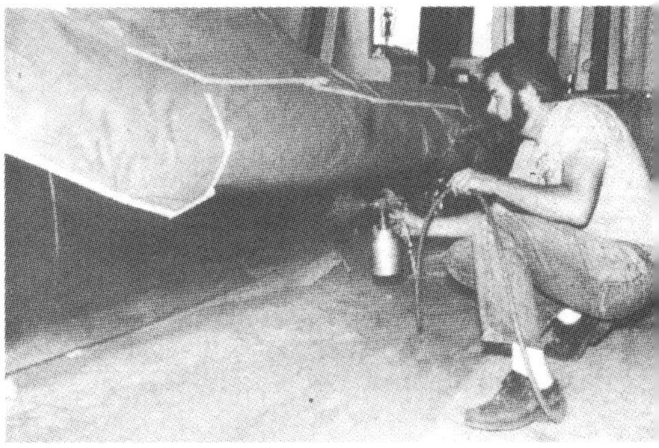

Using a piece of tape as a guide for the body line, the doors are blocked using a "long board" sander. For best (flattest) results, don't use too much pressure and take long, easy strokes.

Bill Carter is shown here (RIGHT) **shooting a little clear over the black lower panel. *Do not* do this without a respirator; Carter wanted you to see his mug. . . .**

visible under the hood. Earl's also supplied the braided stainless brake lines.

For your car, Earl's sells complete kits for any application, with either Tu-Braid or Econo-Braid hose. The photo sequence shows the assembly technique for the Auto-Fit hose ends used on Econo-Braid hose; Tu-Braid simply slips over your existing hoses and is finished at each end with an Econo-Fit hose end.

With the drivetrain installed, the next step was to hook up the Mr. Gasket shifter, which is a gated, no-mistakes street shifter. It utilizes the stock shifter cable and drops right into the console. To shift, you pull the finger grips up and either lift or press down on the shifter handle to negotiate past each gate. It doesn't do anything accidentally, like popping into neutral; you need to be deliberate with this one.

Any decent street machine should have high-performance lurking everywhere you look, and the Aero Camaro delivers. Inside, the new threads were tastefully assembled by Denny Nish Interiors, using cloth the same color as the original vinyl, with added interest from stripes running vertically through not only the seats but onto the back shelf, concealing the speakers for the stereo. Lining the stripes up proved to be the trickiest part for Denny, who is a veteran upholsterer for everything from drag boats to motor homes. The old plastic trim parts in the interior had faded in varying degrees, so Ditzler supplied the dye and primer for recoloring them to match the upholstery. Actually, this is a three-step process, with a cleaning stage, a primer, and finally, the finish coat. Clears are also available for added abrasion resistance.

Tunes? Pioneer to the rescue with one of their ultra-sensitive Supertuner III

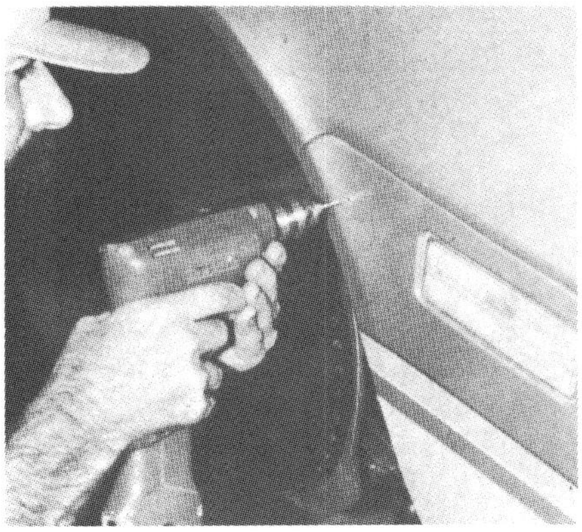

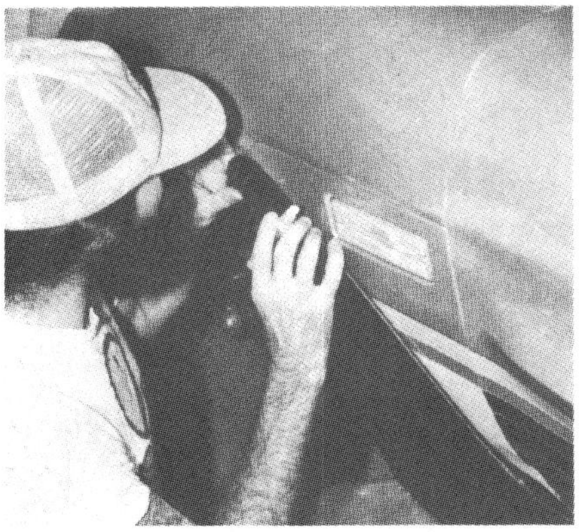

After marking the location of the necessary holes for the front spoiler kit, the holes are drilled. The hardware supplied with the kit is used to attach it. If done carefully, the fit between the side skirts and body can be done perfectly. Most aren't.

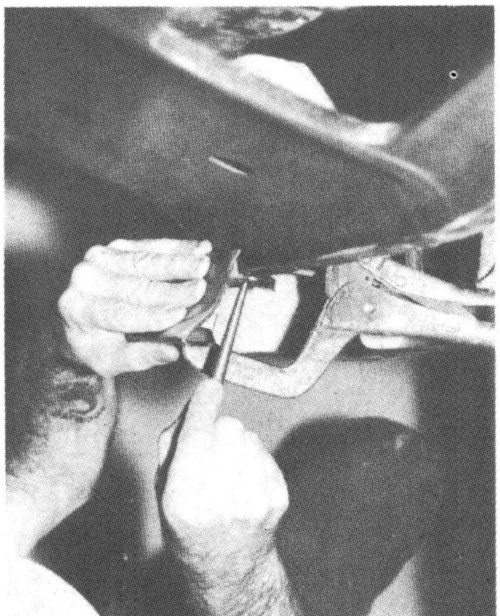

While the spoiler assembly is held temporarily in place with welding clamps, a punch is used to align the holes for the bolts that will attach the center section to the subframe.

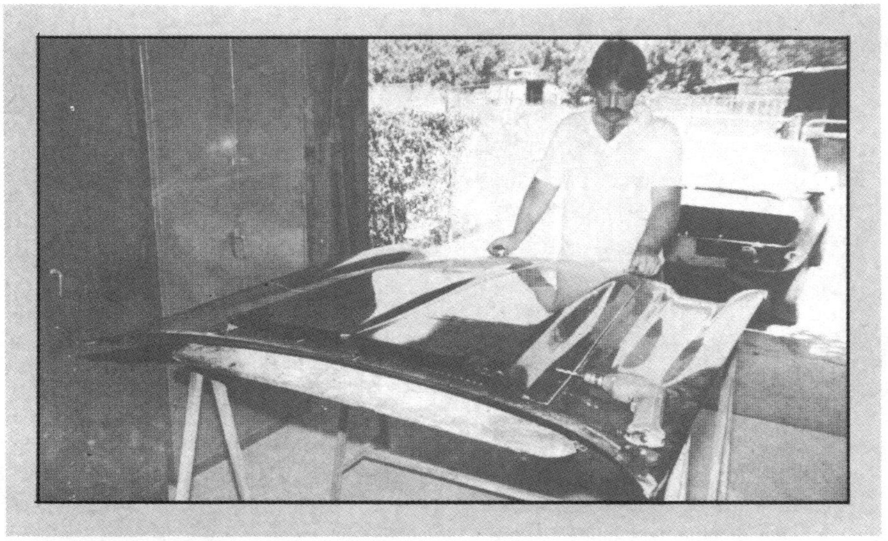

The A&A Fiberglass scoop is trimmed to length with a cutting disc. After the scoop is bonded on, a sander is used to blend the contour to the hood.

decks, an outboard amplifier, and four excellent speakers: two TS-6907 6x9's for the rear deck and a pair of 3½-inch drivers for the front, mounted on a bracket fabricated from the old speaker frame and some perforated strips. The bracketry and installation were the handiwork of Lee's Stereo Installations in Van Nuys, California.

Pioneer's KE-A880 deck is loaded with all the features you might want in a deck, such as 18 station presets, local station seek, Dolby, Music Search, tape EQ, Stereo/Mono switch, fader and preamp outputs, and 10 watts per channel. For a little extra low-end, a Pioneer amplifier was utilized. Model GM A-120 cranks out 30 watts per channel, safely, thanks to triple protection circuitry and excellent heat sinking. The amp has provisions for three types of inputs, also: RCA preamp, speaker level, and the new Centrate connectors. The amplifier was mounted inside the console, but Lee suggests that if your amp is larger, one of the best mounting locations is under the passenger seat in order to keep the cables from the deck as short as possible. Long cables carrying a preamp signal can pick up engine noise more easily than short ones.

As you can see in the photos, the stock Camaro instrument panel was slightly reworked for the installation of Auto Meter's small-faced gauges. The panel face is plastic ("No kidding . . . I never would have guessed . . . , " you're saying) and can be surgically modified easily with a cutoff wheel or hacksaw. The remaining holes aren't exactly round because of the cone-shaped bez-

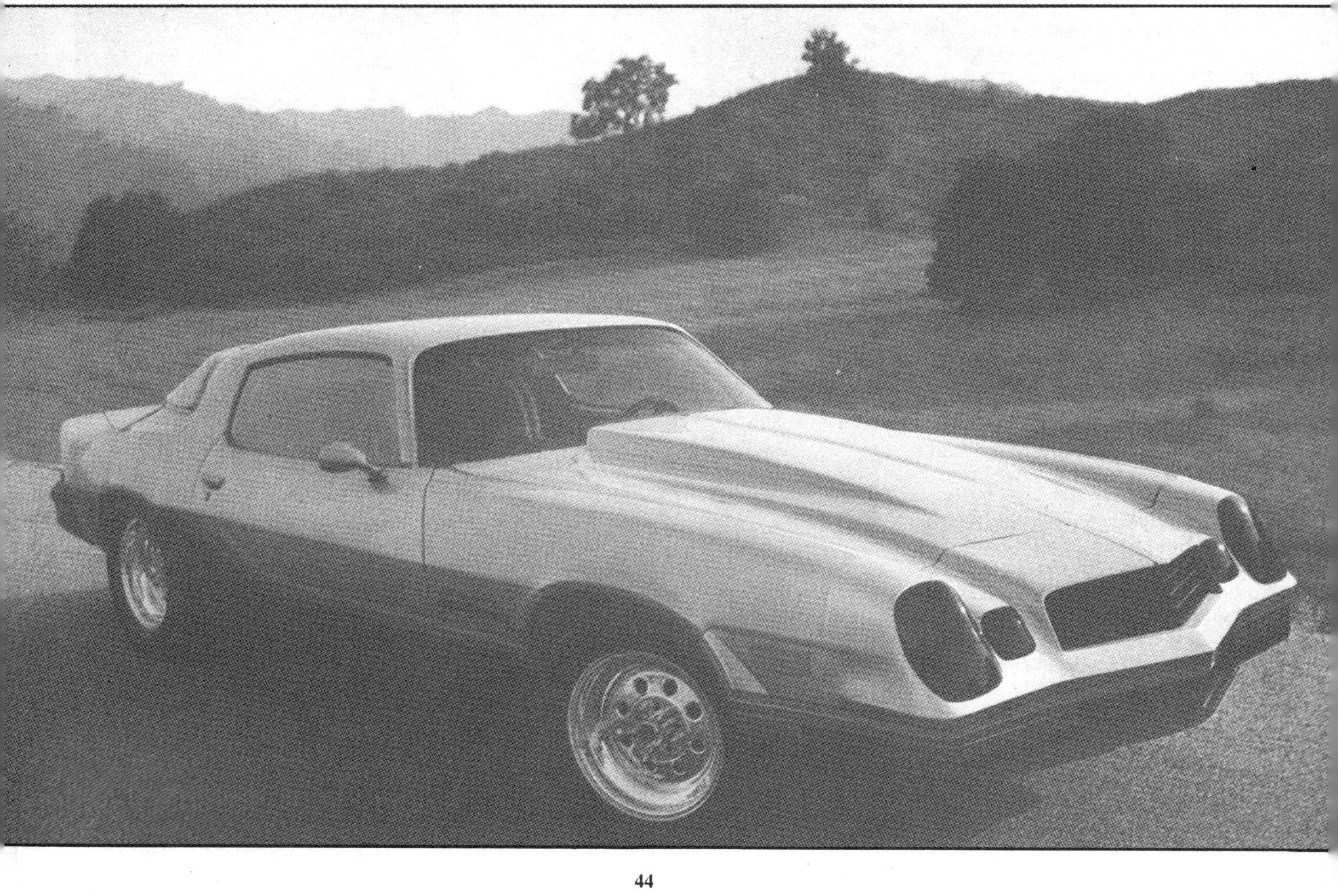

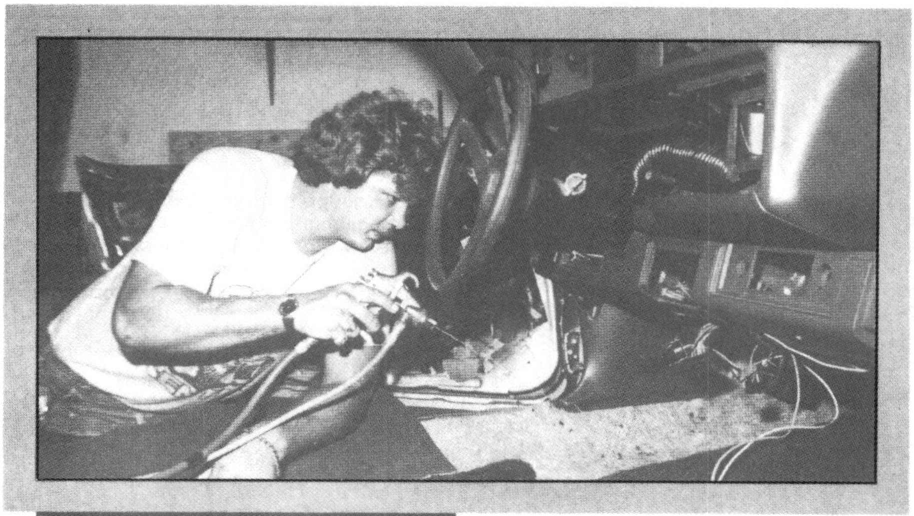

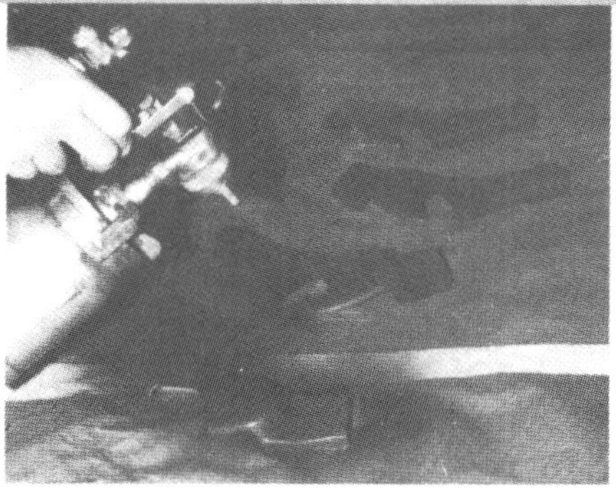

Denny Nish and son will upholster almost anything, although they prefer boats, rods, and street machines to more bizarre projects. Denny (Jr.) is blasting on some spray adhesive to hold everything down. These guys are excelent . . . check out the color shots.

Ditzler's vinyl finishes spray on like paint; different clears, primers, and cleaners are available for preparation of the pieces or added abrasion resistance.

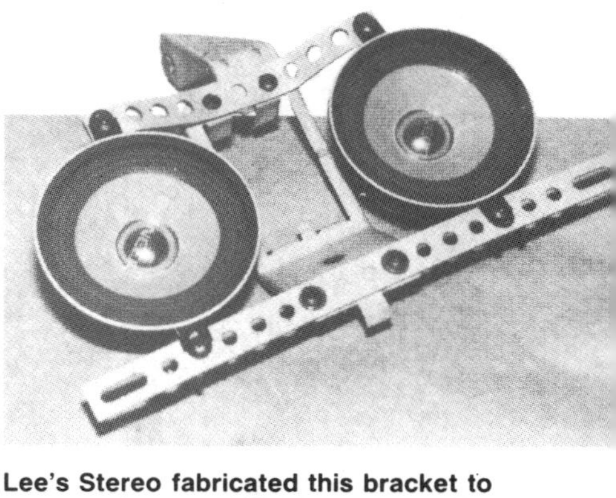

Lee's Stereo fabricated this bracket to support the two 3½-inch speakers. The assembly bolts in place of the stock 4x10 and uses part of the OEM speaker frame.

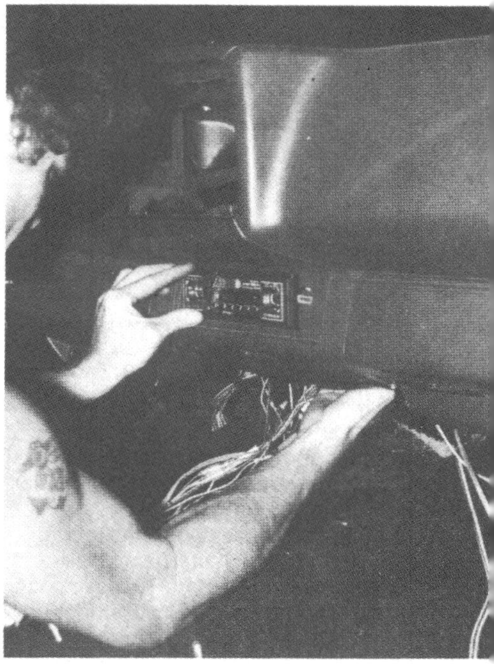

Installation of the Pioneer KE-A880 deck is fairly straightforward. The outboard amplifier will mount beneath the console, about where Lee's forearm is.

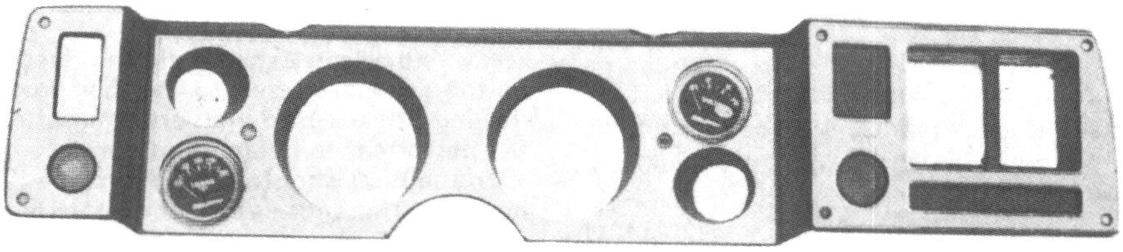

Stock instrument panel was replaced with step-up version from a Z28, modified to accept the Auto Meter gauges. By removing the cone-shaped tubes behind the front plate, the gauges can be flush-mounted.

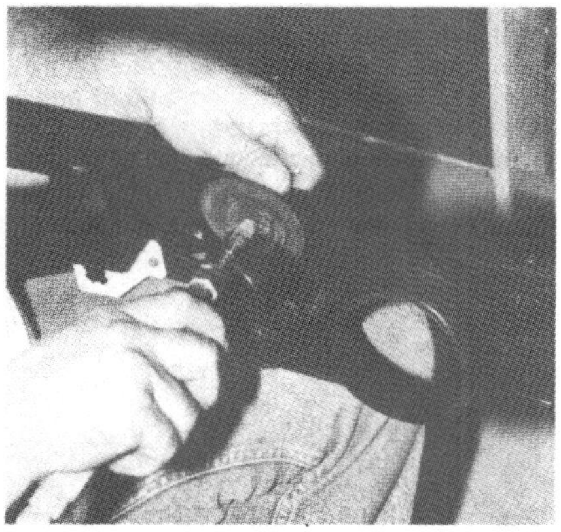

This shows how the panel is modified to accept the gauges. Although the remaining holes aren't exactly round, the gauge bezels are wide enough to cover the problem.

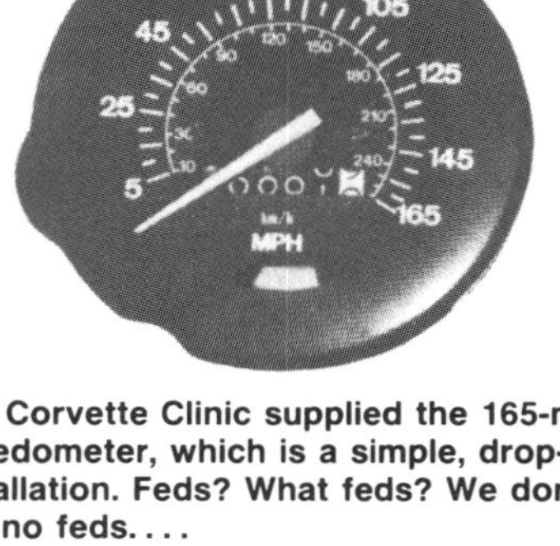

The Corvette Clinic supplied the 165-mph speedometer, which is a simple, drop-in installation. Feds? What feds? We don't see no feds. . . .

els that formerly extended from the front face back to the original gauge faces. Auto Meter's gauges have a large enough chrome bezel to hide the oval openings in the panel.

After seeing one of the Corvette Clinic's 165-mph speedos, the Performance People at Mr. Gasket contacted them to see if one could be made for the Camaro. "No sweat," came the reply. An exact replacement with new dial numbering appeared at Rick Leonard's Special Interest Auto within a few days, and

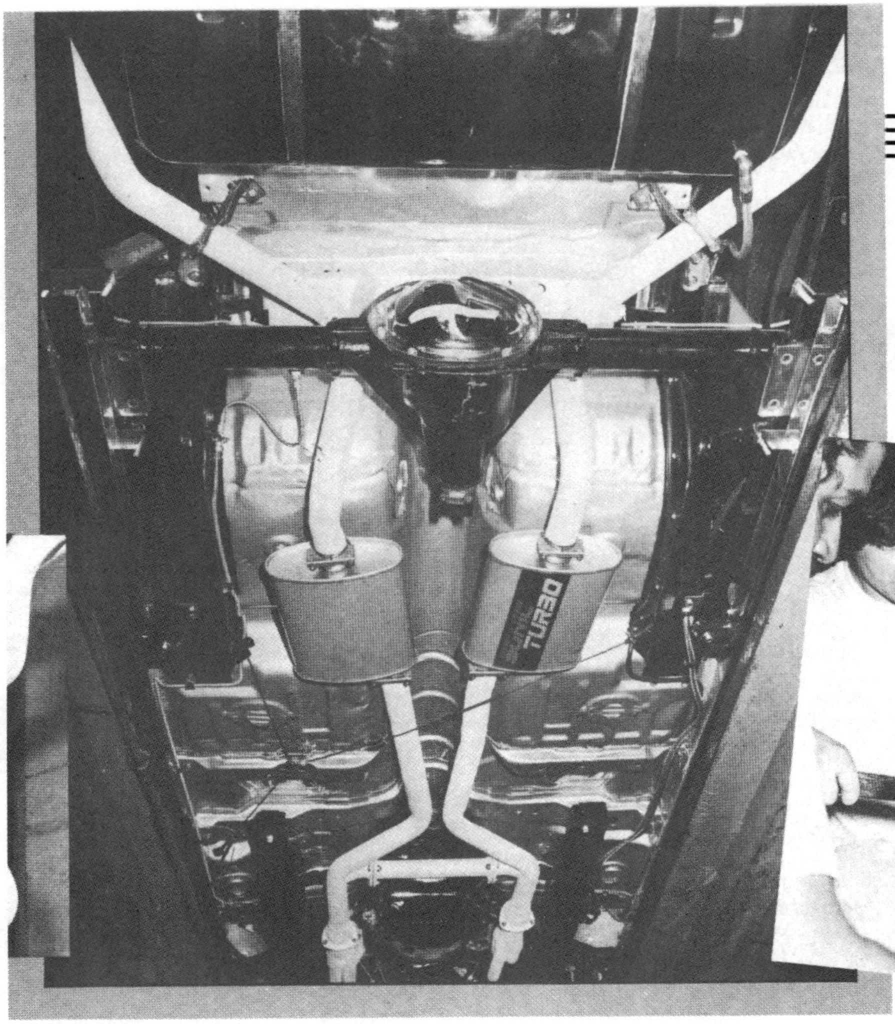

A little nudge here and there.... Using a body hammer as a mandrel and a 24-ounce convincer, you can make any size dent you want for clearance. Small dents won't interfere with header performance, kinks will.

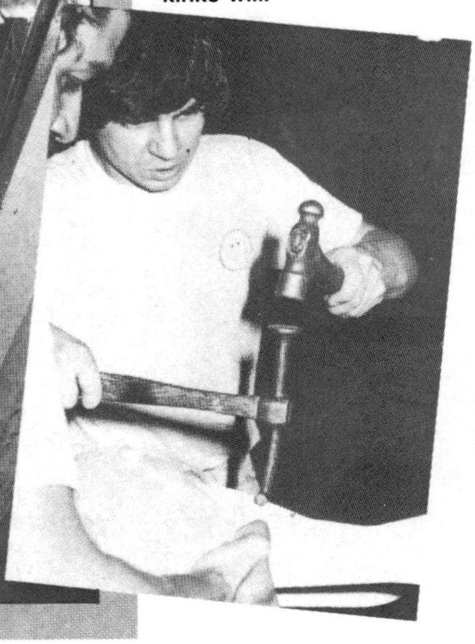

(ABOVE) **American Exhaust did the deed on the exhaust system, using their free-flowing Sonic Turbo mufflers (winner of the performance category in our recent Turbo muffler shootout) and Cyclone headers. The entire exhaust system is alumi-koted, except for the mufflers.**

M&M Welding flame-sprayed the aluminum onto the exhaust components. As you can see, the metal is liquified and sprayed in one operation. Note the feed wire extending to the gun (BELOW).

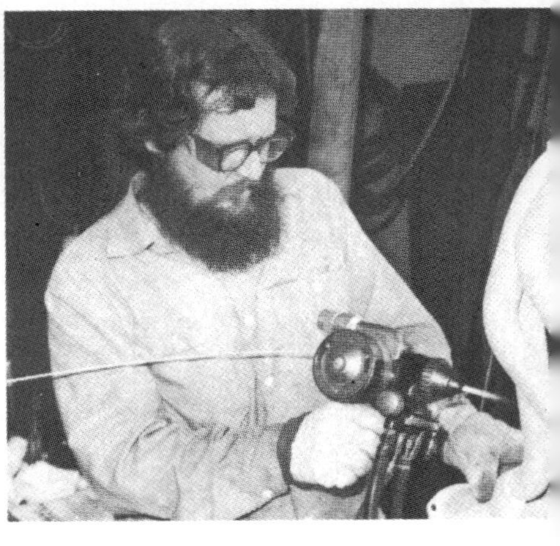

there it is in the dash.

Finishing off the driver's station, the stock X-ribbed steering wheel was discarded in favor of a Nardi leather-wrapped model, coincidentally the same Carmine Red as the rest of the interior.

There were still some remaining details under the car, too. Although the engine and trans had been installed, the exhaust system had not. Sooo . . . one of the Mr. Gasket companies happens to be Cyclone/Blackjack/American Exhaust, and naturally, they were given the task of assembling some hot tubes for the Camaro.

American Exhaust markets Alumi-Koted headers and Sonic Turbo mufflers. For this car, the entire exhaust system was to be "koted" in the interest of corrosion protection. The process has been around for years, used in the aircraft industry to build up abraded areas, such as in compressor seals for jet engines. Although the aluminum is soft, it offers excellent protection from rust, which is the worst threat to the lifespan of your exhaust system.

The aluminum is literally sprayed on the surface in a molten state. Called flame spraying, metal spraying, or plasma spraying, depending on the metal being applied and the method, the metal bonds itself to the base material, forming a coating of variable thicknesses. On the headers, the coating is fairly thin, but it doesn't have to be thick to work.

American Exhaust's Sonic Turbo mufflers also won our recent Muffler Shootout (July '84 HRM) in the sheer performance category. They aren't the quietest in the world, but they don't suck up much horsepower, either.

The final detail under the car was a set of Lakewood chrome traction bars, finishing up the rear axle assembly.

What happens to the Aero Camaro now? Mr. Gasket intends to put it on the show circuit for the next year, which will give all of you the opportunity to see it in the flesh. And don't be surprised if it shows up at the HOT ROD Super Nats!

For those of you who would like more details on the construction of Mr. Gasket's Aero Camaro, this project has also been featured in a book of its own, with extended pictorial coverage of the transmission rebuild, suspension and brakes, painting tips, and fiberglass bonding. The book is HOT ROD Magazine's *How to Build a Street Machine,* on sale in January at newsstands, or from Petersen Retail Sales for $2.95. **HR**

Engine swapping certainly isn't anything new in a car crafter's repertoire. In fact, it's an activity that's never gone out of style. Twenty-five years ago it might have meant inserting a brand new 265-cubic-inch Chevy in a '40 Ford, but the basic premise has always been the same: Put something with more power or reliability into a car that just doesn't have "enough" of whatever it is the car crafter has been after.

For a long time engine swapping's popularity waned, probably due to the fact that the overhead valve V-8 engine in stock form already offered plenty of power and reliability. As the aftermarket industry began designing bolt-on and interior pieces for the new generation of engines there just wasn't the need there had been in the past.

Today there may not be a need for engine swapping, but it's certainly become a regular occurrence among the street machine set—with some major differences. For one thing, most of us aren't trying to stuff V-8's into cars 20 and 30 years old. What we're trying to do is simply update our existing machines from the Sixties and Seventies. But it really goes beyond that. Today there are three generations of Camaros and Mustangs, and who knows how many different engine combinations. With big and small-block engines, manual and automatic transmissions, the list is literally endless. Yet one factor does stand out. If your machine's getting a little tired under the hood, yanking that oil encrusted "thing" out of there and replacing it with a sparkling clean, ready-to-go, 100,000-mile, blazing the tires at every stoplight new engine may be a lot easier than you think.

Rick Voegelin, our resident Camaro expert (he grows 'em in Northern California, honest), recently blurted out in a staff meeting that engine swapping was the way to go for more power, more fun and more everything. Of course Rick tends to get slightly enthusiastic about these things. From the following guides for engine swapping in Camaros and Mustangs, however, it seems that once again he's quite right.

By Rick Voegelin

Powerplant Transplant

Part Numbers to Put Your Camaro Engine In Its Place

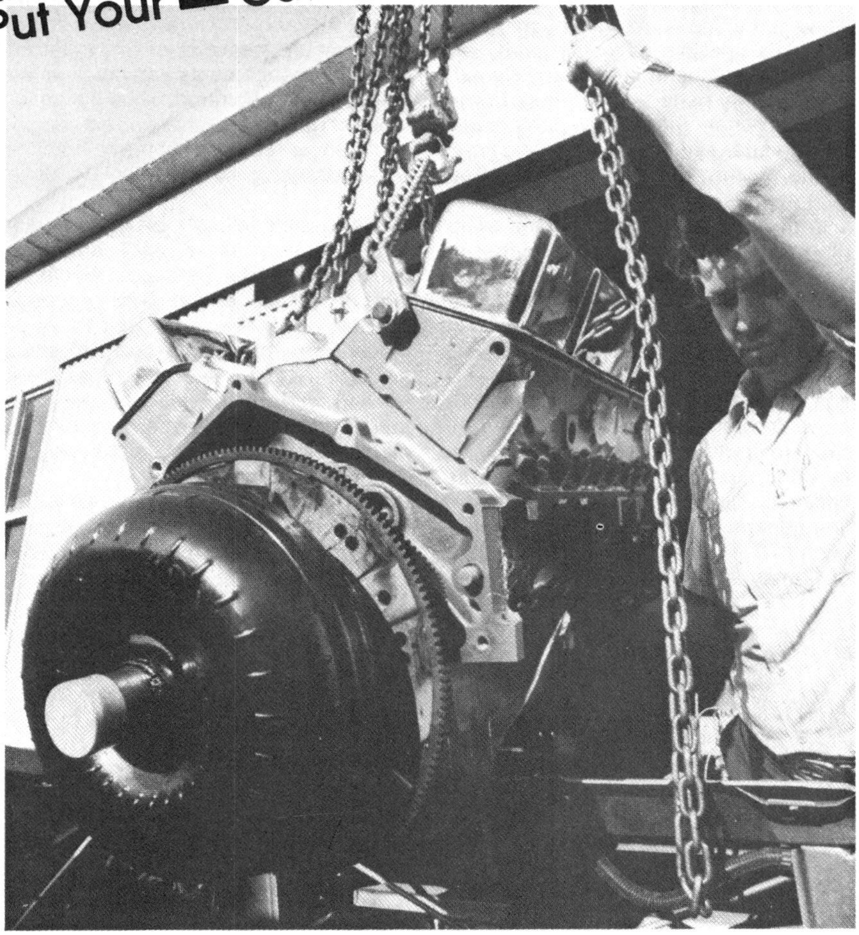

The Fifties was the golden age of oddball engine swaps. No matter how strange the body or bizarre the powerplant, *somebody* made an adapter to bring it all together. Buick straight eights in Henry J's, flathead Fords in '57 Studebakers, and Lincoln V-8's in Jeeps were just a few of the favored engine exchanges. But in the Sixties, when the automakers themselves started putting very big engines in reasonably small bodies, the practice of engine swapping became a lost art.

Yet now there are signs of a Fifties revival in areas beyond popular music and fashions. Engine swapping is making a comeback—in a less eccentric form, thankfully. For the facts of automotive life in the Eighties are that you can't always get what you want, unless you're willing to build it yourself. This time around, however, the swaps tend to respect automotive purity, keeping the powertrain and body all in the same corporate family. Chrysler's Direct Connection group has spread the word to the Mopar clan on how to mix and match A, B and Hemi engines with various chassis; Ford fanciers are well aware of what it takes to build better Boss machines. And for Chevrolet partisans, we've assembled the following guide to Camaro engine swapping.

Powerplant Transplant

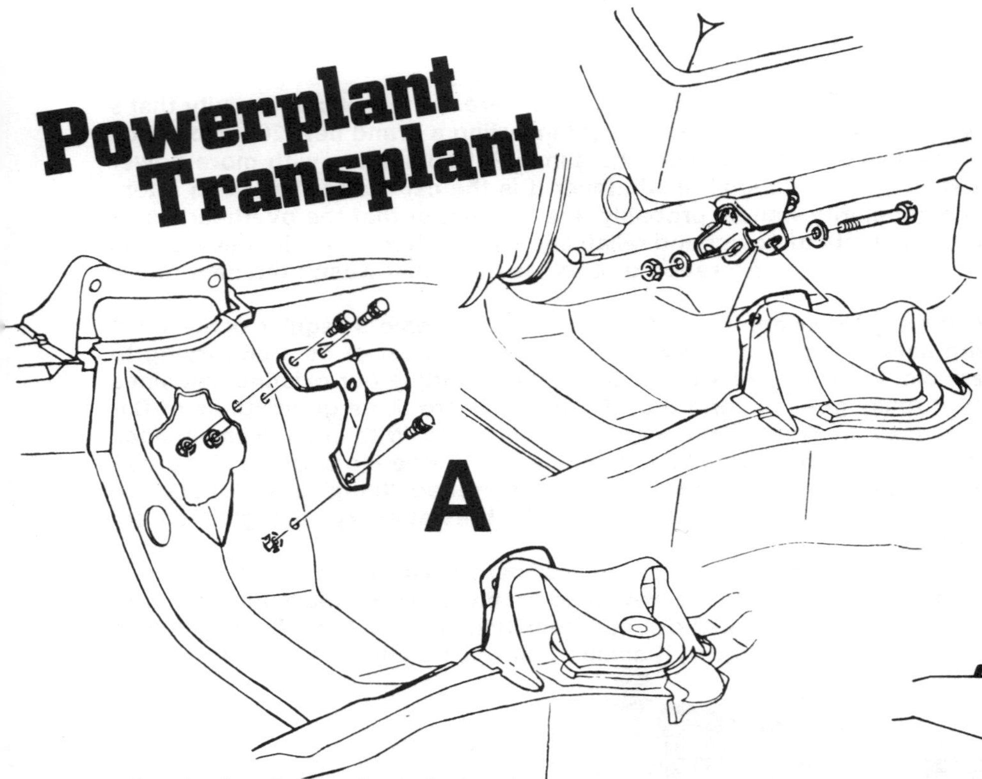

If you've long harbored a desire to put a street Hemi in your Z/28, you'll have to look elsewhere. But if the times call for you to put a fuel efficient small-block where a consumptive big-block has long resided, or to substitute an LS-6 Rat motor for an anemic inline six in order to terrorize the local Pro Gas circuit, then we've got some numbers for you. For, in truth, it's possible to install nearly any Chevrolet engine in almost any Camaro using factory parts. And in those few instances when the Chevy parts books fail, the aftermarket manufacturers come to the rescue with their specialized engine mounts.

In most cases, Camaro engine swapping requires nothing more elaborate than a set of wrenches, a chain fall, and an obliging and sturdy tree limb—although a rented engine hoist may be more secure. Exchanging engines in a Camaro chassis is truly a bolt-in proposition, thanks to Chevrolet's system of removable brackets and mounts. And the Camaro crafter's task is simplified by the fact that there are really only three distinct chassis designs to deal with—the first generation ('67-'69), second generation ('70-'81), and third generation ('82) of Camaros.

All Camaros use a three-point engine mounting system, with a pair of mounts on the sides of the engine block and a single mount on the transmission tailshaft. This wide triangle supports the entire powertrain assembly. To accommodate the various widths, lengths, and motor mount locations of Chevrolet engines, different mounts and brackets are bolted to the subframe. Thus it's important to have both the correct engine mounts and the right subframe brackets when changing over a Camaro drivetrain.

It *is* possible to bolt a big-block engine onto a pair of small-block subframe brackets in some instances, but problems may arise as the work proceeds. The big-block brackets used in early-model Camaros, for example, offset the engine toward the passenger side slightly—and without that minor relocation, there's insufficient header and exhaust manifold clearance around the steering box.

Surprisingly, there are even some differences between engine mounts used with apparently identical blocks. For example, the mounts commonly found on 302 and high performance 350ci engines in first generation Camaros are often wider and lower than mounts used with standard 307 and 327 blocks. Similarly, the mounts specified for high-performance LT-1 small-blocks in 1970-'72 are actually big-block components, installed in deference to the engine's considerable torque output. But many of these variations have disappeared from the parts books, and now it's a matter of using the pieces which Chevrolet still stocks. The part numbers

There are two basic types of Camaro motor mounts. Camaros built through 1971 use stamped steel brackets bolted to the subframe and rubber-cushioned mounts on the engine block. In later years, this system is reversed: steel brackets bolt to the engine, and the rubber cushions are installed on the subframe.

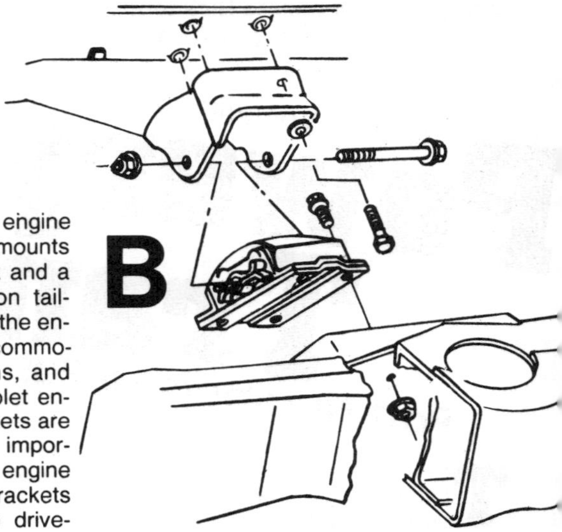

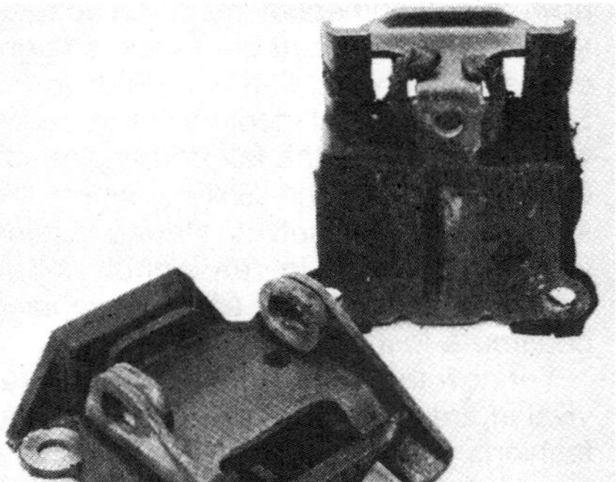

Revised early-model engine mounts have interlocking steel safety tabs to prevent engine movement if bond between rubber and metal breaks.

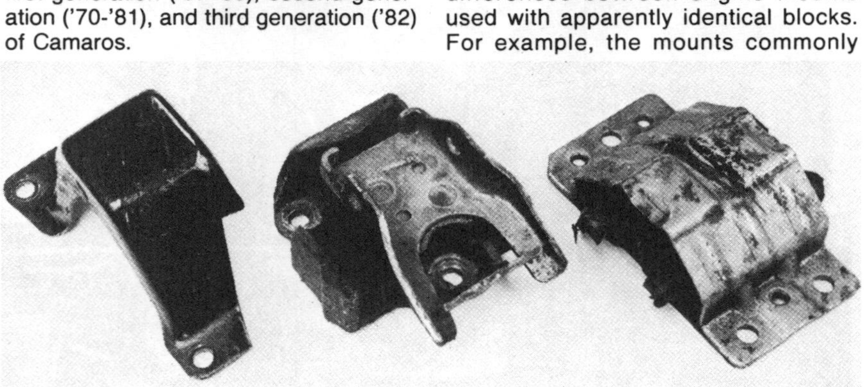

Stamped steel bracket (left) is typical of design used through 1971 with rubber-cushioned engine mount (center). In late-model subframe mount (right), rubber is completely surrounded by steel shell, and is much less prone to failure.

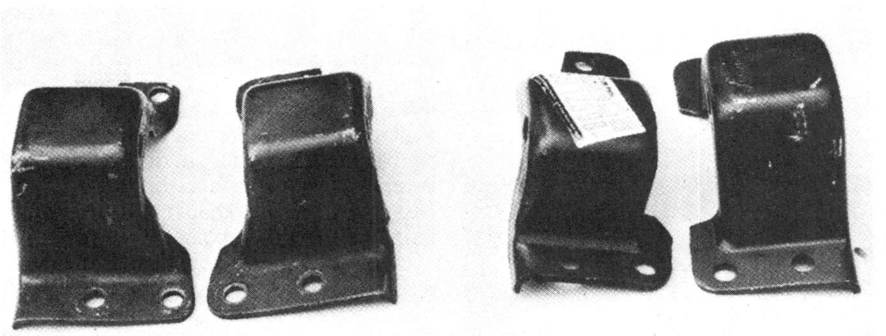

There are subtle differences between early-model subframe brackets for small-blocks (left) and big-blocks (right). Big-block brackets locate engine mounts farther forward and slightly offset toward passenger side.

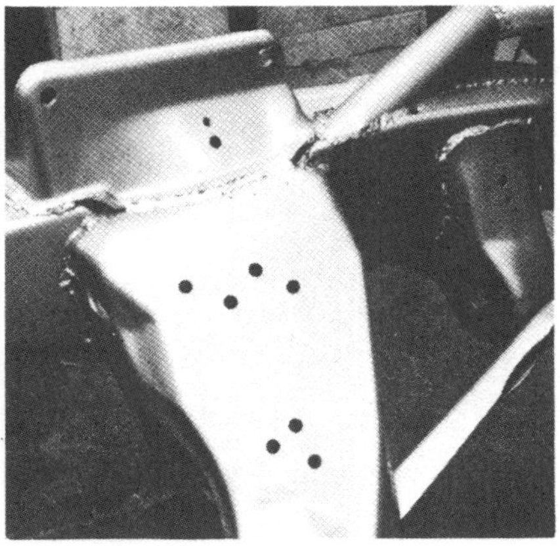

Variety of holes in subframe crossmember accommodate mounts for different engines. New holes may be required for engine swaps.

in the accompanying chart represent the latest listings—but of course these are subject to change without notice, as they say.

Any Camaro built before 1973 should have its motor mounts inspected carefully, since the bond between the mount's rubber cushion and steel bracket may have failed. Even though the engine may *look* like it's mounted securely, torque reaction under hard acceleration may cause it to move enough to bind up the accelerator, clutch or shifter linkages. It's possible that missed shifts are the result of broken mounts, not a lack of driver coordination. Replacement motor mounts have interlocking steel tabs which restrict the engine's movement even if the rubber-to-metal bond breaks.

Chevrolet engineers completely redesigned the Camaro engine mounts in 1972 to eliminate this problem. In the new scheme, the stamped steel brackets attach to the engine block, while rubber cushioned mounts are bolted to the subframe. This simplified the engine mounting hardware, since the same frame mounts could be used with big-block V-8's, small-block V-8's, *and* six-cylinder engines. It's still possible to use the earlier type of stamped steel brackets, but it may be necessary to drill holes in the subframe crossmember to accommodate them.

Solid steel motor mounts eliminate the flex found in factory rubber-filled mounts, which is why they're favored by many racers and serious highway runners. These simple U-shaped brackets are lighter and easier to install than their OEM counterparts, and serve to stiffen the chassis by adding the rigidity of the engine block. Naturally, without the insulation of the original rubber cushions, more vibration and noise is transmitted to the body structure—not a consideration on the race track, but a drawback on the boulevards.

As the Pro Street movement gathers momentum, more Camaro owners are opting for Pro Stock-style engine plates to carry their powerplants. Front engine plates are typically constructed from ¼-inch thick aluminum, and are bolted to tabs welded to the subframe alongside the upper A-arm mounts. In their favor, engine plates simplify maintenance, allow more room for header tubes, and look undeniably racy. As with solid motor mounts, they also transmit more noise

Camaro Engine Swap Chart

CAMARO V-8 MOTOR MOUNTS

Year	Engine	Mounts	Brackets
'67-'68	302, 350	6258154 (2)	3955183(L), 3980938(R)
	327	3990914 (2)	3955183(L), 3980938(R)
	396	6258154 (2)	3912597(L), 3912598(R)
'69	307, 327	3990914 (2)	3955183(L), 3980938(R)
	350, 302	3990918 (2)	3945507(L), 3945508(R)
	396	3990918 (2)	3980941(L), 3980942(R)
'70-'71	LT-1, 396	3990918 (2)	3982307(L), 3964886(R)
'70-'72	307, 350	3990914 (2)	3964890(L), 3980930(R)
'72	LT-1	357717 (2)	332648 (2)
'72	402	3990918 (2)	332648 (2)
'72-'75	Small-block	9762326 (2)	332648 (2)
'76-'81	Small-block	459021 (2)	332648 (2)
'82	Small-block	14039437 (2)	334970(L), 14039436(R)

ENGINE SWAP COMPONENTS

Item	Part Number	Application
Starter Nose	1968122	Aluminum, for 13-inch dia., 153 tooth flywheel
	1893864	Aluminum, for 14-inch dia., 168 tooth flywheel
	1969309	Cast iron, for 14-inch dia., 168 tooth flywheel
Converter Flex Plate	3991402	13-inch dia., 153 tooth, Powerglide or Turbo-Hydro
	3991403	14-inch dia., 168 tooth, Powerglide or Turbo-Hydro
	3992069	14-inch dia., 168 tooth, 454ci only, Turbo-Hydro

Powerplant Transplant

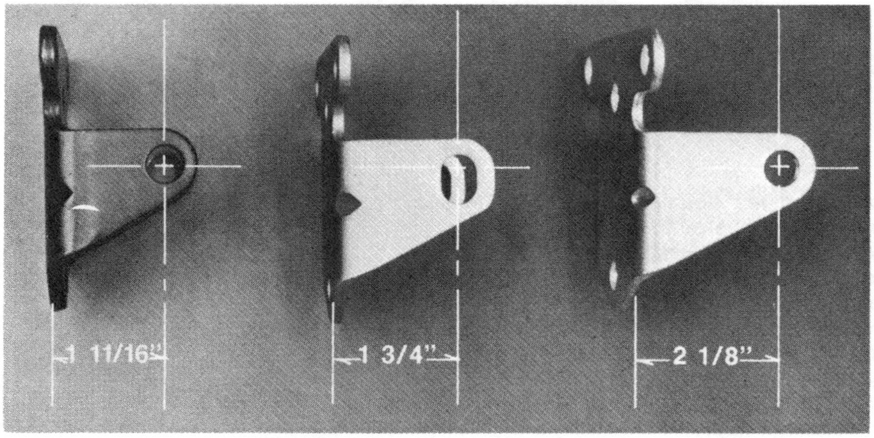

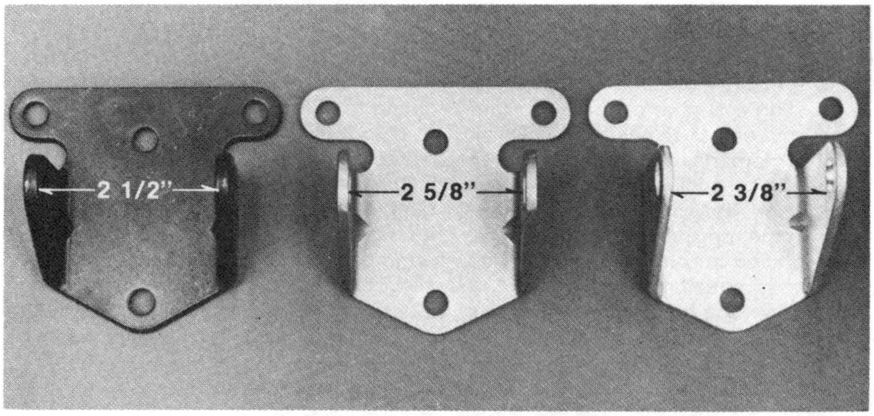

Solid engine mounts are favored by many racers because they eliminate drivetrain movement. These B&B Performance engine mounts illustrate dimension differences among early-model mounts. Mount on left is typically used with 307 and 327ci engines in '67-'69 Camaros; center mount is for 302, 350, and 396 in '67-'68 chassis. Long mount on right fits 302, 350, and 396 in '69 Camaro.

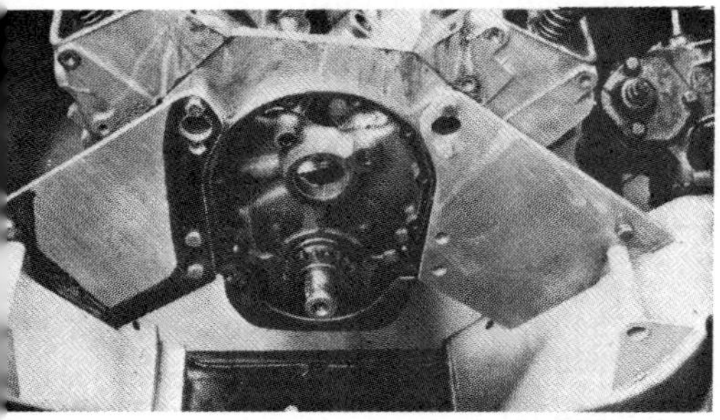

and vibration from the powertrain to the passengers.

The work of changing a Camaro powertrain doesn't stop when the engine's bolted to the subframe. The accessories have to be right for the new engine. A six-cylinder radiator can't cope with a high-compression big-block, nor will a starter intended for a small-diameter converter flex plate work with a large diameter flywheel. The accompanying photos and parts lists point out the components that are required for a re-engined Camaro. If all-out performance is more important than a factory-original installation, then a car crafter's tasks are simplified. Installing a set of headers, for example, will eliminate the hassles of finding the proper exhaust manifolds and pipes for the new engine.

Upgrading the suspension system is also necessary for a successful swap. Most factory big-block engine packages also include heavy-duty chassis underpinnings to cope with the extra weight and horsepower of a Rat motor. A 10-bolt rearend, standard springs, and drum brakes that were perfectly adequate for an inline six aren't really up to handling a healthy V-8. Similarly, installing a small-block where a big-block used to reside can leave the car with its front end up in the air unless the front coils are exchanged for more compliant springs.

Savvy GM performance enthusiasts know that what works for Camaros usually applies to Firebirds as well, since both makes are F-bodies under their sheetmetal skins. The same brackets and motor mounts used to install various Chevrolet engines in Camaros will work perfectly on Firebird subframes, opening up possibilities for 427 Turbo-Jet 'Birds—or Ram Air IV Camaros, for that matter. (Thanks to General Motors' "corporate engine" policy, it's even possible to build an Oldsmobile-powered Camaro, using the mounts designed for a late-model Trans Am!) Installing a Chevrolet engine in a Firebird is somewhat more involved, since the starter motor is on the opposite side, and the bellhousing bolt pattern is different for Pontiac engines. Some extra lengths of wire and a Chevrolet bellhousing will eliminate those dif-

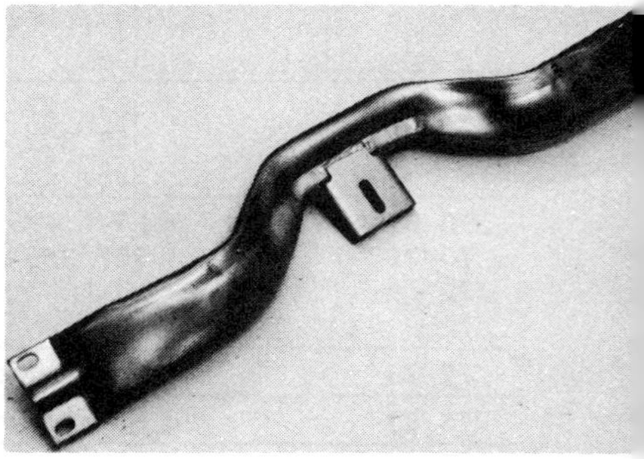

The '70-'81 transmission crossmember is tubular, has tab with slotted hole to accommodate different transmission tailshafts.

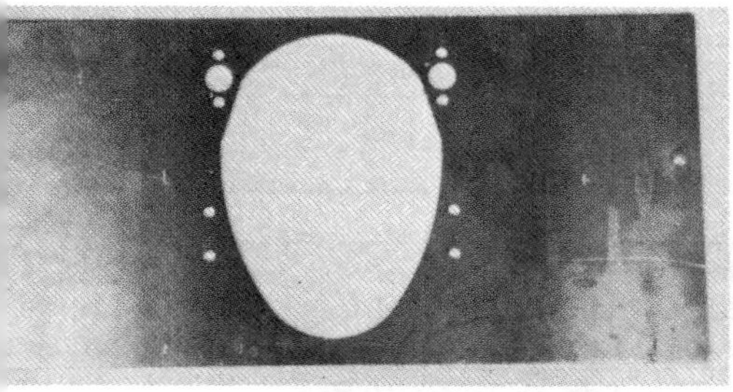

Many Camaro enthusiasts are eliminating factory mounts altogether by replacing them with Pro Stock-style engine plates. Moroso plate is ¼-inch aluminum.

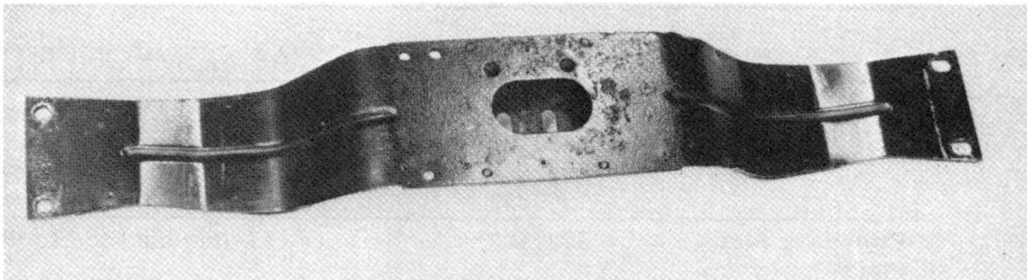

Transmission crossmember for '67-'69 Camaros is wide and flat, kicks up to clear exhaust pipes.

ficulties, however.

Over the years, Chevrolet engineers have given the world some remarkable Camaros, from all-aluminum big-blocks in 1969 to fuel injected small-blocks in 1982. But they've also missed some astounding combinations over the years, too. How about a 454 in a late-model Z/28—or perhaps a 229ci V-6 in a lightweight '70 Camaro for frugal performance? These and other intriguing packages are all possible. We've supplied the part numbers and the impetus. Now it's up to you to turn the wrenches.

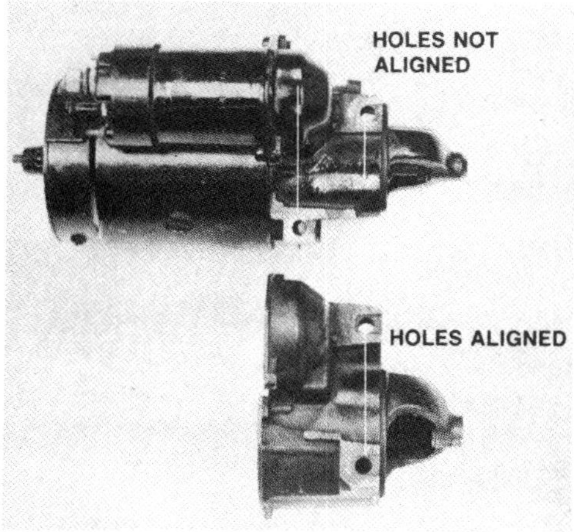

Depending on the flex plate or flywheel used, starter nose may have to be changed to complete swap. Noses for 13-inch diameter flywheel have holes in line; 14-inch flywheels take starter nose with holes offset.

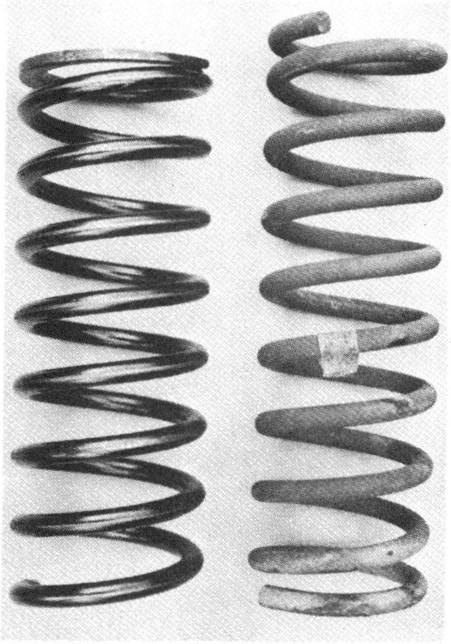

Difference in weight between small-block and big-block engine is approximately 150 pounds, depending on equipment. Front coil springs should be changed to restore proper ride height.

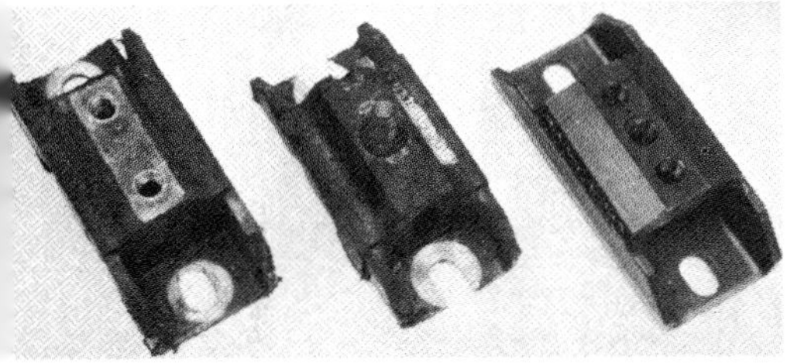

Early-model '67-'69 Powerglide/4-speed transmission mount (left) has two threaded holes; later '70-'81 version has single stud. Solid trans mount on right replaces both to eliminate drivetrain flex.

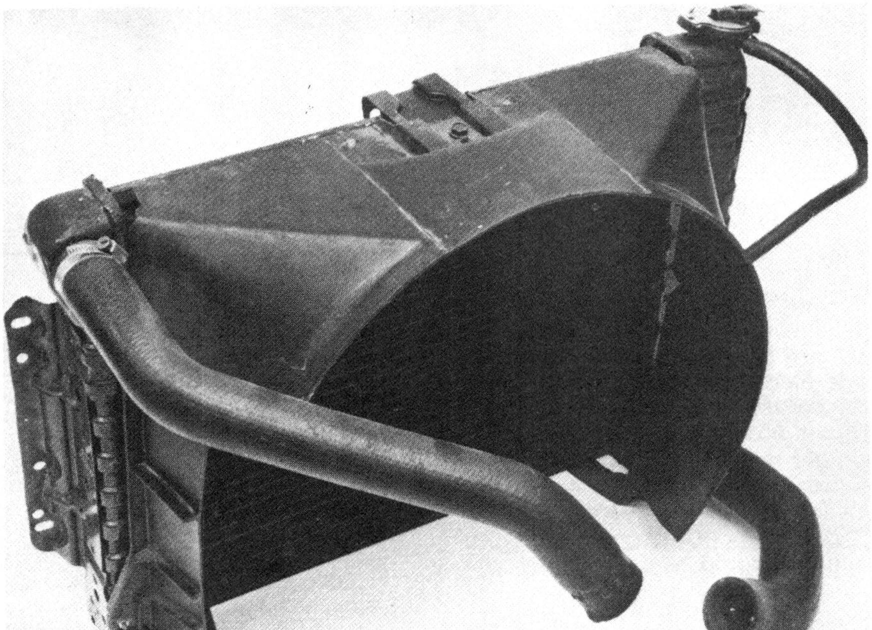

Cooling system should be upgraded to high performance or big-block powerplants. Heavy-duty radiators have double row of cooling tubes.

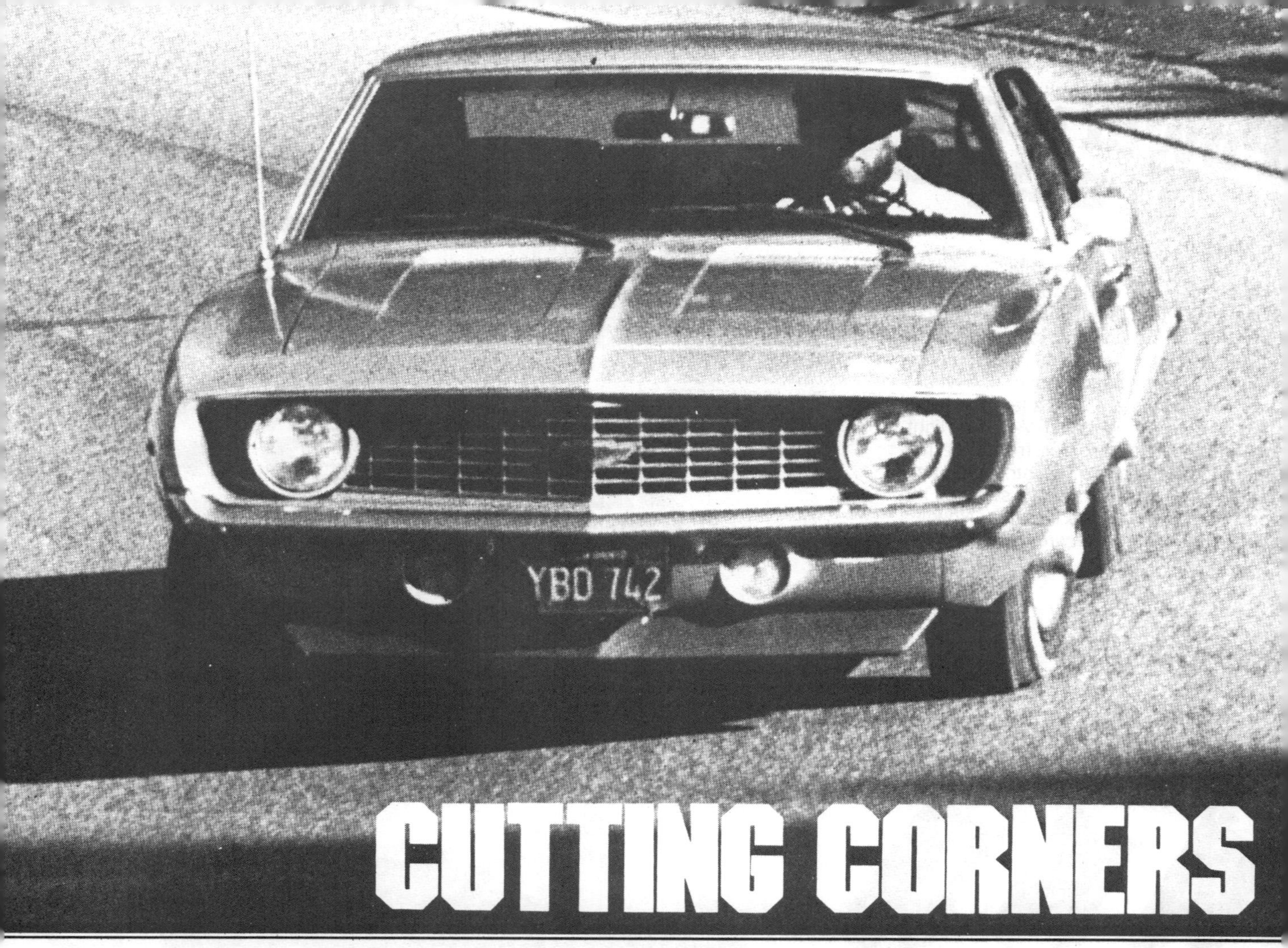

CUTTING CORNERS

High-Speed Handling for Early Camaros and Firebirds

By Rick Voegelin

The silver bullet streaks across the overpass and disappears behind a grove of eucalyptus trees. It explodes back into sight as the sun catches its metallic sheetmetal, and as it flashes past the designated light pole, the electronic stopwatch starts its relentless count. The Camaro's tires claw at the asphalt as Herb holds a tight line around the off-ramp centrifuge. The car's rear end gets a little loose as its tires approach the limits of adhesion; Adams catches it with a quick correction. At the second pole, we hit the "Stop" button and catch a glimpse of the Camaro as it tucks in behind a lumbering 18-wheeler. The LCD display reads, "5.4."

We have just set a world's record for the Highway 101/ Highway 156 skidpad.

Once a week for the past month we have ventured out to this 390-foot section of interstate highway system in the interest of applied automotive science. As I crouch behind a pine tree with telephoto lens and stopwatch, Herb

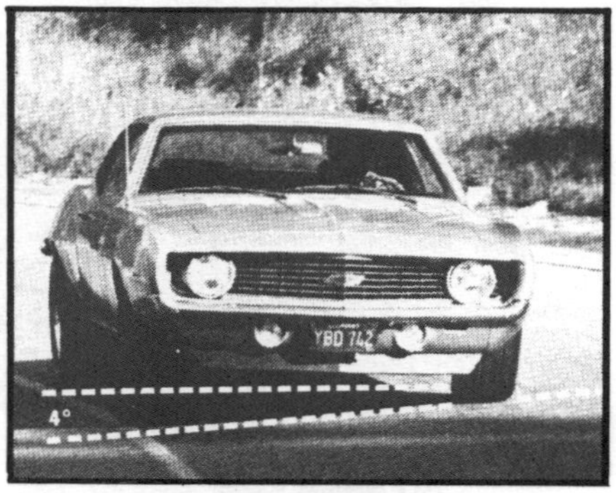

Initial testing was done with E70x14 tires and an OEM ⅝-inch diameter swaybar. Note severe 4-degree body roll and outside front tire heavily loaded.

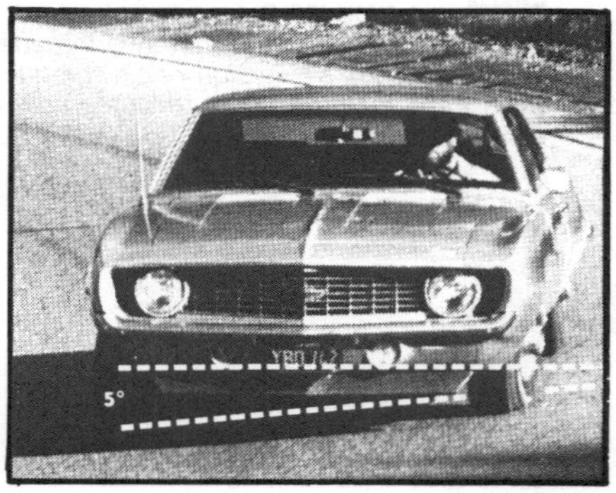

Installing Goodyear Wingfoot P255/60R15 radials on 8-inch wide VSE steel rims increased cornering speeds, but also increased body roll because of greater side forces generated. Roll has increased to 5 degrees in this photo; note camber change on outside front tire.

Adams waits for a break in the traffic. And when the turnoff from Monterey is all clear, it suddenly becomes our own high-speed test track, where the limits of Camaro handling are stretched a little farther with each test session. The evidence of our efforts is as plain as the black tracks which circle the transition road. And the progress of our efforts is apparent in the notations in our logbook: The early Camaro is nearly a half-second quicker through the course than when we started.

Our weekly expeditions are part of CC's own ERA program—creating Equally Roadworthy Automobiles. The object of the campaign is to grant the early-model '67-'69 Camaros and Firebirds the same rights on the road as their newer counterparts. For, in truth, the first F-bodies have been denied equal opportunities at turning corners by all the attention lavished on late-model Trans Ams and Z/28's. If you want a drag racer or a street runner, you get an early Camaro. But if you want a handler, a car that will carve up a canyon

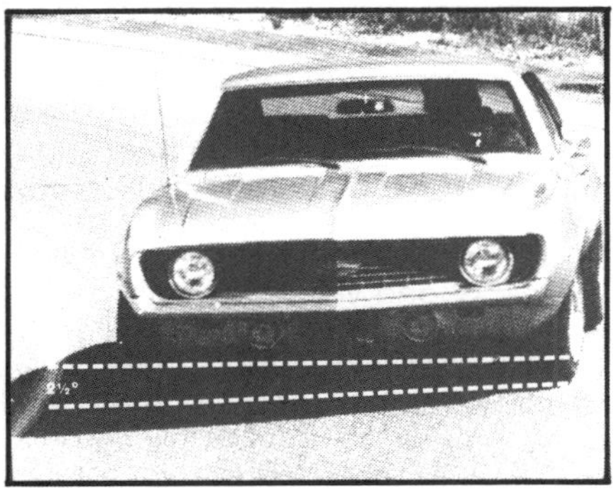

Adding VSE 1-inch diameter front and rear swaybars cuts body roll to 2½ degrees, and increases cornering speed 2½ mph. Car now exhibits a trace of oversteer, with rear end just starting to come around. Lessened body roll keeps outside front tire closer to perpendicular to road.

road with amazing grace, your only option is to surround yourself with post-'70 sheetmetal. True, General Motors did develop some heavy-duty underpinnings for the first generation of ponycars, but they were better suited to the high-banks at Daytona than the off-camber turns on Racer Road. And finding a swaybar designed for the original Trans Am (the racing series, not the car) is only slightly less difficult than locating 30 cent-a-gallon gas.

Our guide through the uncharted territories of early Camaro/Firebird handling is Herb Adams, a man who leaves no corners unturned. Adams is a self-confessed dropout from the GM corporate grind who prefers life on the lunatic fringe to the daily office routine. So now he lives in a solar-powered house overlooking the Pacific and conjures up components for his line of Fire Am Pontiacs and Cheverra Camaros. Quoting from a decade-old "Hi Riser" profile, Adams is credited with "maintaining and bolstering Pontiac's performance image." Now that performance is a treasonous thought in some quarters of Detroit, Herb is happy to be an outsider who can indulge his automotive whims. Witness his participation in this latest form of Camaro madness. Who needs the GM Proving Grounds when there's this lovely, wooded off-ramp there for the using?

Although Adams' appearance has changed greatly since his days with the Advance Design Group, he still has an engineer's perspective. Before tackling the assignment of making an early F-body turn corners, Herb laid out a systematic schedule of changes. Adams is never without a letter-size sketch pad, and within minutes of proposing the project, he had several sheets covered with notes, numbers and plans. "First we'll establish a baseline," Herb suggested. "Test the car just as it is to get an idea of the capabilities of a Sixties-style suspension. Then we'll add improvements one at a time and test the results—but without getting so radical that a guy would have to cut up his car. There'll never be any more of these early F-bodies, and I see more and more of them being preserved by people who appreciate them." Shortly thereafter, I came to know terror as Herb offered an over-the-road

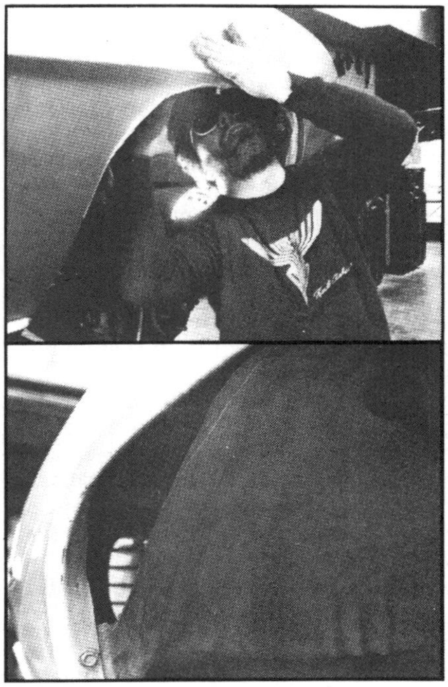

Wide Wingfoot tires required trimming inner fender panels. After cutting out portion of inner panel where it joins fender lip, Adams rolled edge with a body hammer, protecting outer skin with a piece of Masonite held behind fender.

HANDBUILT SWAYBARS

Fabricating a swaybar is a relatively simple procedure which permits you to take out all your frustrations on a hapless piece of steel. If the bar diameter you want is unavailable—or if the aftermarket doesn't produce bars for your particular car—the answer is to take torch in hand and roll your own. A cast-off swaybar from a similar chassis can provide a model for the dips and turns you'll have to form, or a length of wire or tubing can be used to make a template.

The basis for most swaybars is 4140 chrome moly steel. Ask for annealed steel at your metal supplier; he'll charge approximately $2 a foot, and it takes about 6 feet for each bar you want to build. The trick to bending the bar is to use a big vise, a bigger hammer and plenty of heat. Heat the bar with an acetylene torch until the outside begins to turn yellow hot. Adams advises, "When you think you've got it hot enough, heat it some more." To form a 90-degree bend, heat a 3-inch section of the bar stock. When the material starts to look translucent, make the bend with the aid of your trusty hammer. If the metal's not hot enough, it will start to crack open like a banana, which means you've ruined that attempt. After you've made a satisfactory bend, let the bar air-cool—don't quench it with water, and for the sake of your fingertips, don't pick it up until you're sure it's cool.

Attaching the bar to the body can be the most challenging portion of this handling exercise. Adams flattens the

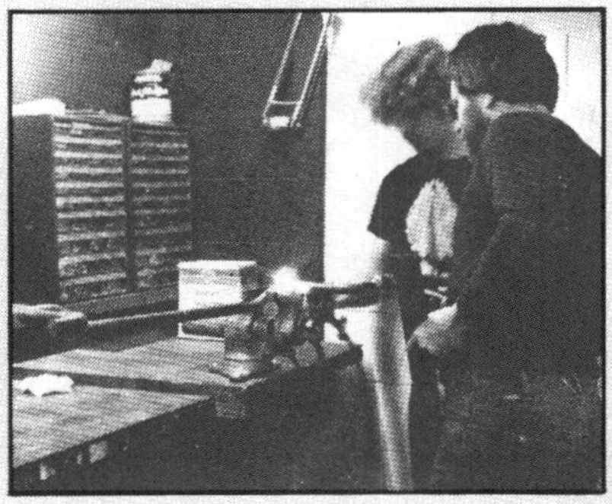

Forming custom swaybar requires plenty of heat and plenty of patience. Secret to bending large diameter bar stock is to heat metal until it becomes translucent. Handbuilt bars must be heat-treated before they're ready to use.

ends of his bars and drills them to accept a spherical rod end. An alternative plan is to pirate the end links and bushings from a factory bar. Rubber or urethane bushings are the hot tip for attaching the bar to the body on street cars; your Pontiac or Chevy dealer's parts bins are a good source for this attaching hardware.

Homemade swaybars have to be heat-treated before trying them out on the nearest off-ramp. Heat-treating is best left to a professional, so forget about stuffing your new bars into the kitchen stove. The bars need to be hardened to 38-40 on the Rockwell C scale. Expect to pay about $20 a bar for this service. Reinstall the bars, check for sufficient clearance, and get ready to take on Mario Andretti.

demonstration of an early Camaro's handling deficiencies.

The original F-bodies have always been considered acceptably stable machines. They don't generally turn upside down without extreme provocation, and there are hundreds of 130 mph Camaros on the nation's dragstrips. But as Herb sliced through a favorite series of mountain switchbacks with our stone-stock '69 Camaro, the F-body's faults were as apparent as the 200-foot drop below us. We'd covered these roads minutes earlier in a fully-prepared Fire Am with reassuring precision; the contrast was astonishing as our Camaro test car pitched and yawed, its front end oscillating in a wild series of leaps and jumps, with Herb cranking in more and more steering as the car tried to slew off the fast line. It was not a pretty sight.

All early F-bodies were not created equal, and ours was less equal than most. With monoleaf rear springs, a 10-bolt rearend, a standard suspension and 140,000 miles of use, it was no road racer. Since the supply of original Z/28's, Super Sports and Trans Ams is finite, Herb regarded the transformation of this low-key Camaro into a freeway flyer as a challenge—the automotive equivalent of turning Gilda Radner into Bo Derek.

The first expedition to the off-ramp test facility established the car's performance baseline. The car was as close to the way God and GM had intended it to be as we could make it, right down to the OEM-type E70x14 bias ply tires on 6-inch rims, inflated to 30 psi. And despite the car's lack of manners in transient situations, Adams found it surprisingly well-behaved in steady-state cornering. "The car has mild understeer," Herb reported, "which is what you'd expect from a stock suspension. In fact, the tire feel and handling are really very acceptable—it's not terrible, by any means. After the fourth pass, though, I could hear the front tires howling." From my vantage point on the hillside, it was clear why the tires were protesting: the outside tread would roll under the rim as the car cornered on the sidewall like some crazy four-wheeled motorcycle.

The first step was to install wider tires and wheels. Although filling an automobile's wheelwells with rubber is a time-honored practice, Adams' concern was more pragmatic than artistic. "You're always better off with the widest tires you can fit inside the sheetmetal," Herb observed. "Wide rubber is the foundation for all subsequent suspension work. Even if the car can't make use of the tires' cornering power yet, later improvements will require wide rubber." Herb's choice was a set of P255/60R15 Goodyear Wingfoot radials, mounted on 8-inch wide VS zero-offset rims. The Wingfoot is as close to a racing tire as you'll find on the street—in fact, Adams runs Wingfoots on his IMSA racer at Sebring and Daytona. The zero-offset wheels were intended to keep the tires and sheetmetal at a discrete distance by tucking the rubber wel inside the Camaro's fender line—a strategy that worked, but only up to a point.

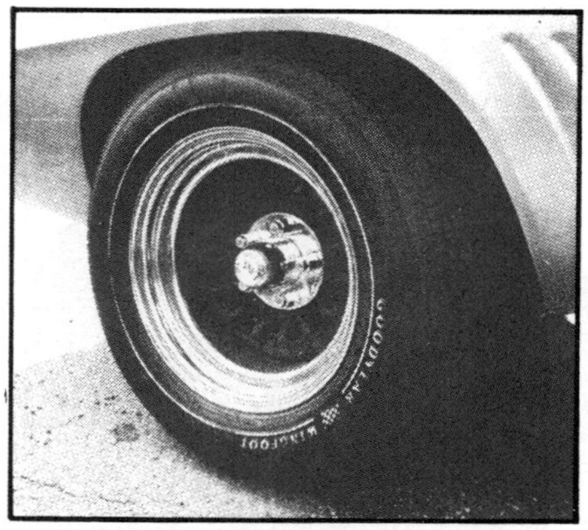

Eight-inch wide VSE rear wheels have ¾-inch positive offset to provide proper clearance; front wheels are zero-offset items. Some deft hammer work on inside of rear wheel lip was all that was required.

The Wingfoot's impact is both visual and visceral. The tires' tread modeled after a Formula One rain tire, and is about as subtle as a thunderstorm. You feel the tires, because the stout sidewalls and wrap-around tread (which are responsible for the tire's cornering capabilities) aren't very yielding to aberrations on the pavement.

Our return engagement at the off-ramp produced an immediate improvement. The circuits produced three 5.7-second clockings. Yet eve though the times were faster, Adar had to work harder. The additional bite provided by the new tires actually increased understeer, sinc the car's limited power prevented using the throttle to toss the back end around. Cornering speeds increased 3 mph, but an occasion puff of smoke from the front wheelwell indicated that the tires and the sheetmetal were still a bit too cozy.

Back at Adams' mountain-top shop, telltale traces of rubber on th fender liners indicated where the clearance was insufficient. A crescent of sheetmetal 6 inches wide and 2 feet long was trimmed from the inner panels where they join the fender, using a sabre saw and electric shears. Herb then rolle the fender lip with a body hamme protecting the outer fender from damage with a piece of Masonite Checking the tire clearance as the front suspension went through its travel revealed that there was now enough room to accommodate later changes such as cutting the front coils and lowering the front end sheetmetal.

In back, the tire sidewalls had been contacting the inside of the wheelwells as the body rolled. The solution was a pair of wheels with ¾-inch positive offset. Moving the tires out increased the track 1½ inches and visually centered the ti in the wheel opening. Again, the outer lip of the fender had to be rolled with a body hammer to provide the necessary jounce clearance.

Adams then announced that it was time to get serious. "We're going to build a set of swaybars," he declared. With visions of forging presses dancing in my head, Herb selected several lengths of bar stock and began the proven road racer's technique known as "heat and beat" to turn out front and rear anti-rollbars. As Herb and his son lit up the acetylene torch and selected an appropriately hefty hammer, the rest of the crew was busy removing the Camaro's original 5/8-inch diameter front bar. A combination of calculations and intuition indicated that a 1-inch front bar would be a good starting place. Herb went to his well-stocked Fire Am parts bins to provide the hardware necessary to connect the handmade bars to the factory frame. Stock swaybars have considerable built-in compliance produced by rubber bushings at all the connections; Adams goes for a rigid setup with spherical rod ends where the bar attaches to the front A-arms. It turns out that fabricating a swaybar is not the impossible task it might seem to be. The accompanying sidebar provides the details for do-it-yourself swaybar builders. And for those who would rather wield a checkbook than a hammer, mass-produced bars for early F-bodies are now available from Herb Adams VSE (125 Ocean

Continued on page 79

EARLY CAMARO HANDLING TEST

Test	Time	Speed	Comments
1	6.0 sec.	49 mph	**Baseline test with E70x14 bias ply tires, 6-inch rims, 30 psi. "Mild understeer, excessive body lean, front tire too heavily loaded."**
2	5.7 sec.	51 mph	**Installed P255/60R15 Goodyear Wingfoot tires on 8-inch wide zero-offset VSE wheels. "Understeer increased but speed improved. Additional tire clearance needed."**
3	5.5 sec.	53.5 mph	**Installed VSE 1-inch diameter front and rear swaybars, trimmed front inner wheelwell, mounted rear tires on 8-inch VSE rims with ¾-inch positive offset. "Car much more neutral handling. Additional cornering force requires more driver input. Front suspension apparently deflecting from loads."**

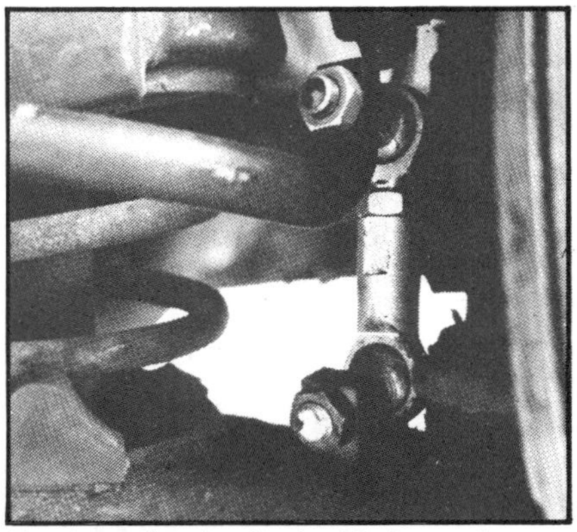

Front swaybar is attached to lower A-arm with spherical rod end. Stock rubber bushings are used to connect bar to frame.

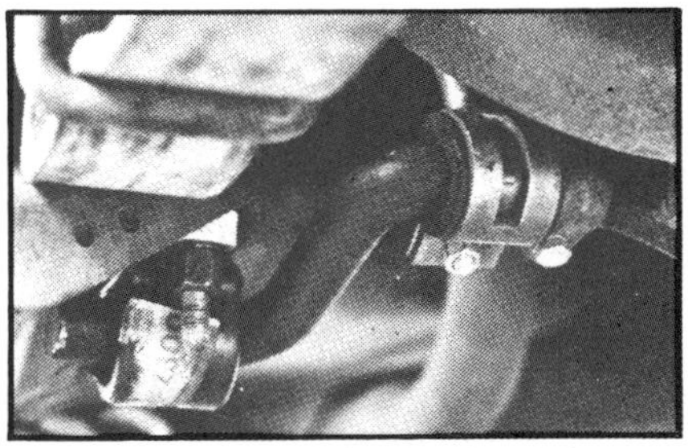

Rear VSE swaybar is located behind differential. Two brackets must be welded to rear subframe to anchor rubber-bushed connecting links.

HANDLING HARDWARE

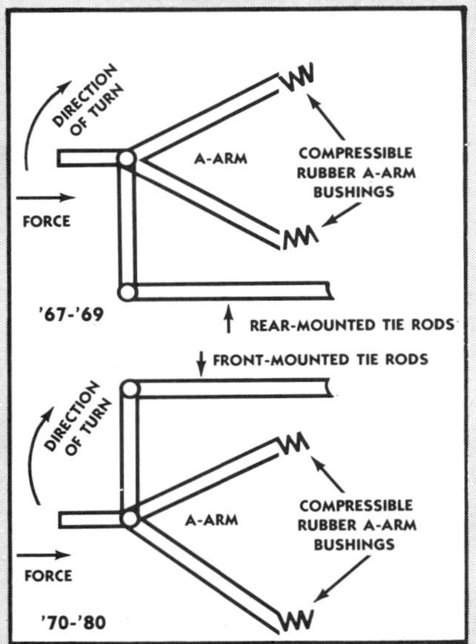

This sketch illustrates a crucial difference between the early and late F-body Camaros and Firebirds. In the '67-'69 models, the tie rods are behind the A-arms, which causes "deflection oversteer" in hard cornering. In the '70-'80 versions, the tie rods are in front of the A-arms. As the bushings in the A-arms deflect, the car gradually understeers, a more manageable handling characteristic.

When General Motors unveiled the '70½ Camaros and Firebirds, the changes went beyond the new cars' swoopy sheetmetal. Few chassis parts are interchangeable between the first generation of F-bodies ('67-'69) and their late-model relatives. Understanding the revisions made in the front suspension explains why the late-model cars are generally regarded as better handlers than the original F-bodies.

The crucial difference is the location of the steering linkage. In the early cars, the steering box and tie rods are behind the A-arms; on the later cars, the box and links are in front. The two tie rods and the drag link form a very rigid bar which is relatively unaffected by the forces of cornering. The rubber bushings which attach the A-arms to the subframe will deflect under load, however. The bushings are, in effect, springs. This combination of compressible bushings and rigid steering links creates some strange reactions at the limits of cornering power.

In a late-model F-body, when the bushings deflect, the tire under load toes out. The driver reads this as understeer, so he cranks the steering wheel over a little farther to maintain his cornering line. The change is gradual and easily controlled.

Now consider what happens in the early-model cars, with their rigid steering linkage behind the front wheel spindle. As the car enters a corner, the tires bite and the bushings deflect. But because the rear steering linkage doesn't compress, the tires toe in. Suddenly the car is turning more sharply than the driver anticipated, an oversteer condition. To correct, he has to turn the wheel away from the corner, which tends to unload the bushings. Thus the driver is constantly sawing back and forth on the steering wheel as he hustles the car through the corner. The solution? Understanding the forces at work and sharpening up your driving skills will go a long way, but taking the deflection out of the rubber A-arm bushings will buy you more cornering power. After all, solid bushings are less expensive than an '80 Trans Am.

MORE CAMARO madness

Cutting Corners, Part II: Handling Hardware for Early F-Bodies

By Rick Voegelin

Nineteen eighty-one is here, and the time is right for touring in the streets. Jagger and the band were right when they told us you can't always get what you want—but if you try, you can get what you need. If what you need is an outlet for your automotive emotions, then what you want to build is a corner burner. Six hundred horsepower is a delight, until the bill comes due at the gas pump. All it takes to drift around an off-ramp is 200 horsepower, however. It can even be regular-fueled, low-compression horsepower at that. It takes a mental shift to start thinking in terms of turns rather than elapsed times, and to pay attention to camber angles instead of angled spark plugs, but that's the long and winding road we're traveling these days.

Consider one man's response to the road-racey early-model Camaro parked at his shop's door. He surveyed the sway bars, the cut-down coils, the decambered front tires, the lowered subframe and the Wingfoot-wrapped 8-inch rims. And then he said, "You know—it doesn't cost a lot of gas to go fast around corners." This from a man who makes his living building cylinder heads for Pro Stocks, the last word in gasoline-burning performance. You see, the revolution has been won; we're down to the mopping-up operations now. The Arabs and the oil companies did what no mega-dollar ad campaign or stacks of sporty-car magazines could accomplish: namely, convince horsepower-crazed Americans that automotive performance is more than straight-ahead, flat-out, take-it-to-the-limit speed. Duntov knew it all along; he told us at a CAR CRAFT banquet half-a-dozen years ago. "Driving 100 mph is fun," Zora said, "but taking a 25 mph corner at 45—well, that's better still." Life in the fast lane can be full of sweeping left-handers.

But we stray from the story at hand—Part Two of how a mundane Camaro was turned into something very special for the Empty Eighties. Under normal circumstances, a 307-powered ponycar wouldn't even register on a scale of 1 to 10. But things aren't normal when gas costs

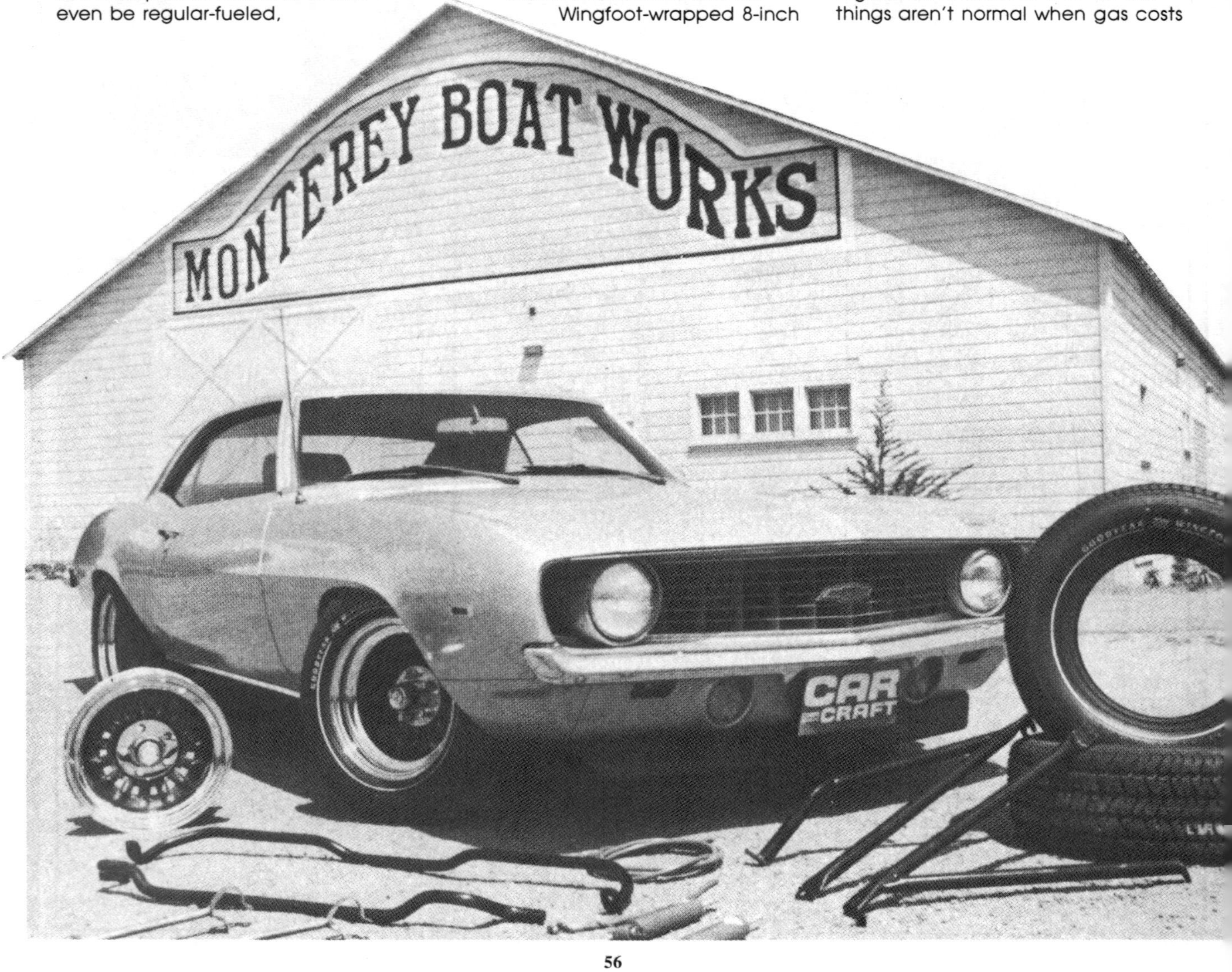

a buck-thirty a gallon—and in these times, a clapped-out 307 can be as much fun to drive as an LS-7, providing the terrain is right and the car is set up to handle it. In the first installment of this continuing saga ("Cutting Corners," July '80 CC), we laid the groundwork with sway bars, wide wheels and tacky tires. Now the work of turning our sow's ear into a silky grand tourer begins in earnest.

We built the kind of car that sends you looking for the thin, tortured lines on the highway map. Never again will you choose the interstate if there's a winding secondary road that takes you in the same direction. Our guide through these uncharted territories was again Herb Adams, the bearded ex-GM engineer who marches to a different automotive drummer. It was Adams who entered his street-driven Tempest in the Trans-Am, and his street-driven Firebird at Daytona. Herb volunteered both his expertise and his equipment in these early F-car experiments. The results were pleasing enough that Adams has added many of the parts to his Very Special Equipment catalog (Herb Adams VSE, 125 Ocean View Blvd., Pacific Grove, CA 93950).

Through the days of turning wrenches and turning corners, we evolved a strategy of automotive survival for the Eighties. Whenever the urge to acquire a new econobox (and its accompanying inch-thick payment book) asserts itself, resist the impulse. Instead, go look for an early Camaro/Mustang/Cuda/Firebird/Challenger/Nova. Even if you have to pay two grand for one without rust, do it. Then invest another thousand making it "right," including a selection of handling and performance equipment. You'll end up with a car that knocks down 20 real miles per gallon, costs $15 a year to register, and appreciates in value with every passing day. But if you want to adopt this survival strategy, get started now. It isn't very original, and plenty of other people have already undertaken the task.

Setting up a car for high-speed handling is as challenging as modifying one for straight-line performance. Things you know will work, don't. Conversely, minor modifications that you think are insignificant can yield great gains. We let the stopwatch and cornering speeds be the judge of our efforts, and furthered our automotive education in the process. Take the front tire flog, for instance. It's a tenet of road work that you use the widest tires that will fit—and that's what we did at the outset. But after trimming the front coils and lowering the body over the subframe, the P255-60R15 Goodyear Wingfoots were simply too cozy with the front end sheetmetal. The only alternative was to bolt on shorter P245's. The car went slower—in fact, the difference was much greater than anyone thought reasonable. But Adams realized our crucial mistake—the brand-new front tires hadn't been broken in as thoroughly as the larger tires, and thus weren't performing to their potential. One week and 800 miles later, we scheduled a return engagement at our favorite off-ramp. The second time around, the smaller tires turned in times nearly as quick as the wider rubber.

Improving a car's handling characteristics doesn't mean downgrading its comfort and driveability. High-output engines tend to be temperamental beasts, but max-effort suspensions are pussycats. As long as you stay away from incompressible springs and rock-hard shocks, ride and comfort aren't compromised with a high-performance suspension. Adams advocates stock springs and a limber suspension that "works" to keep the tires on the pavement when the road gets rough. A suspension that requires a kidney belt has no place on a road car.

So there's hope. It's possible to have your V-8 and drive it, too. The price differential (including depreciation and insurance) between a resurrected ponycar and a chic econobox will buy a lot of gasoline. And if you factor in the fun-per-gallon, the homebuilt supercar's advantage is formidable. We've reached the turning point in high-performance. Now it's time to set up your machine to clip the apex.

HANDLING TEST

One Good Turn . . . Part II

What's it like out there? Funny you should ask. The seat of one's pants is a respected, albeit inaccurate, gauge of cornering forces. Looking for a more objective measure of handling improvements, we made repeated expeditions to a nearby off-ramp for timed runs. Those with total recall will remember that our initial foray to the clandestine test track found the Camaro circulating the course at a leisurely 6.0-second pace. By the time we added Wingfoot radials, 8-inch wide VSE wheels and prototype VSE front and rear sway bars, the time around the off-ramp was down to 5.5 seconds and the speed had increased 4.5 mph. Picking up the story from that point, we rejoin the Camaro improvement course.

Short P245-60R15 Goodyear Wingfoot front tires provided extra fender clearance required for this kind of off-ramp running. Yes, you can find automotive happiness with only 100 horsepower!

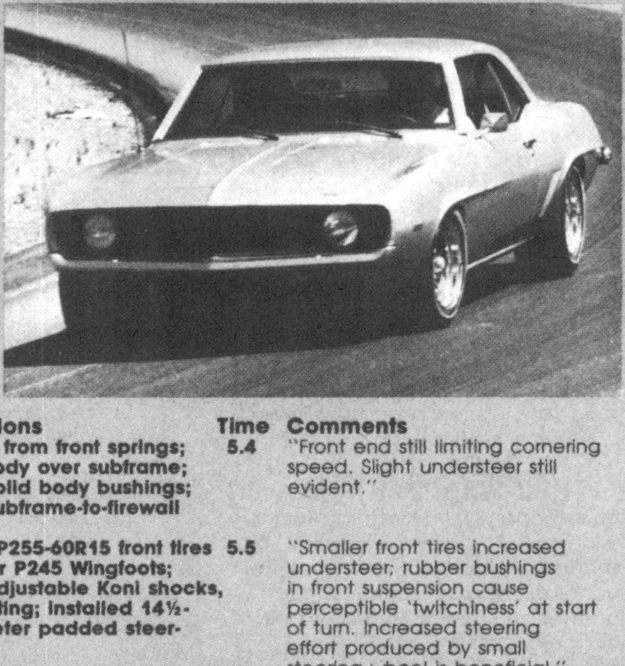

Test	Modifications	Time	Comments
1	**Cut ½ coil from front springs; lowered body over subframe; installed solid body bushings; installed subframe-to-firewall struts**	5.4	"Front end still limiting cornering speed. Slight understeer still evident."
2	**Replaced P255-60R15 front tires with shorter P245 Wingfoots; installed adjustable Koni shocks, middle setting; installed 14½-inch diameter padded steering wheel**	5.5	"Smaller front tires increased understeer; rubber bushings in front suspension cause perceptible 'twitchiness' at start of turn. Increased steering effort produced by small steering wheel is beneficial."
3	**Relocated battery to trunk; realigned front suspension to 1-degree negative camber**	5.3	"Understeer greatly reduced, although still a trace evident. Solid suspension bushings would minimize deflection steer. Decambered suspension dramatically improved stability."

MORE CAMARO madness

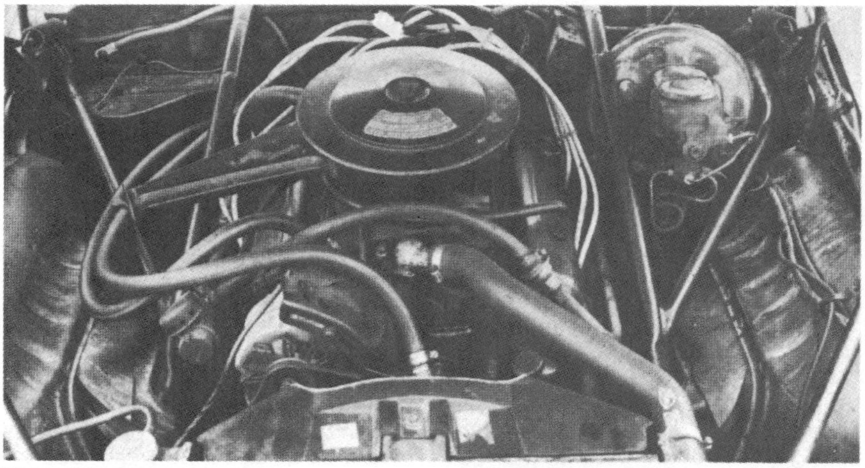

STRUTTIN' YOUR STUFF

All GM F-bodies (the Firebird/Camaro line) are schizophrenic. They've got a full frame in front and a unit body in back. Where these two major components are joined, there's a considerable measure of flex—and a subframe that deflects under load is the root of evil handling. Adams' cure for this unwanted motion (which can be as great as ½-inch under heavy cornering loads) is to fabricate and install these tubular braces, which he terms a "front end structure kit." These struts attach to the upper A-arm mounts and the firewall, and work in conjunction with solid subframe bushings to make the F-body stop behaving like a split personality.

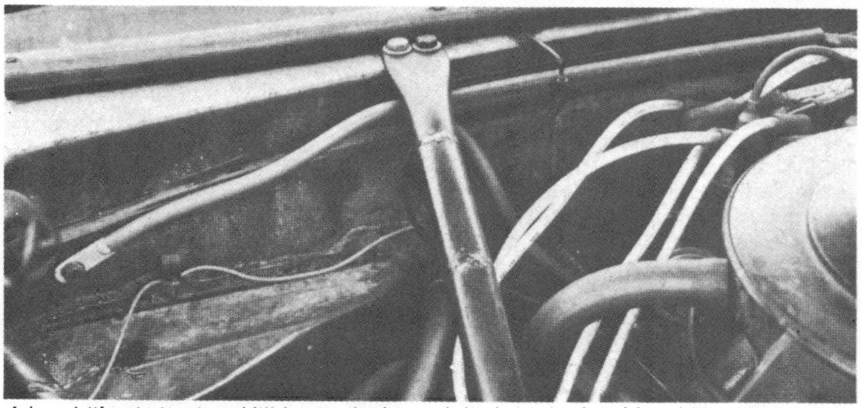

Adams' "front structure kit" is attached to unit body at cowl and hood hinge bulkhead. Ground wire runs from firewall to engine block to provide current flow for trunk-mounted battery.

LOW-DOWN BUSHINGS

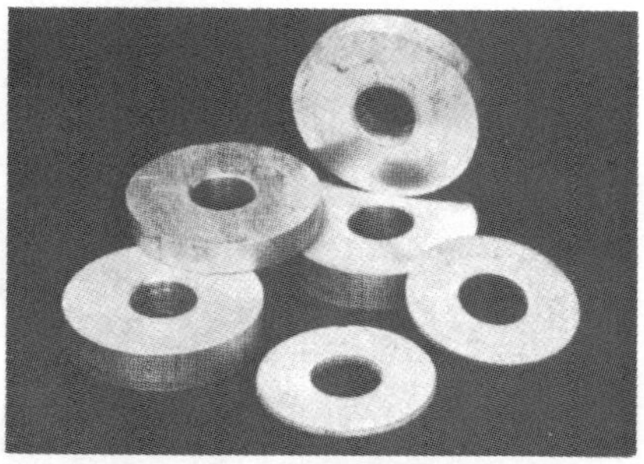

Lowriders have the right idea, but the wrong reasons. A low center of gravity is a joy forever—or at least a joy for handling. These solid aluminum body bushings which Adams turns out have two functions: they lower the body relative to the subframe and they reduce the flex at the body/subframe junction. Quite naturally, they also augment the low-down look that's part of the cafe racer style. They're a bolt-in replacement for the stock rubber biscuits. These solid bushings don't reduce suspension travel or noticeably increase ride harshness. They do require trimming the fan shroud for clearance and realigning the steering column.

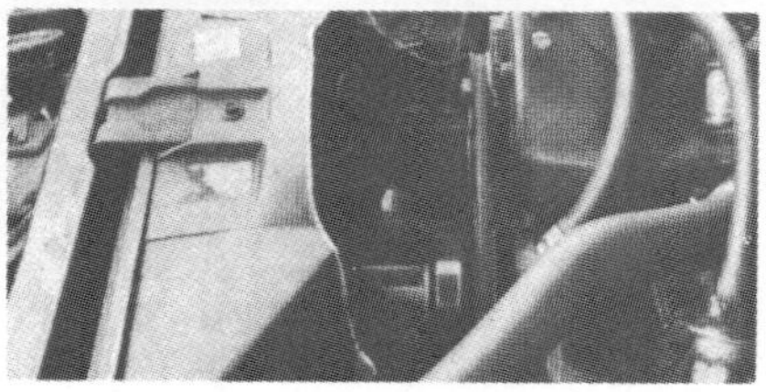

LEATHER-WRAPPED RIM

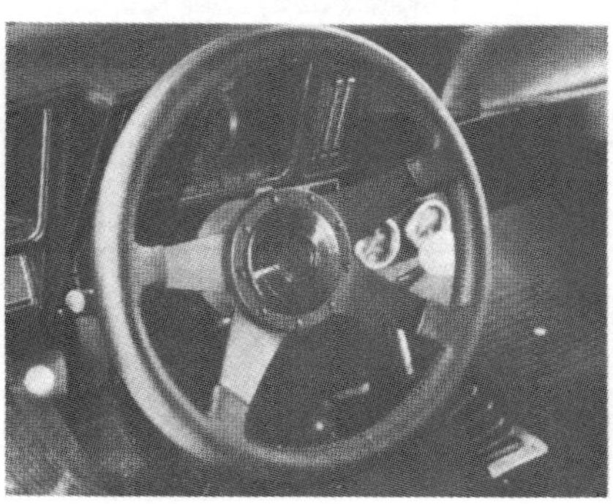

One important part of going fast is feeling fast. A small-diameter soft-rimmed steering wheel is a vital psychological asset. This 14½-inch diameter VSE French import has a mechanical contribution to make as well, since it increases steering effort without the work of installing a special high-effort steering box. It also feels good on sweaty palms, which certainly counts for something.

TIRELESS EFFORTS

Wide tires and narrow fenders can't coexist. That's the lesson we learned after lowering the front end sheetmetal and trimming the front

WHEN BETTER BARS ARE BUILT

Bigger is not necessarily better, especially when designing sway bars. There's a belief abroad in the land that if a 1-inch sway bar is good, a 2-inch thick bar has to be terrific—but it isn't necessarily so. For example, the front sway bar that Herb Adams built for our early-model Camaro is a relatively skinny 1-inch diameter steel rod. That sounds rather anemic compared to the 1 5/16-inch bars he uses on late-model Firebirds and Camaros. But as we pointed out last July, there are

coils. The P255 Goodyears we initially fitted were just too close for comfort after the front end came down. Swapping shorter P245 Wingfoots (Wingfeet?) on the same 8-inch wide zero-offset VSE steel rims eliminated most fender-to-sidewall interference. (The inner fenderwells had already been trimmed in the course of our earlier handling exercises.)

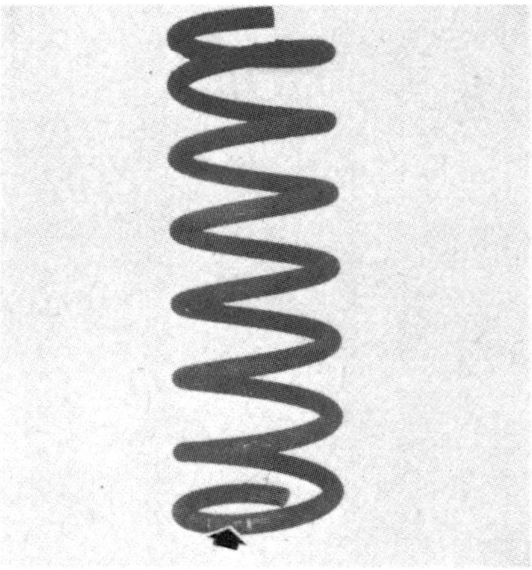

SHORT SPRINGS

Each front coil was relieved of ½ coil with the aid of an acetylene torch. After trimming the springs, the bottom coil is heated and bent to permit the spring to seat properly in the lower A-arm. Cutting the front springs does three things—it raises the spring rate approximately 10 percent, lowers the front end, and modestly reduces front suspension travel. Adams likes to retain at least ¾-inch clearance between the lower A-arm and the factory jounce snubber.

some crucial differences between the early and late-model F-body suspensions. When designing a high-performance sway bar, there are two crucial dimensions to consider: overall bar length and the "swing arm" length (the distance from the frame attachment point to the lower A-arm link). An early-model F-body sway bar is 4 inches wider than a late-model version (38.0 versus 34.0 inches), and the swing arm length is significantly shorter (6.5 versus 12.0 inches). The outcome is that a calculated force of 748 pounds is required to deflect a 1-inch diameter early-model bar 1 inch—virtually the same force needed to twist a thicker 1 5/16-inch late-model bar the same amount. Naturally, testing and fine-tuning will determine the optimum bar diameter after the engineering data has pointed you in the right direction. Clearly, though, selecting the right size sway bar is more than toting home the heaviest piece of steel down at the metal supply.

BAR ASSOCIATIONS

These are production versions of the handmade bars originally fitted to the test Camaro. The goal of any sway bar combination is to achieve neutral handling. With P.255 tires on all four corners, there was a slight tendency toward oversteer, which prompted Herb to build a ⅞-inch diameter rear bar to replace the original 1-inch version. But fitting smaller P245 tires in front increased understeer, and we elected to remain with the original 1-inch rear bar.

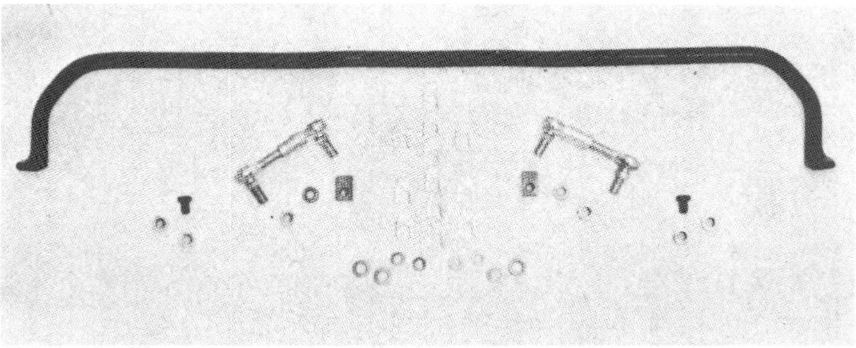

Front swaybar uses spherical rod ends to attach bar to lower A-arm. Extra shims were installed between upper A-arm and frame to produce one-degree negative camber.

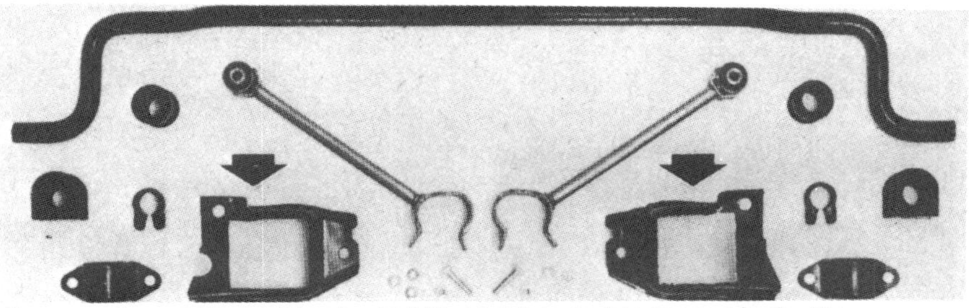

Rear swaybar employs rubber-bushed Firebird links to join bar and body. We experimented with raising the rear spring eye by modifying spring hangers (arrow); the early F-body doesn't provide enough clearance, so we returned spring to stock location.

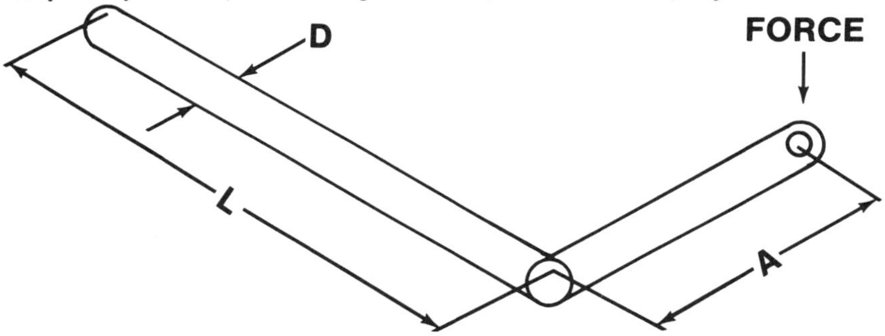

Two important dimensions in swaybar design are overall bar length and "swing arm" length. That's why the 1-inch diameter bar on early F-bodies has the same effect as the thicker 1 15/16-inch bar on late-model Camaros and Firebirds. (D = bar diameter, L = bar length, A = swing arm length)

BACK DOOR BATTERY

Part of the Camaro handling plan was a weight redistribution program. Moving the battery to the trunk is a quick way to change the weight distribution nearly 1 percent. We opted for a permanently sealed battery to prevent corrosion in the trunk floor and to forestall a buildup of fumes in the unvented area. A large diameter copper cable runs from the battery's positive terminal to the starter solenoid; the ground post is attached to the body shell with the original ground wire. A second heavy ground cable connects the firewall to the engine block.

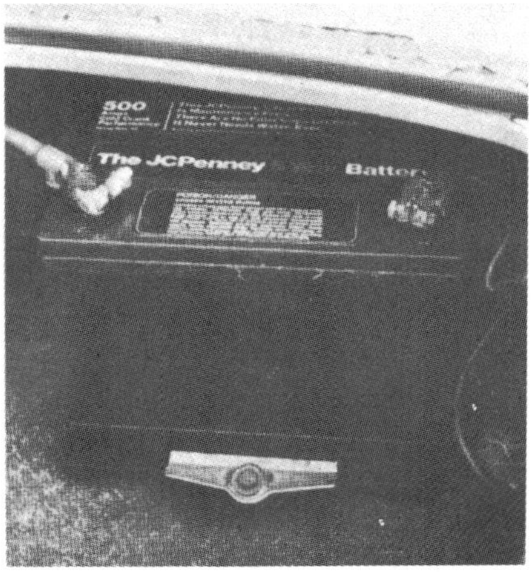

MORE CAMARO madness

RUNNING IN CIRCLES

"Come on down to Laguna Seca tomorrow. We're testing the Fire Am, and maybe we can complete your automotive education." The speaker is Herb Adams, and the invitation is irresistible. Laguna Seca is a road course nestled in the hills behind California's Monterey Bay, and a Fire Am is Adams' personal vision of what a race car should be—a reliable engine, a simple suspension, and a whole lot of fun. In fact, Adams enjoys baiting the megadollar teams with his homebuilt racers, usually driving out to the course in the car he intends to race.

This Saturday was no different. He'd driven the 20 miles out to Laguna in his subtle blue and white 'Bird—the one with 2-foot high numbers on the doors and an exhaust note that would make Mick Jagger cringe. Adams wanted to get some time on the new big-block Chevy cam they'd installed the night before. "Want to go for a ride?" he asked idly. He didn't have to ask twice.

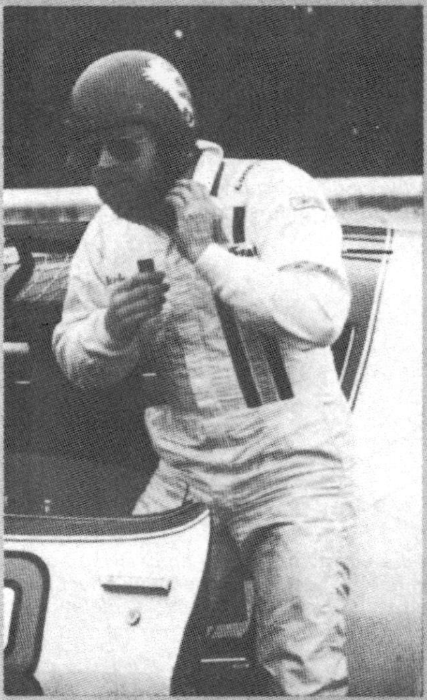

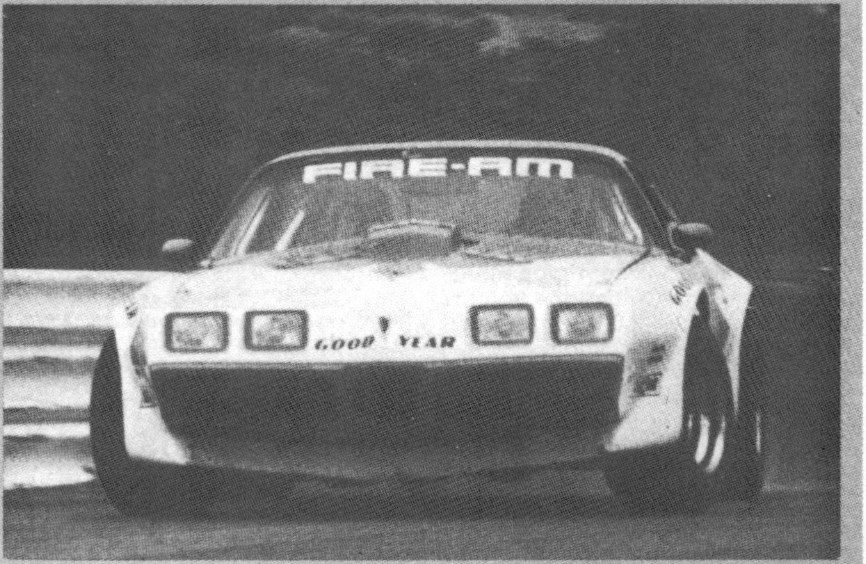

Herb introduced me to the art of high-speed cornering. Drag racing was never like this; each lap is the equivalent of a dozen full passes, with a handful of roller coaster rides thrown in for good measure. The side loads are amazing. As the Firebird flies into a sweeping left turn, the helmet seems to pick up an extra 30 pounds as you struggle to keep your head upright. The car skitters as we catch some dirt going under the spectator bridge, then drifts as Herb shoots through the uphill portion of the course. At the crest, the road disappears from under the Fire Am's long hood; we spiral down the Corkscrew like a paper airplane down an elevator shaft. I'm reassured that this is Adams' home course, since we are flying blind, with no indication of which way the road will dart after the next turn. As we take the slow left-hander at the entrance to the pit straight, the big-block's torque overcomes the tires' grip, and the car's rearend swings around like a 3500-pound dirt tracker. Herb cranks in more opposite lock, the Firebird sluices back to the right, and we are once again at full speed down the straight, like a drag racer with a rolling start. As Herb sets up the Fire Am for the first corner, I find myself wondering what would happen if you took a Pro Stocker through the first turnout at 100 mph.

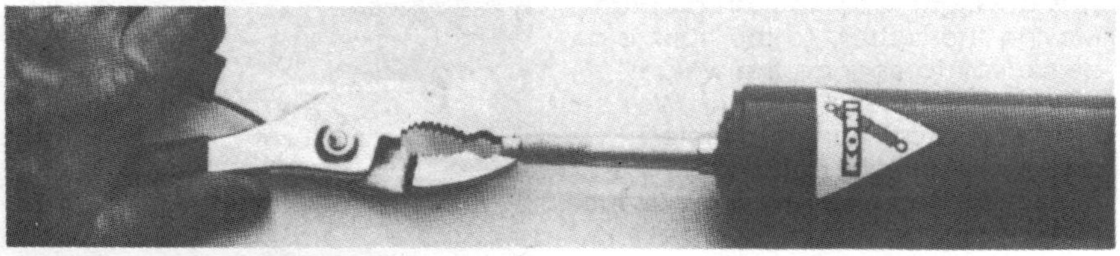

SHOCK TREATMENTS

Although shock absorbers have little effect on steady-state cornering, they do have an impact on transient maneuvers and over-the-road feel. We followed Adams' recommendations and installed Koni shock absorbers. These were adjusted to their middle setting, which produced a firm ride without undue harshness. An added, unexpected bonus is the multi-language instructions that accompany these Dutch beauties, just in case you ever want to install shocks in Italian.

EASY ALIGNMENT

One no-cost modification that pays high-priced dividends in improved performance is a decambered front end. "Camber" is the inward or outward tilt of the front tires. Adams realigned our Camaro's front end by adding shims to the upper A-arm mounts to produce 1-degree negative camber (which tilts the top of the tire "in"). This allows the full tire tread to stay in contact with the pavement when the suspension bushings and spindle deflect under severe cornering loads.

More Travels Down the Long and Winding Road With Herb Adams and His Magic Camaros

Flat-Out F-Body

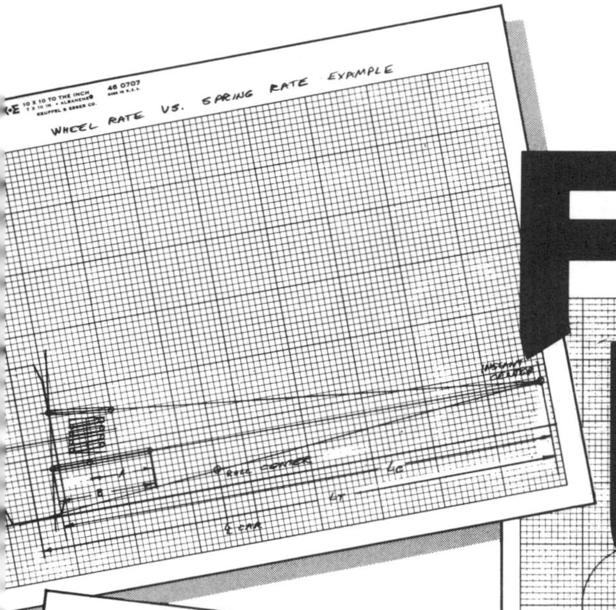

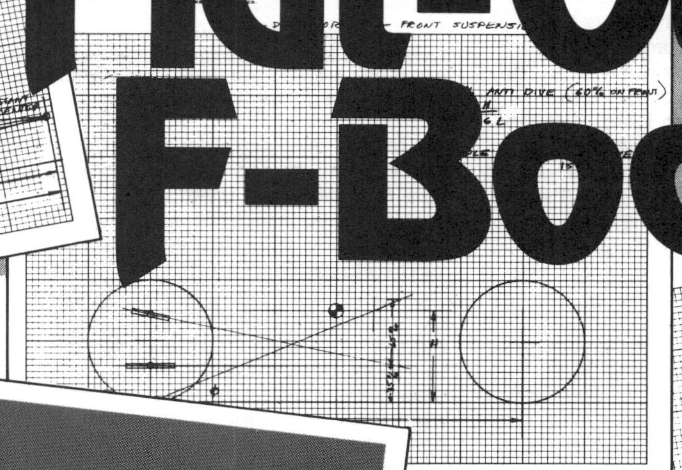

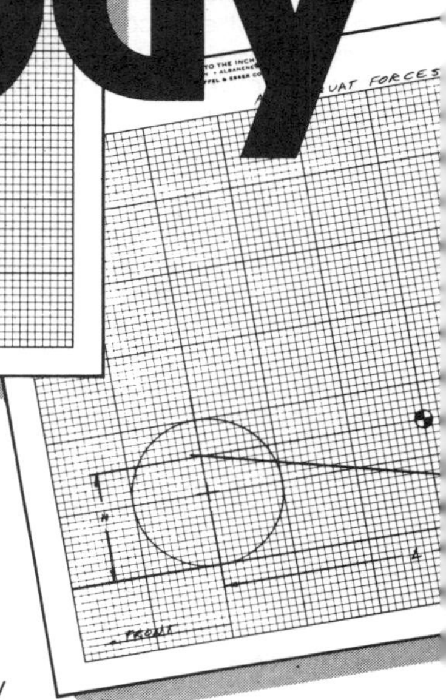

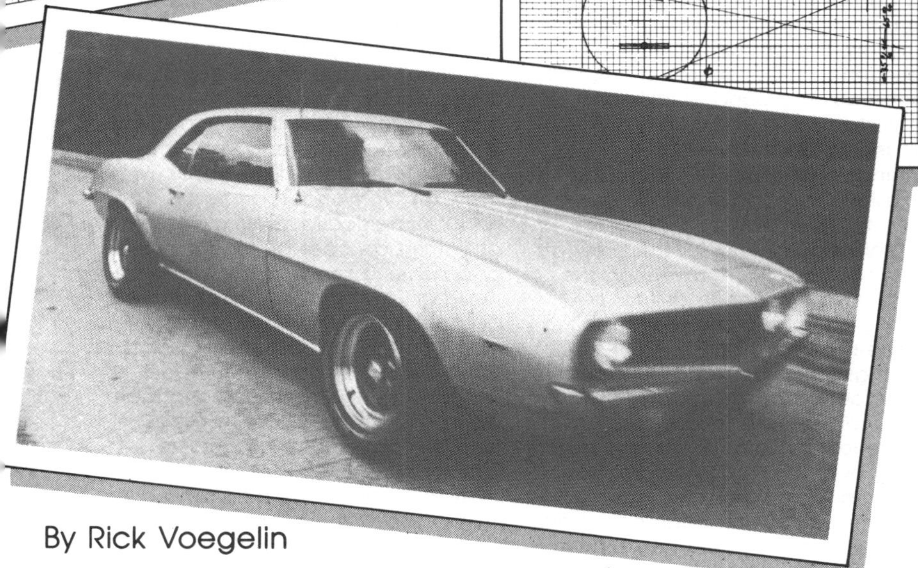

By Rick Voegelin

Street cars, like rambunctious children, should be seen and not heard. They should be obedient, mindful of your will, and respectful of your wishes. Street cars that don't behave are an intrusion on your life; rough-running, rough-riding, gas-swilling transportation can turn the normally pleasant pastime of driving into a chore to be endured. The first two installments of Herb Adams' Camaro/Firebird handling exercises were directed at giving the early F-body the most refined of road manners. (See "Cutting Corners," July '80 CC and "More Camaro Madness," Jan. '81 CC.) The suspension improvements that the crew at Herb Adams Very Special Equipment performed—adding brawny swaybars, fitting wider tires, installing supple shock absorbers, and bringing the '69 Camaro development car closer to Mother Earth with a judicious lowering program—did nothing to compromise the car's civility. But this time around, as Herb Adams, former GM engineer and current road racer, outlined the last steps in his F-body handling flog, it seemed that he had forgotten the First Law of Street Cars: "Don't mess with your transportation." As Herb described his intended plan of action, it sounded more like a race car buildup than a street car upgrade.

At first glance, his intended modifications seemed certain to rough up the edges which his earlier efforts had so carefully smoothed. "Solid suspension bushings," the list read. "Brake cooling package" and "shaved tires," it continued. "NASCAR-style wheels," it went on. But now, having seen, driven and reflected on Adams' efforts, it's apparent that a car set up for high-speed handling is still entirely streetable. It turns out that solid suspension bushings won't necessarily knock the fillings out of your molars, but they will make an early F-body more predictable, stable, and flat-out fun to drive. Brake pads that will withstand an all-out charge down a canyon road will still stop the car when you pull out of the driveway on a cold morning.

One area that Adams immediately targeted for improvement was the '69 Camaro's deflection oversteer—in layman's terms, the front end slop that raises anxiety levels when cornering close to the limit. Rubber suspension bushings are perfectly adequate for ordinary driving, but they have too much compliance when the suspension loads are high. The problem is apparent in any production car, but it's especially vexing in an early F-body which is likely to have accumulated over 100,000 road miles. And, through a simple twist of fate, the early model '67-'69 Camaros and Firebirds are much more vulnerable to deflection-induced handling ills than their late-model brethren. As Adams pointed out in earlier installments, the early and late-model F-bodies are related in name alone. Their totally distinct front suspension systems have quite different characteristics under duress. In a

Flat-Out F-Body

late-model Trans Am or Z/28, the suspension geometry induces understeer; the driver cranks the wheel over a little harder and continues on his merry way. The early Camaros and Firebirds behave differently. Push them hard and the geometry shows a predilection toward oversteer when all the rubber in the front end compresses. The driver steers away from the turn to correct, making the process of getting around a corner with haste a series of jabs and feints at the steering wheel.

The most effective remedy for this condition—short of a total front suspension redesign—is to eliminate the deflection that is the root cause of handling inconsistencies. In his all-out Daytona and Sebring sedan racers, Adams specifies aircraft quality spherical bearings at all the critical suspension pivot points. In the interest of economy and a smooth ride, Adams opted for polyurethane bushings in this instance. The front suspension was completely disassembled, the factory rubber bushings pressed out, and new urethane bushings installed in their place. New upper and lower ball joints were also installed since the A-arms were already out of the car. Similarly, the rubber bushings in the front spring eyes of both rear leaf springs were put aside in favor of harder urethane replacements. All the other suspension components—VSE 1-inch diameter front and rear swaybars, Koni shocks, and VSE front structure kit—remained the same.

Well, it didn't work. The car became, in Herb's terse evaluation, "Terrible." It bucked and bounced and in general behaved like an ill-mannered race car, not a silky street GT. The problem was soon diagnosed as too much friction in the plastic bushings. Like a motorcycle with too much drag in its front fork seals, the urethane bushings prevented the suspension from reacting to irregularities in the road surface. Over rough pavement, the front A-arms would stay in the jounce position; with the suspension virtually resting on its rubber jounce snubbers, the car was effectively unsuspended. Definitely not a good way to cover ground quickly and comfortably.

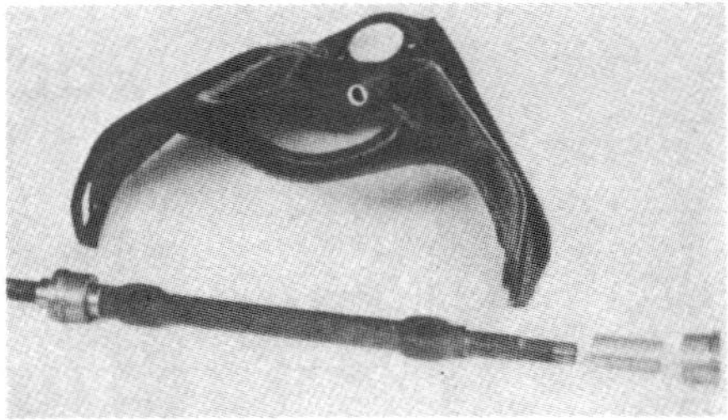

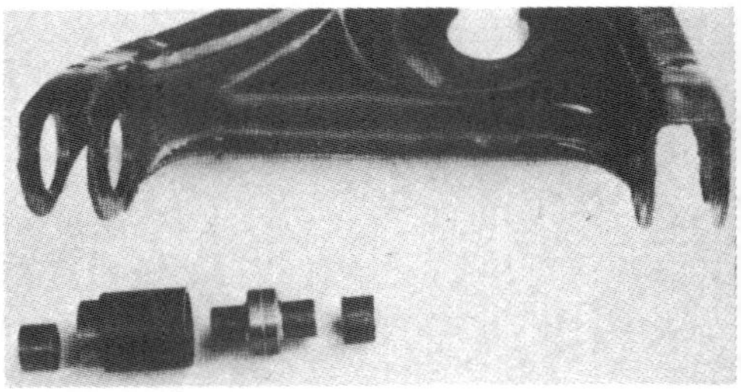

Eliminating deflection sometimes requires drastic action. Adams installed solid steel bushings in the Camaro development car. These VSE bushings (pictured here with late-model A-arms) use steel-on-nylon bearings for the upper pivots and caged spherical bearings at the lower control arm pivots.

Opening up the urethane bushings enough to eliminate this undesirable friction would have given the old devil Deflection a foothold again. And so, the crew at Herb Adams VSE went through the entire front suspension drill for a second time. This time they installed a race-bred solid bushing kit—steel bushings riding on nylon sleeves at the upper A-arm pivots, and hardened steel spherical bearings for the lower A-arm pivots and front spring eyes. The entire assembly is an exercise in close-tolerance machine work, with clearances held to a thousandth of an inch between the sleeves and bearings to forestall slop, deflection, and things that go "bump" in the suspension. After prototyping the various components and hardware on the '69 Camaro development car, Herb added an early-model solid bushing kit to his burgeoning catalog of Camaro-Firebird performance parts (available from Herb Adams VSE, 125 Ocean View Blvd., Pacific Grove, CA 93950).

One test drive and it was clear that the car was back on its good behavior. And wonder of wonders—the harsh ride that everyone anticipated simply failed to appear. In fact, unless a passenger is told that there's nothing between him and the control arms except hardened steel, there's nothing in the Camaro's demeanor to tip him off. The combination of bushings and sleeves seems nearly frictionless; the tires are free to follow each divot and glitch in the asphalt. The only time the driver is reminded of the solidity of the Camaro's undercarriage is when he lackadaisically steers along a row of Bott's Dots freeway lane markers or catches a seam in the roadway head-on. In these instances, there's no doubt that there's no room for deflection in the suspension pivots. On a fast two-lane country road, though, there are no objectionable traits—it's just the driver, the car, and the pavement, all in close communication.

At this point in Herb's Camaro ministrations, there was scarcely a speck of rubber left in the underpinnings to muddy the messages coming from the tires. Aircraft spherical rod ends connected the front swaybar to the control arms, aluminum biscuits were sandwiched between the subframe and body, and tubular steel struts tied the subframe into the body structure. Yet for all these unyielding connections, the car still rode comfortably on all but the worst road surfaces. Perhaps a Manhattan pothole would break the car in half,

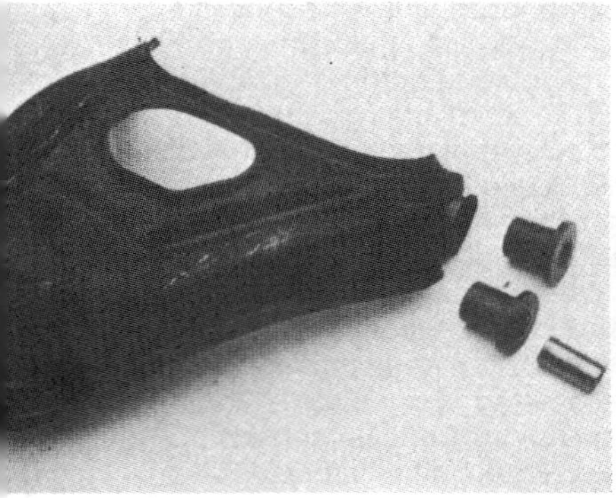

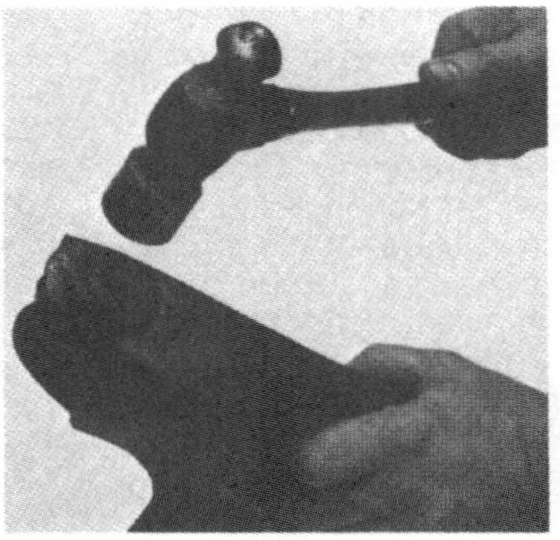

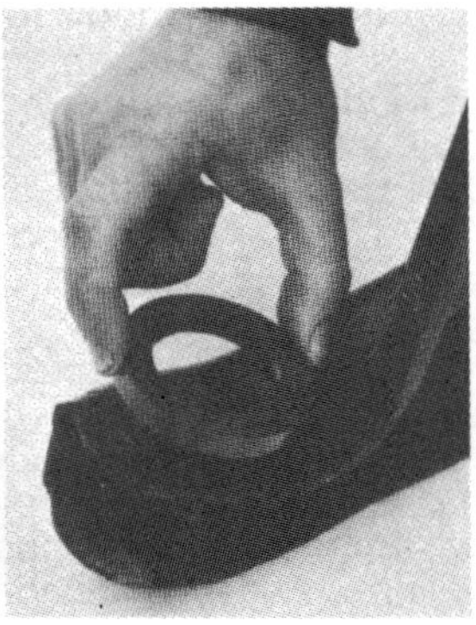

Adams' initial efforts at reducing deflection in early Camaro front ends relied on urethane plastic bushings. These replacement bushings are installed by tack welding sleeves in the control arms, then pressing in the urethane and an inner steel sleeve. Unfortunately, the results of all this work were disappointing.

Flat-Out F-Body

but nothing on the interstates upsets its composure.

Journeying to Herb's favorite test track revealed a startling development. With Herb's practiced hands at the wheel, the Camaro proved no faster around a sweeping off-ramp than when it had been handicapped with the spongy rubber bushings. But with a less skillful driver strapped in place, the car went unquestionably faster than the novice driver had ever been able to coax it to go. The moral of this lesson is clear: A good driver can make up for a car's shortcomings with sheer skill, but give a less talented driver a good car and the automobile will make up for his lack of ability. Herb could compensate for the rubber bushings' vacillations without conscious effort, though eliminating their mushiness didn't help his times. But the predictability, firmness and control that the solid bushings provided allowed novice drivers to complete their rounds with more confidence—and ultimately, more speed.

The fires of competition burn hot in Adams' soul, and he still wasn't satisfied that he had extended the limits of early-model F-car performance as far as they would go. Since he'd taken the chassis setup to the bounds of streetability, he looked to improved traction as the source of faster cornering speeds. Throughout the entire F-body flog, the Goodyear Wingfoots (P245x15 front tires and P255x15 rears) had proven themselves capable of providing traction close to that of a racing tire while retaining the durability and wet pavement adhesion of street rubber. For Adams' final assault on the skidpad, he selected a very special set of Wingfoots—special only because two-thirds of their tread depth had been removed! The simple facts of tire life are these: New tires aren't as fast as old ones. As the tread wears down, the tire's ability to hang onto the pavement increases, because there's less tread wiggle. Adams had shaved the treads on a set of used Wingfoots for just such an occasion. And since he was fast approaching race car cornering forces, he also bolted on a set of VSE NASCAR-style chrome wheels, with thick steel centers, nine vent holes and full 360-degree welded construction.

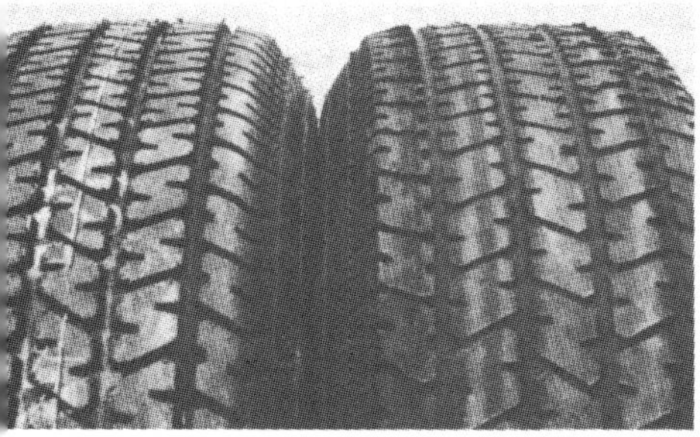

What's the difference between the Wingfoot on the left and the one on the right? About two-tenths of a second on the skidpad! Tire on the right has had approximately two-thirds of its tread depth shaved off. This autocross trick improves cornering traction by reducing tread flex—but you wouldn't want to do this to brand-new tires unless your father owns a tire store.

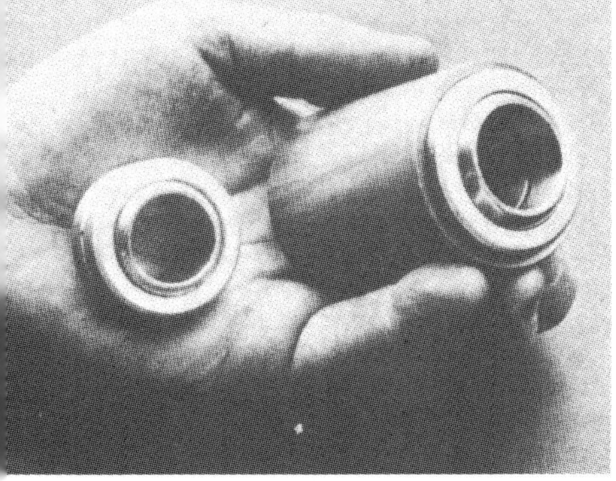

The leaf spring front eye bushings are also replaced with spherical bearings in Adams' war on suspension sloppiness. The inside diameter of the steel sleeve is honed to provide a close tolerance fit for the hardened bearing.

What a difference a little less tread depth can make. It's apparent why the winning autocross racers insist on shaved tires: The shallow-tread Goodyears reduced the Camaro's time around the skidpad by a full two-tenths of a second, producing the quickest time recorded during the entire year-long development program. Not that the stopwatch was necessary to gauge the improvement. Just listening to the car travel around the circuit without tire squeal, and watching how smoothly it negotiated the corners, was clear evidence of the improvement. The effect of the shaved tires was so remarkable that Herb did back-to-back tests twice to verify the results.

The portion of any car's life when it is driven to the brink of Edge City is short when compared to the hours devoted to commuting, errand running and just plain driving. Accumulating thousands of miles on a car reveals all its little quirks and idiosyncrasies, like discovering that the love of your life puts catsup on her scrambled eggs. In this instance, two areas proved to need extra attention. The effectiveness of the VSE front structure struts was clearly proven when their attachment points at the top edge of the firewall began to flex the sheetmetal. Adams cured this condition by fabricating four reinforcing plates to augment the factory's sheetmetal. Similarly, the forces transmitted from the rear suspension to the body convinced

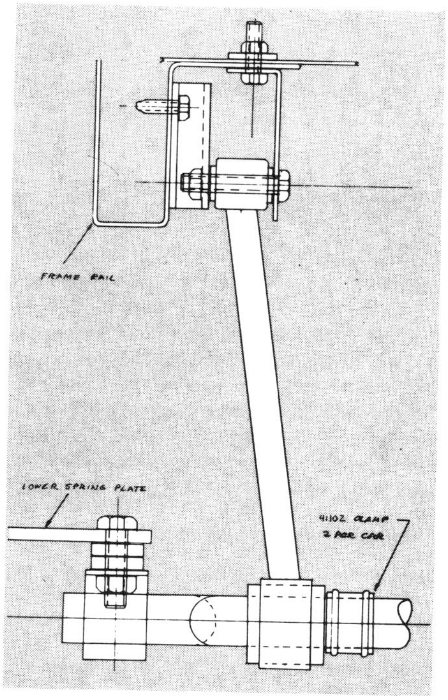

This diagram shows revised rear swaybar mount that straddles upper link. There's not much metal in the back half of a Camaro, so stressed mounts must be carefully designed to dissipate the forces of cornering.

Sometimes observing the 55-mph limit takes a little reminding. Adams recommends this Escort radar early warning device for drivers whose right foot occasionally grows heavy on the highway.

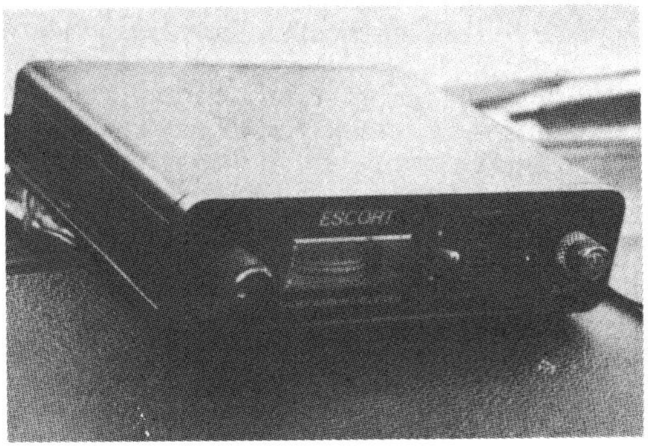

Flat-Out F-Body

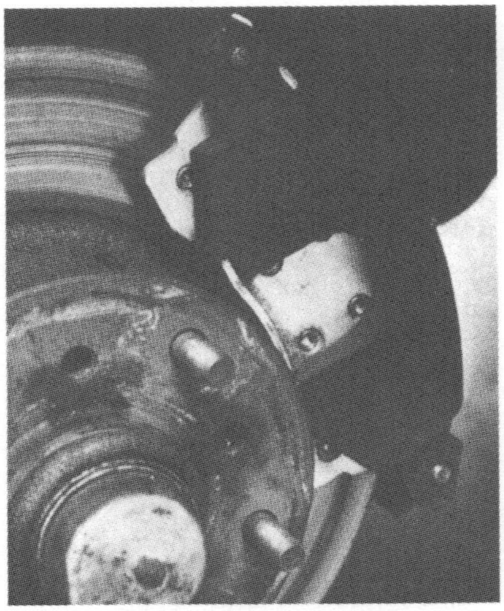

Experimental Raybestos front disc brake pads contain no asbestos and are designed for high-performance driving with minimal fade after repeated hard stops.

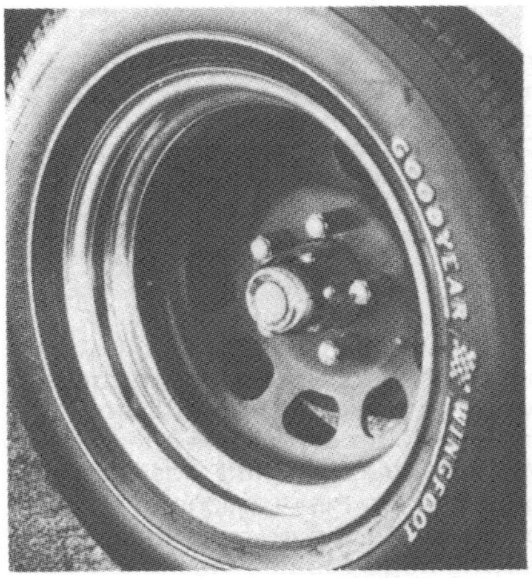

Race-styled VSE steel wheels help brake cooling with nine vents. These rims feature full 360-degree welds around the center section and 3/4-inch offset for tire clearance with P255x15 Wingfoot radial tire.

Herb that the rear swaybar mounts needed to be strengthened. He designed extra braces which straddle the upper swaybar link. Then there was the matter of sheer overexuberance while driving. Nothing could cure that, but an Escort radar detector mounted on the dashboard made one more aware of the possible consequences of "excessive" speed.

An all-around performance car has to do more than accelerate and corner. It has to slow down occasionally. And if these occasions happen to occur in rapid succession along some winding mountain road, then the braking system had better be able to cope with the load. Herb applied a double-barreled approach to improving braking efficiency: high-performance linings designed for fade-free stops, and plenty of fresh air to help the brakes keep their cool. The VSE crew installed experimental Raybestos disc brake pads in front. These new linings are unique since they are made without asbestos, which is generally considered to be a good thing for the environment; but more to the point, they offer improved fade-resistance, which is a good thing for a healthy F-body. The rear drums and backing plates were vented with a hole saw to hasten the circulation of cooling air through the brakes. The VSE crew took care not to remove any of the pads that support the brake shoes while performing this operation.

From the very beginning, it had been Adams' intention to develop a package of components that would make the early F-body the equal in handling of its late-model counterparts. The test of his efforts was to come when he arranged a showdown between a state-of-the-art VSE-built Cheverra '79 Camaro and the decade-older '69 test mule. To no one's surprise, the '69 Camaro matched the Cheverra's performance perfectly on the test track. To highlight how far the early-model had come, Adams calculated its cornering power, expressed as a "g" force. (One "g" equals the force of gravity.) Both the late-model and the early-model Camaros generated a lateral acceleration of .85g. To get an idea of how fast this really is, consider the cornering power of some rather pricey competition as reported in the automotive press: A BMW 320i corners at .74g, a Ferrari Dino 308 registers .77g, and a Porsche 911SC circles a skidpad at .79g. In other words, put a VSE-equipped early Camaro alongside a car costing eight to 15 times more money, and the homebuilt F-body will drive around the high-priced competition in the corners. No wonder that Chevrolet performance rivals apple pie as America's favorite—especially when chefs like Herb Adams add the extra spice. ■

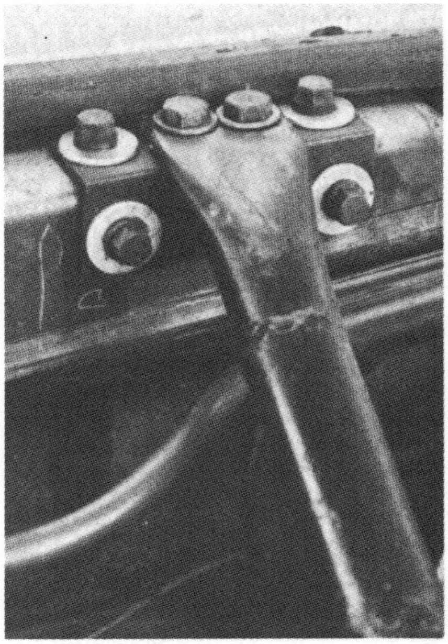

Effectiveness of the VSE front structure kit was proven when struts began to fatigue firewall sheetmetal. Herb's crew fabricated these steel reinforcements to help distribute the load.

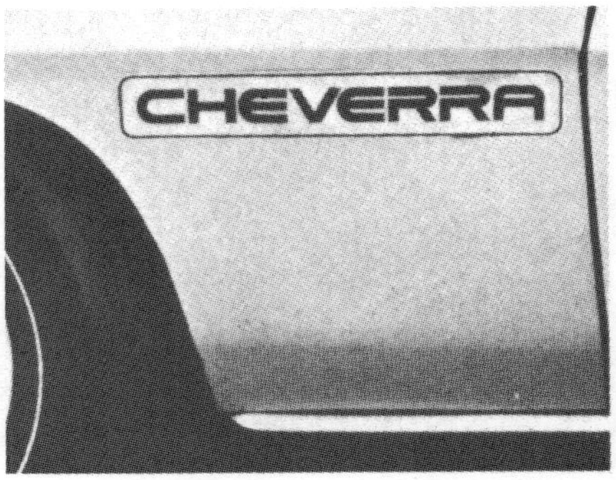

What's a Cheverra? That's Herb Adams' name for a Camaro that's been upgraded to supercar status. With cornering power superior to high-dollar foreign machinery, a Camaro deserves appropriate identification.

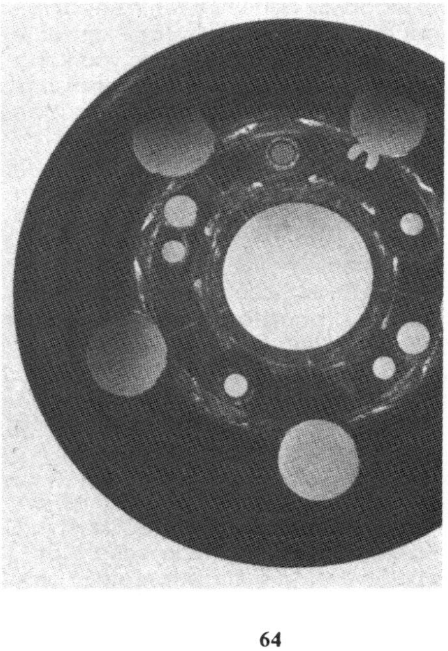

Adams improved ventilation of rear brake drums with carefully spaced holes in backing plates and drums.

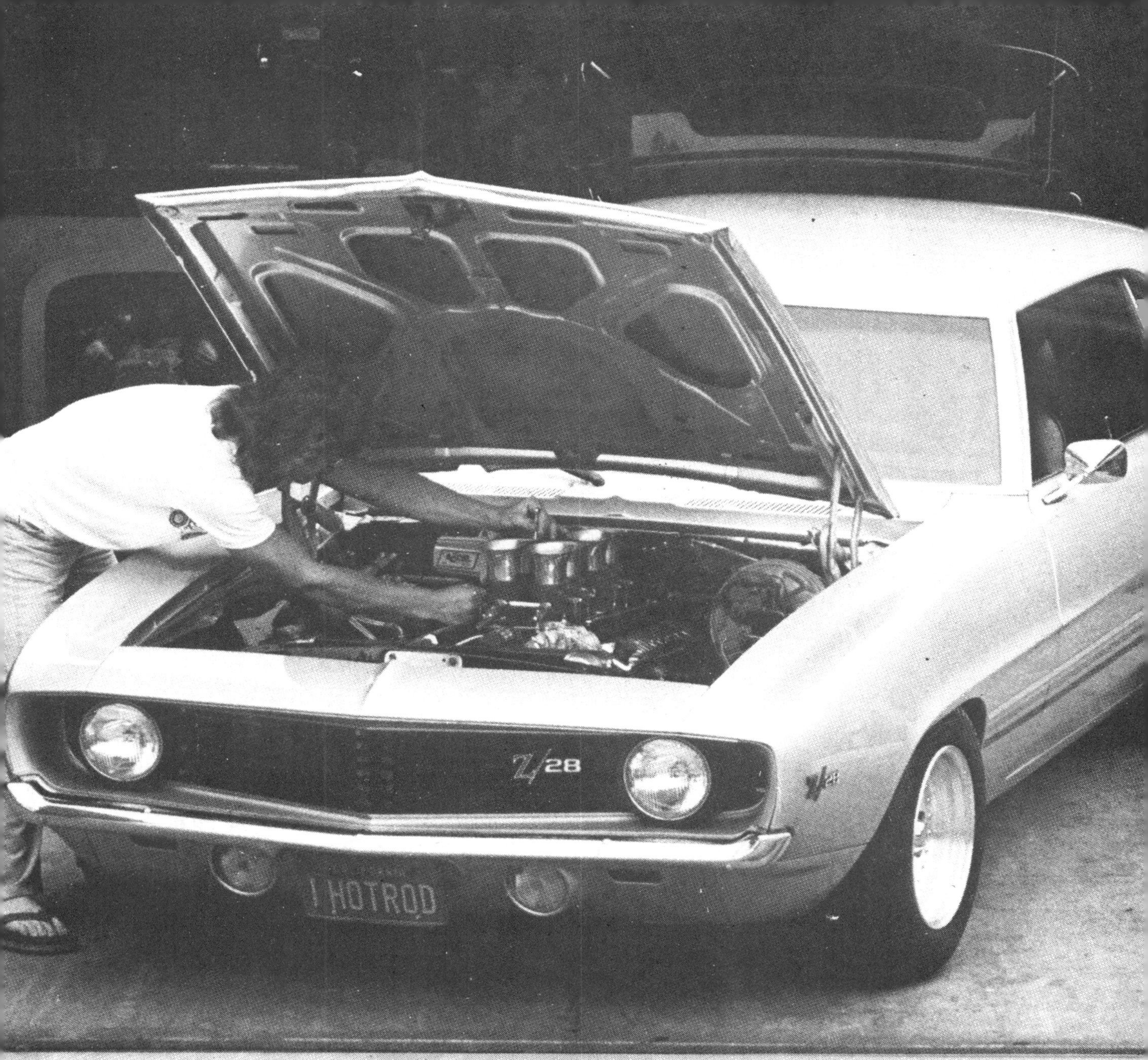

HOW TO BUILD A STREET MACHINE

By Gray Baskerville

'THE BEST WAY FOR A
BEGINNER TO START IS TO
UNSCREW IT, THEN
CAREFULLY SCREW IT
BACK TOGETHER."
—Lil' John Buttera

Don't tackle any project without adequate preparation. Ask for help or advice from someone who has had experience with your kind of project. A lot of guys who don't know what they're doing point with pride at a botched-up job because they did it themselves—and it looks like it! You must take one step at a time to realize the full monetary and common-sense rewards from that first experience. Remember, if you do fail, don't be afraid to fix it, because we all foul up."

Ah, yes, the wisdom gained from building 200-plus cars in the last 20 years is something Little John Buttera can impart to our readers in general, and to his son Chris in particular, for that's what this yarn is all about: how a 19-year-old kid built his first street machine.

"You know, kids are crazy and do dumb things," mused the mustachioed Lil' John the other day. "Take my kid Chris, for instance. He had this perfect '69 Z28 Camaro and wrapped it around a telephone pole one night while showing off in front of his pals. He's your typical kid who likes cars and the type who can identify with late models because that's what he and his pals mess around with. About a

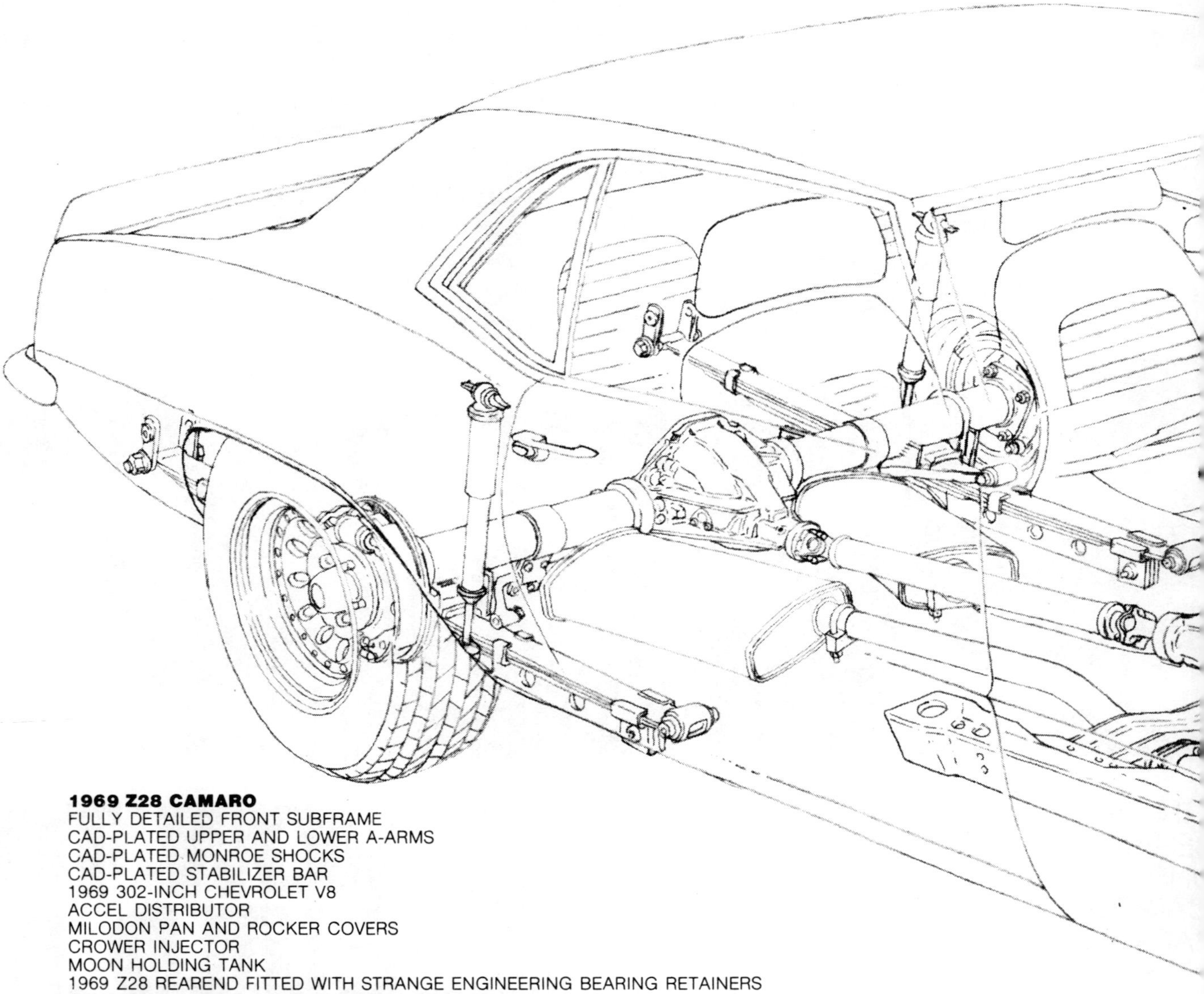

1969 Z28 CAMARO
FULLY DETAILED FRONT SUBFRAME
CAD-PLATED UPPER AND LOWER A-ARMS
CAD-PLATED MONROE SHOCKS
CAD-PLATED STABILIZER BAR
1969 302-INCH CHEVROLET V8
ACCEL DISTRIBUTOR
MILODON PAN AND ROCKER COVERS
CROWER INJECTOR
MOON HOLDING TANK
1969 Z28 REAREND FITTED WITH STRANGE ENGINEERING BEARING RETAINERS
CENTER LINE WHEELS 7 x14 FRONT, 9x15 REAR
FIRESTONE TIRES P205/70R14 FRONT, SS RADIALS LR60-15 REAR
SIDE MARKER LIGHTS REMOVED
REAR FENDER LOUVERS REMOVED
STOCK 1969 Z28 INTERIOR
OPTIONAL HOOD
LIFT BARS
KENNY MOORES HEADERS AND EXHAUST SYSTEM

year ago he bought his Camaro from an old guy who liked fast cars and who had it equipped with all the factory hot rod options. Chris got it for $2500, put $1000 worth of wheels and tires on it, and then promptly bent it in half. He couldn't afford to turn his back on his investment, so he went out to the garage and began to fix it.

"You might say that Chris sort of got forced into it. I wasn't going to let him walk away from his bent Camaro, even though he had no real automotive background—except from hanging around me or what he had learned from bench racing with his pals at the local Jack-In-The-Box. That crash turned out to be a blessing in disguise, because instead of taking a $3500 bath, Chris learned how to take a car apart, clean up the various components, detail them, then put the entire package back together again by himself. He got himself a six-year education in six months by becoming aware of how the wiring was run, how the upholstery went in, how the front and rear suspension worked, how the window glass was attached, how the fenders bolted on and how an entire car was constructed."

In a body shell, that is what this article is all about. We're simply going to take a car apart, rebuild or refinish the various assemblies, then reassemble it in the age-old tradition of building a street rod. We know many street machiners will simply add some aftermarket wheels, tires, headers and/or chrome goodies to their otherwise

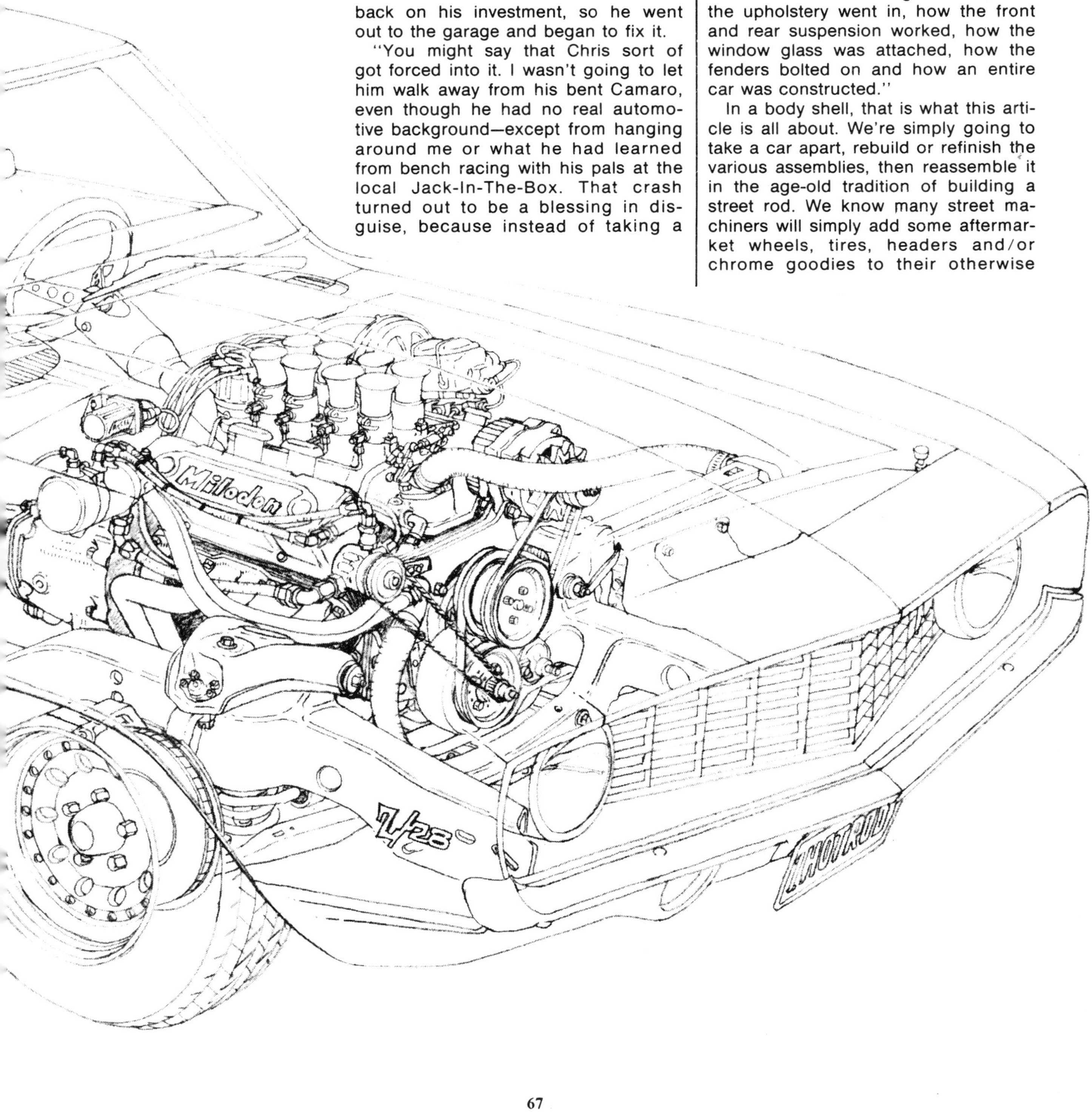

"stocker," and consider it done. We also realize that others will go the full blower route, using the eye-catching appeal of a polished 6-71 Jimmy huffer, wheezing through the hood of a flamed and fuming shagnasty, to build a car that looks more at home in the staging lanes at the local drag strip than huffing down "burger lane." However, Chris' Camaro is what we like to call the middle ground. His Z28 combines the best of the bolt-on world with some of the most eye-catching performance tricks—never forgetting, however, that he must not adversely affect the car's potential resale value.

In essence, Chris' six-month learning process involved nothing more than finding a '69 Camaro "parts" car that had no more than a near-perfect body. He could then strip it down to the bare shell, sell off what he didn't need, take the high-performance factory components from his bent original, detail them, and carefully reassemble the total package—and eventually wind up with what you see on these pages.

UNSCREWING IT

Because Chris unscrewed his original '69 Z28 with a telephone pole, all that was left for him to do was to cannibalize it for its high-performance options and store them away for future use. "The secret here," according to his fidgety father, "is to remember where everything goes. Now this isn't easy, so Chris took a Polaroid of each assembly and made notes on how it came apart; then he bought a '69 Camaro Service Manual, so he could see and read how those components went back together again. He then took everything out of the wrecked Z28; and that included the headliner, upholstery, wiring harnesses, window glass, trim—you know, everything that could be used to turn his parts car into another Z28."

SOCKING IT AWAY

It's all well and good to remember how something comes apart and goes back together again, but if you can't find that fastener or rod end, you're in trouble. "If Chris learned one thing," recalls Lil' John, "it was how to keep track of all the pieces. He used grease pencils (or China markers), labeled boxes and coffee cans and stored them in such a way as to make it easy for him when it came time to massage each component. By doing it that way, he didn't run into any memory problems.

MOTOR RESPONSE

"It was both Chris' and my thought that, from a psychological standpoint, his Z28 should have a distinctive mo-

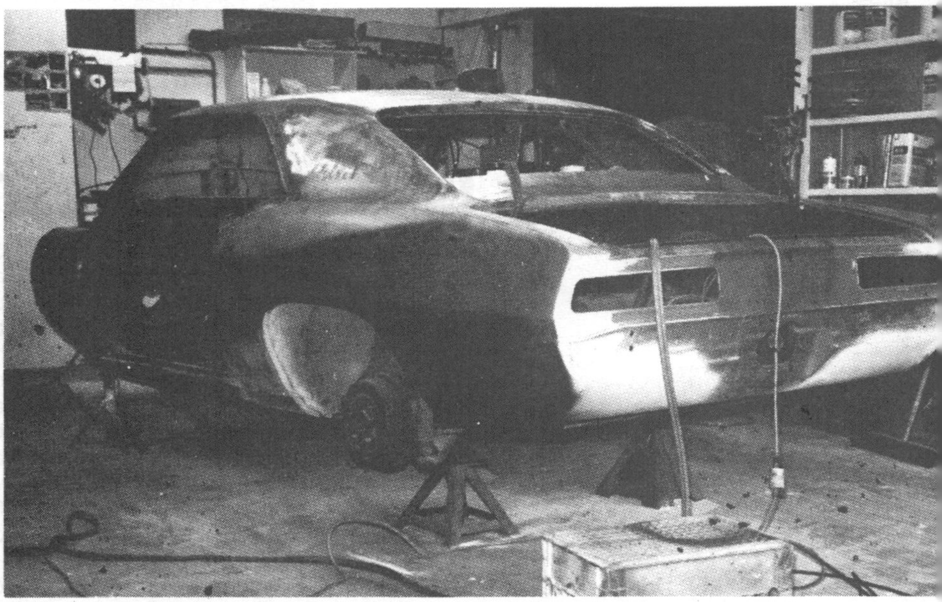

Second Camaro was completely disassembled (including wiring harnesses, upholstery, sheetmetal and exterior brightwork) by Chris Buttera before the "parts" car was primered and painted by Terry Finch.

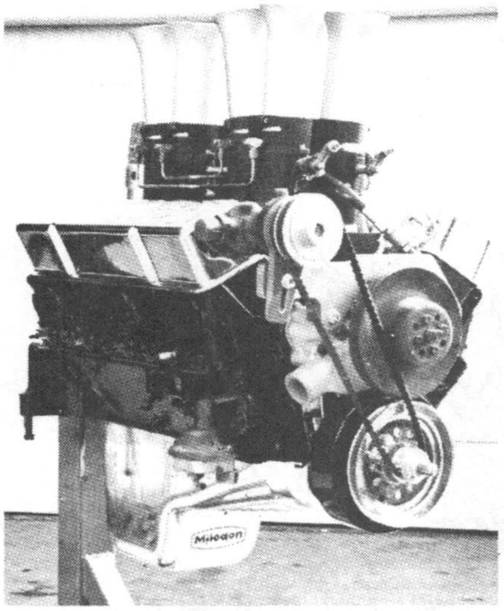

The 302-inch Z28 engine retains all the high-performance factory-supplied hardware, augmented by a Milodon oiling system, an Accel ignition package and this somewhat unique Crower-built induction assembly.

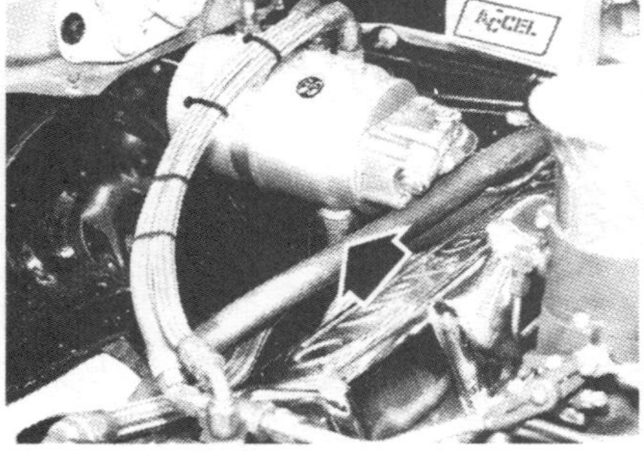

The two-quart Moon receiver tank was fitted with a removable Holley float bowl (and float assembly) to control the fuel level. Fuel is pumped into the top of the tank so gravity can be used to help prime the injector pump. Bottom line (arrow) goes to pump, while the top lines are bypass returns.

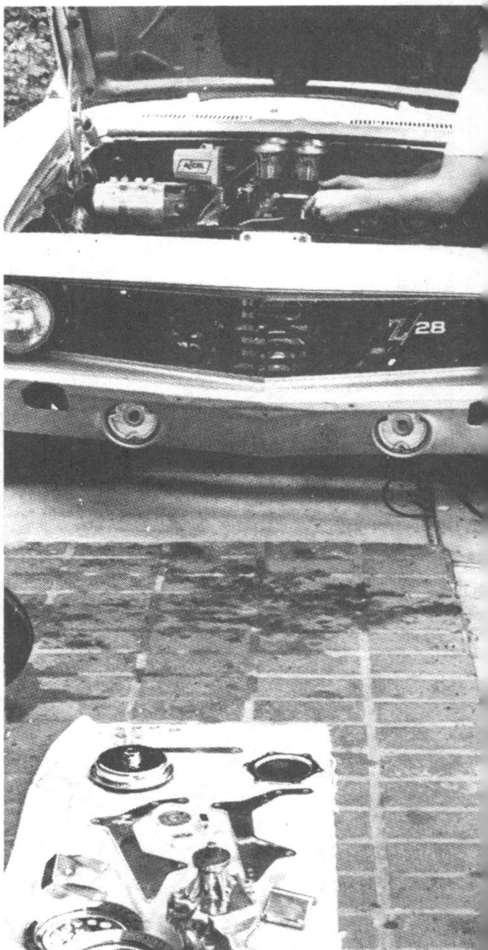

However, Chris decided to spend some money dressing up the engine compartment by polishing or plating the water pump, fuel pump and alternator mounting brackets, pulley and timing cover.

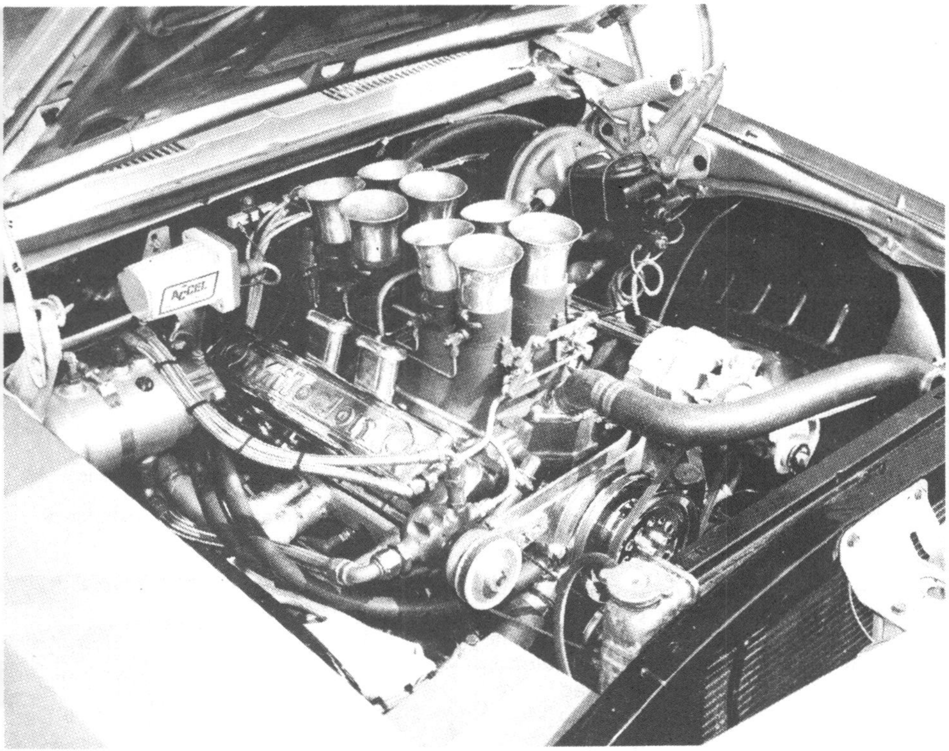

Injector stacks have been cut down for underhood clearance. Other than some basic detailing and the addition of a pair of Milodon rocker covers, the underhood package is basically factory stock.

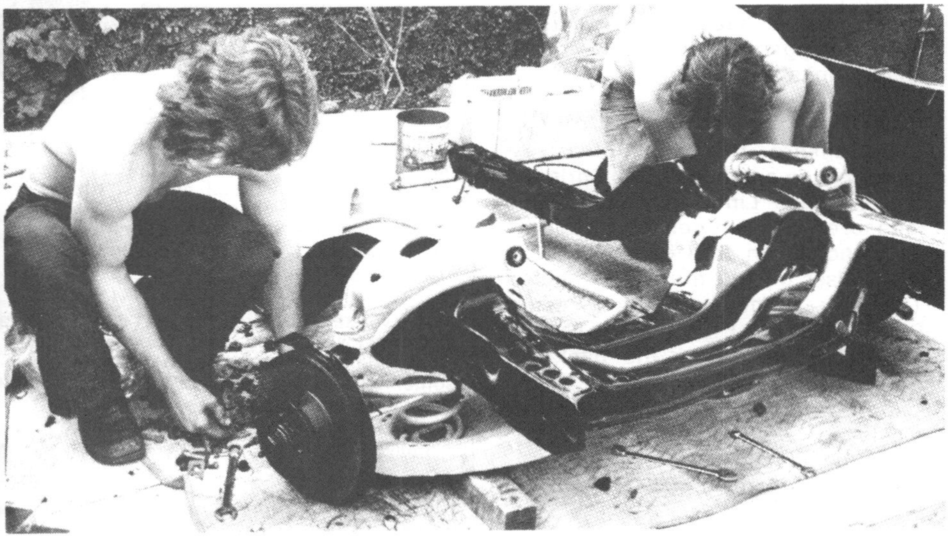

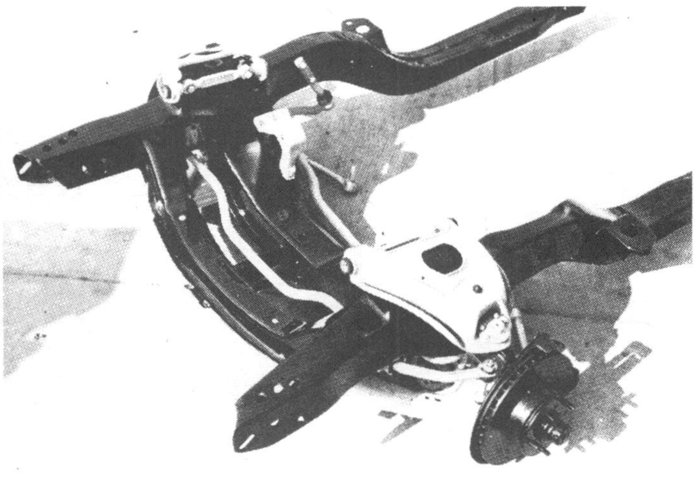

Now any novice street rodder can do this to his front suspension. Here Chris (left) begins to assemble the Camaro's underpinnings. Note that the welds on the front subframe have been ground and filled with "bodyman in the can" and then painted with high-gloss black acrylic enamel.

tor. The idea, therefore, was to build a performance engine to fit the Z28's image of high performance. The engine out of the original car was good; it had all the right components; and when we made a leak-down test, we found the 302-inch small-block Chevy V8 was in great condition. I guess Chris could have taken the engine apart and had it balanced and blueprinted, but why spend the money when you don't have it?"

Consequently, the little 302-incher was simply steam-cleaned, repainted a glossy, spray-can black and fitted with four basic elements: an Accel distributor, a Milodon oiling system, a pair of Kenny Moores headers (Positive Performance Headers, Fullerton, California) and one streetified-type set of Crower port injectors.

"The reason we decided to equip the 302 with an Accel distributor is simple. We wanted to fire all the fuel from the injectors, and under compression that could be difficult with a less powerful ignition system. As far as I'm concerned, a good ignition is the most important part of a performance engine.

"As for the 'Dooner' oiling package, I felt that a high-capacity pump with a big oil pan was a plus in the engine-cooling department. Oil is a coolant too, and by lowering the oil temperature, you can lower the temperature of the engine.

"The reason behind the exhaust headers is this: While your out-of-the-box-type headers are good, Chris wanted something distinctive, something that didn't allow the rest of the system to fall below the rocker panel line and kill the Camaro's profile by having anything hanging down along the bottom of the car. We wanted to make the exhaust system as unobtrusive as possible and still keep its performance image. The balance of the custom exhaust system is made up of Corvair-type turbo mufflers which combine a nice shape with a mellow tone ["Why tee off your neighbors with a set of raspy pipes?"—Lil'] and two 2½-inch-i.d. tailpipes shaped to keep the exhaust system hidden from view."

But the big deal was, and is, the fuel injectors. "I guess Chris could have bolted a blower on his engine, but that would mean he would have had to do away with the original hood line, because you gotta hack 'n' slash the hood so the blower will clear the sheetmetal. We decided to add a set of port injectors and still retain the clean performance image.

"The only problem we ran into was finding a cheap and simple way to prime the injector pump. The way we solved that item was to take a two-quart Moon aluminum tank and bolt a Holley removable float bowl to one side of the tank. It was simply a matter of hooking up a Carter competition-type electric fuel pump near the stock gas tank and running a fuel line from it to the firewall-mounted 'receiver' tank. What happens here is that the electric pump fills the receiver tank until the float shuts it off. A No. 12 steel-braided line is run from the bottom of the tank to the injector pump, which always keeps the injector pump primed. Two return lines (one a high-pressure relief valve on the pump and the other the bypass jet) are rerouted back to the top of the receiver tank. It's all very simple and works just as easily.

"The injector itself is strictly a bolt-on item. The only thing you have to do when ordering a system like this is to tell the Crower people how their injector is going to be used, and they'll program it for your particular application. They'll flow it for you, send you the right port nozzles and bypass jets. Chris' unusual induction system cost $750 for the injectors, pump, belts, pulleys and jets, plus an additional $170 for the receiver tank and about $50 for the steel-braided fuel lines, although he could have saved himself $40 if he had used rubber hoses. However, he's got something different, with a heck of a lot of visual clout per dollar."

DRIVELINE DISCUSSION

The only other thing Chris did to the engine compartment was to use a Deuce Factory thermostatically controlled electric fan (because the injector pump pulleys and belts interfered with the stock radiator shroud) and to fully detail and paint the inner fender panels and firewall—but we'll get to that later. Meanwhile, the stock Z28 clutch and flywheel package was replaced by a Schiefer "street/strip" single-disc model used in conjunction with a steel flywheel and a pair of new U-joints, because "you can't touch one part of a car without touching other parts." To round out the balance of the driveline, Chris retained the stock axles (although he checked them for twist) and added a set of Strange Engineering bearing retainers to keep the wheels on the axle housing in case he snapped an axle. The stock 4.11 "posi" was taken apart to make sure all the bearings were good; and the brakes were completely rebuilt using good, aftermarket, NAPA-type linings and rebuilding kits.

As Lil' John points out, "Essentially all the changes Chris made to his car were done for the sake of improvement, not for the sake of change. By upgrading the Camaro with longevity and safety in mind, you can compensate for a power increase without having to worry about undue breakage."

SUSPENSION EVALUATION

Except for one coil cut out of the front springs and the addition of a pair of high-performance Monroe tube shocks, the front suspension on Chris' Camaro is box-stock. "We could have made the front end lower, but Chris didn't want to use small front tires, and he didn't want his new Firestone P205/70R14 radials rubbing on the inside of the fenders." However, all the various suspension components were either cad-plated or spray bomb-painted, and the front-end assembly was equipped with all-new tie rod ends and ball joints.

Like the front suspension, the rear suspension was simply taken apart; stripped of old paint, road grease and dirt buildup; then either cad-plated or repainted. Chris bought a traction bar kit from a friend and used it, along with a set of new Monroe Road Handler shocks and a chrome-plated Milodon differential cover. The axle housings were sanded, then painted with a combination of Rustoleum and out-of-the-bomb semi-gloss black enamel paint. Again, all the suspension bushings were replaced with new factory units.

WHEELING AND DEALING

To ensure that distinctive Z28 performance-oriented look, Chris opted for Center Line "Indy" rims which measure 7x14 inches on the front, 9x15 inches on the rear, with a 3½-inch offset on the front and the rear. The tires themselves are Firestone P205/70R14s on the front and LR60-15 SS radials on the back. The front skins are 25 inches tall, with a 6-inch tread width, while the rear meats are 27½ inches tall, with a 9-inch tread. According to Lil' John, "I figured I had one crazy kid, so if he

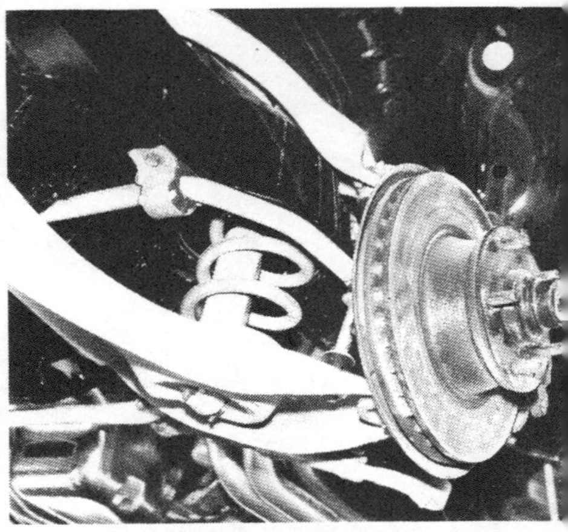

As you can see, the upper and lower A-arms, shocks, springs (cut down one inch) and stabilizer bar have been cad-plated to contrast with the gloss black subframe and floor pan.

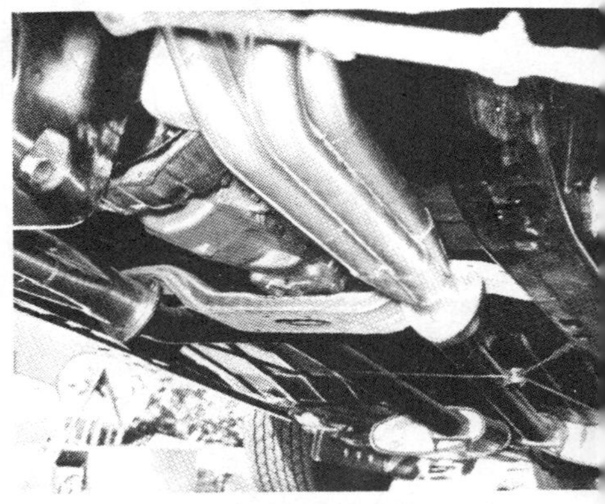

See how the Kenny Moores-massaged headers are tucked up inside the subframe so the entire exhaust system doesn't hang down below the rocker panel line. Again, note how the trans mount has been cad-plated and the floor pan painted a contrasting black.

As we've stated before, this is a '69 Z28 Cama that any novice could build. The wheel/tire combination is appropriate, the paint layout contemporary, and there are just enough body modifications (like the removal of the side marker lights and rear fender louvers) to visually clean up the exterior of the car.

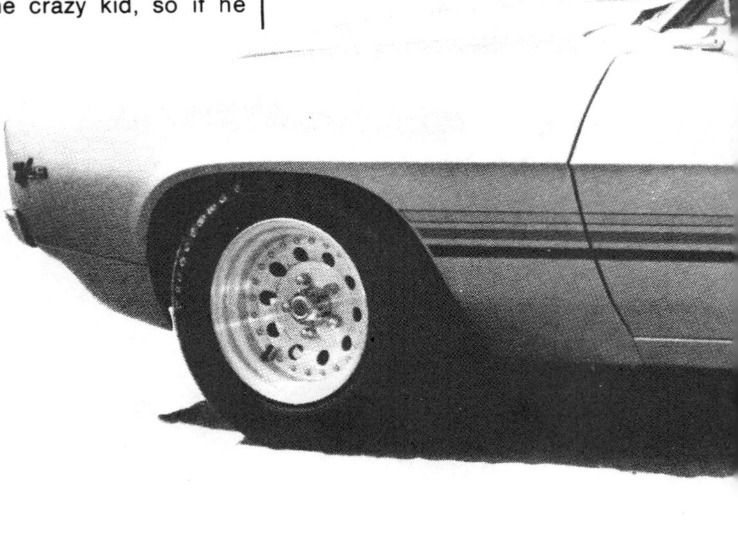

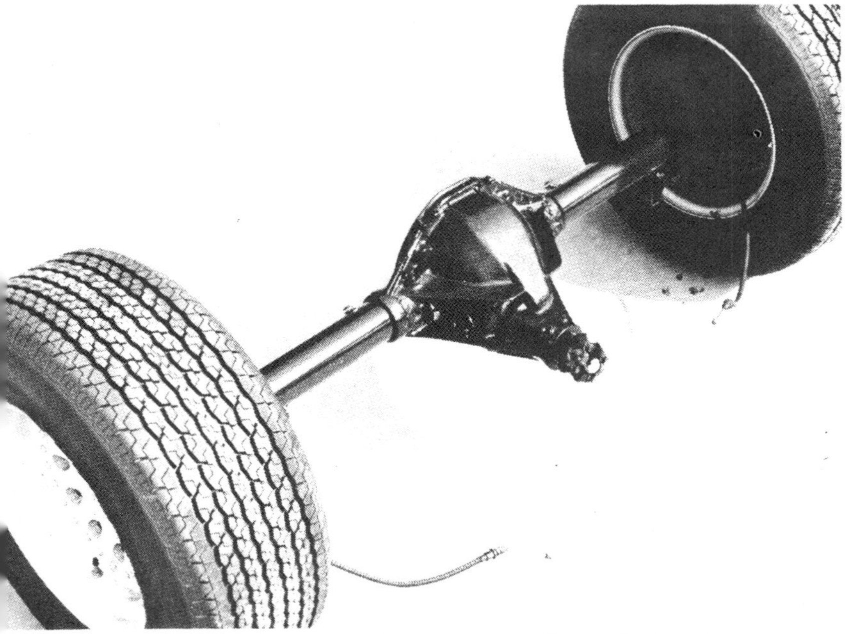

earend housing has been detailed and equipped with Strange Engineering earing retainers to keep the wheels in place should Chris break an axle. Buttera lso checked both the "posi" bearings for wear and the axles for unwanted twist.

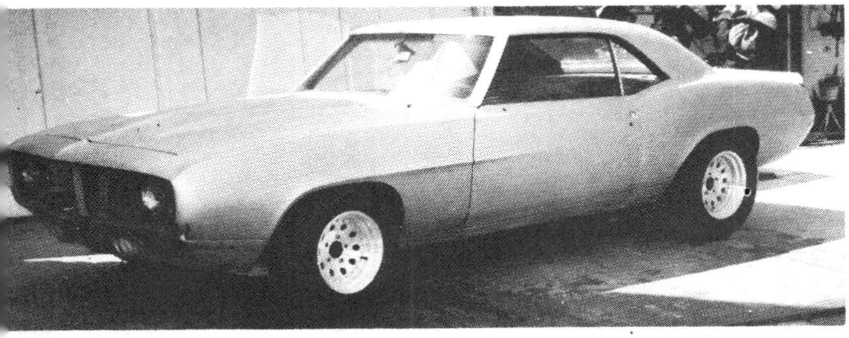

his is how Chris' Camaro appeared before the body was rubbed out. Check out uper body fit and subtle rake.

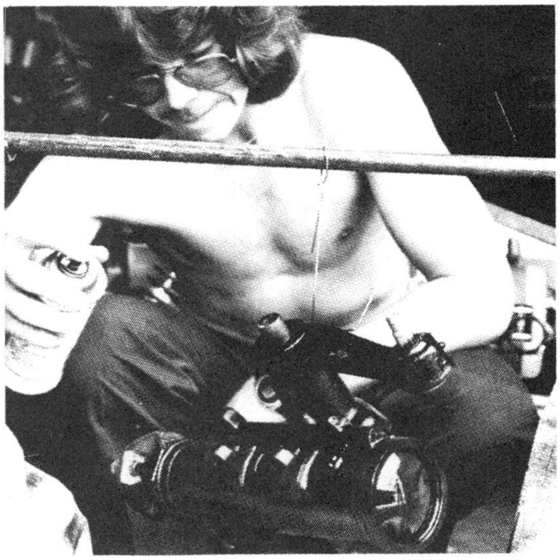

Like we've said, half of Chris' Camaro was painted with a spray can. Here he squirts the power steering box hanging on a backyard-style support system.

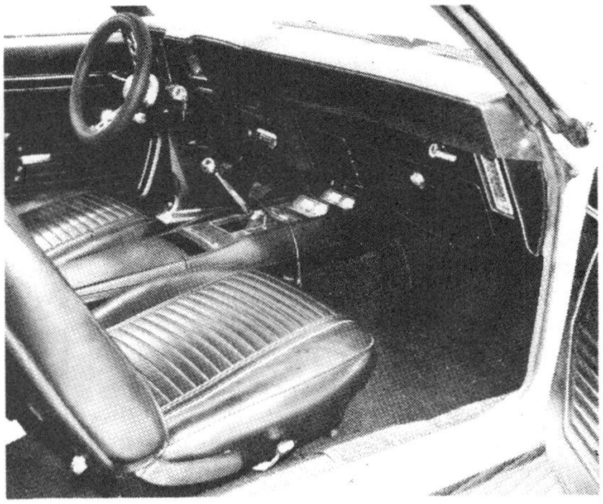

While the all-black interior is completely stock, it does feature a Hurst shifter and linkage, as well as an aftermarket wheel.

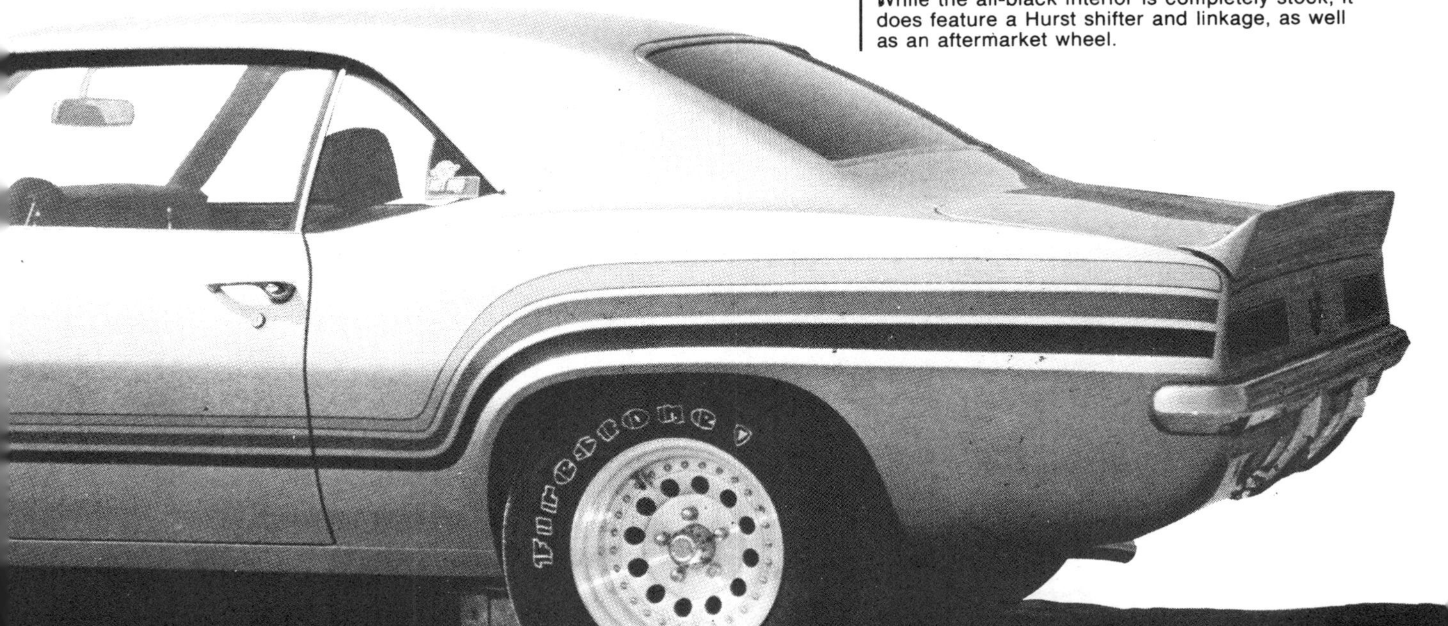

went out and cut a tire in the middle of Resume Speed, Iowa, he could find a replacement without too much trouble."

FRAME UP

As you might recall, Camaros, like most late models, rely upon combining a unibody with a front frame section that bolts to the firewall. Well, Chris unbolted the front frame section; had it sandblasted; and then, with a lot of help from Terry Finch, detailed the welds with an air grinder, a file or two and some plastic filler, before having painter Terry cover it with black acrylic enamel laced with a hardening agent that promotes a high gloss and durable finish. The rear frame section was also wire-brushed, cleaned, detailed and painted a semi-gloss black with a spray bomb.

BODY, PAINT, UPHOLSTERY AND SOUL

As we have already written, Chris bought a '69 Camaro parts car for its body and not its used-up engine or running gear. Consequently, little was needed in the way of straightening out the body, but it did need a thorough stripping. It's harder to do it this way, but the only means to ensure a first-class paint job is to do what Chris did: remove all the trim, tear it apart and take it down to bare metal. It's hard, dirty, no-good, son-of-a-gun work; but if you want it straight, you've got to play it straight.

To make sure his particular Camaro would be clean, Chris had Steve Davis fill in the side marker lights, as well as remove the louvers in the rear fenders, because they would interfere with the paint layout. Every salvageable part of the "bent" original was retained, because Chris didn't know what he could or could not buy. For instance, the Z28 had its own interior, and securing gennie replacement could have been a problem; but it wasn't, and all the OEM (Original Equipment Manufacturer) rubber is still available from Chevrolet.

The paint layout is based on the stock Camaro silver, set off by Dennis Ricklefs graphics, encompassing red, yellow and blue panels separated by red, yellow and blue pinstriping. "The thought behind this Terry Finch-squirted paint scheme," Buttera points out, "was to accent the Camaro's body lines without destroying the performance look of the car. A paint job like this could have cost Chris $1500, but he saved himself a lot of money by doing all the dirty work."

The interior of the Z28 is bone-stock and in keeping with the philosophy that this particular car is a learning process, not a radical rod. It's a first step, a getting your feet wet and eventually ending up with a brand-new old car. The rear kick panels, or inside trim panels, are brand-new, and one seat has a new cover. Other than that, what you see here is the original '69 Z28 upholstery. The interior also sports new rugs, the best of the window glass from both cars, and although the stock wiring looms were retained, Chris had to rewire both the ignition and fuel pump circuits.

The old man is proud of this particular car, because his son Chris persevered, put in a lot of time, and can now see the value of his work. "Everyone likes it. It's a tribute to showing people you can have something nice without having to spend a lot of money or buy a lot of equipment. It's a nice car to his peers, but Chris hasn't destroyed its resale value. It's a resto-machine, and he's gained the knowledge that he can do it. There are times when car building is boring. It's a dirty job. It's removing 10 years of oil, lube and road grime. But he proved to himself he could do it."

Yes, Chris Buttera certainly did—and how! **HR**

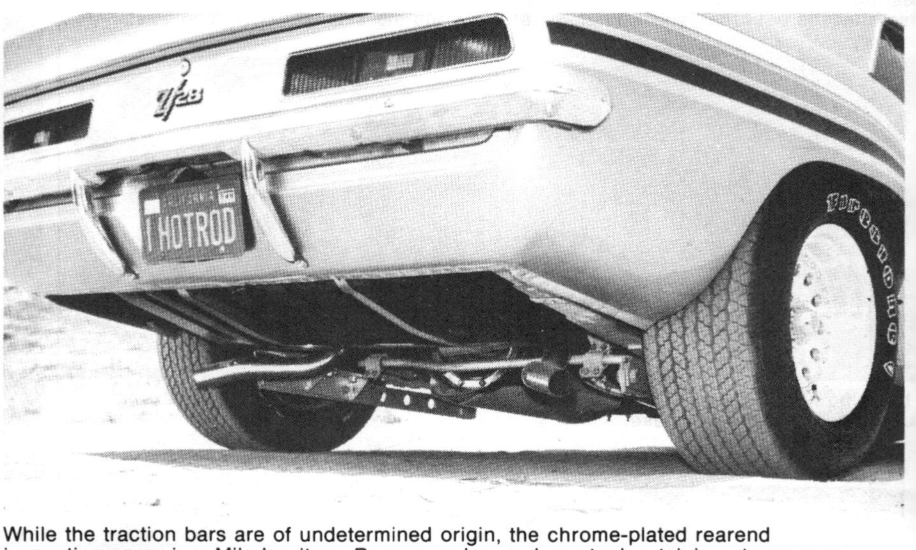

While the traction bars are of undetermined origin, the chrome-plated rearend inspection cover is a Milodon item. Rear sway bar and gas tank retaining straps were also cad-plated.

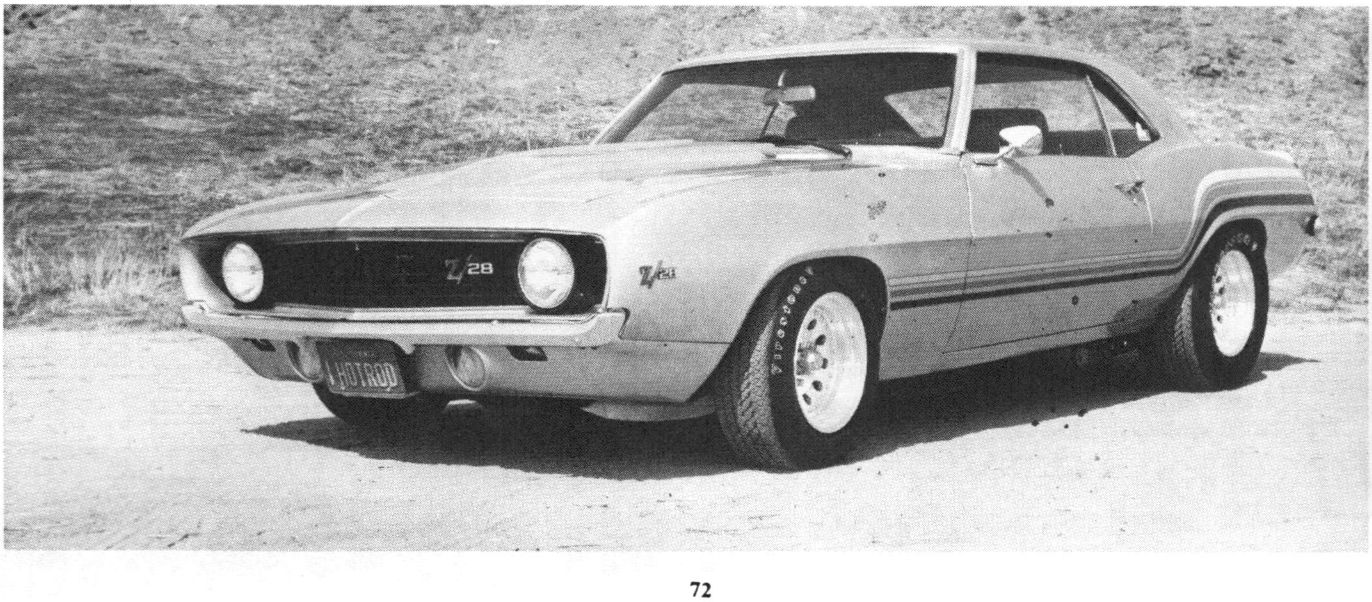

HANDLING HANDBOOK

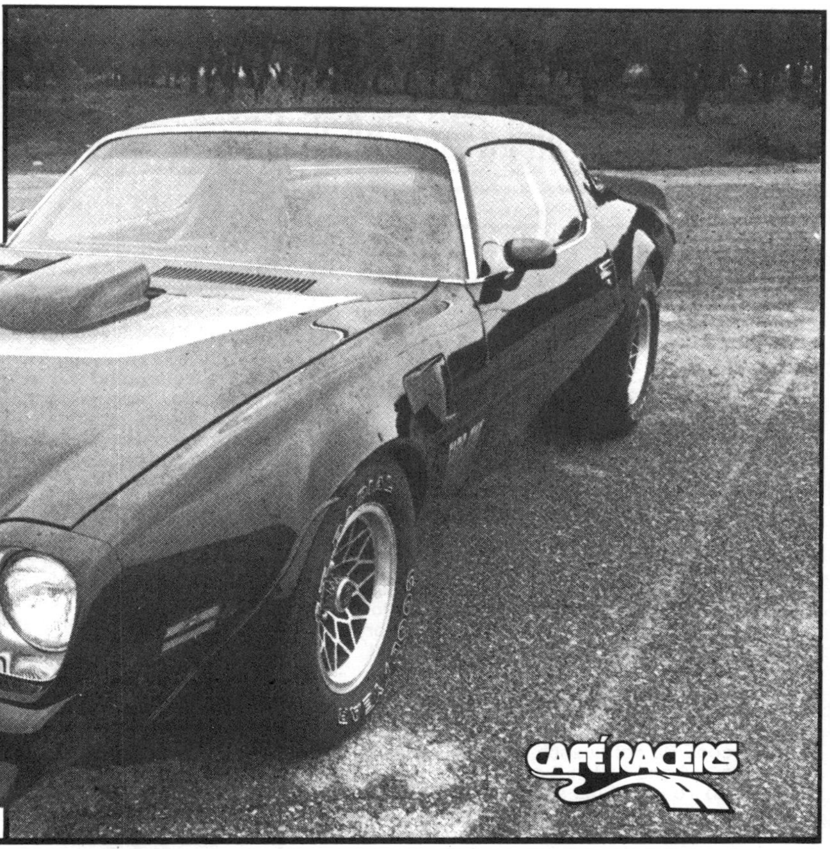

By John Baechtel

THE WHOLE EARTH CATALOG FOR FIREBIRDS & CAMAROS

All you Trans Am owners must be feeling pretty smug these days. For years you've heard that the Trans Am is the best handling car in America, and by now you're probably quite convinced that it's true. Unfortunately, your car's lofty position has been jeopardized—and by a former Trans Am engineer, no less. But Herb

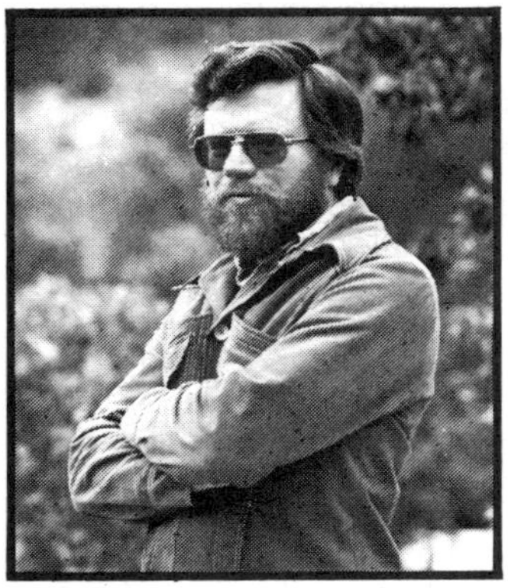

Adams is no traitor. He helped develop the original Trans Am handling package back in the Sixties, and nursed the concept through some difficult times. Who could be better acquainted with the compromises and inadequacies that still remain in the production vehicle? The current package is good and Herb knows it, but he also knows how to make it a whole lot better. His new company, Herb Adams VSE (Very Special Equipment), was formed to provide Trans Am owners with the hardware and technical assistance to turn their cars into true street driven road racers.

In September, 1977, we showed you one of Herb's tricked-up Firebirds with many of these pieces installed. They're now available through the mails from Adams' catalog. But Herb is pragmatic enough to suggest that these parts are not for everybody. Some of them have a very noticeable effect on the car's ride and comfort features. In order to take maximum advantage of this hardware, you have to be willing to live with more noise, more maintenance and a stiffer ride. The true enthusiast not only accepts these discomforts, he revels in them. If you're only interested in the T/A image, put your money into a stereo system. If you see yourself as Niki Lauda's successor and want to experience an absolutely phenomenal ride, read on.

Nearly everything Adams has to offer can be used on a Camaro or a standard Firebird. This opens the way for some interesting chicanery. Imagine an unadorned Firebird that has been tricked up with these pieces. An enthusiast who would go that far would undoubtedly be capable of putting together a respectable powerplant,

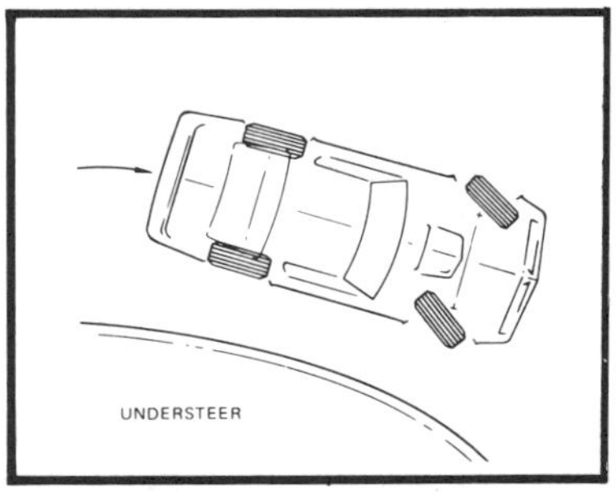

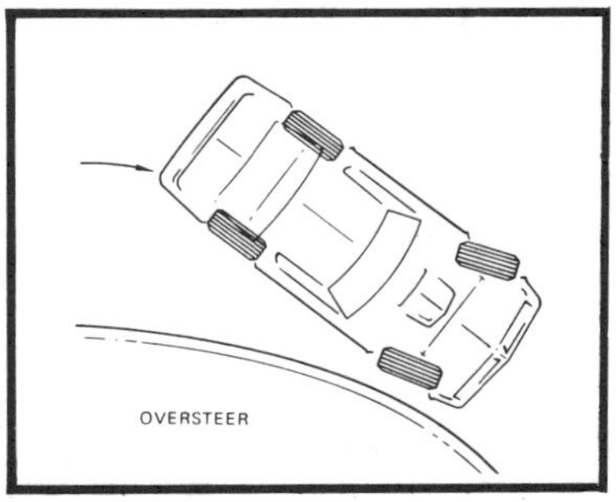

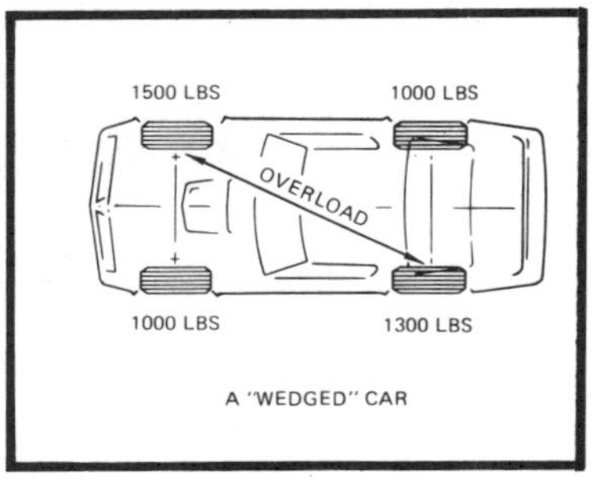

These drawings illustrate the conditions of understeer, oversteer and wedge. In an understeer condition, the front tires are pointed into the turn more than the rear tires. In an oversteer condition, the front tires are pointed out of the turn more. A wedged car compounds the problem by understeering in one direction and oversteering in the other.

HANDLING HANDBOOK

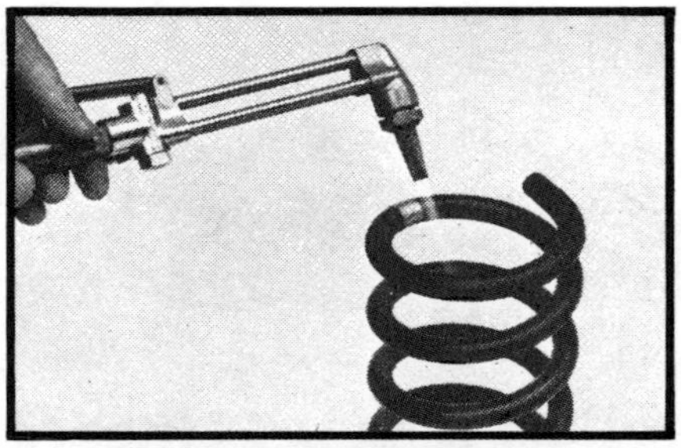

SHORTENING FRONT SPRINGS

(1) Remove the front wheels, shocks and shock mounting nuts.
(2) Disconnect the lower ball joint. Use a puller or rap on the mounting boss to free it.
(3) Using a spring compressor or a pry bar, cautiously remove the spring and rubber cushion. Discard the cushion.
(4) Shorten the spring by cutting off ¾ of a coil. Cutting it with a torch is acceptable, but it must be allowed to air cool. Do not quench. Heat an additional ½ a coil and flatten it against the pavement.
(5) To install the shortened spring, place the cut end into the spring pocket in the frame. Align the spring so the lower end will cover the outside shock mounting hole. Push the bottom of the spring toward the car.
(6) Jack up the control arm and guide the ball joint back into the upper arm. Secure it and reinstall the other parts.
(7) An alternate method is to heat the springs in the car, causing them to sag. Run a torch along the bottom coil of the spring. It will slowly nestle into the control arm socket. This must be done carefully because it is very easy to lower the car too much.

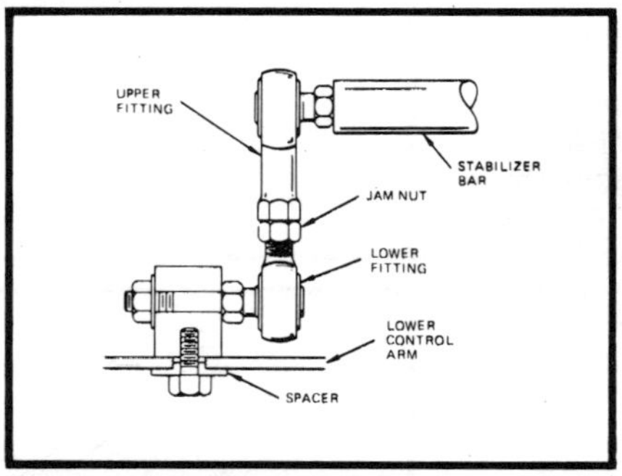

The front stabilizer bar is a bolt-on piece, and features adjustable links for fine tuning the suspension. The lower mounting block should be tightened in the location of the old stabilizer bar link, with the hole parallel to the threaded hole in the new bar.

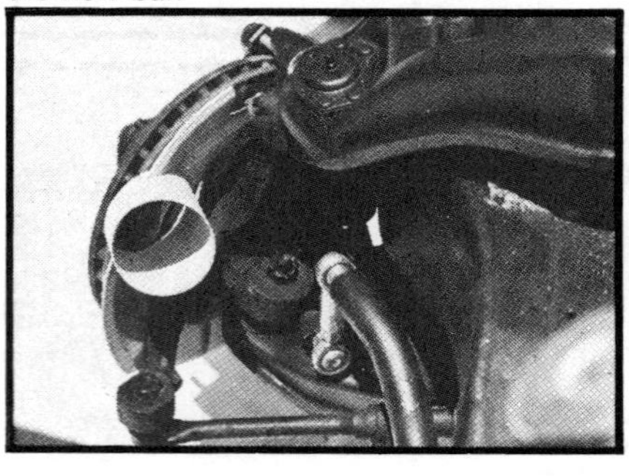

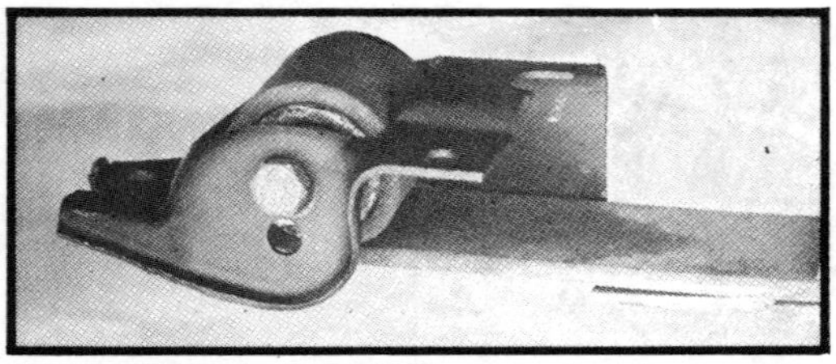

To cure the car of a twitchy rearend on turns, you need more anti-squat in the rear suspension. This is done by modifying the front spring eye brackets in the following manner.

(1) With the car resting solidly on jackstands, support the rearend with a jack and remove the shocks and upper link stabilizer bar bolts.
(2) Remove the front spring eye bracket mounting bolts (three on each bracket). Lower the rear axle until the brackets clear the body. Be careful not to stretch the brake hose on the rearend housing.
(3) Remove the large bolt from each bracket and remove them from the springs.
(4) Redrill the holes for the large mounting bolts exactly ¾-inch above the existing holes.
(5) With a torch, carefully cut out the top of each mounting bracket until you can mount the spring in the new hole. Cut only as much as you need.
(6) Reinstall the brackets and springs.
(7) Connect the stabilizer bar and reinstall shocks.
(8) If you are not able to modify the brackets yourself, they are available from Adams.

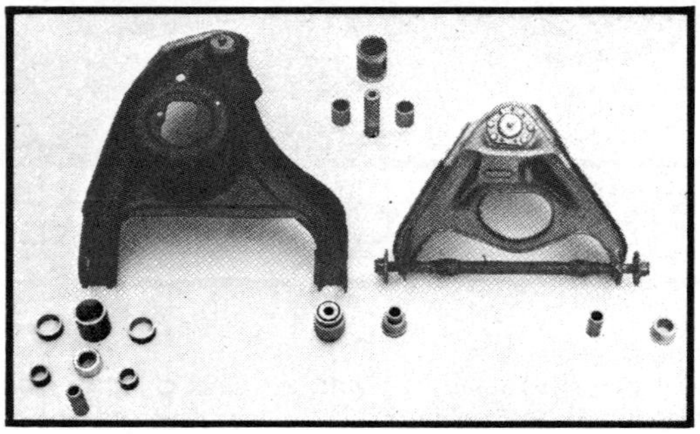

The solid suspension bushing kit is a serious step. Designed to eliminate deflection in the suspension, it will also eliminate your spine. Herb suggests that you do not install these pieces unless your intentions are really serious.

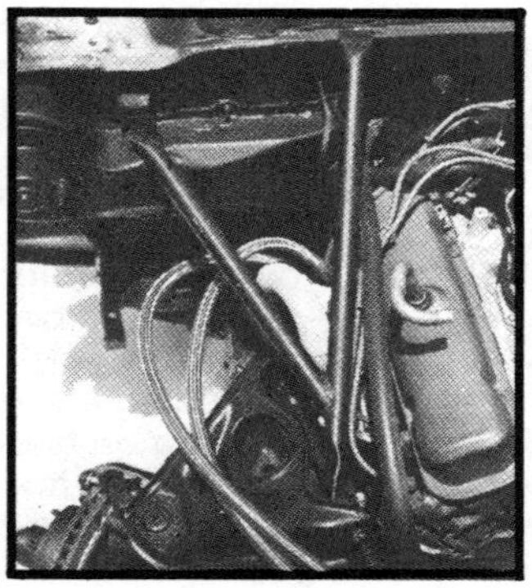

Dash support struts brace the firewall to the front subframe. They are included in the front end lowering and structuring kit. It also contains new body mounts to lower the body in relation to the frame.

To install the recommended wheels and tires, you will have to modify the wheelwells for clearance. Use the following procedure.

(1) Place the car on jackstands and remove the tires.
(2) Remove rear seat belts.
(3) Use a large hammer to remove the bump at the front of rear wheelwell (Photo A overleaf). Rework the entire area until it conforms to the shape of the rest of the wheelwell. Be sure the bottom edge turns away from the tires.
(4) With a large Vise-grip, bend the rear wheel opening flange upward toward a 45-degree angle. Do this all along the top half of the wheel opening. Finish it off with a hammer.
(5) Remove the front fenders.

and the combination would walk all over any unmodified Trans Am he might encounter. Trans Am owners may soon be forced to look into these parts just to keep up. Let's examine some of the handling characteristics of these cars, and see what beneficial changes can be made.

Herb describes these characteristics in simple terms and offers straightforward changes and adjustments to alter them. Basically, the car's handling is affected by how the tire contacts the road, how heavily it is loaded, how perpendicular it is and where it is pointed. In nearly every case, the largest possible tire patch is desirable. Bigger tires and wider wheels are the first order of improvement. Herb recommends HR-60x15 tires on eight-inch rims with no offset. He feels the Goodyear G.T. Radial is the best choice, but fender modifications are required for these tires. This first step will give a marked improvement even if no other changes are made.

Controlling body roll is the next step. The more body roll you encounter, the more the tires run cambered (not perpendicular to the road). Specific cures are offered in the form of revised camber settings and the use of special front and rear stabilizer bars. These bars reduce roll and are stiffer than factory bars. Modifying the rear spring mounts as illustrated will also help reduce roll steer by increasing the amount of anti-squat in the rear suspension. Firebird owners are no doubt familiar with the car's tendency to get loose in the rearend. "Anti-squat" uses rear axle torque to press the tires against the road surface with greater force.

In most situations, front and rear roll must be closely balanced to control understeer and oversteer. The front and rear stabilizer bars must be used together. They will provide

HANDLING HANDBOOK

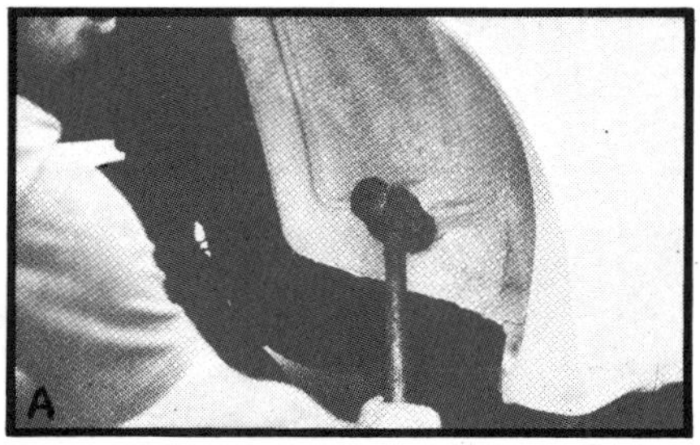

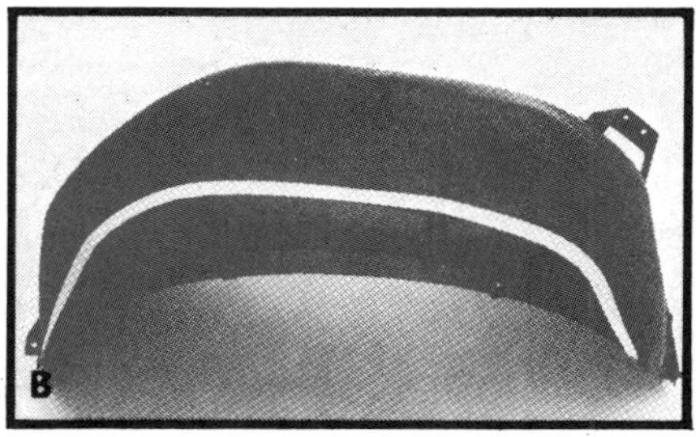

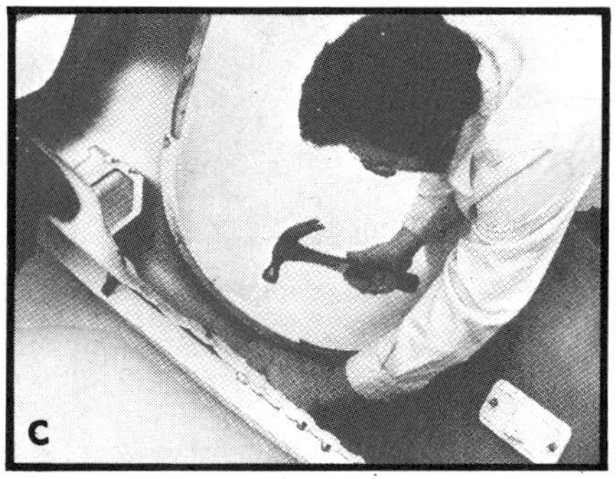

(6) Remove the inner fender panels and use a saber saw to trim the wheel opening as shown in photo B.
(7) Reinstall inner fender panel.
(8) Lay the front fenders on some soft material, and bend the inner wheel opening flange upward at a 45-degree angle in the area where the inner panel was trimmed (Photo C).
(9) Reinstall front fenders.

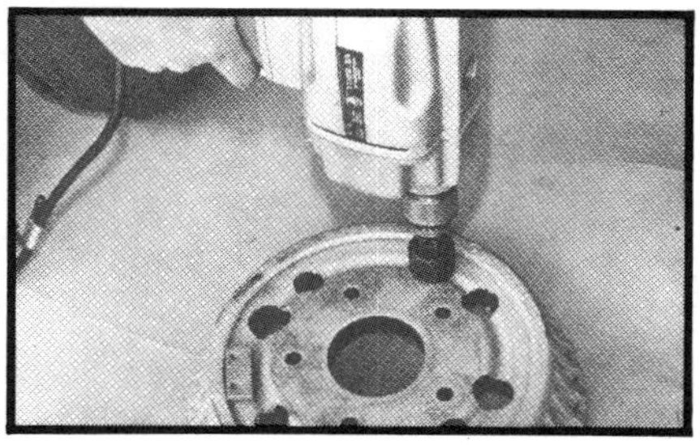

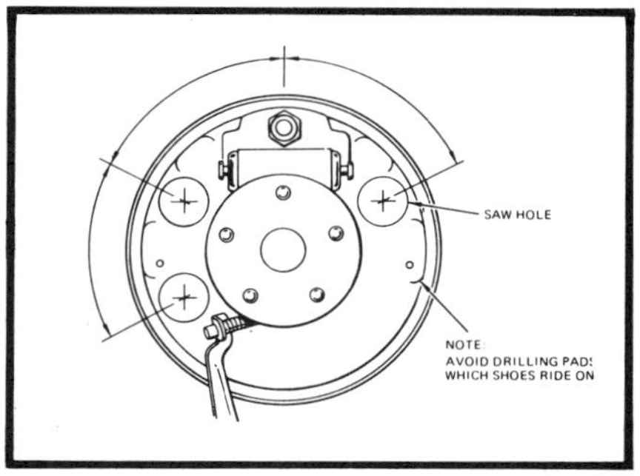

Herb supplies a brake cooling kit for the front brakes, but you can make your own for the rear brakes.

(1) Place car on jackstands.
(2) Remove tires
(3) Remove rear brake drums and ventilate them with a 1½-inch hole saw. Avoid drilling at any mounting points.
(4) Remove the brake shoes from one side and ventilate the backing plate by drilling three 1⅞-inch holes as shown in the drawing.
(5) Following a reverse order, reassemble the brakes with new linings.
(6) Perform the same operation on the other rear wheel.
(7) Reinstall the brake drums and adjust the brakes.
(8) Install tires and go.

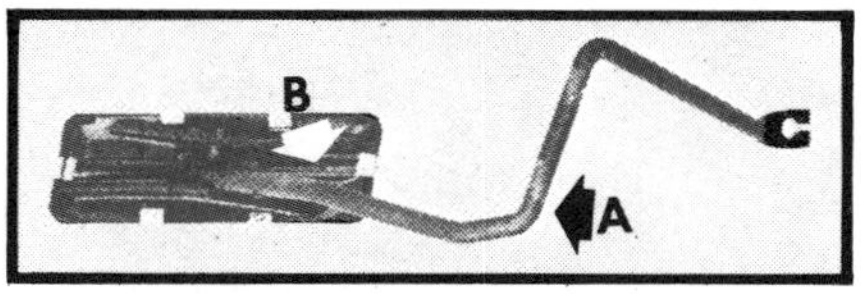

Another simple modification you can make is the throttle pedal. By moving the pedal closer to the brake pedal, you can use the classic "heel and toe" throttle technique.

(1) Remove the throttle pedal and note how the pedal is mounted to the shaft.
(2) Remove the pedal from the shaft.
(3) Bend the rod to the left, concentrating the bend at point A in the photo. This will move the pedal closer to the brake pedal.
(4) Bend the pedal mounting point back to the right, concentrating the bend at point B. This brings the pedal back to a near vertical position.
(5) Continue to alternate bending at point A or B until the pedal is where you want it. After reassembly, be sure to check for wide-open throttle plates when the pedal is pressed to the floor.

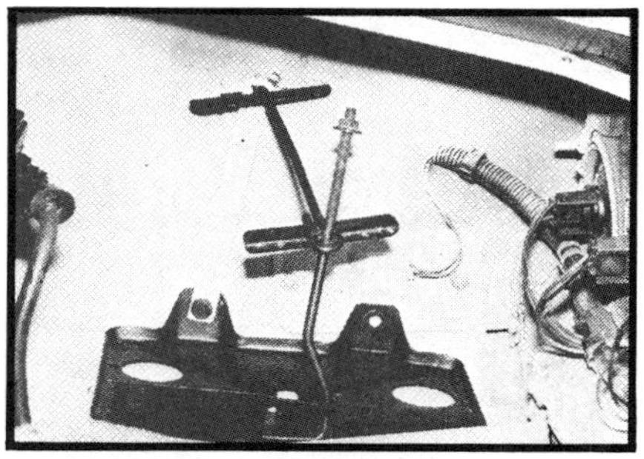

The battery relocation kit lets you move the battery from the engine compartment to the trunk. Since the Firebird has so much weight on the front end, this is a good plan. It amounts to nearly the same thing as moving the engine back ten inches.

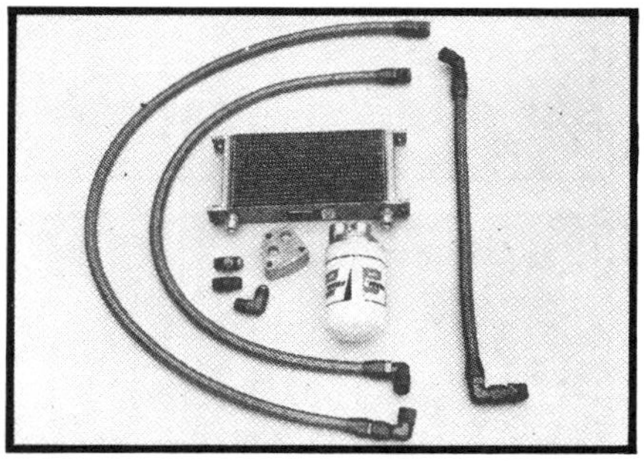

For regular street driving, Herb offers the wet sump kit. It features a specially baffled pan, relocated oil filter and oil cooler. Braided steel lines are included to hook it up. Stock Firebirds can already corner hard enough to empty the oil pan and cause starvation, so the wet sump kit is a must-have item.

nearly neutral steer which promotes high-speed, steady-state cornering. If the car oversteers in one direction and understeers in the other direction, the car is "wedged," a condition where the tires on one diagonal of the car are more heavily loaded than the other. Fire-Am front stabilizers have threaded front links, so this condition may be adjusted out. Weight distribution and suspension geometry are the other factors which affect the car's handling. Moving weight off the front suspension reduces the car's understeer tendency. A battery relocation kit moves 50 pounds into the trunk, and the addition of fiberglass body components would give a substantial advantage.

Herb feels the factory suspension geometry is quite good, but it has too much deflection due to the rubber bushings. The cure is to add steel front and rear suspension bushings. This modification is serious business, and should not be performed unless you are ready to live with the stiff and noisy ride it will cause. It will also generate a cornering force in excess of .85g, and that's a bunch.

Many of Adams' modifications are of a "budget" nature, and could be performed right in your own driveway without spending a dime. We're going to show you some examples that will get you started in the right direction. The catalog serves as a sort of textbook and contains the complete picture. If these samples pique your interest, you can get the catalog by offering up five bucks to: Herb Adams VSE, 1036 San Carlos, Pebble Beach, California 93953, (408) 372-2767. In addition to all the handling information, you will find valuable tips on chassis stiffening, exhaust and intake systems and wet and dry sump systems to control the oiling problem that even a stock T/A can experience on a really hard cornering maneuver. Adams' Whole Earth Catalog for Firebirds and Camaros could help put your Camaro or Firebird on the serpentine road to the Garden of Eden. ■

BOLT-ON PERFORMANCE: CAMARO Z/28

UNLEASHING ALL Z KING'S HORSES

by John Baechtel

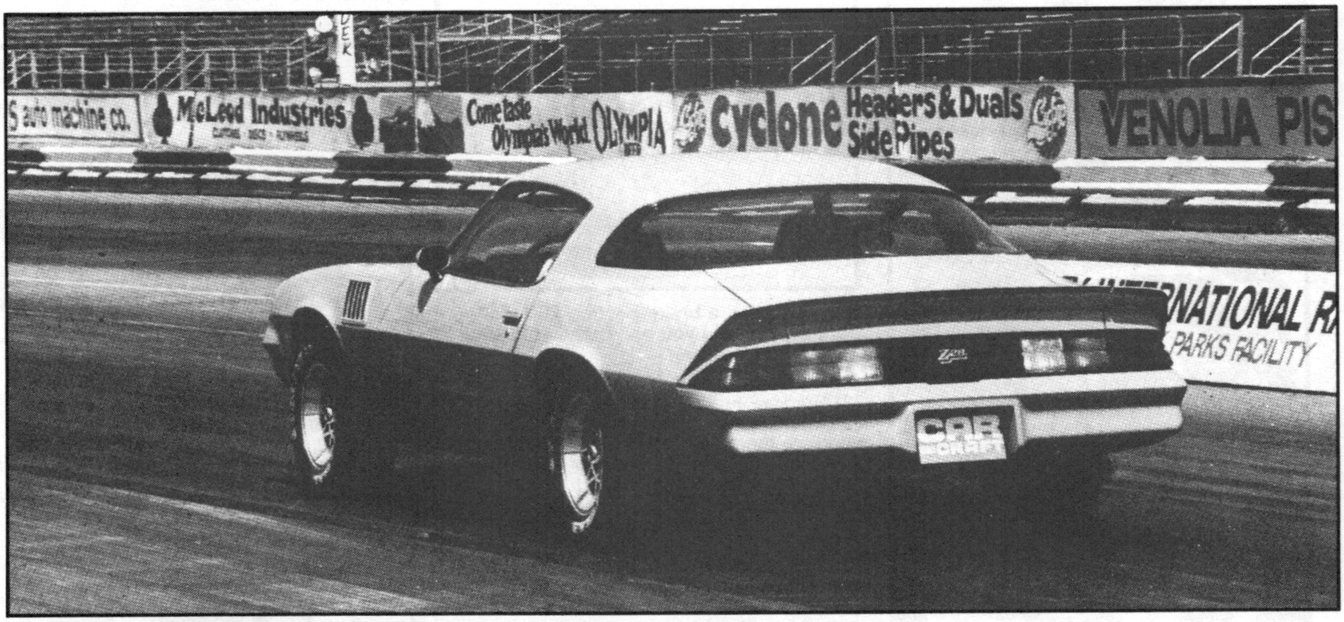

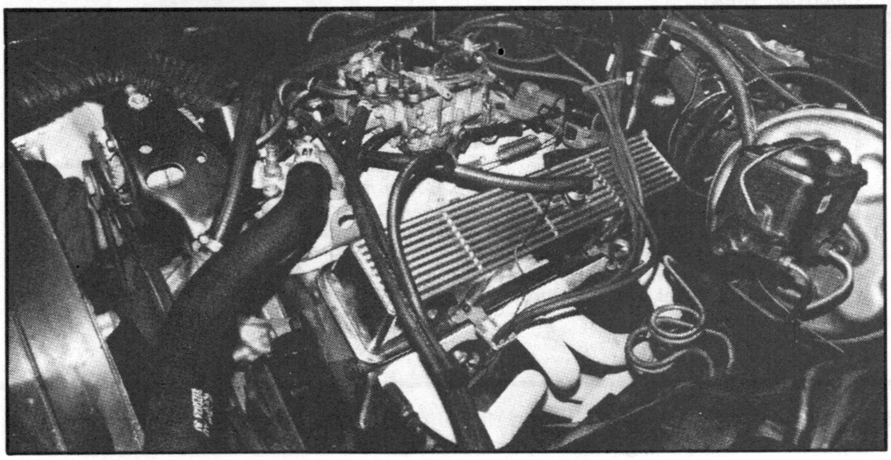

The Z/28 is one of the few cars left that can bring back the good old days. Our modifications put it solidly in the fourteen-second bracket, and proved it is still very much a fun car.

Mr. Gasket valve covers, BlackJack headers and a Team G intake manifold provided the performance increase we were looking for. They also imparted a suitably racy look to the underhood area.

Elsewhere in this issue we expound at length on the virtues of bolt-on speed equipment. We advocate a common-sense approach to modifying late-model street machines, and rather piously guarantee it to work. We feel certain it is a logical plan, but we're not so smug as to think we could lay it on our readers without backing up our claims. So, we arranged to test our hypothesis before making it public—not only to ensure our integrity, but to supply readers with "real world" information. To enliven the research (and, coincidentally, make more fun for ourselves) we secured a car whose performance reputation is well respected: a '78 Z/28 Camaro. Equipped with a four-speed transmission, the Camaro seemed a suitable challenger, especially since the factory had already taken care of the first step for us. The car was already equipped with a 3.73:1 rear axle ratio, which gave it a responsive feeling even in stock form. Previous experience in other Chevrolets equipped with the LM-1 350ci V-8 made it easy to discern the value of stiffer gearing. The stock engine in this car is dramatically more responsive in everyday driving, which reaffirms our faith in good old-fashioned torque multiplication. The Z/28 is the only current production car that offers a performance axle ratio—and for that, we award it a big checkmark in the plus column.

Offsetting that good point was the horrible condition of the posi-traction unit. The clutches were constantly locking up and producing an alarming growling and clunking on every corner. The condition was most noticeable after the car had been driven on the freeway; it appears that extended driving builds enough heat in the rearend to cause a breakdown of the gear lube. The clutches are not sufficiently lubricated, and they start to grab. The condition can be temporarily remedied by locating a large empty parking lot and driving the car in numerous tight circles to the right and then the left. This loads the axle assembly and works the lube into the clutches, but it certainly doesn't cure the problem. A dose of 140 weight gear oil might help, but the best plan would make use of the special GM Posi-lube and additive. This condition is not uncommon with new GM offerings. Owners are advised to avoid the problem by immediately flushing and refilling the rearend with a top

'78 Z/28 CAMARO HORSEPOWER READINGS

RPM	2500	3500	4000	4000 (air cleaner off)
Stock	126	165	174	182
Headers & dual exhaust	130	180	192	200+
Manifold & ignition wires, custom air cleaner	132	194	200+	

'78 Z/28 Camaro track results

	E.T.	MPH
Stock	15.65	91.57
	15.61	91.02
	15.53	91.80
Advance curve kit installed	15.51	90.78
	15.45	91.37
Modified car running with mufflers	15.09	94.80
	15.01	95.91
	14.84	96.88
	14.80	97.08
Modified car with open headers	14.79	96.70
	14.74	96.70
	14.71	96.92

quality lube designed for use in limited-slip differentials.

To help flog the car, we enlisted the cooperation of the performance people of the automotive division of the W.R. Grace Company, who graciously offered the use of their testing facilities located at Brooks Racing Components in Garden Grove, California. Our plan was to install headers, dual exhausts and an intake manifold, but before modifying the car, we baselined its performance at the track. Its average e.t. was 15.60; its best pass (15.45) was recorded after the addition of an advance curve kit from H-O Racing Specialties. Speeds remained constant in the 91 mph range. The car had enough power to blitz the tires, but it was obviously suffering from an obscenely small profile cam and terminally weak valvesprings. Racing a small-block Chevy that's out of breath at 4500 rpm is decidedly distasteful, but it had to stay that way in keeping with our plan.

The car was then delivered to Art Chrisman at BRC. His first move was to check it out on the chassis dyno. The LM-1 engine is factory rated at 175 hp; it dutifully cranked out a maximum of 174 hp at 4000 rpm. With these stock figures faithfully logged, the first major modification was initiated. Mr. Gasket's aluminized BlackJack headers, PN B5032, were installed along with a max-effort dual exhaust system. The stock exhaust system employs a single catalytic converter under the passenger seat, with a "Y" connection splitting exhaust flow to two small mufflers. Chevrolet

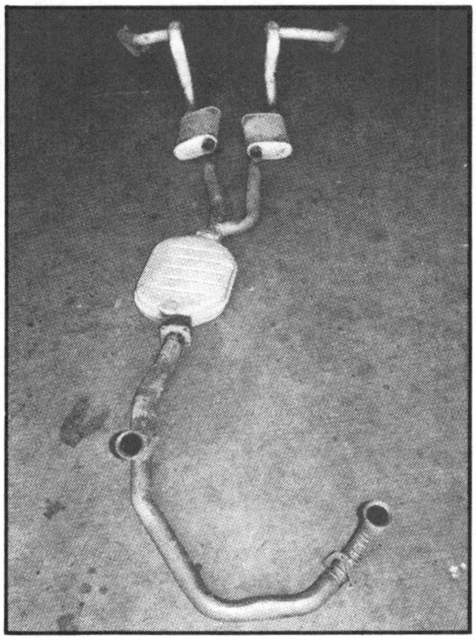

The factory system is obviously restrictive, but as we discovered, the catalytic converter was not the problem. The small diameter single pipe ahead of the converter is the real restriction.

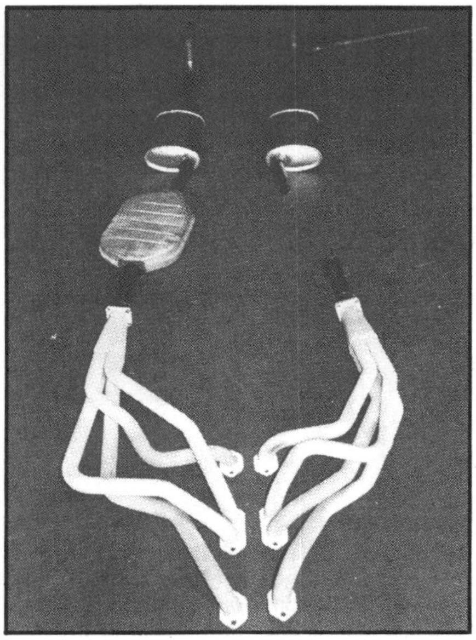

The custom dual exhaust system utilized the existing catalytic converter on the right side. A second converter was added on the left side to ensure that all exhaust gases were properly treated. Dual turbo mufflers were laid in behind the converters.

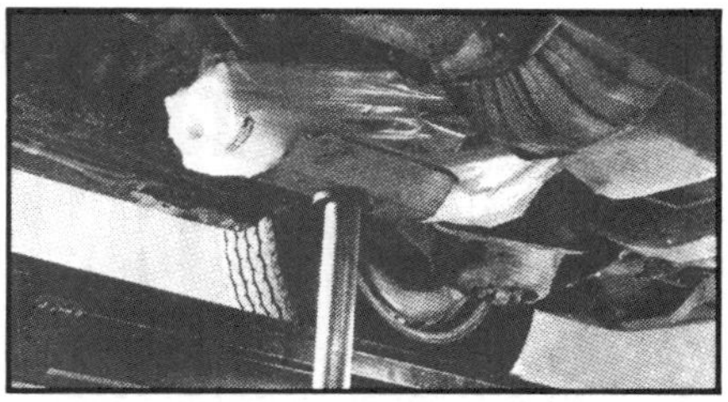

All BlackJack headers come wrapped in a heavy plastic bag. By cutting small holes for the flanges to come through, the headers can be installed with the bag intact to prevent them from getting greasy or chipped.

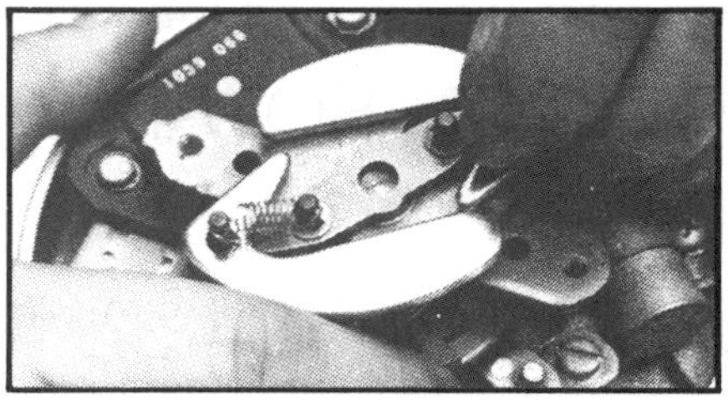

H-O Racing's advance curve kit enabled us to quickly alter the factory settings. Performance picked up when the 36 degrees total advance was all in by 2200 rpm.

engineers found this layout superior to a single exhaust system, even though all the exhaust gases pass through a single converter. We felt it could be improved with a true dual exhaust system—but to avoid eliminating the converter from the system, we used two converters instead, one for each bank of cylinders. Behind them, we connected Mr. Gasket Turbo mufflers, and ran the tailpipes clear to the back bumper.

The result was an unrestricted but quiet exhaust system with a social conscience. Now we all know that removing catalytic converters is against the law, but we also recognize that the temptation to do so is strong. But, as we will see later, the effort and risk may not be worth as much as you think.

Back on the dyno, the car immediately pegged the dial. BRC's chassis dyno only reads to 200 hp, a figure the car easily reached. Tire slippage was not yet evident, and significant gains were recorded at all rpm levels. The value of the exhaust system work was immediately evident: a 26 horsepower improvement was indicated, and the true figure was probably more. This gain was recorded with the dual cats and mufflers connected. The exhaust system was clearly a good one, but it included the assumed restriction of the catalytic converters.

Satisfied with the improvement, Mike Chrisman and Dave Bekaert of BRC undertook the manifold installation. The new Team G Street

CAMARO BOLT-ON

Ram intake manifold was selected for this test, and the installation was straightforward. Particular care was taken to see that the seal between the manifold and cylinder heads on the lower half of the ports was leakproof. Sucking oil out of the lifter valley is a common problem in many manifold installations, and we strived to avoid it. Stock-type beaded gaskets and a quality sealer will normally ensure a good seal. The swap was completed without difficulty, and the stock Quadrajet was reinstalled for evaluation. Before taking the car back to the dyno room, the engine compartment was dressed up with the addition of Mr. Gasket custom valve covers and a chrome air cleaner. These pieces complemented the aluminum-coated headers, and provided just the right amount of

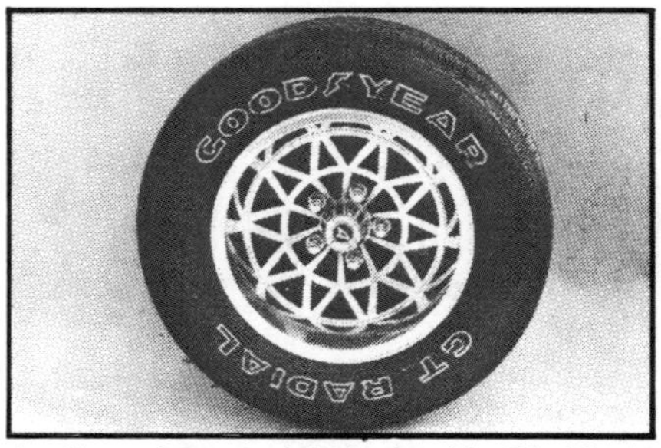

Appliance's new eight-inch wide Dia-Mag's were fitted with Goodyear GR60x15 Custom radials. The combination looked fabulous and performed superbly. The Goodyear tires offered considerably more bite than the factory tires.

BOLT-ON HOW-TO: **INTAKE MANIFOLD**

Installing a new intake manifold on our Z/28 presented an opportunity to point out some of the problems you might encounter when putting one on your own car. Intake manifold manufacturers have recognized the importance of retaining emission control equipment, and have included provisions for mounting the various anti-smog devices. If you're using the stock carburetor, most of the hardware will be reusable on the new manifold. Some items, like throttle cable brackets, may require minor adjustments, but it is generally a direct swap. Before you begin, make sure you have everything you are likely to need for the swap: this includes an intake gasket set, a thermostat gasket, distributor gasket, carburetor base gasket, carburetor mounting with nuts and washers, Teflon tape, thread adhesive, a tube of silicone sealer and an assortment of hand tools.

The first step is to accurately mark the

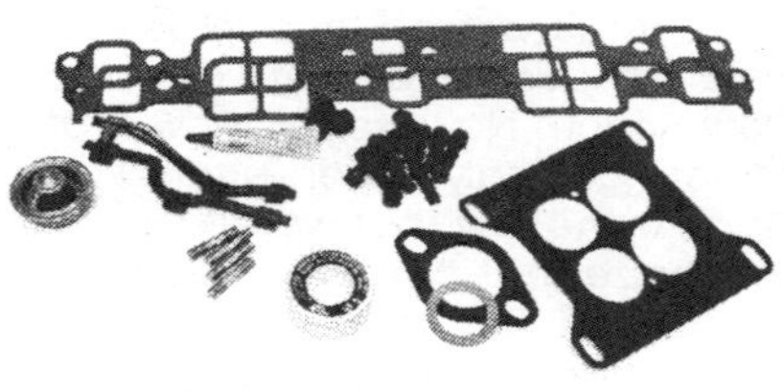

location of all the stock hoses and fittings. Take a Polaroid picture or tag each item as you disconnect it. Drain the radiator to lower the water level in the block below the manifold and cylinder heads. Mark each plug wire and make a sketch of the cylinder numbering. (Chevy engines are numbered 1,3,5,7, front to back, on the driver's side and 2,4,6,8 on the opposite side.) Remove the distributor cap and wires for clear access to the distributor. Note the direction the rotor is pointing, and also the approximate position of the vacuum advance diaphragm. Relate their positions to the face of a clock, with 12 o'clock being straight back toward the firewall. If the rotor is pointing to two o'clock, note this on your sketch; then draw in the approximate position of the vacuum advance assembly. Remove the distributor by pulling firmly upward. As it rises, the rotor will turn counterclockwise

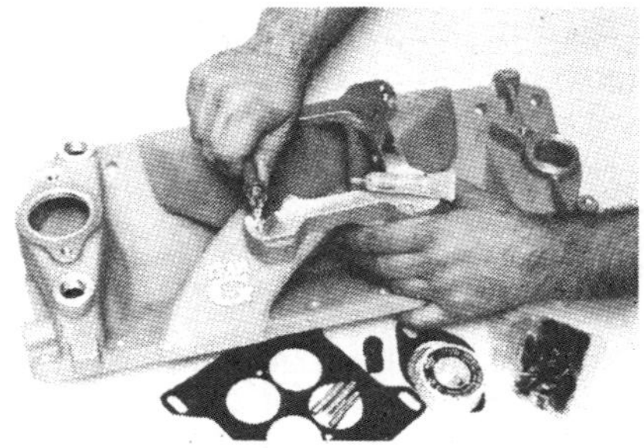

slightly when the distributor gear disengages the cam gear. When reinstalling the distributor, be sure to lead your mark by a similar amount. If the rotor was pointing to two o'clock, start it at approximately one o'clock and watch it rotate as it drops in. If the engine hasn't been cranked, the oil pump drive will engage and you can rotate the vacuum advance assembly to its approximate position to fire the engine.

Once the distributor is out, remove the manifold bolts and carefully pry it loose. Lift it slowly, and watch for debris that might fall into the engine. A common problem in manifold installations is the lack of a good seal between the heads and the manifold. There are three steps to prevent this problem, which usually shows itself by fouling spark plugs with oil sucked out of the lifter valley. First, thoroughly clean the mating surfaces. Place two shop towels down in the lifter valley, and arrange them so they cover

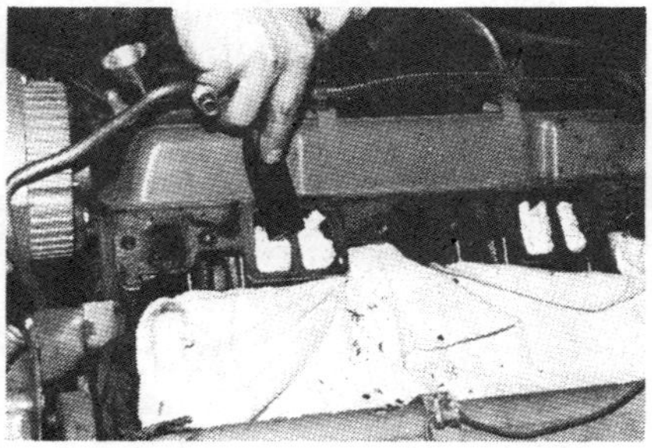

as much as possible. Stuff a crumpled paper towel in each intake port to prevent any debris from falling in. Then, use a good scraper tool (the kind that holds a single-edge razor blade) to remove every trace of the old gaskets. Work carefully, and allow the bits and pieces to fall on the shop towels. Remove the towels carefully, and inspect for any bits that may have escaped your precautions. Use a clean towel and some lacquer thinner to wipe down the surfaces.

Circle the intake and water passages with a thin bead of silicone sealer and set the new gaskets in place. Apply more sealer to the upper surfaces and put an extra dab at each corner where the gaskets join. Lower the new manifold into place, being careful not to disturb the gaskets. Try not to move it around any more than is necessary to start the bolts. Snug down the bolts and then check the end gaskets before the bolts are tightened. Make sure the end gaskets

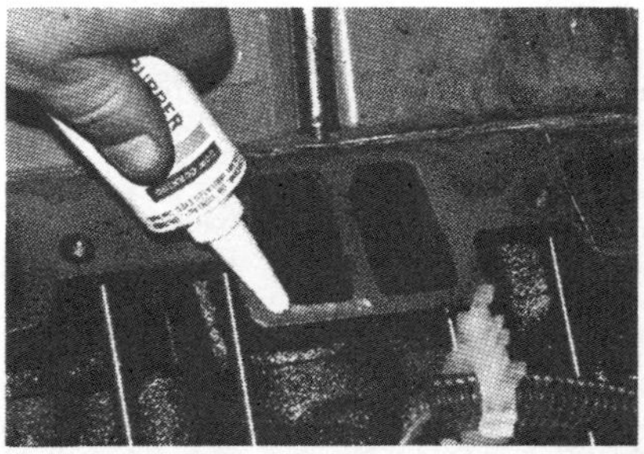

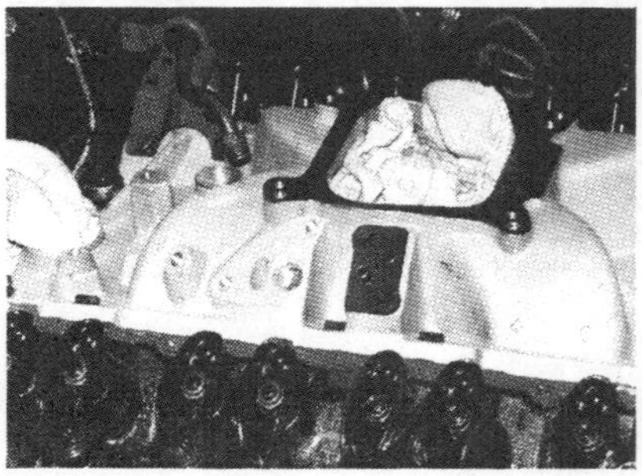

are not sliding out. The bolts should be torqued to 35 foot-pounds, starting with the center bolts and working in a circular pattern toward the ends of the manifold. Using small flat washers under each bolt will help spread the load on the manifold flanges. Install the distributor as previously described and hook up the hardware. Refill the cooling water. After starting the engine, set the timing and check for fuel or water leaks. If there are no visible problems, hit the streets. Periodically check the bolts and the end gasket seal to be sure there are no leaks.

CAMARO BOLT-ON

dazzle to elicit a satisfactory "Wow!" from service station attendants offering to check the oil.

Dyno performance was suddenly down. We had expected this, but we weren't fully prepared to deal with it. The new manifold was a single plane design with a much larger plenum area than the factory unit. Thus the fuel mixture was painfully lean. Like most people, we had absolutely no Quadrajet metering parts on hand, so we resorted to piracy. A thorough search of the grounds turned up a Quadrajet with richer secondary rods, and a rod hanger that lifted the rods .020-inch higher than the stock hanger. Power increased, but the car ran rich and seemed sluggish. The only remaining alternative was a return to the stock rod hanger to lower the rods back into the orifices a bit farther. This admittedly shadetree approach seemed to work, as power shot way up and the motor ran cleanly. Plug readings showed a normal coloring, and it was decided that we were as close as we could get without scouring area parts stores for more metering components. Our final configuration, utilizing an "L" hanger with "DH" metering rods, ran very smoothly on the street. The open plenum manifold did not significantly detract from low end response, and the overall effect was intoxicating. The Camaro had all the right attributes: it sounded racy, and felt really strong on the street.

It was time to head back to the dragstrip, but first we needed to trick up the car's appearance a bit. Appliance's new Dia-Mag was too pretty to resist; we ringed a set of 15x8 wheels with Goodyear G60x15 Custom GT Radials. Our first impression was that these tires had a little more bite than the stock tires, and we were soon glad they did. The first pass netted a tire-boiling 15.01 at 95.94 mph. The tires were better, but if you simply put the hammer down, you could burn them unmercifully through the top of second gear. A little practice enabled us to launch the car successfully, and it quickened considerably. Competition Editor Asher immediately issued his customary challenge for Low E.T., and we all smiled and silently cursed him because we knew he would win. (We're certain he cheats somehow, but we can't seem to expose him.) Art Chrisman joined us for the day at Orange County International Raceway, and we took turns regaining our youth, reaffirming our masculinity and generally mistreating a thoroughly cooperative car.

We did, however, make an effort to arrive at valid figures from which to judge the car's improvement. It ran best with six degrees initial timing; nine degrees produced detonation and an audible complaint from the motor. Shift points were up slightly, but the Z/28 stopped pulling at about 4800 rpm in first and second gear, although it pulled strongly to about 5500 rpm in third gear. We eventually established a best pass of 14.80 through the mufflers. That constituted an e.t. reduction of seven tenths, a performance improvement anyone can feel.

Reasonably content with our findings, we promptly opened the headers in search of another gain. This got us a modest improvement as the car recorded a 14.71, but it never went that quick again. It always bettered the 14.80 mark, but the average was only four hundreths of a second. Art informed us that much of his dyno work on other cars indicated the catalytic converters did not harm performance. We were pleasantly surprised. While the good exhaust system had obviously helped, the converters were not detrimental to the overall performance. The car might have responded to a little richer mixture, but the plug readings remained unchanged when we uncapped the headers.

Our well-worn office power/speed calculator reveals that it takes a 46-horsepower increase to produce our Z/28's performance improvement. That's quite a gain to make without going into the engine's innards. In keeping with the strictly street spirit of the thing, slicks were not tried, although they would have provided an enormous increase. Our "best buys" plan called for traction bars, and the car could certainly have used them.

There's no question that a coordinated package of bolt-on speed equipment can still be a magic elixir, even for late-model cars. Legitimate fourteen-second times produced by the addition of headers, dual exhausts and a manifold are very respectable. Bolt on some slicks at the bracket races and it will run hard. A switch to a hotter cam and maybe a carb change could transform this smog motor into a real screamer—and on unleaded gas at that. Not all the new cars are dogs, after all. Just think of them as puppies that need a little grooming.

Cutting Corners

Continued from page 55

View, Pacific Grove, CA 93940).

It is one of the immutable laws of high-speed handling that a stiff front bar demands a healthy rear bar to keep the car balanced. Adams and company had to start from scratch to bend up a 1-inch diameter rear bar, since none of the late-model designs on hand would clear the early Camaro's underpinnings. Fitting the handbuilt bar around the differential housing and muffler proved relatively easy, but attaching the bar to the body required some deep thinking. After innumerable trial fitups, late-model Firebird attaching links and a pair of steel brackets welded to the rear subframes produced the desired geometry. Once Adams was satisfied with the bars' fit, they were removed and sent off to be heat-treated.

Road testing the reinstalled bars was a revelation. A run through Echo Valley provided an introduction to the Camaro's new personality. Its wishy-washy response to steering input was banished, replaced by a taut, instantaneous change of direction with every flick of the steering wheel. And the body roll which had threatened to pitch the driver out the side window had vanished.

The stopwatch confirmed our subjective impressions. Returning to our favorite off-ramp, the Camaro was two-tenths of a second quicker and 2½ mph faster through the course. But in spite of the better times, Adams wasn't totally satisfied with the package. "The car is much closer to neutral handling now," our bearded test pilot reported. "The car isn't as 'comfortable' now in turning—it demands more of the driver's attention to keep it on line. Before we added the bars, you just kept cranking in more steering. Now that we've boosted the roll stiffness, the effects of deflection in the front A-arm bushings and spindle movement are becoming apparent. Bars alone don't make for the best handling car. We've come a long way, but we're not nearly as far as we can go."

And with that, Herb's pad was once again filling up with lists of things to do. "Shocks," "Front End Structure," "Reduce Anti-Squat," "Solid Bushings," and "Trans Am Steering Gear" were on the agenda of things to come which I read over his shoulder. Together we mapped out our next travels down the road to high-performance handling. After all, one good turn deserves another.

BOLT-ONS

B&M SHIFTER INSTALLATION

By Marlan Davis

One of the leaders in the production of attractive performance shifters for automatic transmissions is B&M Automotive Products (9152 Independence Ave., Chatsworth, CA 91311, 213/882-6422). One of their newer products is the QuickSilver—a dual-action shifter incorporating both conventional and ratchet operation in one mechanism. You can select any gear desired from Park to Drive, merely by lifting the handle and shifting to the desired detent, or click off crisp, sure upshifts or downshifts without any chance of missing a gear when the ratcheting feature is engaged. The shifter will work with automatic, manual/automatic or full-manual valve bodies utilizing standard or reversed shift patterns. While a universal kit has been out for about a year, B&M recently introduced a special version (part No. 80688) designed to bolt-in in place of stock '73-'80 Camaro and '70-'80 Firebird automatic shifters, using the factory console opening and stock shifter cable. This new kit forms the basis of this feature; it can be installed with a ¼-inch-drive socket set or conventional wrenches, plus a screwdriver, pliers, metal cutters, drill and 5/32 Allen wrench. **HR**

PHOTOGRAPHY: MARLAN DAVIS & AL KIRSCHENBAUM

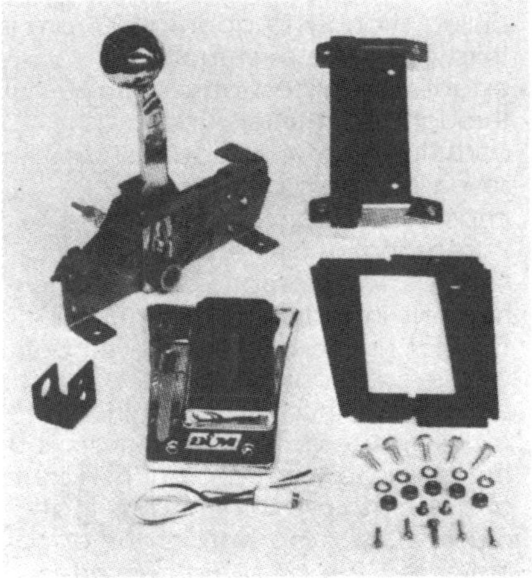

(1) *Kit includes shifter mechanism with knob, mounting bracket, console plate, attachment hardware, shifter detent light bulb, chrome cover, indicator window and cable adapter. Not shown is the complete illustrated six-page instruction brochure.*

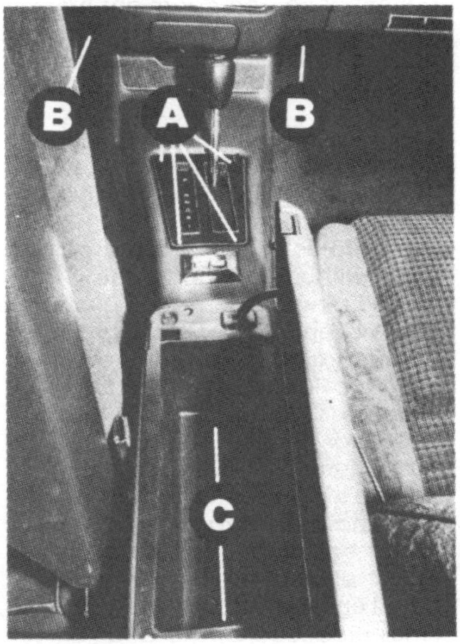

(2) *After disconnecting battery ground cable and placing shifter in neutral, remove four Allen screws located on shifter cover plate (A), and two screws fastening console front to dash (B). Next remove console storage compartment by unscrewing two screws in floor (C) and lifting it out. If car is equipped with electric windows, reach underneath console through space left by removed storage compartment and disconnect power plug.*

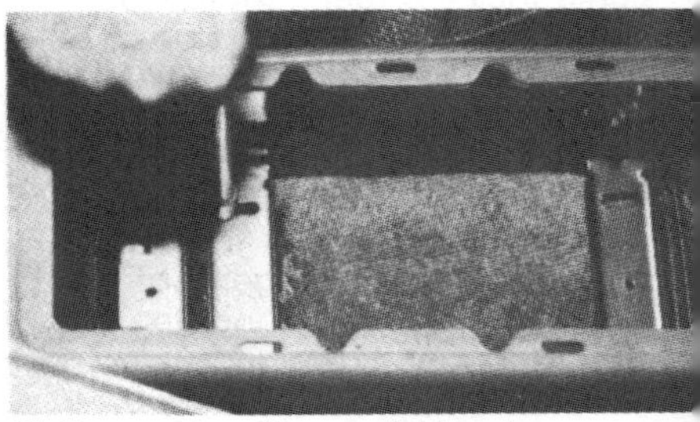

(3) *Remove two (late models) or four (early versions) screws on console-retaining bracket(s).*

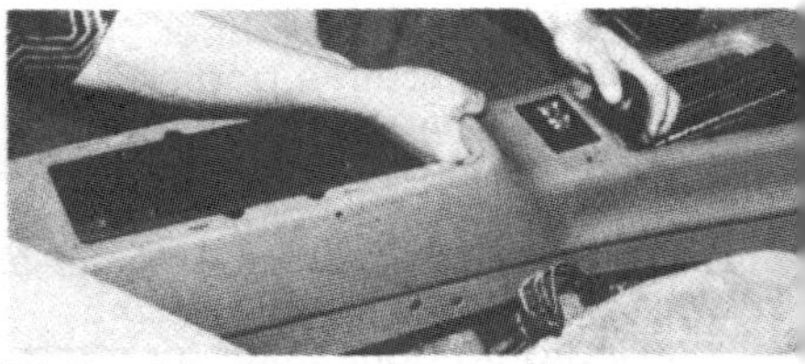

(4) *Carefully lift console out while allowing shifter cover plate to slide down through console hole.*

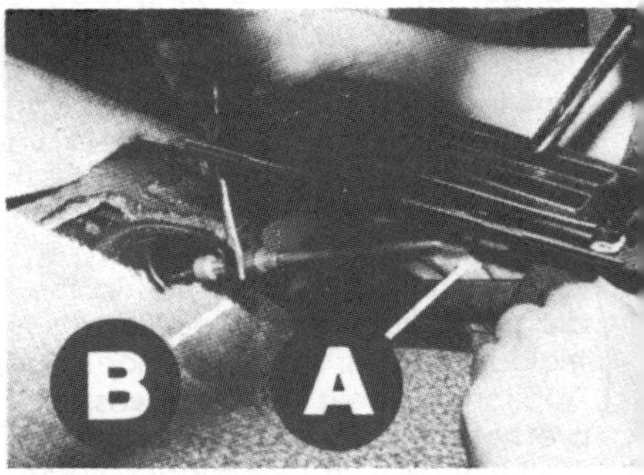

(5) *With stock shifter mechanism now exposed, the two control cable clips can be removed by prying with a screwdriver. They are located at the cable end (A) and at the front where cable joins shifter mechanism (B).*

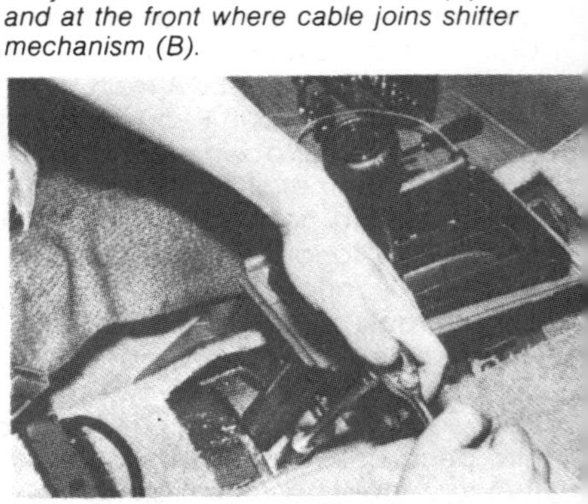

(6) *Now remove four 5/16-inch shifter mounting bolts and lift shifter up and out.*

B&M SHIFTER INSTALLATION

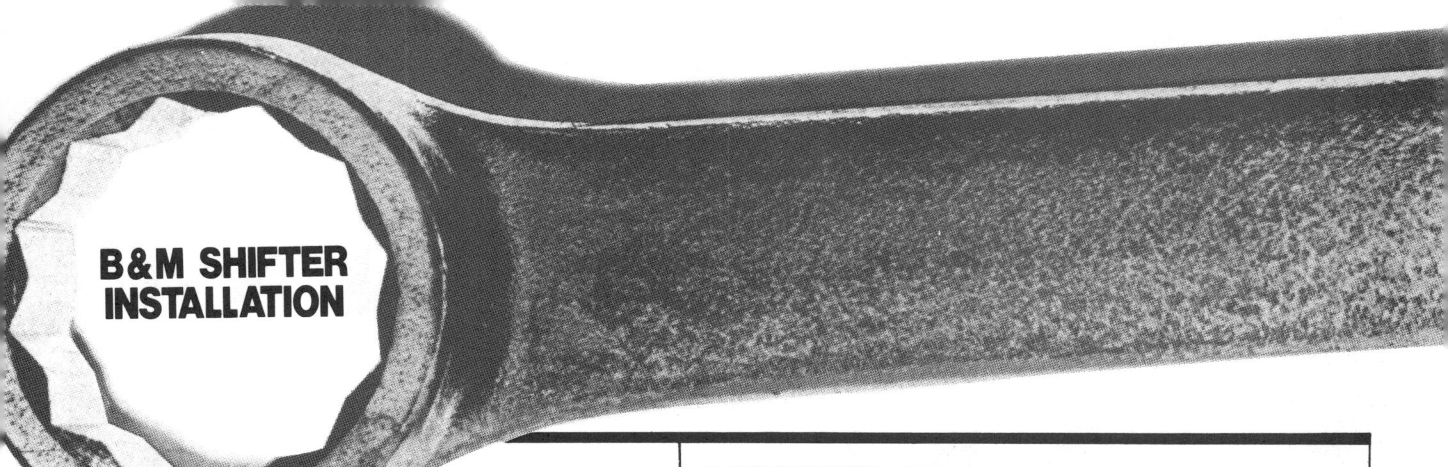

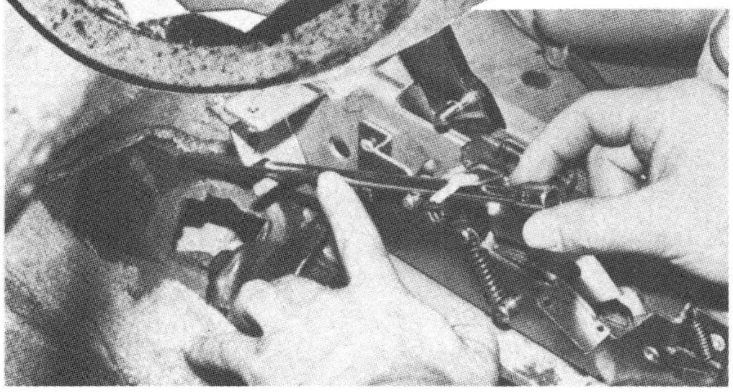

(7) *Remove cable boot cover plate screws and plate. Since cable bolts to new shifter at a slightly different angle than stock, install new shifter mounting plate, using four stock 5/16 bolts; then "dummy" the new shifter into position. Use a long ratchet extension or screwdriver as shown to simulate new cable hookup. Stock hole must be enlarged to meet point where tool contacts floor.*

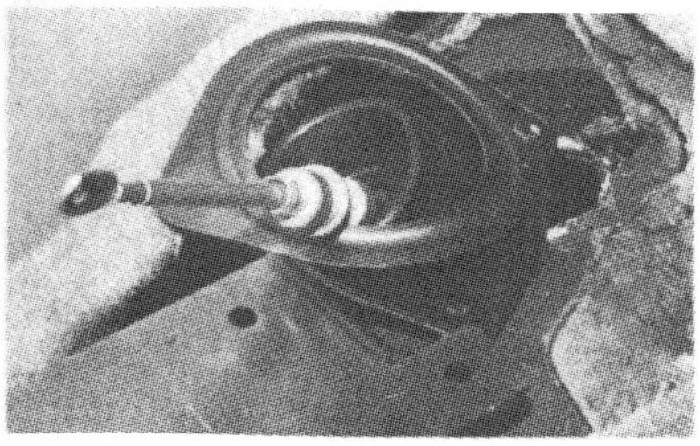

(8) *With shifter mechanism out of the way, use a hacksaw or sheetmetal cutters to enlarge cable hole. Usually, enlarged portion should be about 1 inch forward and to right of existing hole. File off any sharp edges after completing cut.*

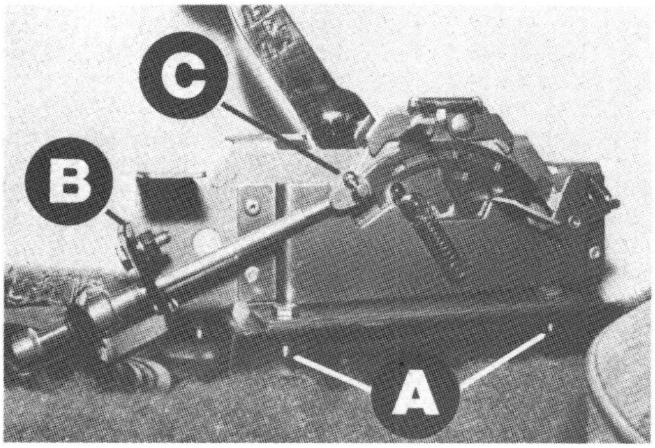

(9) *Now permanently bolt shifter to mounting bracket, using four ¼-inch bolts, lock washers and nuts (A), verifying adequate cable clearance. Align cable adapter (B) with U-shaped notch on shifter and attach it with ¼-inch bolt, lock washer and nut. It can go on either side of notch, depending on how long cable must be to slip its end link eyelet over shifter cable pin (C). Sometimes eyelet may have to be deburred to fit. Finally, reinstall stock cable clip and relocate cable boot and cover plate, using a No. 32 bit to drill new holes in floor for stock screws.*

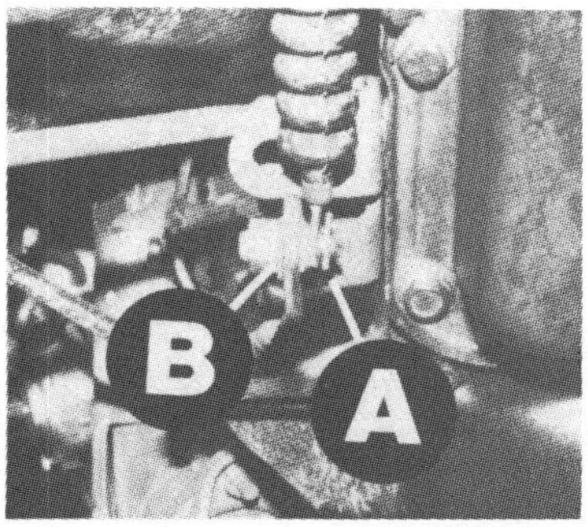

(10) *To check for correct shift linkage adjustment, put shifter in first gear and raise vehicle to obtain access to trans cable mounting bracket. Remove cable clip (A) and pull eyelet off. If it can't be reinstalled without binding, adjust by loosening nut (B) and moving cable pin until correct fit is obtained; then tighten nut and reinstall clip.*

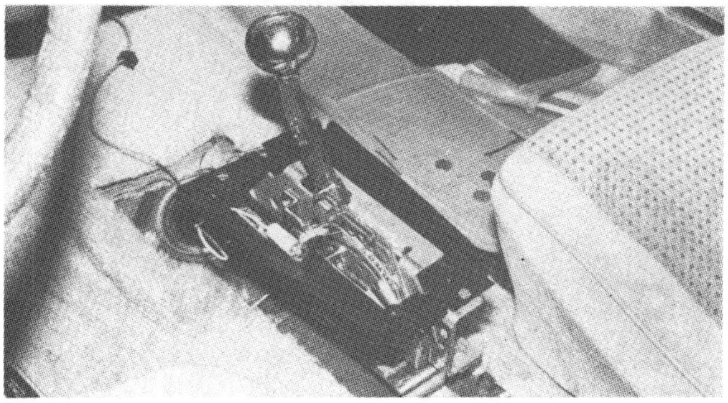

(11) *Using two No. 10 self-tapping screws, install new console plate with wide end forward. Mount detent illumination light with another No. 10 screw. Connect one wire to ground, the other (if a Firebird) to stock power wire after cutting off old socket. Camaros must string a new wire back to the instrument panel light switch.*

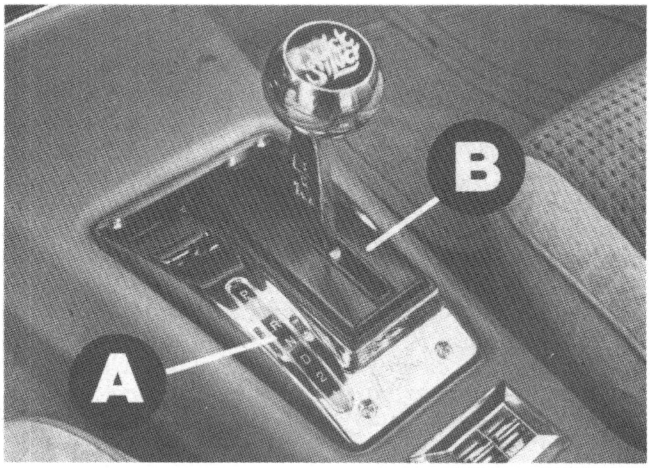

(12) *Reversing removal procedure described in Steps 2-5, reinstall stock console over shifter. Press indicator window (A) and black boot (B) into place on chrome cover; then slip cover over handle (after removing knob) and fasten it to console, using four No. 6 self-tapping screws. Reinstall knob, hook up battery cable—and you're ready to get your act into gear.*

CHEVERRA A POOR MAN'S PORSCHE

HOW TO BUILD A KILLER CAMARO IN 52 HOURS

By Dave Wallace

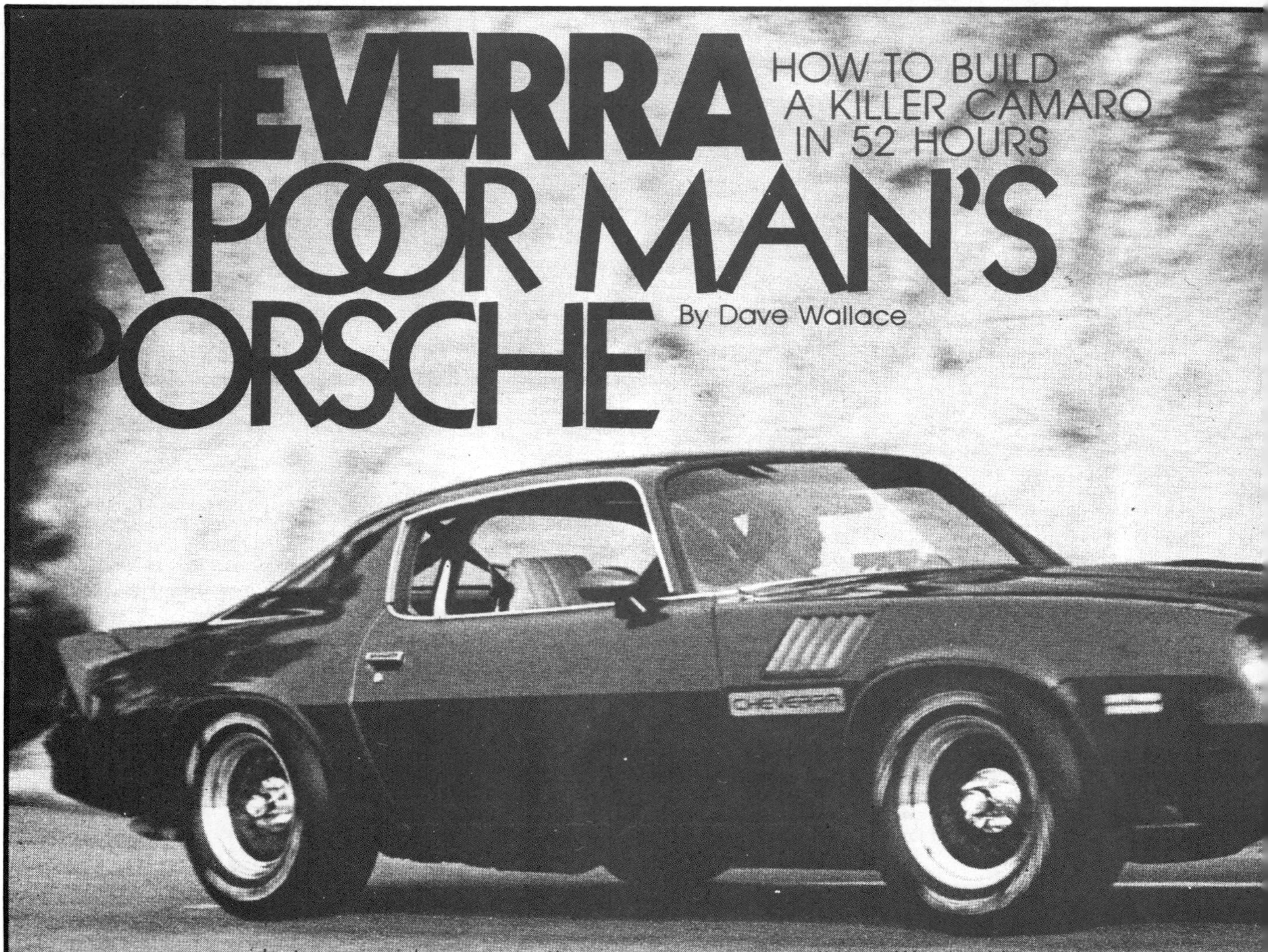

Company presidents are quick to promise miracles when it comes to the "easy," "no-hassle," "do-it-yourself" installation of their particular products, right? For example Herb Adams, the former Pontiac engineer and current president of Herb Adams VSE (Very Special Equipment), made the comment a few months ago that two guys could turn a bone-stock Firebird or Camaro into one of his company's mean-street Fire-Am/Cheverra road-racing rigs in 26 hours flat. Having spent two glorious weeks with a shop-built Fire-Am (see "Cutting Corners," March '78 HOT ROD), we were skeptical that such a drastic change in handling and braking could be accomplished in such a short amount of time, and we told him so. Adams responded with an open invitation to make him prove the point, and we were only too happy to oblige.

HOT ROD accepted an invitation to watch two guys bolt one of these babies together, without special equipment or modifications, in somebody's garage at home. In this case, the garage was well-equipped, to be sure, and the same two-man crew had some experience at turning Firebirds into VSE Fire-Ams. But this time the car was a Chevrolet Camaro, their very first non-Firebird, and this time the tools would all be your basic "backyard" items: hand wrenches, jackstands, a small drill motor, blocks of wood, crowbars, hammers—nothing exotic. The only "outside" work required at all, according to Adams, would be the final alignment of the front wheels and suspension.

But he was wrong about that: Chief mechanic Bill Fisher went so far as to execute an unscheduled garage-floor alignment job (using nothing fancier than a carpenter's level, two 10-foot lengths of channel iron and one ball of string), just to emphasize the do-it-yourself nature of the Fire-Am/Cheverra conversion process. Right after that he handed me back the keys to a 1978 Camaro Z28 that I had personally driven into that garage exactly 39 man-hours earlier (two men working 19.5 hours each), as advertised.

We should note that the front coil springs had both been removed, shortened (by ¾-coil), and reinstalled by the time I arrived to road-test the stock Z-car, so the entire operation came a little closer to the company's own 52-hour conversion estimate for a Stage I Cheverra. (Two more preinstalled VSE items were the chrome rollbar and trick steering wheel, but neither is considered a Stage I requirement.) But a Cheverra doesn't have to be built, or paid for, all at once; Herb Adams designed the whole program in stages, with due consideration for the everyday Camaro owner who wants to build his Cheverra in stages, weekend by weekend, and keep driving his car in the meantime.

Anybody in a particular hurry, on the other hand, to get out and start mixing it up with the Porsches can save time by turning the whole job over to one of VSE's four authorized installation centers (in California, Maryland, Michigan and New York), and he'll be back on the road in a couple of days. Another deal in the works will possibly bring

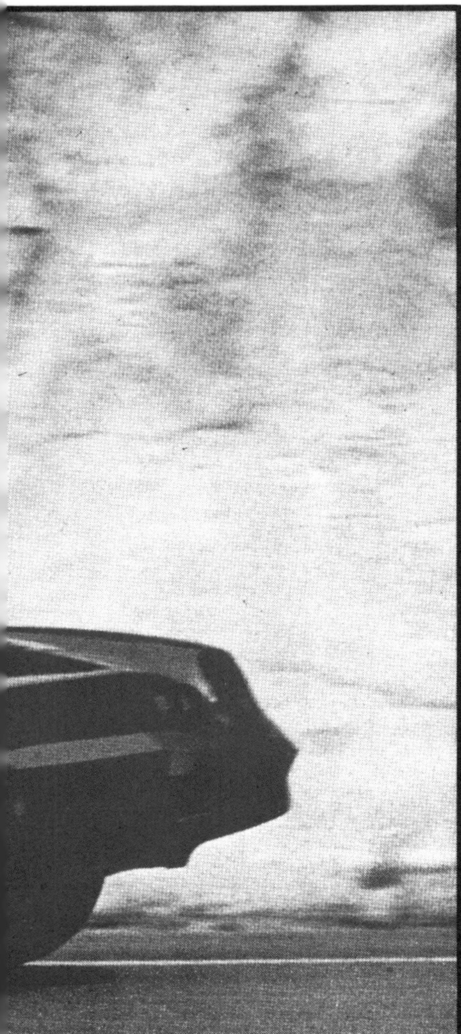

ınning, rolling, complete Stage I :heverras right into the showrooms f Chevy dealers by the end of the ummer. (Details on either plan are vailable upon request.)

But most of these cars will con- nue to fall together over a series f weekends, going together in the riveways and garages of people issatisfied with the handling and/ r braking characteristics of their ıte-model ('70½-up) Camaros. A ıll-color "cutaway" Cheverra post- r, printed complete with parts de- criptions and prices, is now avail- ble for $2 from Herb Adams VSE, 25 Ocean View, Pacific Grove, CA 3950, telephone (408) 372-2767. etter yet, send $5 for the Chever- a poster *plus* a copy of the com- any's master catalog, which fea- ıres detailed Cheverra and Fire-Am ıstallation instructions, valuable andling information and a full run- own on the ever-expanding Herb dams product line. Meanwhile, on't try to stay too close to nother Camaro with "Cheverra" ecals—and that goes for you orsche drivers too! **HR**

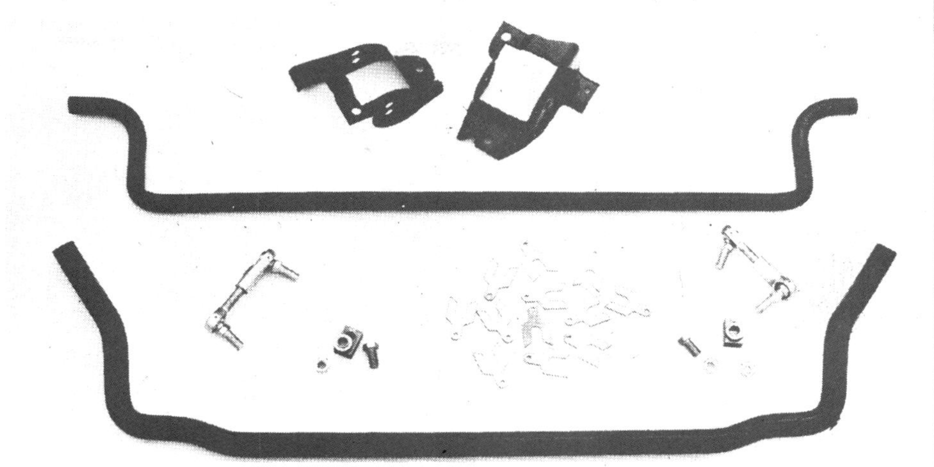

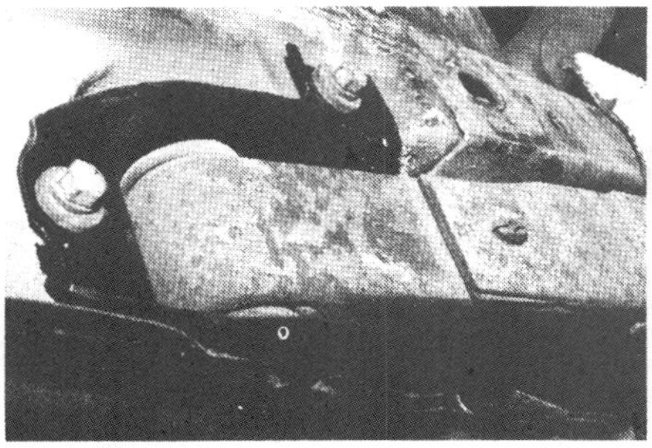

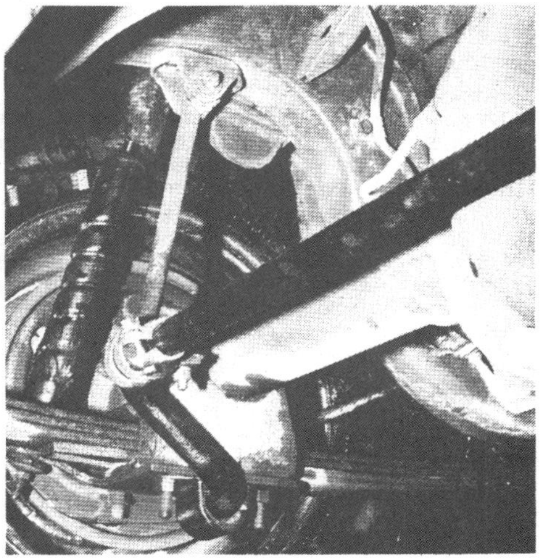

Recommended starting point in the step-by-step transition from Chevy to VSE Cheverra is the hardware shown here: replacement leaf-spring brackets and big anti-sway bars. VSE "wrench" Jeff Odell handled the back half of the project, a true bolt-on operation, in less than two hours. After wedging apart the leaf spring and rear subframe with a common two-by-four, Jeff swapped in the VSE spring bracket. Simple device relocates the front eye of the Hochkiss-type spring ¾-inch higher than stock, effectively lowering the rear of the car. In reducing "roll understeer," modified GM bracket increases "anti-squat" of stock suspension. Result is improved traction for cornering and dead-stop launching. Because Z28-model Camaros come equipped with a rear sway bar, mounting hardware and brackets were already in place for the larger (1-inch-diameter) VSE replacement. Proper mounting holes are predrilled in all Camaros, and VSE provides the rest.

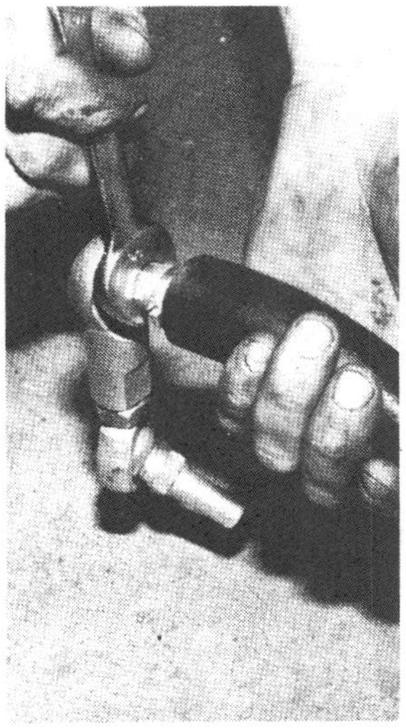

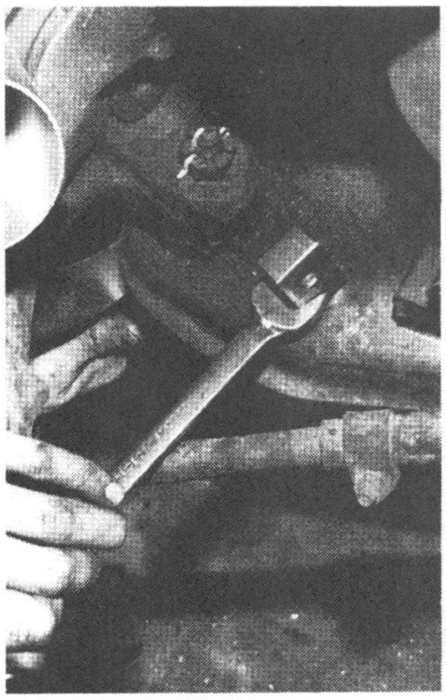

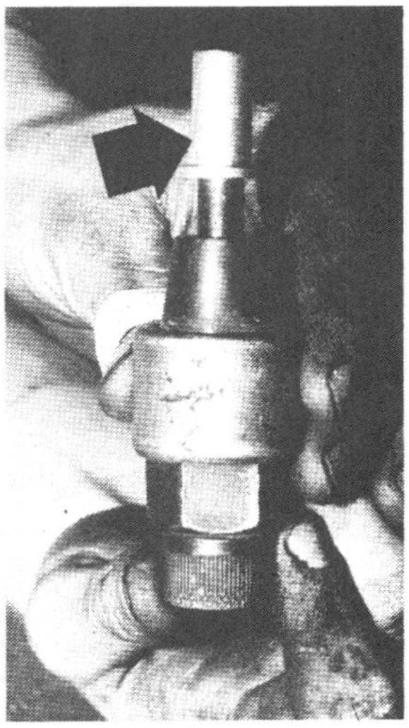

Front sway bar kit does away with rubber-mounted links connecting a stock Camaro's stabilizer bar and control arms. Rubber allows for "lost motion" in front suspension, so VSE provides replacement links made from steel. The kit shortens "swing-arm length" of the sway bar 10 percent and allows adjustment, so Cheverra owner can fine-tune under- or oversteer. However, original Camaro frame holes must be drilled out slightly (to ½-inch) to accommodate the larger VSE mounting brackets, then fitted with replacement "rivnuts" (arrow). VSE and some Pontiac dealers own a trick rivnut-collapsing tool, as shown, but rivnuts can also be tack-welded in place. (No other welding is required in a Stage I Cheverra conversion.)

(photos continued overleaf)

A POOR MAN'S PORSCHE

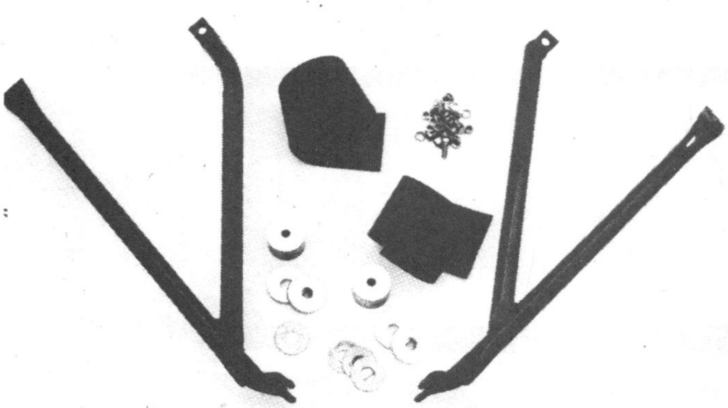

All the preceding pieces combine to produce a markedly different Camaro, especially when installed in conjunction with the recommended VSE battery kit and wheel/tire combination (see "boxes"). However, what really puts the Cheverra into road-racing territory is the VSE Front-End Lowering and Structure Kit, which consists primarily of four subframe-to-body spacers and a pair of subframe-to-firewall struts. Installation of the solid spacers effectively lowers the car around its chassis, while eliminating the deflection of the stock rubber body mounts. VSE's frame struts connect the upper control arm frame mounts to the firewall, where as much as ¾-inch of deflection occurs in a stock Camaro.

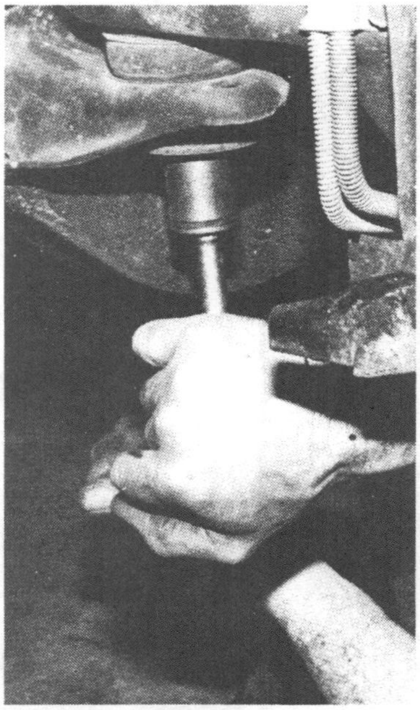

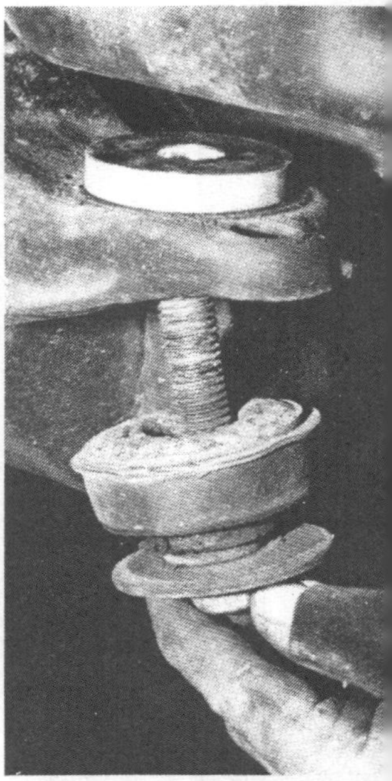

Production Camaros are built with rubber biscuits between the front subframe, body and radiator baffle (left). VSE removes the mounting bolt and rubber mount, both of which are retained and reinstalled (right), along with a solid spacer. Spacers of varying thickness are supplied with every kit.

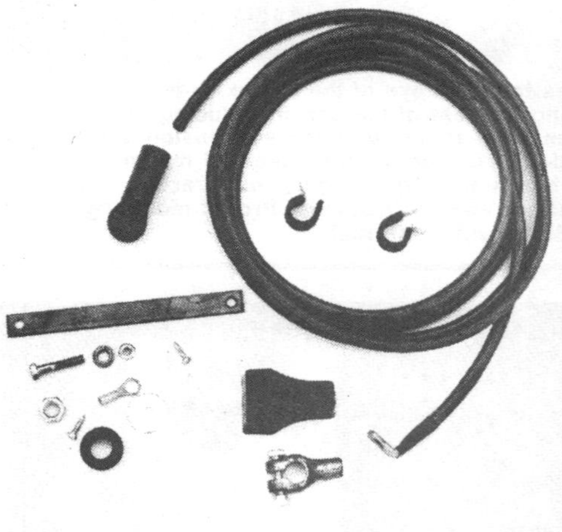

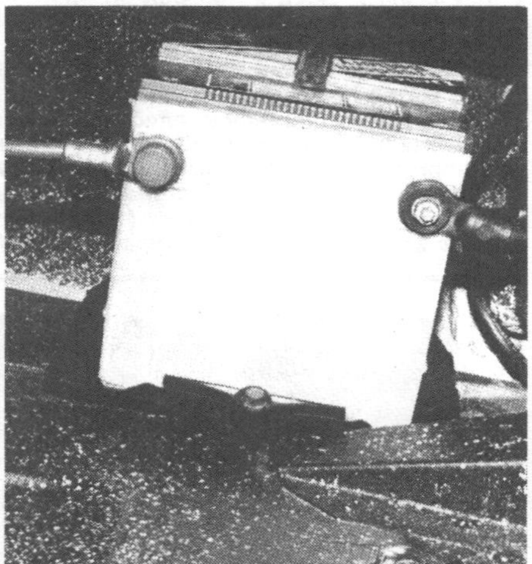

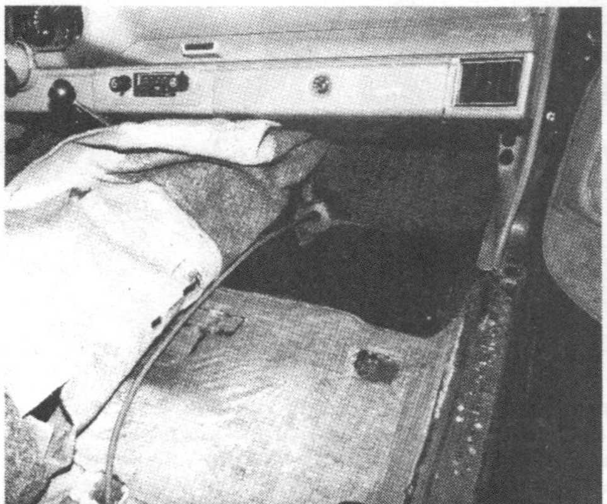

Any Camaro/Firebird can benefit from VSE's battery relocation kit, which makes use of the stock GM mounting tray. This simple operation decreases front-end weight on the nose-heavy GM F-body cars by one full percent, which Herb Adams equates to 10 inches of engine setback! Benefits include reduced understeer, improved braking and better traction. (Spillfree, high-output batteries like the Z-car's Delco Freedom unit are recommended for trunk installations.)

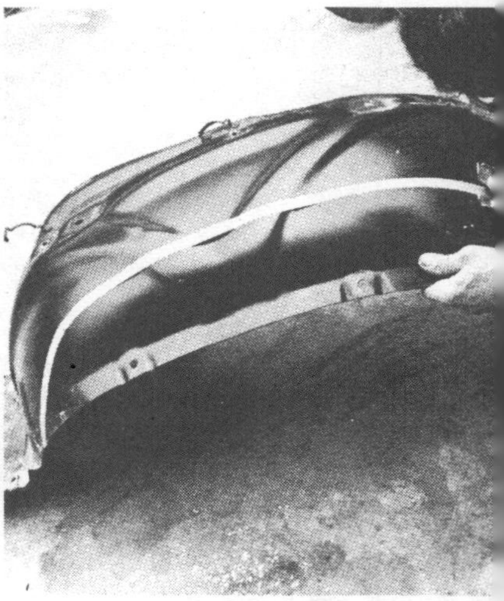

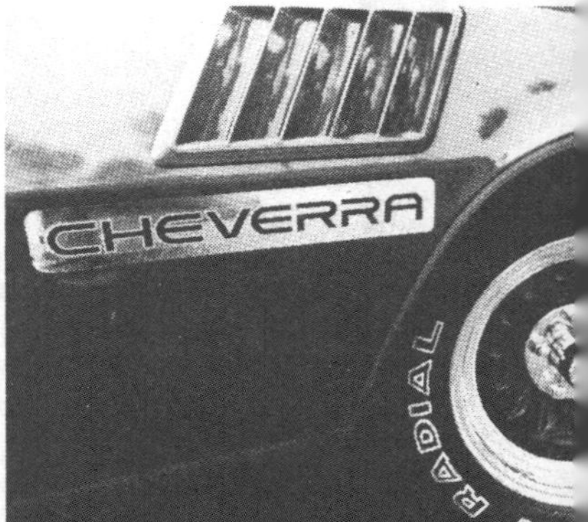

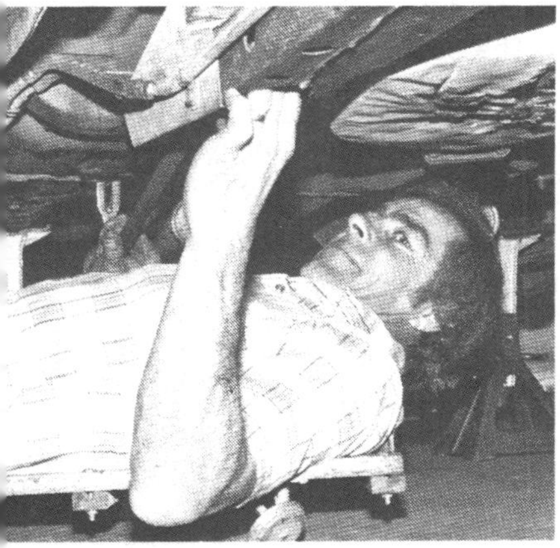

ıt the rear of the subframe is a similar bolt ınd rubber body mount. After removal of ıe Grade 5 mounting bolt, VSE's Bill 'isher pries the frame and body apart with crowbar, then fishes out the rubber ıount. A solid VSE spacer is installed in :s place. Once again the stock rubber ›iscuit is retained and reinstalled with the ame bolt, this time beneath the subframe. ‹ rubber strip is provided to isolate the ›ody from the subframe.

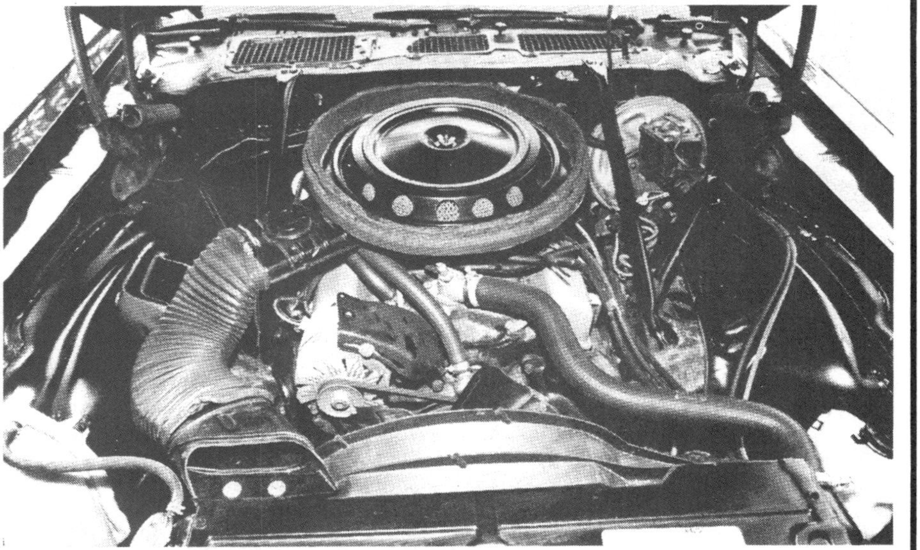

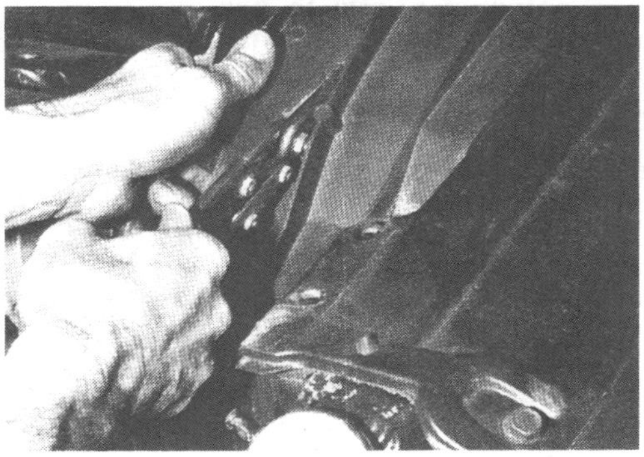

The new, closer relationship between body and frame is reflected in the rubber fan shroud, which must be trimmed as shown to clear the fan blades. Installation of the passenger-side strut requires relocation of the coolant recovery tank to the area formerly occupied by the battery.

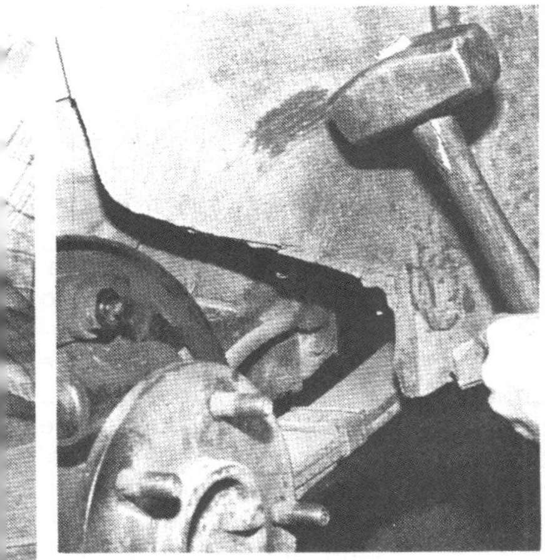

VSE sells a special 15x8-inch, zero-offset steel wheel and recommends HR60 Goodyear rubber, but first you've got to make room. Catalog provides guidelines for trimming the front wheel tubs, as demonstrated here by Bill Fisher, and snipping the outer fender lip. (Tubs receive three cuts each for frame clearance, tire clearance and brake ducts.) Late-model Camaros also require some heavy blows to flatten out this section of the rear fenderwell, but no cutting; fender lip can be folded over with channel locks for outside clearance.

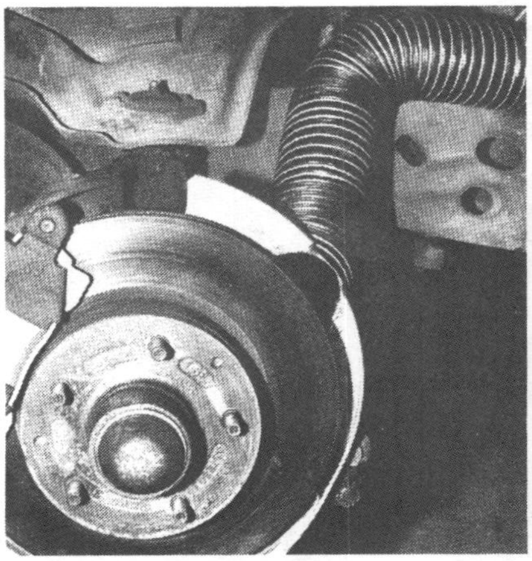

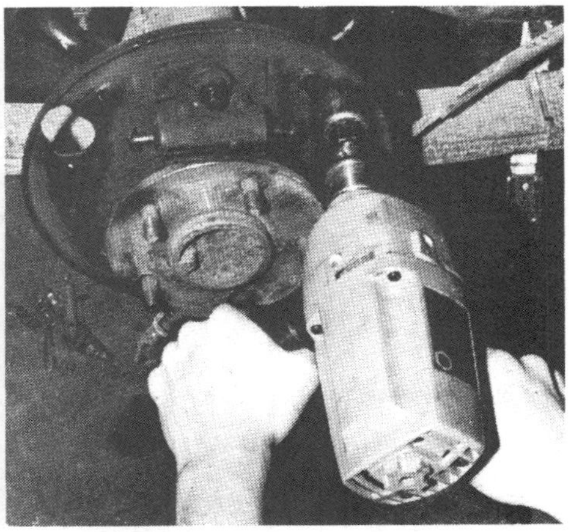

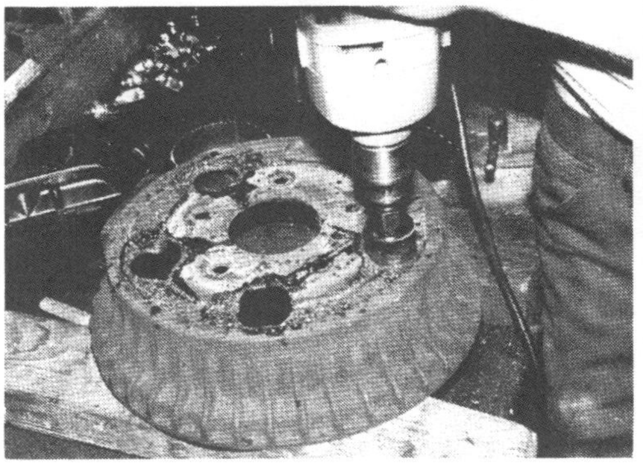

VSE Brake Cooling Kit brings ducted backing plates for the front discs, a section of flexible ducting and instructions for properly ventilating stock rear drum webs and backing plates, as demonstrated by Jeff Odell. Full metallic linings are also available for use with power-assisted brake systems. HR

TWO-WAY STREET

JOHN BUTTERA AND EMMETT GLASGOW BUILD A PAIR OF CAMAROS

By Gray Baskerville

In the inimitable vernacular of Lil' John Buttera, car constructor supreme, "If a vehicle is more than simple transportation, then this is the way to do it." As you might have known, Lil's way of doing it is the street rodder's approach to massaging a machine, which is to take a late-model completely apart, detail it to the last weld, then put it back together with tender loving care.

On the other hand is the route taken by Emmett Glasgow of Street Customs Limited. His way of doing it is radically different. "I want to show what can be done to a late-model by using 'bolt-on' components augmented with minor bodywork."

But there is still a certain amount of agreement here. Both men stress what we call easy parts availability, or for want of a better word, still being able to get all the "stuff." Stuff in this case includes engines, suspension components, driveline assemblies, sheetmetal parts, rubber goodies and trim elements. The packages are as close as your local junkyard, police impound areas, automotive auctions, parts houses, dealerships or your telephone. In other words, you don't need to depend on swap meets, garage sales or other potential ripoffs to give your late-model a shave and a haircut.

Obviously, our intentions are to present two different but valid ways to skin a Camaro. We realize that not all street machiners have the financial wherewithal to buy and then personalize a new or nearly new car. By the same token, not all machiners have the time, patience or energy to indulge in a lengthy ground-up restoration. Consequently, we'll take on both approaches and let you be the judge of which way best suits your individual style. But remember one thing: "Make the cops smile and wave, not frown," counsels Buttera. "You're fooling around with an endangered species, and the more you mess up the more restrictions will be placed on the rest of us."

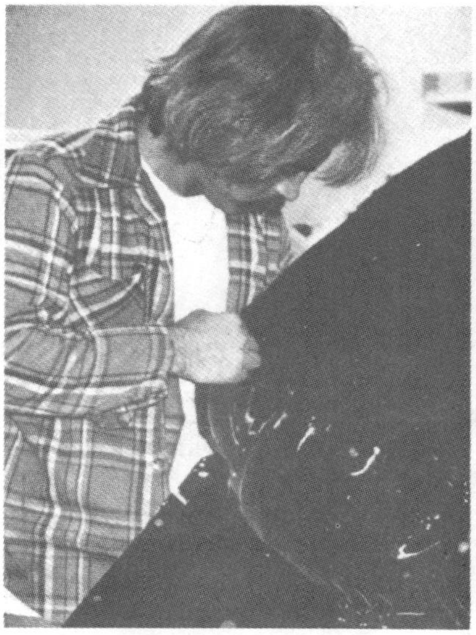

"Combine all the high-performance elements of a wrecked Z28 with those of a cherry '69 coupe and buy whatever else you need from a Chevy dealer, and you can end up with something that will sell for $7500." So sayeth Lil' John Buttera.

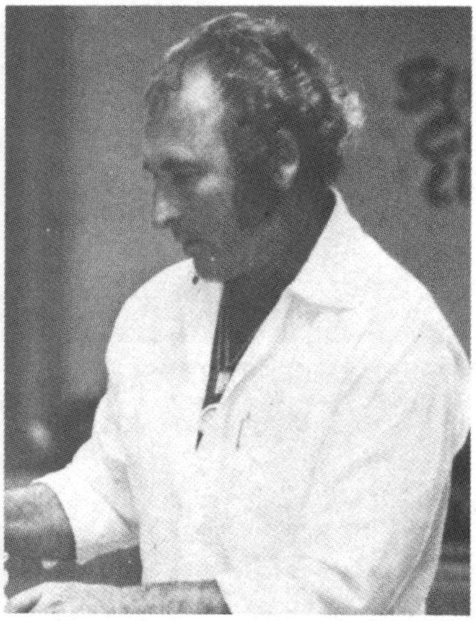

"Sunshine Too is a basic late-model Camaro that shows what can be done to a '78-type Z28 by using bolt-ons and minor bodywork. In fact, my first Sunshine project was one of the most popular series that Lee Kelley ever did." So sayeth Emmett Glasgow.

ROD ON: JOHN BUTTERA

"The best way to build a Camaro," offers John Buttera, "is to find a cherry that needs paint." Lil' did, but his son, much to Buttera's chagrin, managed to wrap the one-owner '69 Z28 around a telephone pole. Like presto-foldo, we got ourselves a story. "Hey, Baskerville," came the dulcet voice from Los Alamitos, California, "would you be interested in showing your readers how to rebuild a Camaro the right way?"

"That depends," came the reply, "on how you propose to do it the right way."

"No sweat," the voice continued. "Come down to my garage and I'll show you."

You can imagine what greeted our eyes as we entered into the cloistered confines of Lil's shop. It was just like building a rod. Parts and subassemblies were scattered everywhere, and the body was up at some guy's house getting painted. "The ideal program," he reiterated, "would be to nail a cream puff, but my kid can't afford to do that now. So my idea is to start with the wrecked '69 Z28, salvage all the high-performance parts, then bolt them onto a clean '69 coupe. What's more, the chances of finding musclecar option parts for a Camaro are still excellent."

Being a hard-core roadster lover, and of the opinion that hot rods don't have fenders or tops, I had to ask Lil' John the reason for building a late-model Chevy in the first place. "Easy," he said, "kids don't want to mess around with old Fords anymore, and attractive musclecars are on their way up in value. In fact, they've doubled in price during the past two years. By playing their 'cars' right, they can make their money work for them, have fun, learn something and come out with a profit if they don't cut corners—or end up with a car that turns people off. If they stick with stock factory colors, a nice lowering job, a good wheel-and-tire combination, and load it up with factory options, they'll be okay. I'll be honest with you," he continued, "the neophyte car builder shouldn't deviate too far from stock. If he does, he's liable to end up with a visual tuna or something that isn't worth anything. Let's face it, Camaros are going to be collector's items, so kids should do things to their cars that can be changed back to stock so the sales potential of the cars won't be ruined. I'll bet you can find a whole lot of guys kicking themselves because they turned their '55 T-birds into full customs."

So the obvious question is what's he right way to turn your own Camao into a practical form of "self-xpression"? That depends on you. If ou've never worked on a car before, ou should take it apart as far as our courage will allow. You should etail it by cleaning up the suspenion components, plating or painting he various subassemblies, straightning out the sheetmetal and making both visually attractive and mehanically sound. Instead of turning project into a jigsaw puzzle that ould eventually lead to frustration nd a subsequent turnoff, make ourself a step-by-step instruction sheet to help you put the car back together. Take notes and make drawings, because they will save you a lot of time and money. Label coffee cans, boxes and crates, and keep all the various parts and pieces physically separated. And don't buy anything sight unseen. Like the man said, "A penny saved is a penny earned."

The way John Buttera saves money is simple. He isn't afraid to do a little horse trading. Within a week after his son "poled" his '69 Z28, Buttera bought a very straight '69 coupe for $750. He then unloaded the parts he didn't need. For example, he sold the engine for $250, the Powerglide transmission for $75, the stock wheels and tires for $150, got $75 for the front frame section, and still has $150 worth of leftover goodies. "I ended up with a good body and a complete interior for $50. By simply taking all the good pieces off my son's '69 Z-car and combining them with the $50 coupe, I figure I could sell the finished car for $7500 to $10,000. There is no reason why you can't have fun and make money doing it." And that's precisely what we had in mind when Lil' John Buttera offered to show us how to build a better mouse trip.

PHOTOGRAPHY: GRAY BASKERVILLE AND BOB McCLURG

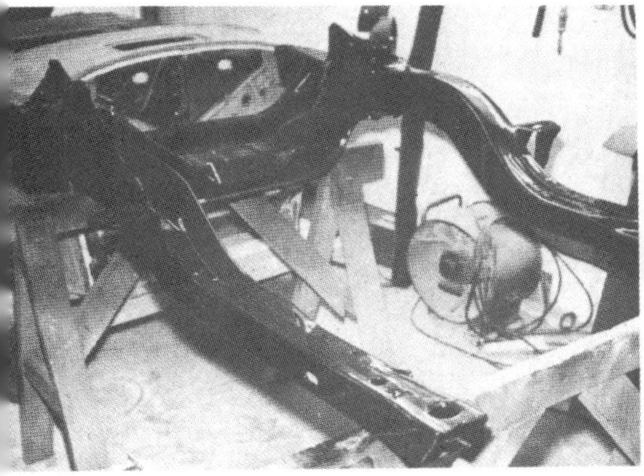

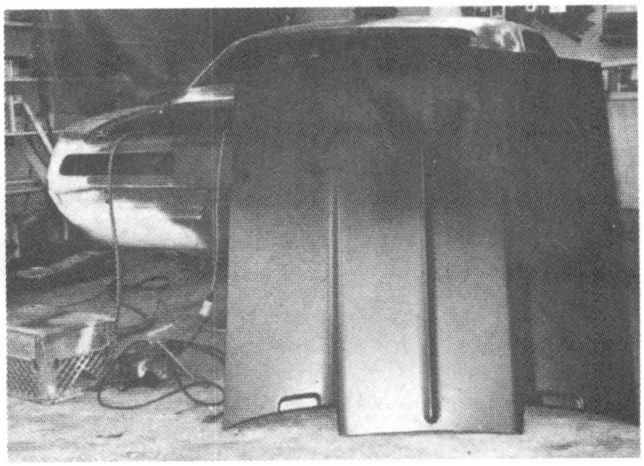

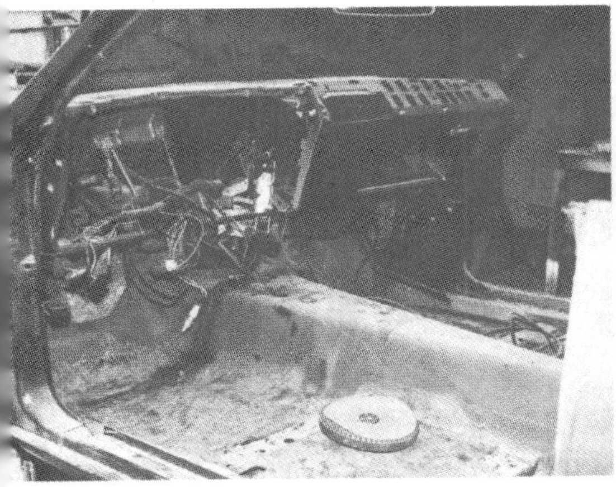

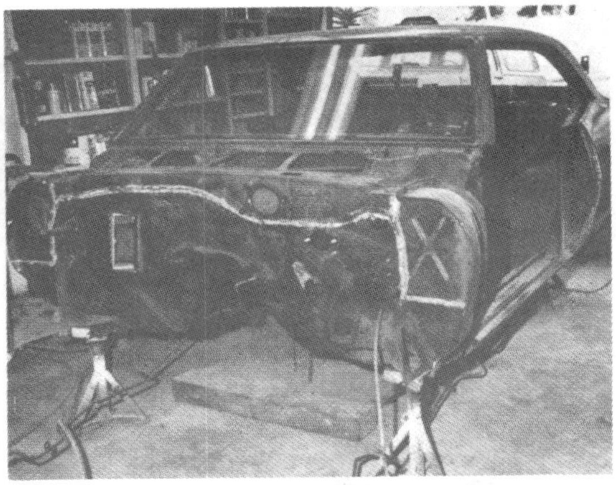

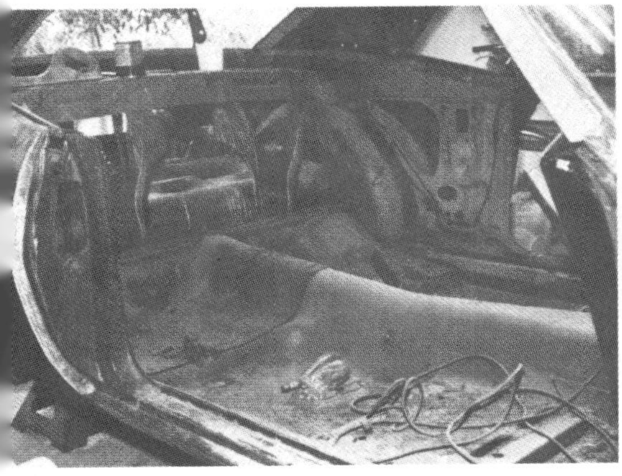

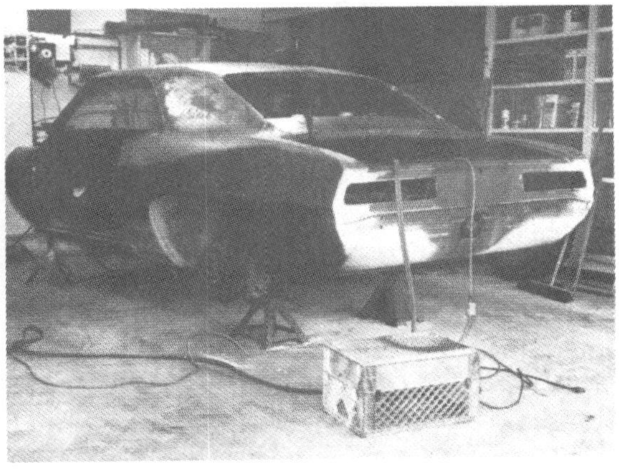

Front frame section has had the welds massaged and covered with plastic filler before being painted with a black polyurethane. It isn't hard to do this to a late-model, and it will increase the value of the car. What Buttera means by "parts are easy to get" is this hood. You can still purchase high-performance-style sheetmetal from Chevy. Now this is how a street rodder would put a late-model together. He would strip it down to the last bolt, then reassemble it to fit his tastes. Interior will be repainted and rewired to reflect current thinking. The only way you can successfully detail a car is to make sure all the various assemblies have been rubbed on. Although the '69 coupe body needed minor cleaning up, you'll note that there is no radical sheetmetal work. Gutted interior will be fitted with stock-type upholstery.

(continued overleaf)

BOLT ON: EMMETT GLASGOW

Philosophically, Emmett Glasgow's "Sunshine Too" encompasses a far different approach to constructing a Camaro. Early-model lovers may recall Emmett's original show-winning '67 Camaro "Sunshine." This time, Emmett is dealing with a new and complete car. Second, as the owner of Street Customs Limited, Glasgow is more visually oriented, so Sunshine Too features a mildly customized look. Third, the car hasn't been completely disassembled, detailed and then reassembled. Fourth, the engine and interior were given new life through some fancy brightwork, and last, the whole package was set off by Candy red lacquer, augmented by gold-anodized trim.

To be more specific, Glasgow's '78 Z28 was never wrecked, nor will it employ OEM (Original Equipment Manufacturer) high-performance suspension and brake components. Basically, the '78 is made up of visual elements, including two extra added sets of factory-built Camaro fender gills. One set of gills was grafted onto the front fenders, just below the stock fender gills, and the other set was spliced into the top of the hood. Glasgow carried the bolt-on plan even further by molding a Hooker Industries spoiler and fender flare kit onto the front of the car; then he trimmed the flares to resemble those on a Trans Am Pontiac. To carry the "Poncho" theme further, Emmett attached Pontiac T/A fender pods to the rear fenders and finished off the Camaro by molding on a hood scoop, rear spoiler and an Old Glory square tube grille. And to ensure that the Camaro is as colorful as it is clean, Emmett squirted on a load of

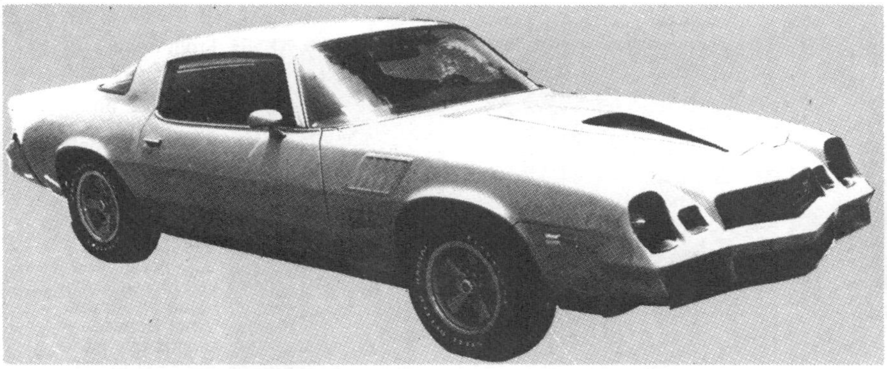

It takes a lot of guts to lay grinder and torch to a brand-new Z28, but that's what customizing is all about. Street Customs' past experiences with early-model Camaros prompted the late-model project, and the results, when finished, should reflect one fine street custom.

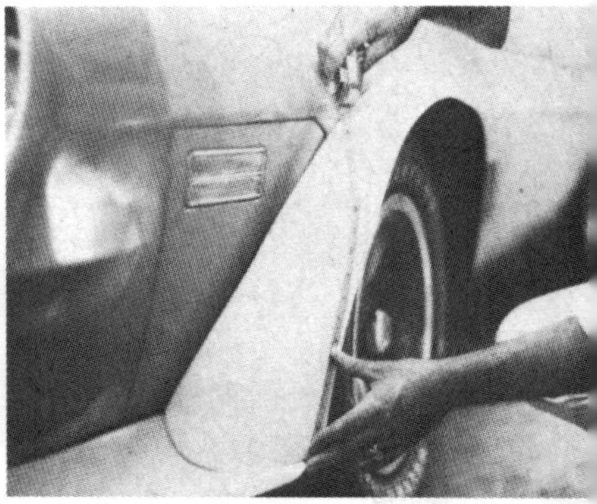

A Hooker Industries Camaro/Trans Am-style front spoiler and fender flare kit was used up front and then trimmed, a la Trans Am Pontiac.

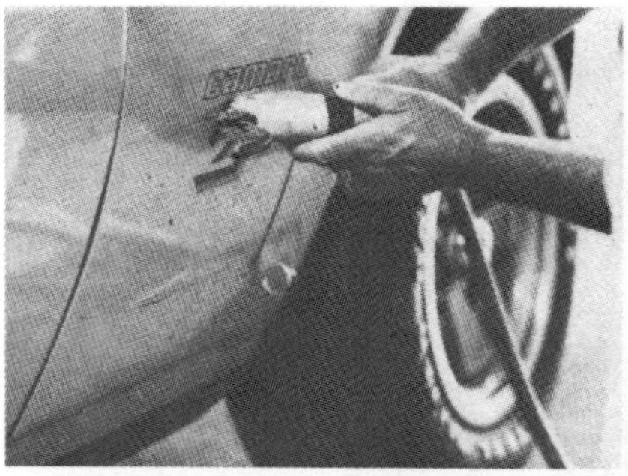

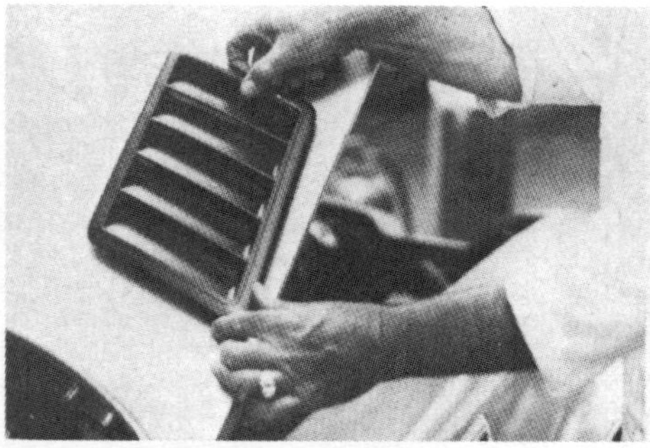

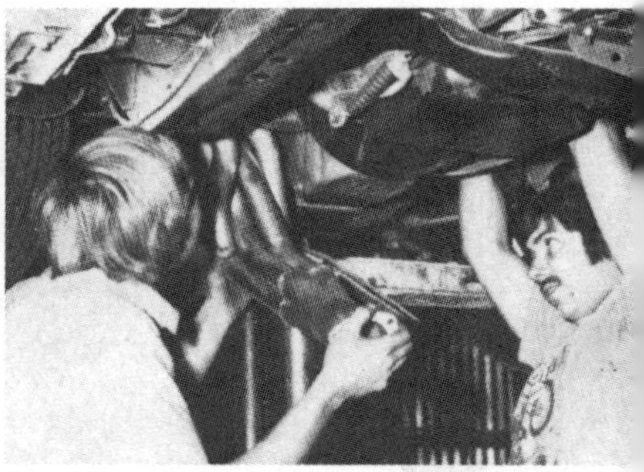

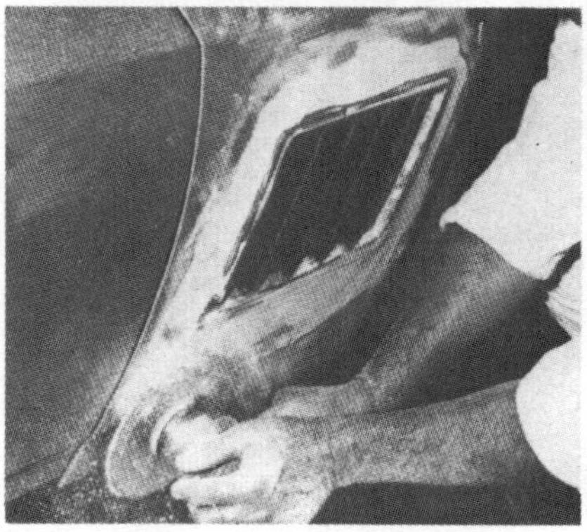

Emmett attacks the Camaro sheetmetal like there's nothing to it. Faint-hearted Camaro lovers and inexperienced types beware; this type of customizing takes an experienced hand and a lot of patience.

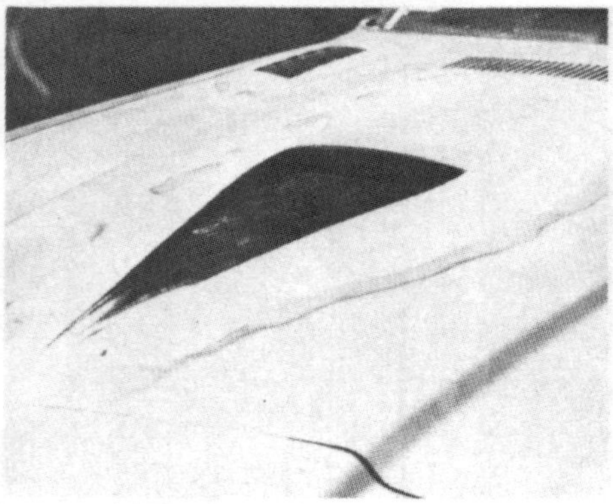

Emmett's side-gill fetish was carried out on the hood as well. This task proved a little sticky, since the hood mounting brackets are below. The Z28 hood scoop was also molded in.

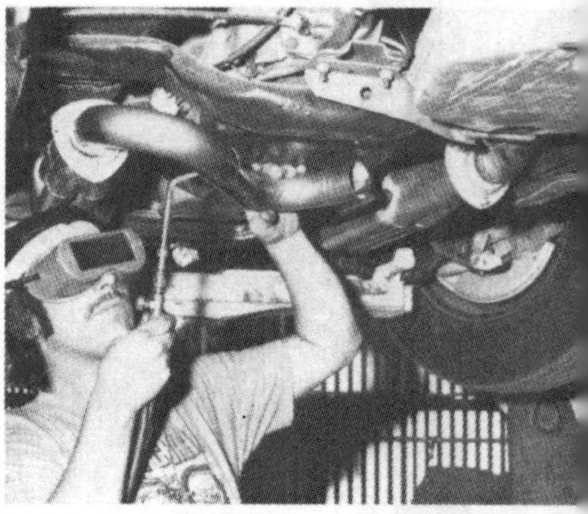

Fitting a set of headers to a late-model anything is a hassle these days, and the authorities have no sense of humor when the project involves a catalytic converter. Here the Hooker Headers crew shows you how it's done legally and right.

Candy Apple red lacquer over a gold-tint base.

Like Lil' John, Glasgow feels that wheels and tires make a Camaro. So he too discarded the solids and replaced them with 15-inch Tru-Spokes, set off by gold-plated knock-offs, and Winston Winner radials.

Unlike Buttera, Emmett decided to put a little "hair" into his Camaro's interior. Consequently, he commissioned "Rafael" to stitch up a mess of threads made out of crushed burgundy mohair, then finished the interior with an A.I.A. GT steering wheel.

The engine compartment, while tame by yesteryear's standards, has got more going for it than a circa '78 smogger. Glasgow had Jack Richardson bolt together something that would look nice and operate on low-lead. In essence, the stock 350 has been given the Holley "Dominator/600-cfm/rocker cover" treatment, augmented by an Accel BEI ignition. To further liven up the engine compartment, Glasgow had every removable exterior part chrome-plated. And let's not forget the Hooker headers. Mating them to the catalytic converter proved to be no simple task.

As you can see, Sunshine Too does indeed shine; in fact, it "Glasglows."

Both cars encompass different but reasonable ways to personalize one's mode of transportation. Nonetheless, there is one thing we would like to stress: Be it a rod-on or a bolt-on, keep it safe and sane. There is no percentage in starting off with a $7500+ winner, having an accident and ending up with a $350 loser because you didn't follow the advice of John Buttera or Emmett Glasgow. **HR**

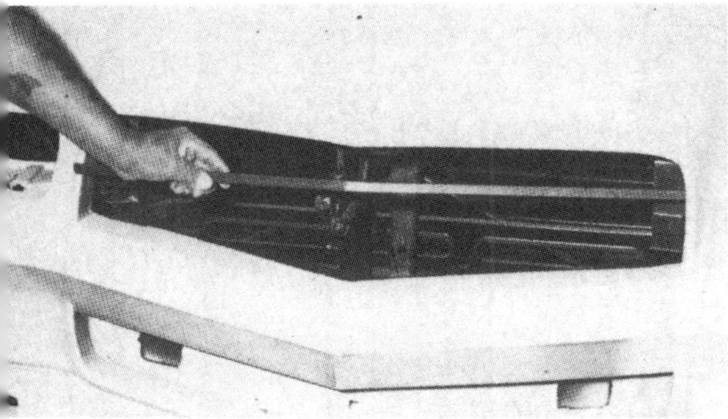

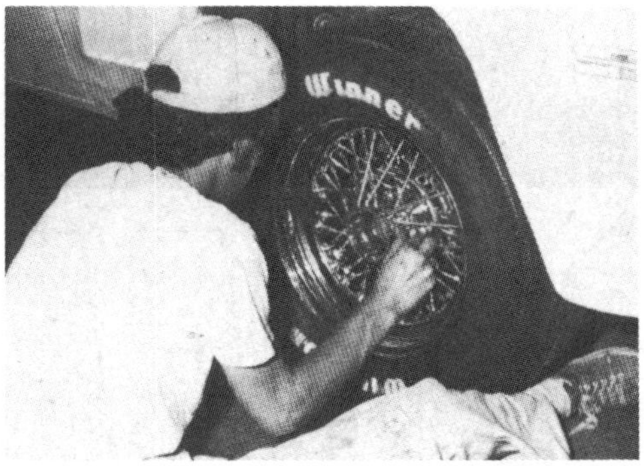

An "Old Glory" square-tube grille settled the custom grille problem. It will be gold-plated. Those Winston Winner Radials are also in good company. They're mounted on a set of Tru-Spoke wires, and naturally, the spinners got the gold treatment too.

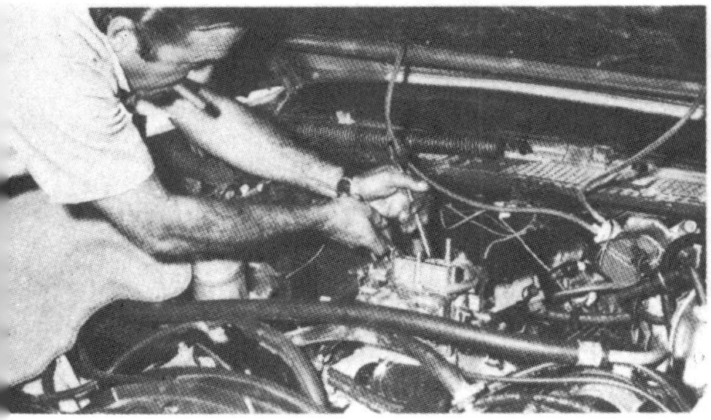

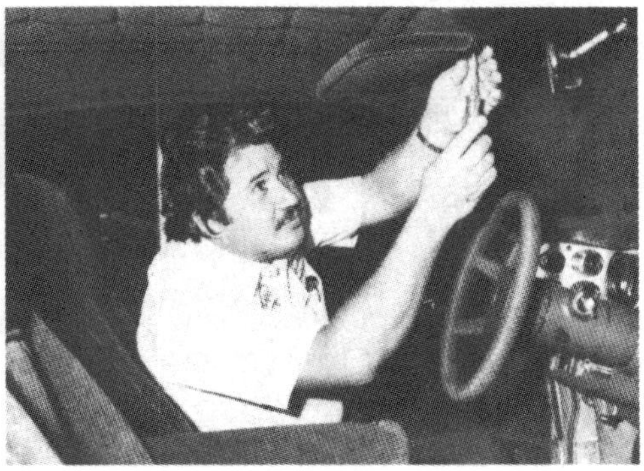

Not just another pretty face. The inside gets a few oats and good looks added, courtesy of Holley, Accel and Jack Richardson. Here super-stitcher "Rafael" puts the finishing stitch to the brandywine mohair interior.

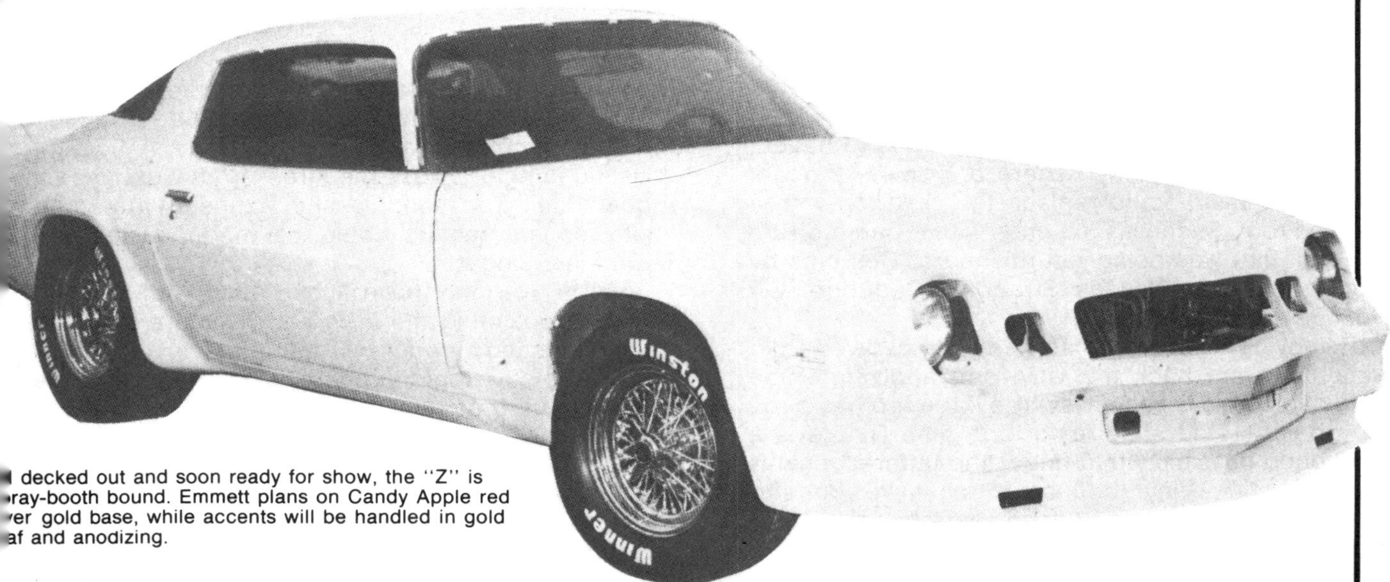

I decked out and soon ready for show, the "Z" is ray-booth bound. Emmett plans on Candy Apple red er gold base, while accents will be handled in gold af and anodizing.

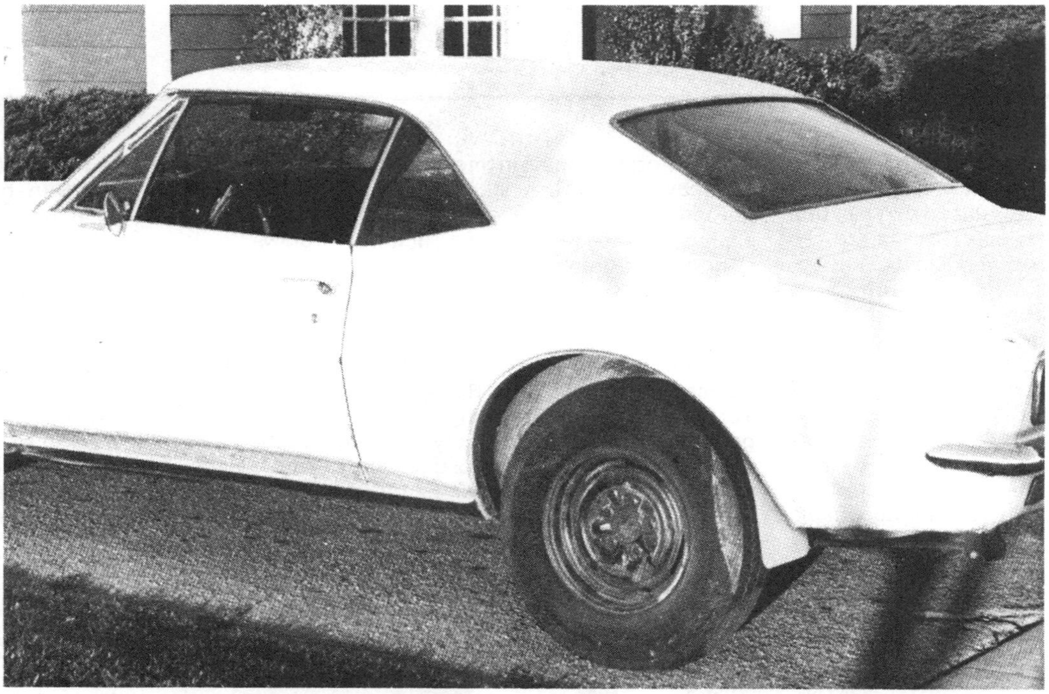

CAMARO
Vinyl Top Replacement

Vinyl tops have been a popular option on new cars for a long time. Many used car owners also had vinyl tops installed on their cars. A vinyl roof in good condition can be a handsome addition to any car. Vinyl roofs were meant to impart a feeling of luxury, which they do, until the vinyl gets ragged. Worn out vinyl tops look far worse than a plain roof. A frayed vinyl top marks a car as a loser that hasn't been well maintained.

One solution to a worn out vinyl top is to have it replaced with new vinyl. There are times, though, when it is desirable to replace the vinyl top with a painted roof. Such was the case with a Camaro Rally Sport that was being refurbished at Customs by Eddie Paul, 124 Nevada Street, El Segundo, CA 90245.

Removing a vinyl top can be more involved than simply peeling back the vinyl and applying some new paint. This is especially true if the top has been in a deteriorated state for a long time. There is a very good possibility that under the tattered vinyl is severe rusting. Vinyl roofs are much more prone to rust damage than non-padded roofs because they can trap moisture. If you have a bad vinyl top, take care of it as soon as possible. The longer you wait, the greater the chance of rust.

The subject Camaro sat outside for three or four years with a bad top. There were areas around the back window where the vinyl appeared "puffy." This, coupled with the fact that the trunk filled with water during the slightest rain storm, was a symptom that should have been spotted sooner. The diagnosis was terminal rustout.

When faced with areas of complete metal deterioration and rustout, the problem can be approached better if the area is sandblasted. Luckily, A-1 Sandblasting is right across the street from Customs by Eddie Paul. The sandblasting exposed the good metal so the repairs could be made easier and would last longer.

Even if you only have a bad vinyl roof with no rustout, repainting the area can be more involved than repainting a non-vinyl roof. If the vinyl roof was installed by the factory, chances are good that the roof is less than perfect. Since most roofs are welded to the lower body panels, there is always some bodywork around the seams. It makes sense for the factory to use the bodies that need more than average work as the basis for vinyl roof equipped cars. The lack of final finishing on the subject Camaro was quite obvious when the old vinyl was removed.

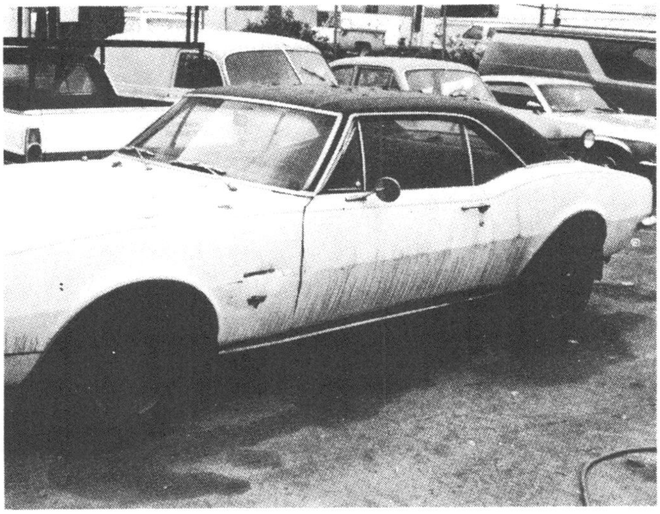

1. One well-worn early Camaro, a victim of too many years in the California smog and salty ocean air.

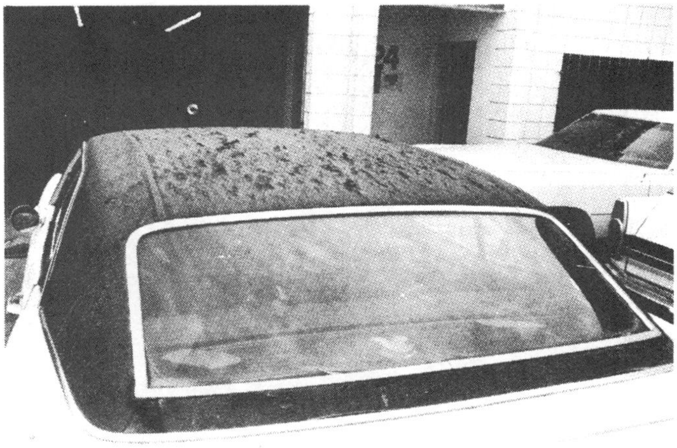

2. The factory installed vinyl top was badly blistered. There were ominous "puffy" sections around the bottom of the rear window. These signs hinted at rustout.

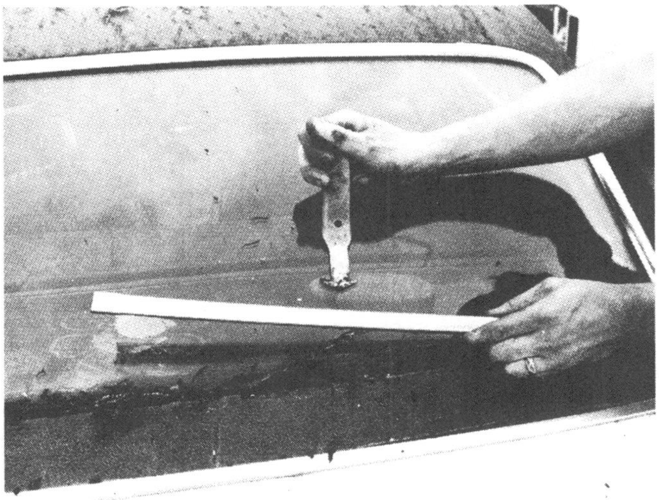

3. The chrome trim was removed using an inexpensive trim removal tool to loosen the clips under the trim. A tool like this is a must if you don't want to damage the fragile trim pieces.

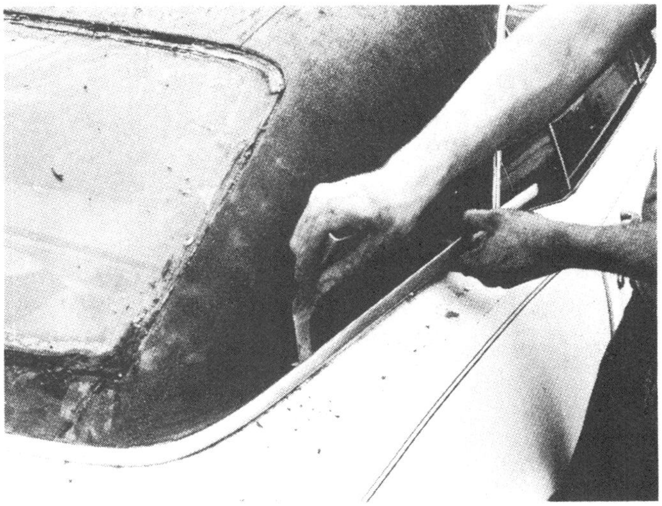

4. Since the lower trim piece that separates the vinyl roof from the painted part of the body won't be needed any longer, you needn't exercise much care in its removal.

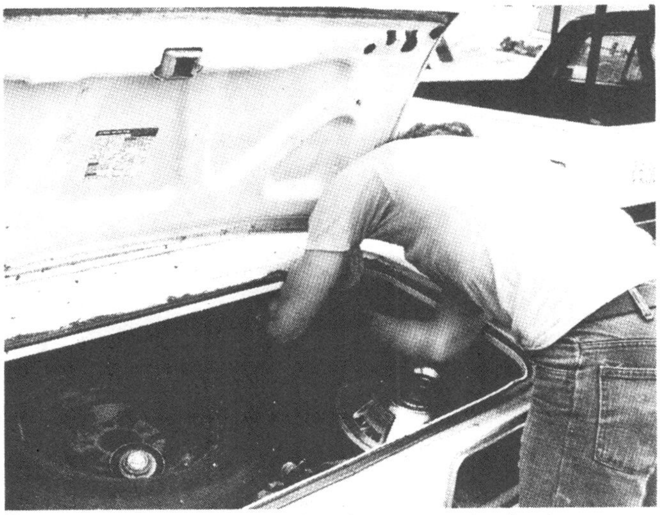

5. The belt line trim pieces often have a couple of nuts holding them in place. These nuts are located inside the trunk area.

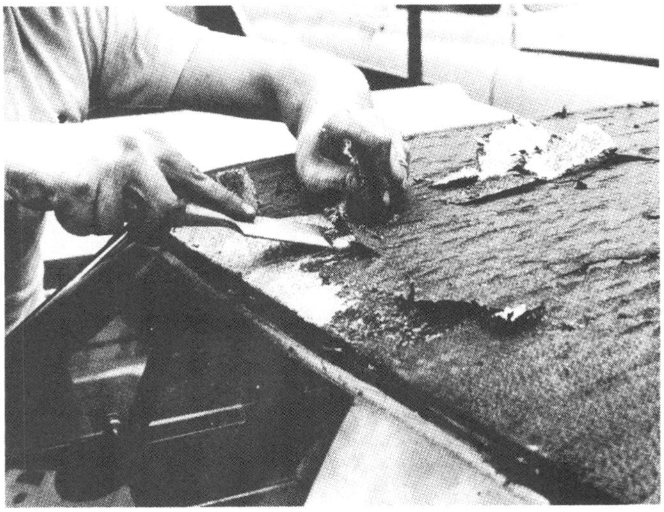

6. Any type of stiff putty knife can be used to remove the old vinyl. Here, a wood-working chisel is being used because of its rigid blade and sloping leading edge.

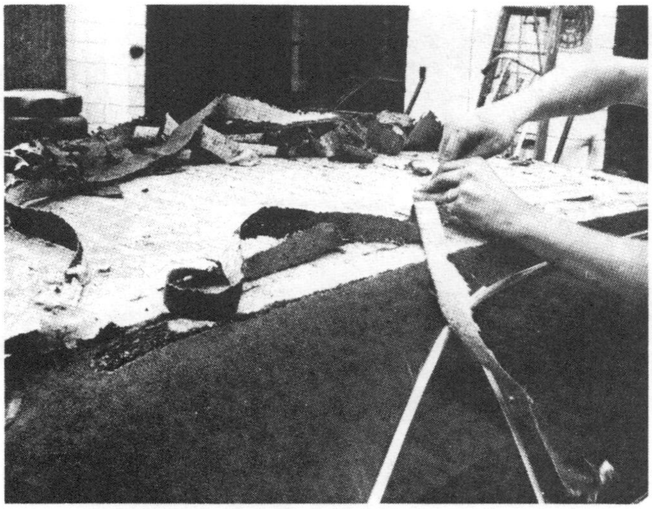

7. Peel back the vinyl and don't worry about the rough surface left by the old adhesive. Don't push the blade too hard or deep or you will risk gouging the metal.

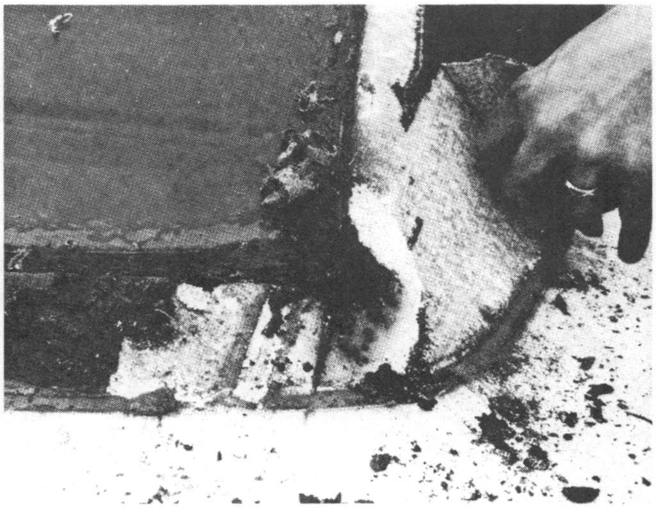

8. Underneath the "puffy" sections around the rear window was rustout. This can be a real problem since this section of metal holds the rear window in place.

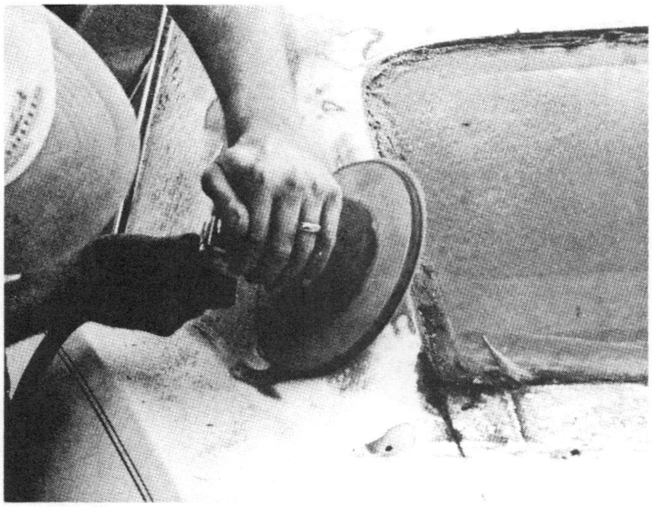

9. A dual-action air sander should be used to remove the old glue residue after the vinyl has been removed.

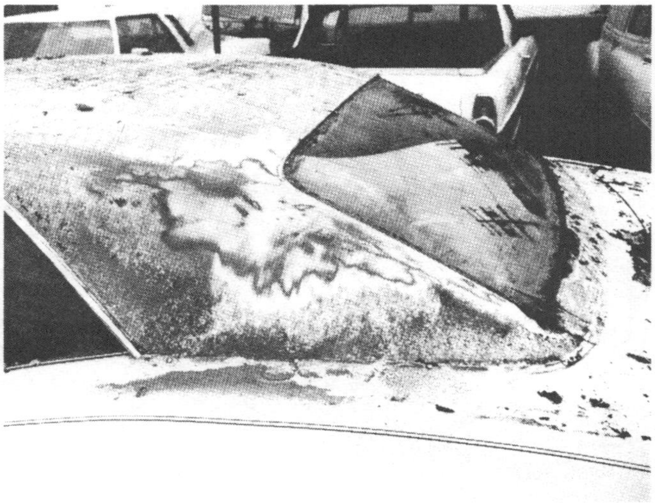

10. Notice the difference between the sanded metal and the areas where the glue residue remains.

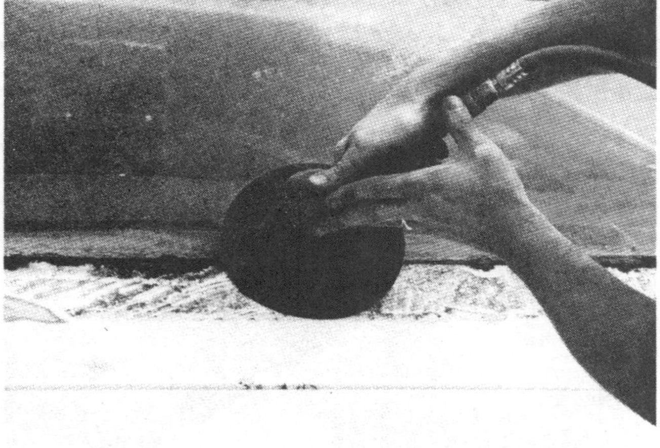

11. An attempt was made to grind away the rust around the rear window with a disc grinder, but the damage was too severe.

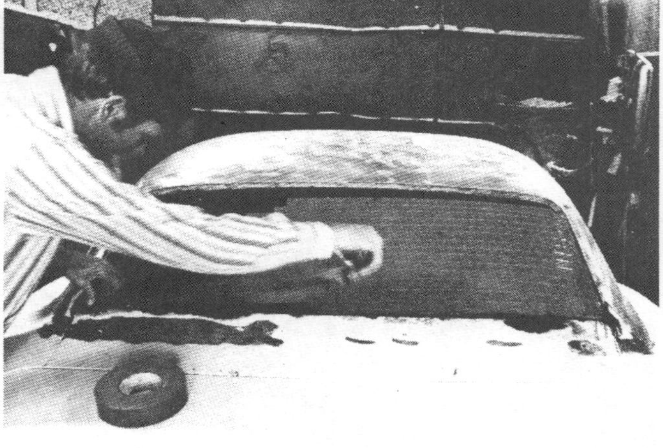

12. The car was taken to A-1 Sandblasting in El Segundo, CA where the rustout areas were blasted clean. Heavy duct tape (or racer's tape) was used to cover the glass.

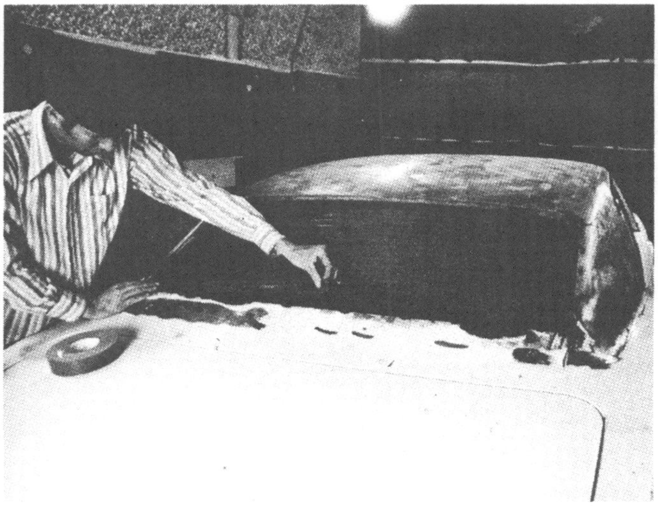

13. Multiple layers of tape should be applied over the glass nearest the damaged areas since sandblasting can permanently etch glass.

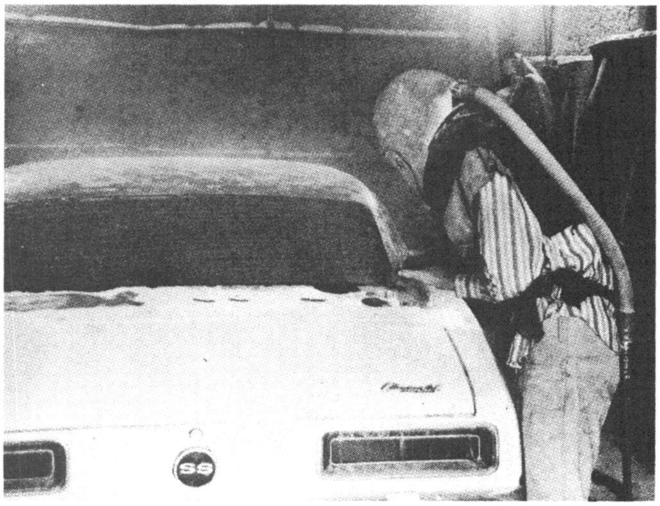

14. Since the metal was quite deteriorated, lower pressure than normal was used during the sandblasting process. The "spaceman" outfit is really protective equipment with a breathing system used in commercial sandblasting.

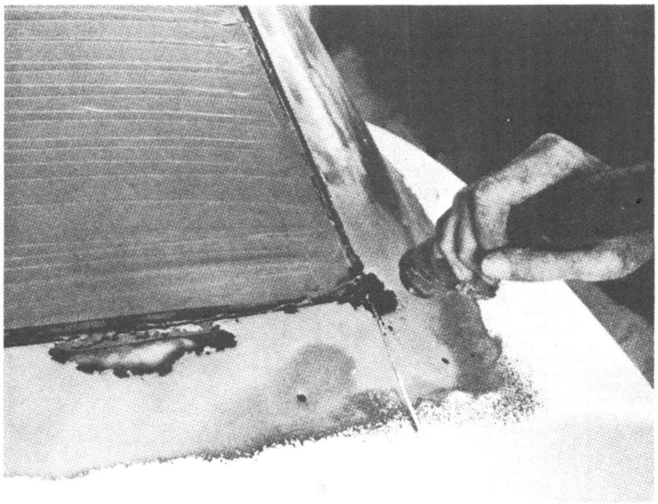

15. The sandblasting removed all the paint and rust. The border area was blasted just in case there was any hidden rust. Notice the size of the rustout holes.

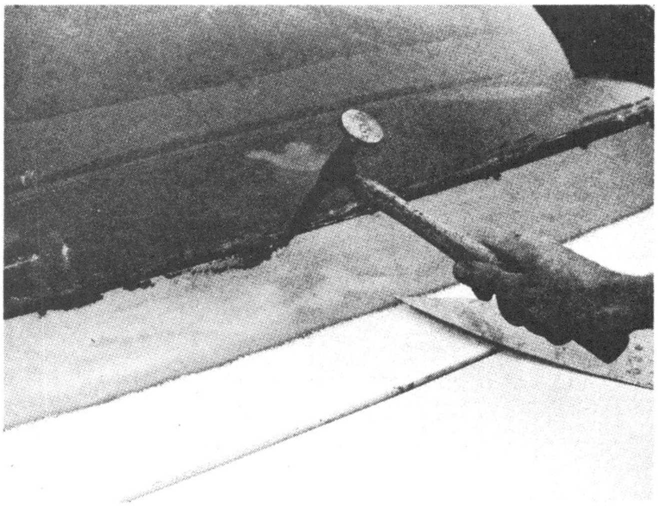

16. To properly repair this extensive damage, a new piece of metal should be installed. It would be costly. Rather than use this method, it was decided to fill the area. Use a body hammer to lower the area to be built up with body filler.

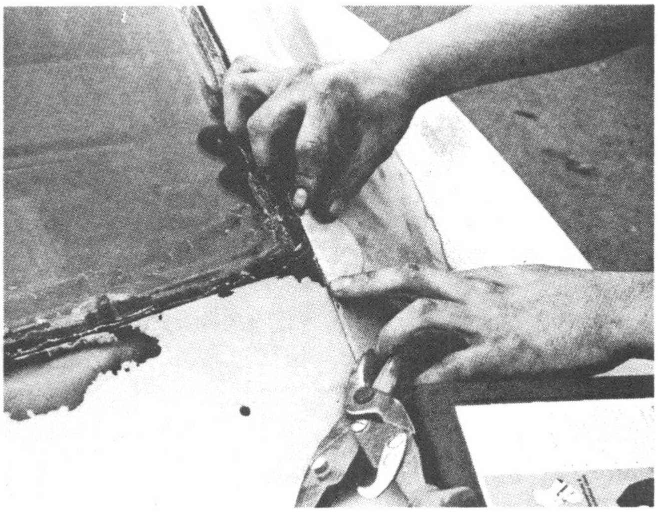

17. A piece of sheetmetal was cut and fitted to each rusted out hole.

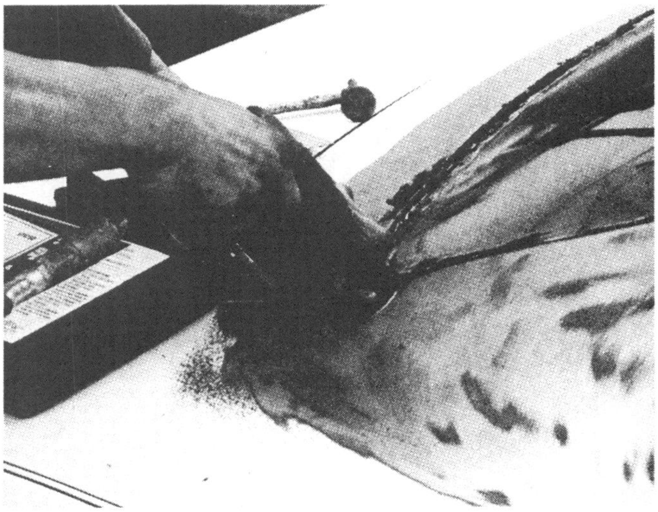

18. Drill holes through both the patch and the underlying metal.

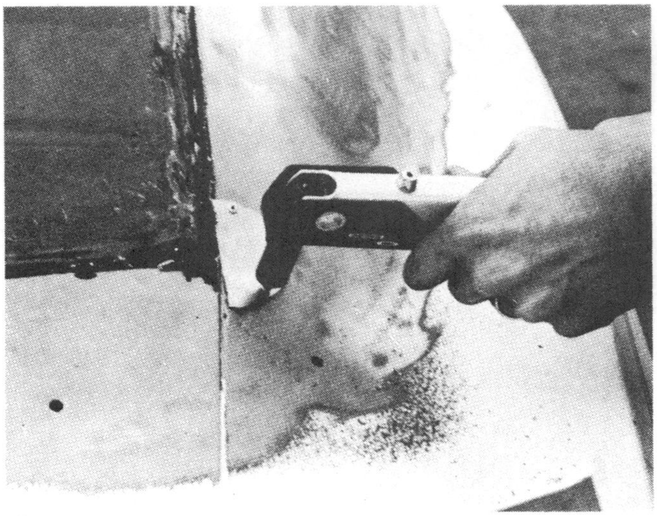

19. Rivet the patch piece in place.

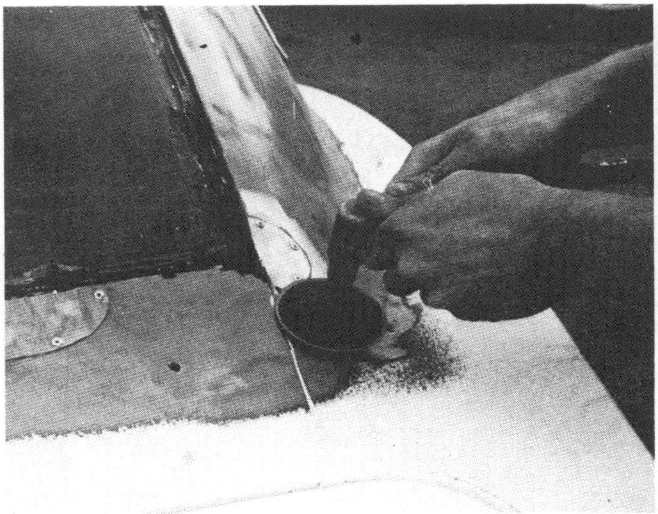

20. Grind the edges of the patch panel smooth, and also make sure that the areas surrounding the repair are cleaned to the bare metal in preparation for the body filler.

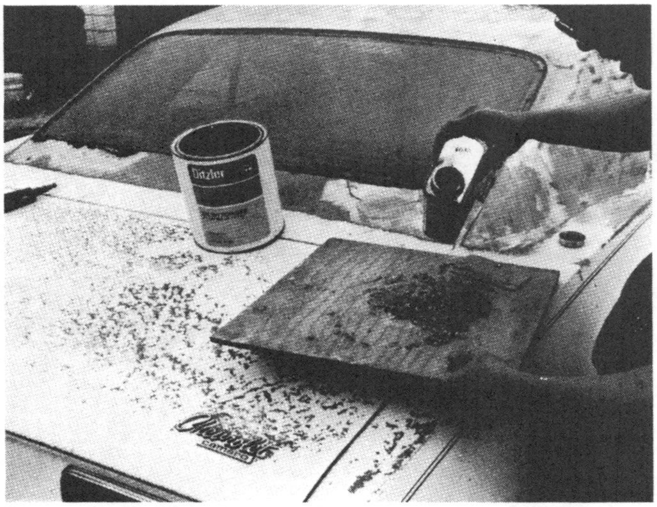

21. Ditzler Alum-A-Lead filler was used because it is superior to plastic fillers in hardness and durability. (It is also more expensive.)

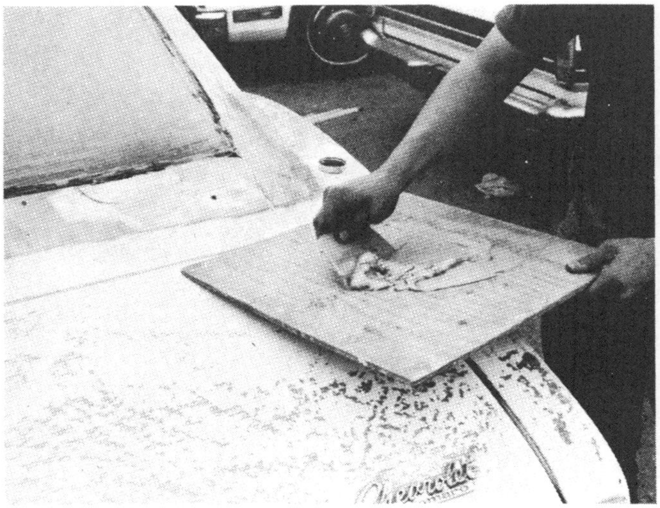

22. The Alum-A-Lead comes in powder form and must be mixed with a special resin before it is applied to the car.

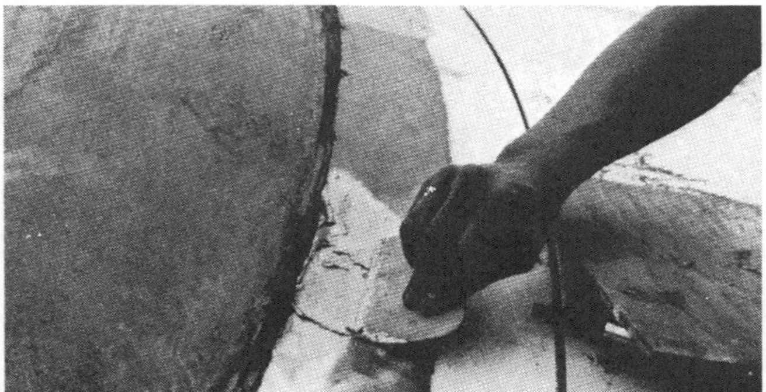

23. Apply the Alum-A-Lead as you would any other filler. Be sure that the patch pieces are slightly below the final surface so that the filler will bring the area back to normal.

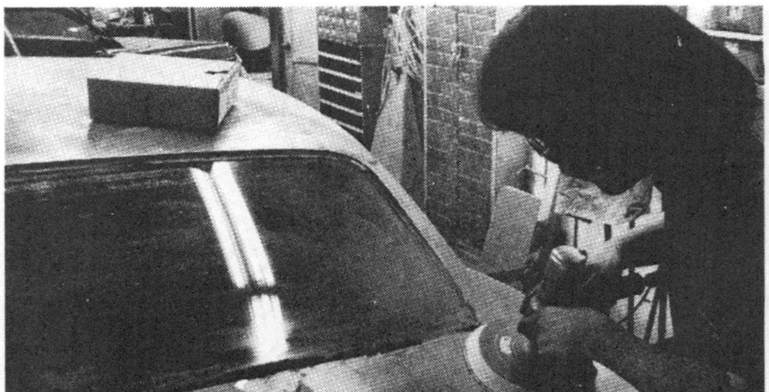

24. After the Alum-A-Lead is dry, it is sanded or filed to the desired shape. (The filler isn't worked with a cheesegrater file like plastic body filler.)

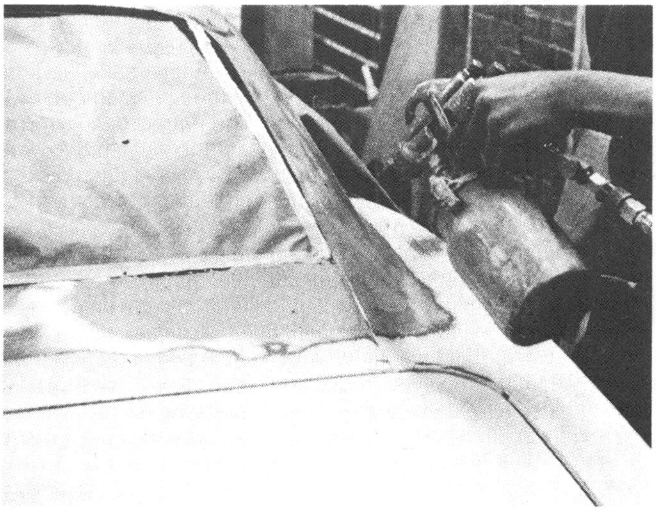

25. The filled areas should be primed and checked for additional low spots. Finish the area with sanding and spot putty.

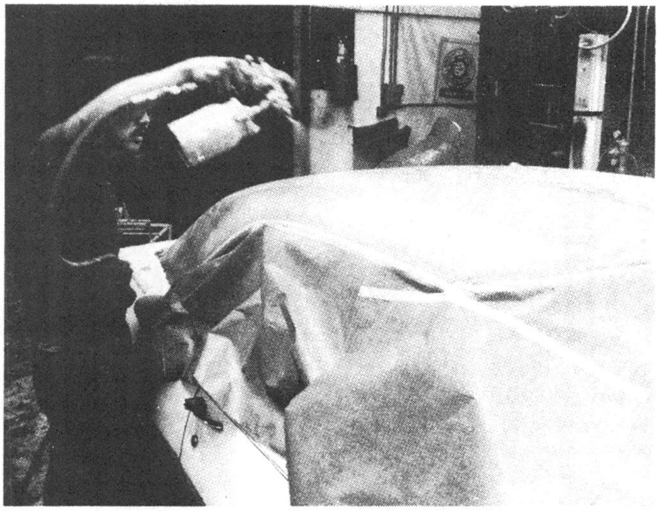

26. After the repaired areas are satisfactory, tape and prime the entire roof.

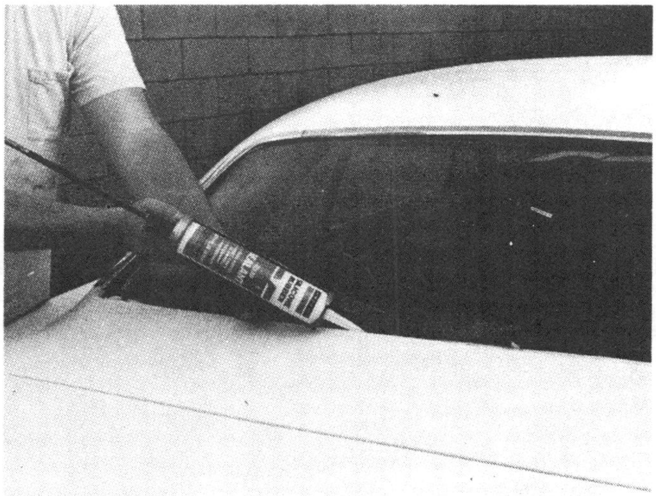

27. Silicone sealer should be used around the rear window to help seal out moisture.

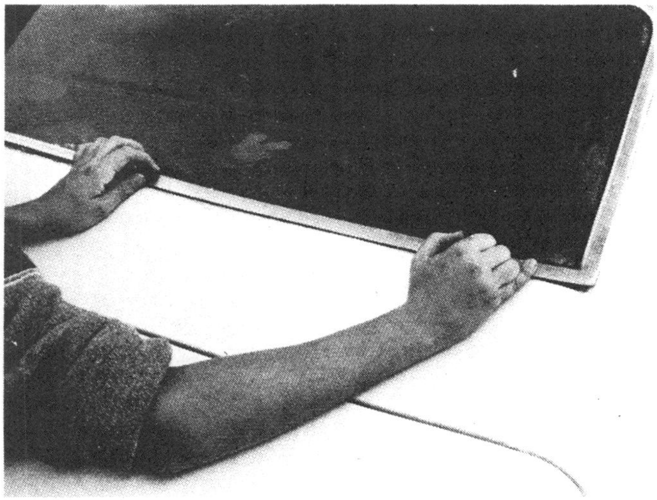

28. Reinstall the window trim.

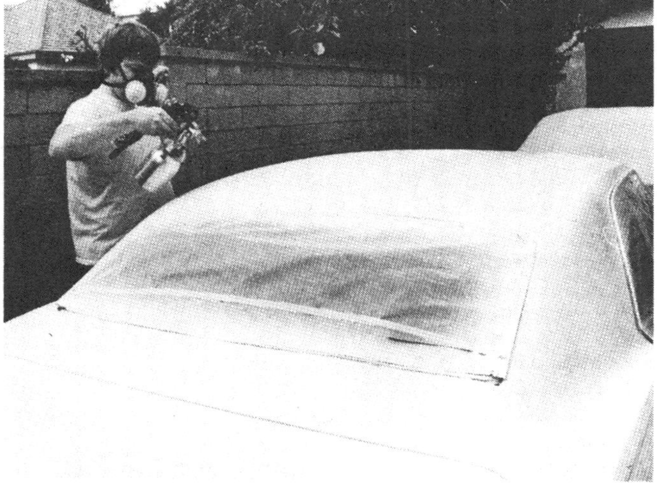

29. Apply the final color after making sure that the entire roof is ready for paint.

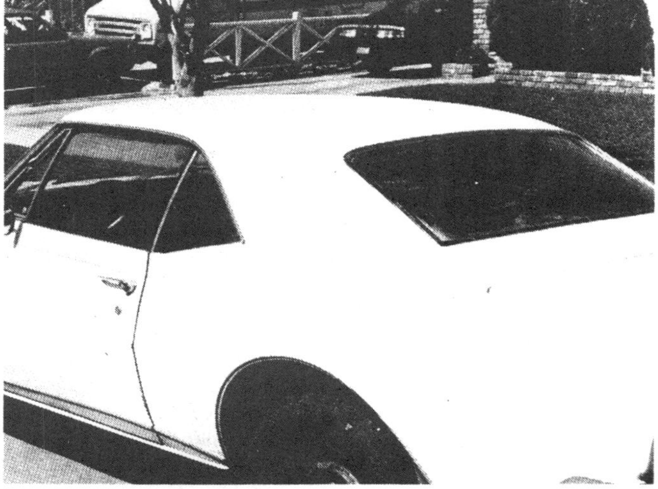

30. The finished product. The rest of the car still needs work, but the top looks great.

Backyard Buildup: '69 Z28

One Hot Rodder's Approach to Restoration

Part 1: The Engine

By Cam Benty

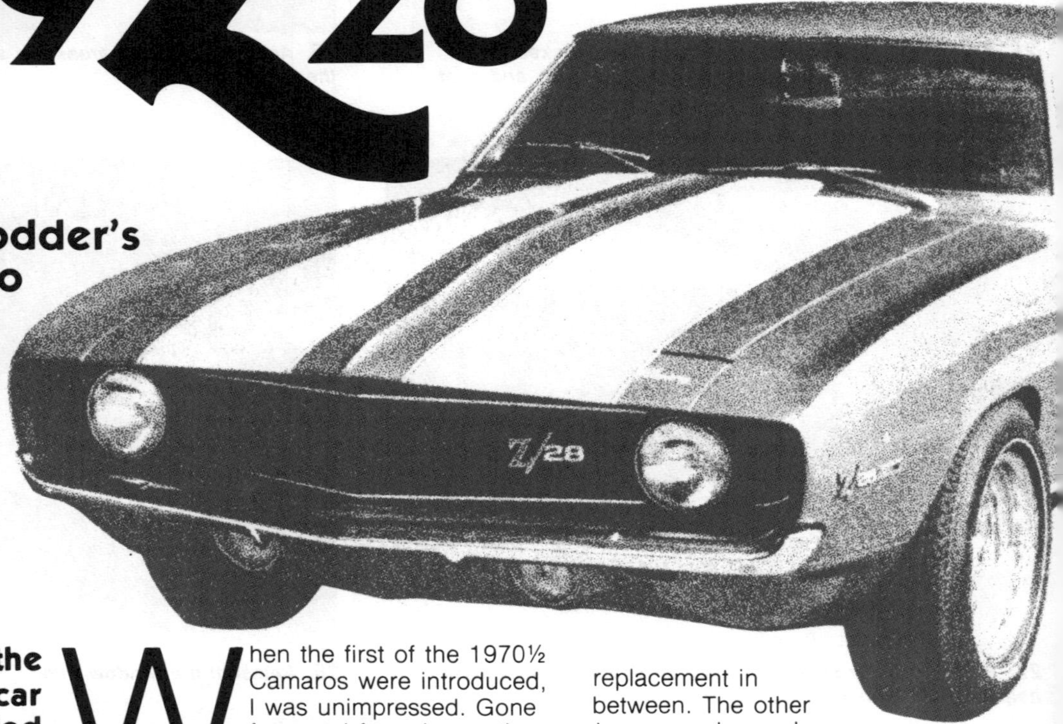

"It was the type of car that notified the locals a mile away that we'd be there shortly!"

When the first of the 1970½ Camaros were introduced, I was unimpressed. Gone forever, I feared, was the styling that I had grown to know and love; gone too, was the optional Z28 302-cubic-inch engine; and, worse still, you could now get an *automatic transmission* on the Z28. What had happened to my race car? What was this "thing" they dared to call a Camaro? It was not until many years later that the current styling in Camaros finally earned my approval.

That same year my father, a true sports car nut, purchased a Rally Sport 1970½ Camaro and spent the following years attempting to persuade me that the change was for the better. The 307-cubic-inch-powered machine certainly had plenty of power, but the car I really longed for was a Z28.

While living in Stockton, California, I had my first close encounter with Z28 power. The sleek '69 Z-car was highly modified, with 12.5:1 compression pistons, a high-lift camshaft, and red and blue candy stripes over a white paint scheme. Owner Dan Meckler and I spent many a day wringing out the Camaro, waking up the surrounding farming community with 7000-rpm blasts. It was the type of car that notified the locals a mile away that we'd be there shortly!

We learned a number of things from that experience—the most important bein the type of treatment Z28s were subjecte to. While Boss 302 Mustang owners tend to polish and baby their machines, Z28 owners thrash their cars constantly. Engine rebuilds take place every 40,000 miles, with constant attention and parts replacement in between. The other lesson we learned was that if you don't adjust and examine the temperamental engine, the performance is lost, never to be recovered without major surgery. Soon t Z28 had a dented front valance panel ar a terminal case of piston ring blow-by, and Dan had had enough and was ready to sell. In a classic example of impulse buying, I purchased the Z28 for $2200 and set about restoring its former potenc

When I bought the car, my knowledge of restoration techniques or the collectability of early Camaros was not extensive, so before tackling the project, researched it. Today the Camaro restoration business is booming, with experts predicting that the Camaro will g the way of the Corvette, and asking pric are going through the roof. For this reason, I decided that the car's interior and exterior should be returned to their stock configurations, but that for driving pleasure, the driveline should be strengthened with the best engine components affordable. The hoped-for result was a car suitable for street driving and road racing, yet one that would also be an investment.

The engine was the first order of business, because the 302 engine was in dire need of reringing. Previously bored-out to .030, the cylinder walls had been severely scored, and there was no

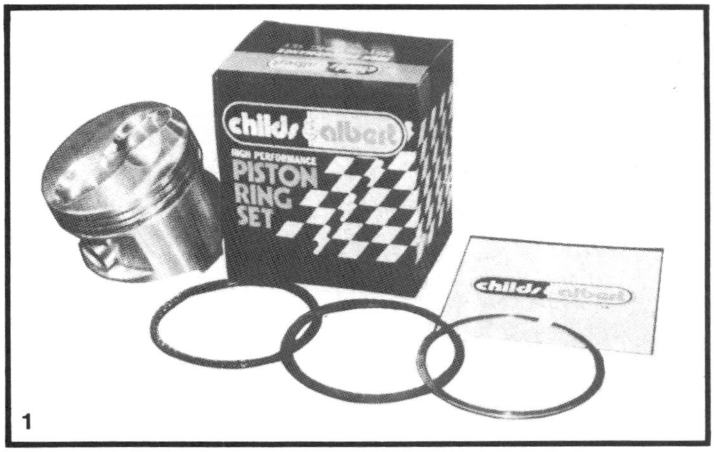

(1) The 302-cubic-inch engine had previously been overbored to .030, but severely scored cylinder walls made it impossible to save the old pistons. BRC made a custom set of .040-over 11.5:1 compression forged aluminum pistons which were fitted with Childs & Albert piston rings.

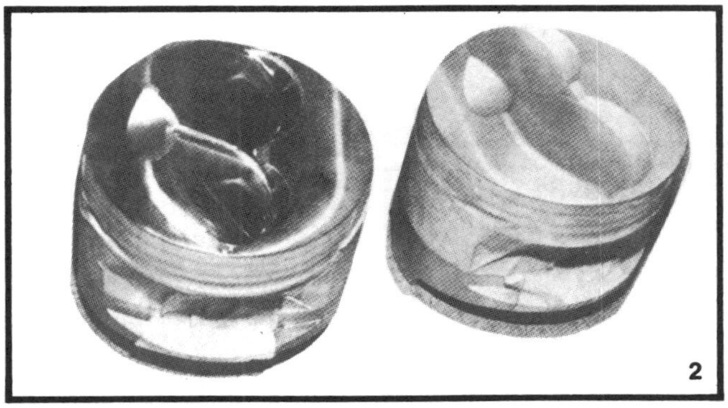

(2) Each piston received a light milling of the dome for better flame propagation in the cylinder. The surface was then glass-beaded smooth, taking care not to harm the piston ring lands.

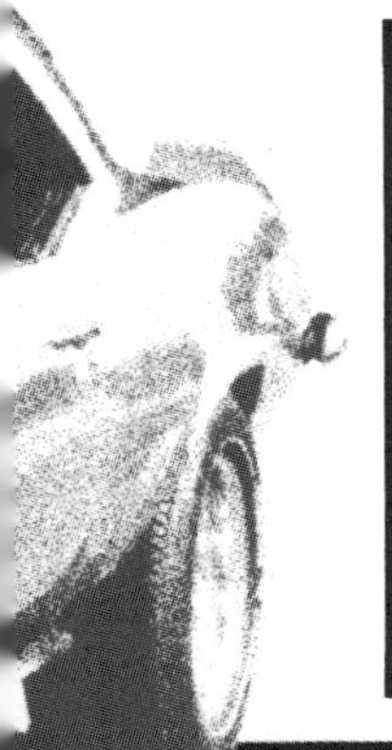

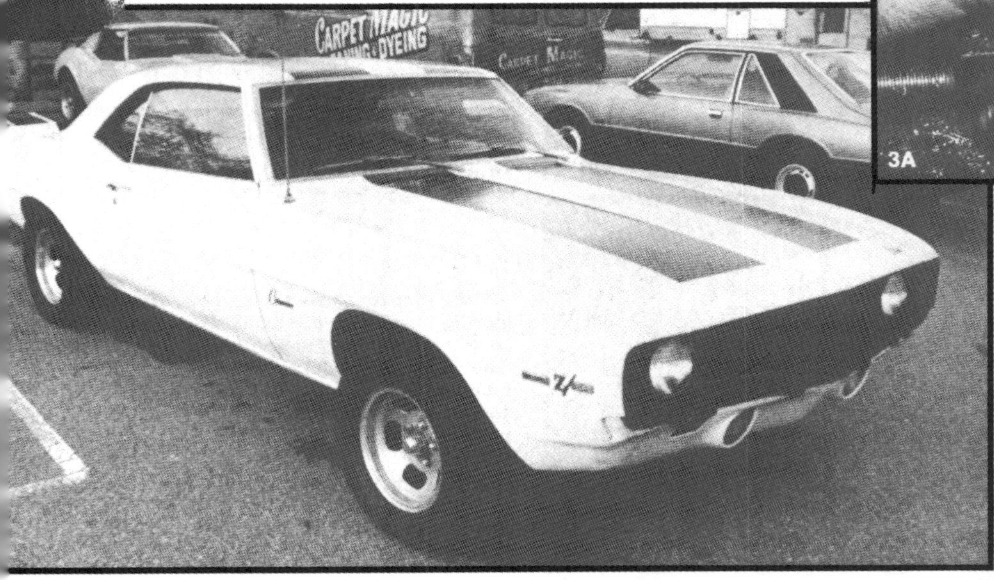

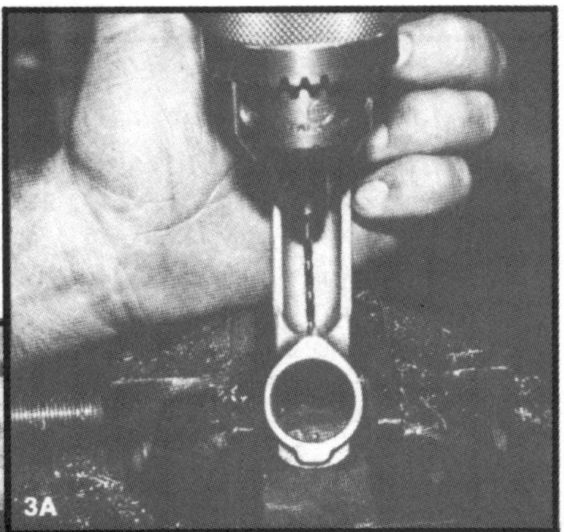

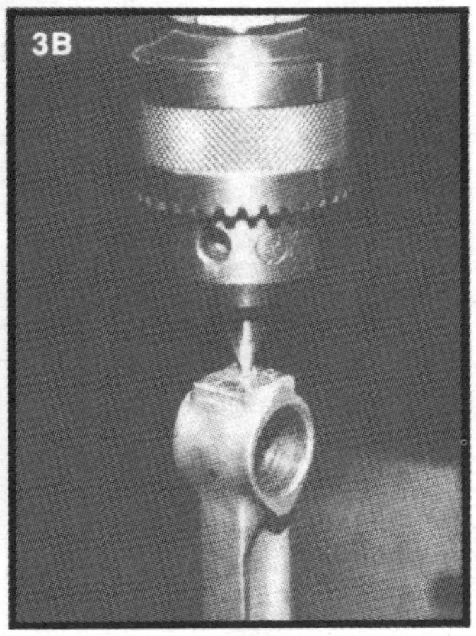

(3A & B) While the stock 302 Chevy engine had the best racing parts available, one point the factory forgot was how to oil the piston pins. To rectify this, we drilled rod ends from both top and bottom to form oil passages.

chance of using the same pistons again. For fear of overheating problems with a .060 overbore, the engine was honed to .040 and BRC custom 11.5:1 compression pistons were fitted to the block. The 11.5 compression ratio was suggested by Dick Guldstrand, a former preparer and driver of racing Camaros in the famous Trans-Am series. These racing Camaros ran the 11.5 compression, as opposed to the standard Z28 compression of 11:1. After all, if you're going to have to use water injection and octane booster anyway, half a compression point won't make much difference. Childs & Albert piston ring expert Bruce Walker set up the ring combination for the small-block. The top 1/16 ring was a torsional compression stainless chrome ring, while the second ring was a 1/16 low-tension ductile iron ring with a taper face. The oil ring, which measured 3/16-inch, was a standard tension circumferential graphite-impregnated ring containing no

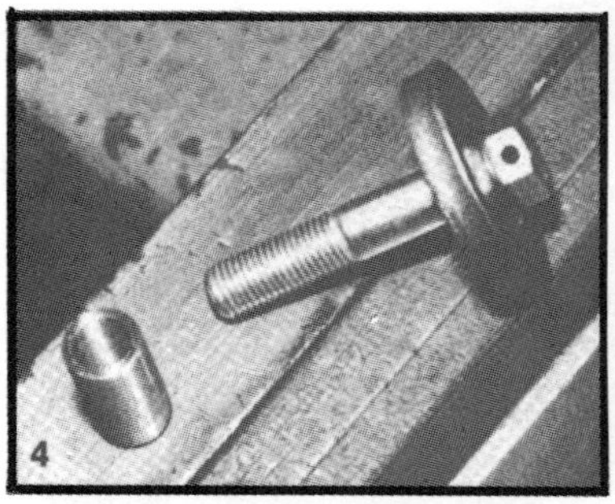

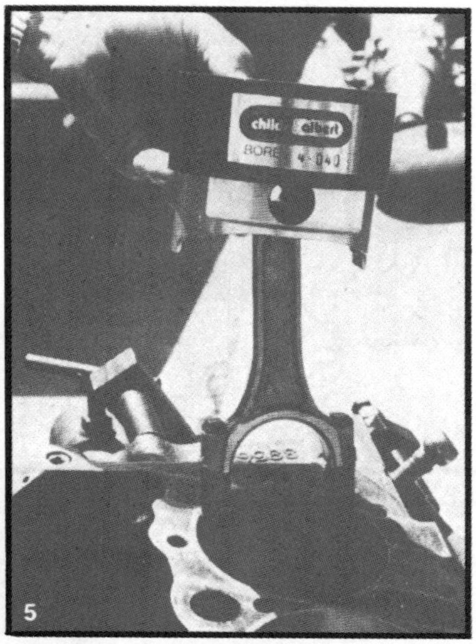

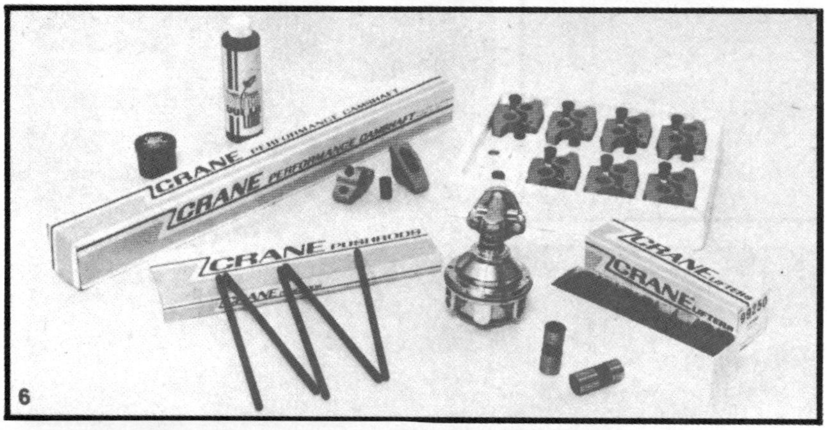

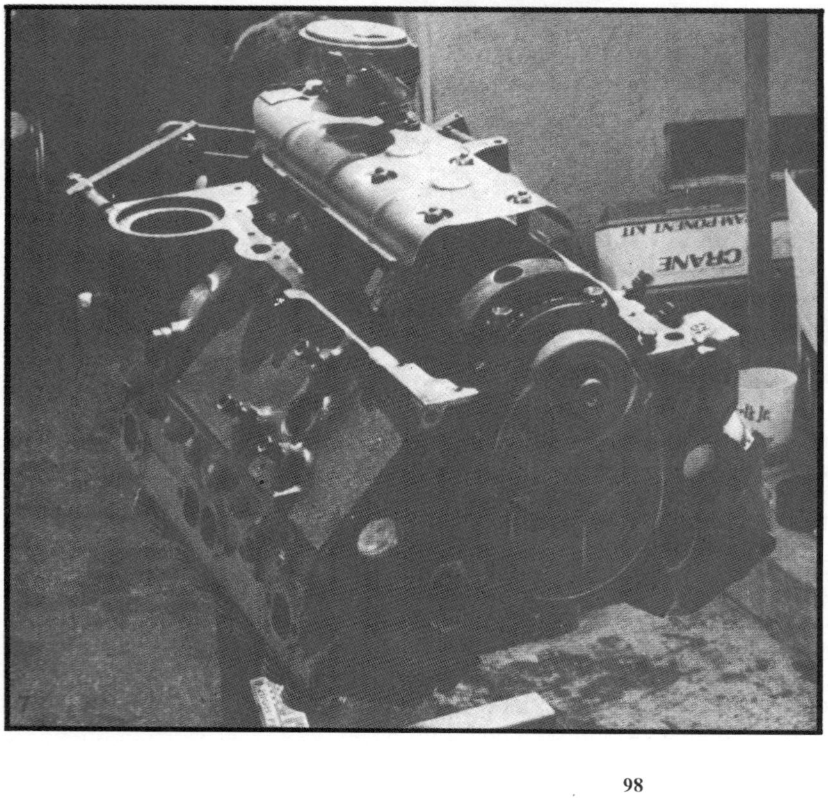

chrome surfaces. The result was that the rings seated almost instantly after the initial start-up.

All the machine work on the 302 Chevy was performed by the experts at Valley Head Service (19340 Londelius St., Northridge, CA 91324), under the guidance of Larry Ofria, who applied a little 1980 technology to the 11-year-old engine. While the 302 came with the best Chevy parts, there were a number of areas which benefited from some light massaging. (A classic example was the piston pin oiling.) All 302 Chevys came with full-floating pins that tended to gall from a lack of oiling, but by drilling the piston end of the connecting rod from the top, and at an angle from the bottom, an oil passage was formed. The rods received further refinement as they were shot-peened and Magnafluxed. The standard Chevy rod bolts were replaced with SPS bolts, which were costly but less expensive than a new engine.

The stock forged and Tufftrided crankshaft had finally shown signs of wear after 106,000 miles and, at the suggestion of Larry Ofria, was ground .010 on the rod and main journals. Since a .010 cut was enough to remove all the Tufftriding, the crank was retreated to make it as strong as the original. To make up the additional journal clearance, TRW .010-over Clevite 77 bearings were used.

In keeping with the general restoration plans, a duplicate of the Chevy first racing design camshaft was installed. The camshaft was a Crane "Blueprint Series" unit which featured an intake lift of .493 and an exhaust lift of .512, while the

(4) The crankshaft snout had been stripped, making it impossible to tighten down the harmonic balancer. Valley Head Service devised a cure by hollowing out an aircraft bolt and rethreading the inside. The crankshaft was drilled and tapped and the threaded insert screwed into place.

(5) This nifty tool from Childs & Albert made installing the eight pistons a breeze. The tapered ring allowed installation of the pistons without the hassle of bent rings or slipping ring compressors.

(6) Crane and Holley supplied important valve and fuel system gear to strengthen the Chevy small-block. The Crane camshaft was a duplicate of the Chevy "140" first racing design camshaft, while roller rockers and pushrods replaced the worn Chevy parts. The Holley high-performance mechanical fuel pump was large enough to handle the fuel needs of even the cross-ram system.

(7) The four-bolt-mained block (available only in '68-'69 302 models) came with a forged steel crankshaft and full windage tray. The oil pump was a TRW high-performance unit.

duration was 299 and 313 degrees, respectively. The heads received a light polish job on the intake and exhaust ports, as well as the 68cc combustion chambers. Bronze valve guides were honed out lightly until the TRW valves slid smoothly in the head. Crane pushrods and roller rocker arms completed the head modifications, replacing the stamped-steel Chevy rockers and worn pushrods.

A Cloyes Tru-Roller timing chain proved to be a very helpful addition to the small-block Chevy. Not only did it give strength to an area which was extremely sloppy previously, but it allowed easy degreeing of the camshaft. The three-position crankshaft sprocket allowed adjustment of the camshaft at four degrees advanced, four degrees retarded or right on time. Experimentation also allowed adjusting the camshaft for extra advance for racing purposes and setting the timing back for street use.

Other additions to the engine included a set of Hooker headers, a Stinger II ignition, Holley mechanical fuel pump and a Rapid Cool oil cooler (used for racing purposes only).

Once the engine was put together, the clutch unit was installed. We decided to replace the aluminum bellhousing with a Lakewood scattershield, giving the driver

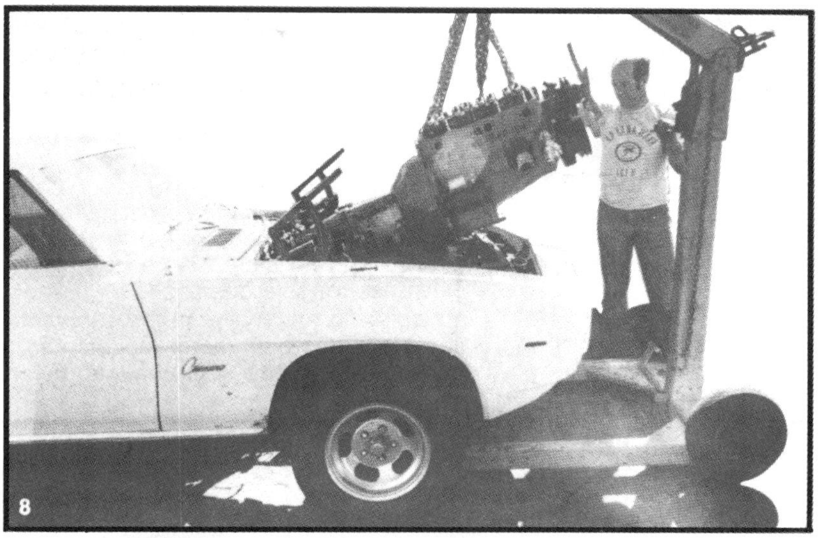

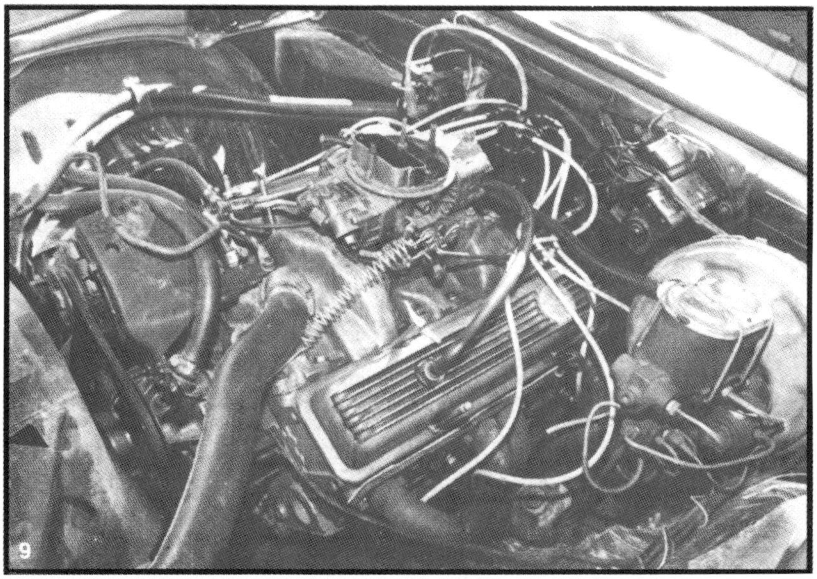

(8) Installing the engine in the car was easy, but all-new motor mounts were used to assure that the high-revving motor would stay where it was bolted. Note the use of a Lakewood scattershield to protect against flywheel mishaps.

(9) The completed engine installed and ready to fire.

Options

The 1969 Z28 Camaro featured one of the longest lists of factory- and dealer-installed racing options ever offered by Chevrolet, thanks to the successful racing career of the Trans-Am Camaros campaigned by Mark Donohue and Roger Penske. The short-lived series, which lasted from 1967 to 1971, ruled that in order for a part to be legal for competition, a certain number of cars bearing these race-bred options be made available to the public.

The option list grew, and hit an all-time peak in '69, when the racing Camaros took the Trans-Am title. The benefactors of the victory were those wary auto enthusiasts who ran down to their local Chevy dealers and purchased their street-legal race cars.

In '69, when you ordered a Regular Production Option Z28 (RPO Z28) the car came with a special Z28 suspension, 302-cubic-inch engine, 4-speed transmission, dual exhausts, quick-ratio steering, 15x7-inch Rally wheels, F70x15 tires, Z28 emblems and rear bumper guards. Rumor notwithstanding, all Z28s did not come with racing stripes, spoilers or full instrumentation. Listed below are the various options available in 1969.

(1) Cross-ram aluminum manifold with twin 600-cfm Holley carburetors. The dealer-installed option cost $500.

(2) RPO JL8 four-wheel disc brakes. This was a factory-installed option which was delivered on only 206 Camaros. The cost of the option was $493.

(3) ZL2 raised cowl induction hood. This hood came on 10,026 Camaros in '69 and was $79 as a factory-installed option.

(4) Steel tube exhaust headers (part Nos. 3942501 and 3942502). Dealer-installed options.

(5) RPO M20, M21 and M22 4-speed transmissions. The Z28 Camaro came standard with either the M20 or the M21 wide- and close-ratio transmission. For $322 the factory would install the M22 "rockcrusher" transmission.

(6) First racing design camshaft (part No. 3927140). Featured intake lift of .493 and exhaust lift of .512.

(7) Transistorized ignition system. Featured ball bearing distributor for reduced friction.

(8) RPO D80 spoiler package. Installed on 19,040 Camaros, the spoiler option was available for $37.

(9) Rearend gearing. Available in ratios of 3.23, 3.55, 3.73, 4.10, 4.56 or 4.88, with 5.13 as a service part.

(10) Eight thousand-rpm tachometer and 140-mph speedometer.

(11) Oiler cooler. Available under part No. 3157804.

(12) A complete line of suspension equipment, including different springs, tie rods, knuckles, ball joints, spindles, etc., for any racing application.

an extra measure of security knowing that the flywheel wouldn't come through the floorboards. Along with the scattershield, a Schiefer clutch disc and pressure plate ensured a strong link to the Muncie transmission. However, one problem did occur with the installation of the Lakewood scattershield: The block plate interfered with the flywheel rotation. Shaving the back side of the flywheel a full .070 gave it more than enough clearance for safe operation.

Replacing the engine in the car went flawlessly, and both motor mounts and the transmission mount were changed before the powerplant was bolted down. The 302 engine tends to be a high winder and places extra stress on motor mounts because of its great movement.

Except for a slight adjustment of the initial timing, the engine started easily; and as soon as the operating temperature was reached, the hot lash on the camshaft was adjusted—so ending part one of the Camaro restoration. Part two gives a detailed look into suspension improvements and final painting techniques, so be with us as we return this Z28 to its former glory in next month's installment. **HR**

Cross-ram

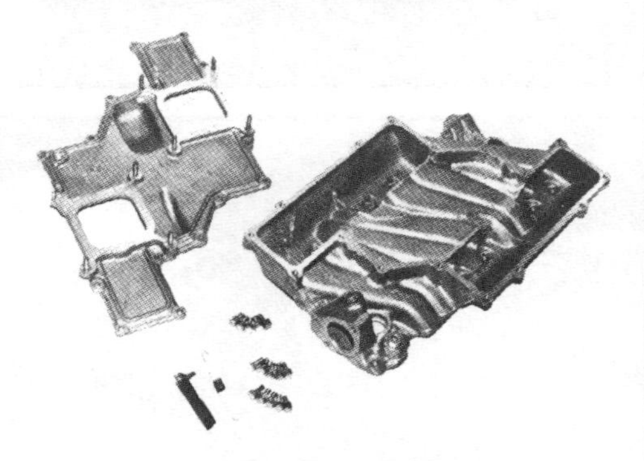

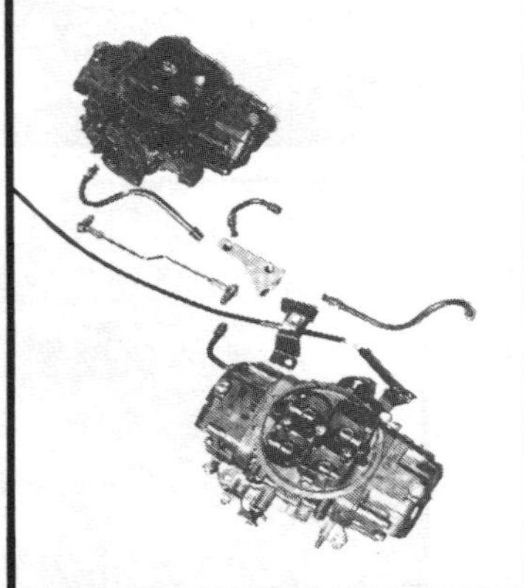

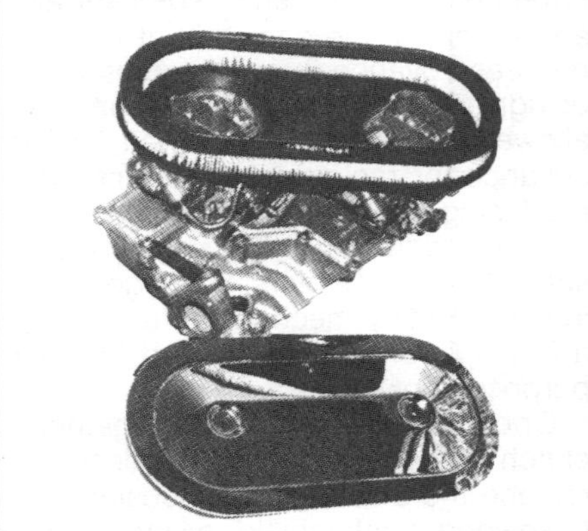

By far the most outstanding performance option available on the 1969 Z28 Camaro was the cross-ram-style manifold with its Holley 600-cfm double-pump carburetors. Many racers feel that this option, which added a full 100 horsepower to the 302-cubic-inch engine, was a major factor contributing to Camaro taking the Trans-Am title back in 1969.

The cross-ram, or two-by-four as the Chevy engineers called it, was quite a departure from the norm. It was—and is—the only dual Holley 4-barrel setup ever offered to the public by Chevrolet. The manifold with accompanying hardware was a dealer-installed option and cost the buyer $500 back in 1969. The number of these manifolds sold to the public is quite a mystery, as many were sold over the parts counter to racers who found out about the system and installed it themselves. Today only the carburetors and the foam ring which sealed the air cleaner base to the fiberglass hood remain available from a Chevy dealer.

The carburetors used on the cross-ram (which cost close to $350 apiece today) were not the standard 600-cfm Holleys, as they retained totally different float bowls and had accelerator pumps on both the primary and secondary circuits as large as those found in the Dominator-series Holleys. The low-torque 302-cubic-inch Chevy engine was always a poor drag-strip performer, and it was even slower with the cross-ram, because of the large fuel flow. At over 3000 rpm, however, the engine responded, pressing the passengers back in the seats and showing its race-track-intended teeth. This was a racing manifold designed for racing purposes, not for motoring around in bumper-to-bumper traffic.

Along with the basic manifold and carburetors, the cross-ram system included an air cleaner unit which ducted directly to a fiberglass hood with an elongated hole or attached box (two different styles were available) on its underside. The 1968 model Camaros also came with a cross-ram system which had a series of ducts leading to a hole cut in the firewall. Air was pulled in at the windshield cowling vents through the ducts and into the carburetors.

According to Chevrolet, the fiberglass hoods used in '69 were built for racing only, but complaints from the Trans-Am series organizers forced Chevy to make the hood available to the public—or stop using them on the race cars. For a while, steel hoods were cut for the race cars; but after enough fiberglass pieces had been made available to the public, the racing Camaros switched back to the 'glass hoods.

While restoring my Camaro, I spent many hours in Southern California swap meets looking for parts. By luck, and nothing more, one swap meet turned up a brand-new cross-ram, complete with 20 extra manifold gaskets, the fuel lines, five-way fuel block, the throttle cable and the bell crank for the foot pedal. After some fast talking and the exchange of a couple of dollars, the cross-ram was mine.

As I said earlier, the total number of cross-ram systems produced is a big question. My best guess, from a variety of different contacts made with Chevy engineers, owners of cross-rams and people at the Winters foundry which cast the Z28 single and cross-ram manifolds, is that no more than 600 manifolds (possibly fewer) were ever made. Because so many were for race-only purposes, the number of handmade air cleaners is something like 300 to 400. The same numbers come to mind with the ZL-2 fiberglass cowl induction hood; however, these pieces are extremely rare, due to their destructibility ratio: Usually the first thing to go was the 'glass hood.

Depending on the condition of the cross-ram system, the price can range from $2000 to $4000 for the complete manifold, air cleaner and fiberglass hood setup. Individual items, such as the fuel block and carburetor tie rod, are getting high dollars as well. It is a matter of how much the buyer wants the part—and lately there have been quite a number of desperate buyers.

Remember that while the cross-ram was nice for the track, it was a bother on the street, and there are a number of more recent modifications that can be performed on a small-block Chevy with more manageable performance gains. Today the cross-ram remains to remind us of the days when you fed your machine 36-cent-per-gallon Chevron Custom Supreme and got back all the performance you could handle.

Backyard Buildup: '69 Z28

PART 2: Suspension and Paint Modifications

By Cam Benty

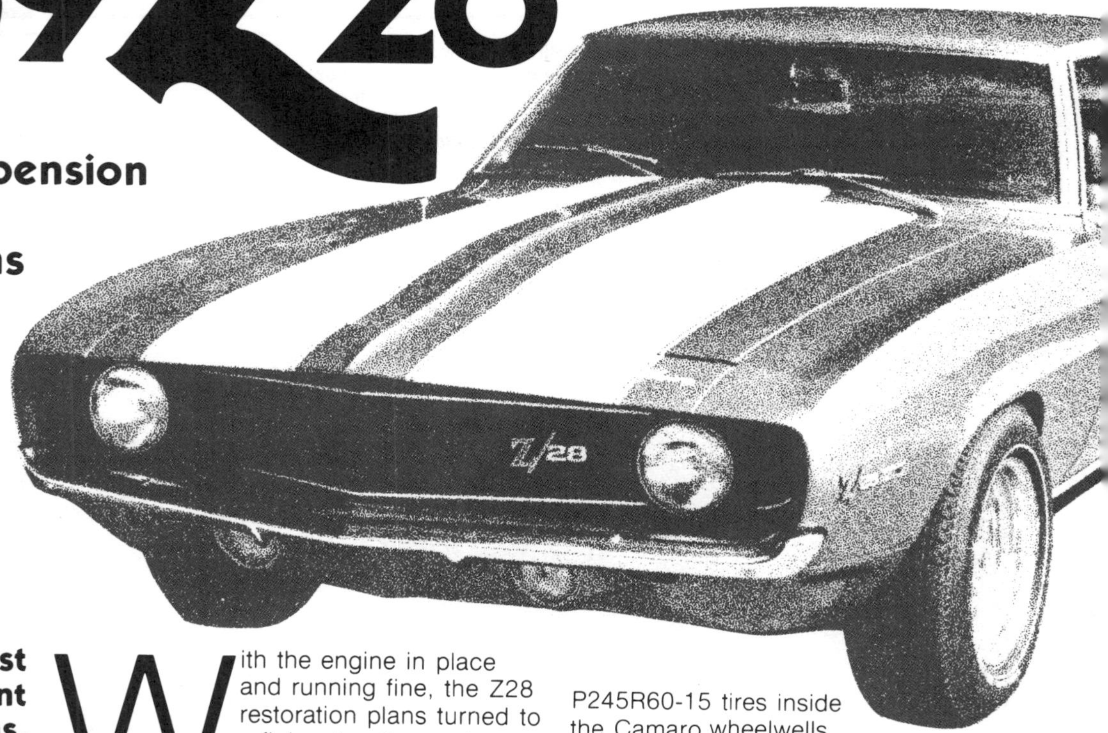

"The most frequent question was, 'Is this really the same car?' "

With the engine in place and running fine, the Z28 restoration plans turned to refining the Camaro's handling characteristics and repainting the body. The driveway was the site for these modifications, as it was for the engine building. After all, there's no place like home. Right!

While Z28s were the handlers of the Camaro line, they still left a lot to be desired in terms of cornering attitude and understeering tendencies. In an effort to correct these problems, Guldstrand aluminum subframe bushings were installed in the joint between the Camaro subframe and the body. Each side was done separately to avoid shifting of the frame, which could really affect the car's handling. The bushing installation involved undoing the two large bolts attaching the subframe to the car, removing the rubber bushings, and replacing them with solid bushings. The change produced incredible tightening of the front end, but made the ride a bit rougher and lowered the front end about ½-inch.

Bilstein shock absorbers were installed at the four corners and produced a drastic change in the Camaro's handling. The gas-filled shocks controlled the understeering tendency and worked well with the big-block Camaro front springs, which were cut three-quarters of a coil to give the correct ride height.

The final handling trick was the addition of Goodyear Wingfoot tires and American 200S rims. The 15x8-inch wheels proved the perfect offset to house the P245R60-15 tires inside the Camaro wheelwells. The Wingfoots greatly improved the Camaro traction, most noticeably in hard corners.

We spent more than a month painting the Camaro in the original style. The dark blue interior mandated the use of Chevy's "Marina Blue," which was found on stock Z28 Camaros. A helpful representative at Ditzler tracked down the color as number 2083 (or Le Mans Blue Poly) and a batch was made up. Only white and black stripes were available in '69, and my choice was white—as simple as that. Lacquer paint was used because of its quick drying properties and our ability to sand or buff out the various imperfections. The painting was by far the most exhausting step!

To fix the front end, the previous owner had purchased fresh sheetmetal, which we fitted into place. The curved fender section attaching to the front valance panel was distorted and had to be pulled

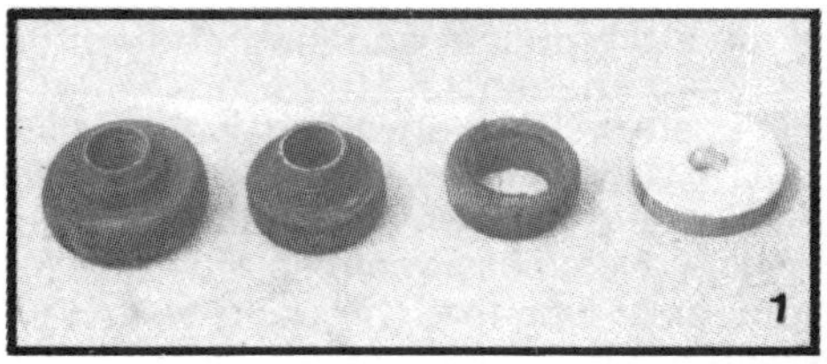

(1) All three rubber rings on the left were replaced by Guldstrand aluminum subframe bushings similar to the one shown on the right. The solid bushings dropped the front end ½-inch and tightened up the front-end flex. A total of six bushings were installed in the Camaro suspension.

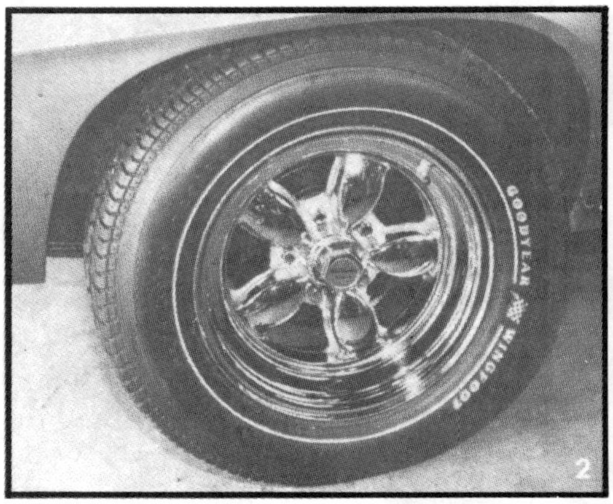

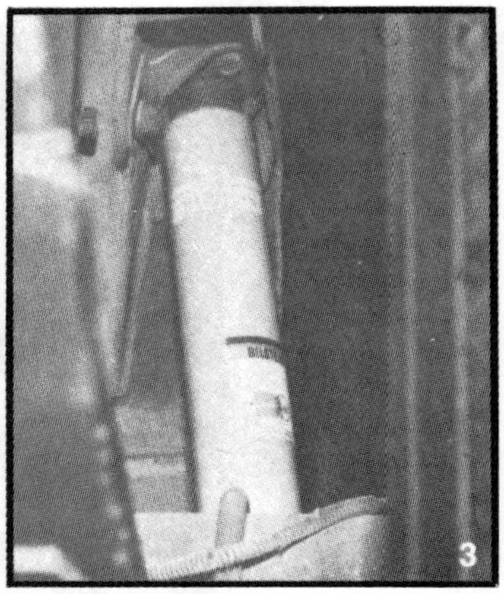

(2) Possibly the most noticeable improvement in the Camaro's handling came after the installation of the Goodyear Wingfoot tires mounted on the 15x8-inch American 200S wheels. The wide P245R60-15 tires required some inner fender lip bending to clear.

(3) Bilstein shock absorbers were installed at the four corners, greatly improving the car's cornering ability. To install the front shocks, the hole in the lower A-arms had to be enlarged with a cutting torch.

(4) Past experience in racing the Z28 Camaro warranted the use of a Rapid Cool oil cooler. The cooler was mounted in the front grille and tapped oil at the oil filter mounting tab.

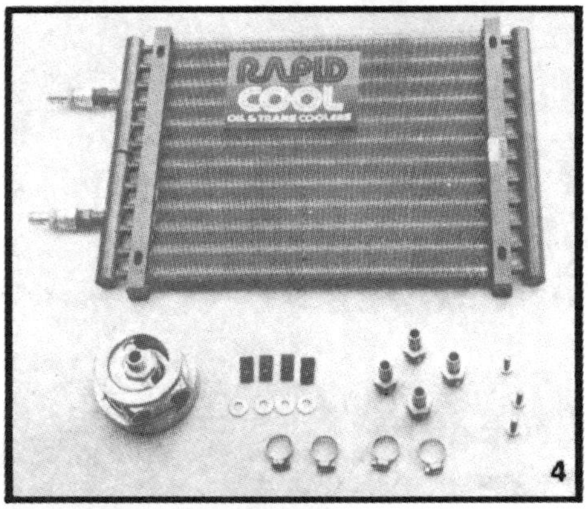

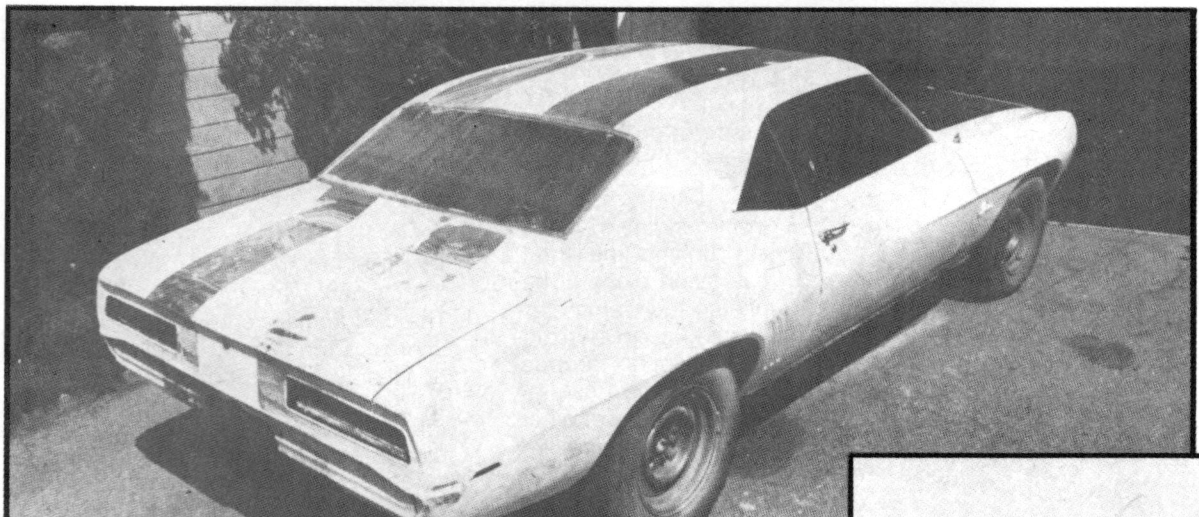

back into position. To make it entirely smooth, the panel was spread with a small amount of Bondo and sanded to the same contours as the opposite fender curves. The only other sheetmetal change on the Camaro was the addition of a steel cowl induction hood, purchased from the local Chevy dealer for $210.

The rest of the Camaro body was extremely straight and, aside from a few paint chips, needed little correction. The

(5) Although the engine was running fine, the exterior was still in a sorry state. The blue and red candy stripes were painted over a gold base that was extremely tough to cut through.

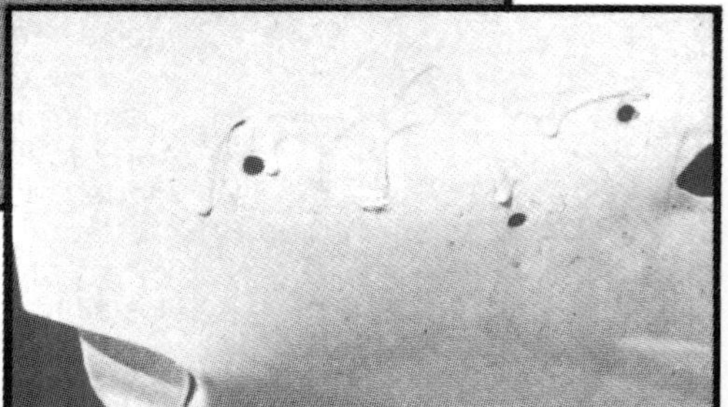

(6) The original paint was white. However, a terrible repainting job, which didn't include removing any emblems or checking for runs, meant the entire car had to be sanded smooth.

hardest chore in preparing the body was removing the candy red and blue stripes. While the first layer of color came off easily, the gold base was extremely thick and required air-sanding with 150-grit sandpaper before any headway was made. Rougher sandpaper could have marred the surface of the metal, which may have shown up in the final paint.

Several layers of light gray primer were used to build up the surface, as a large expanse of bare metal had been exposed when we were grinding the stripes. Other areas had cracked, and they too were taken down to the bare metal. Each layer of primer was sanded smooth before the next coat was applied.

With the variety of weather conditions facing our driveway paint job, the correct ratio of thinner to paint was always a question. A constant adversary was wind, which caused the paint to "orange peel" at times and threaten to run at others. We thinned the primer 150 percent, which was drier than necessary but allowed sanding off the dry areas instead of dealing with runs. The blue paint was a different story, however, because although we started thinning (per the Ditzler instructions) at 150 percent, we progressively thinned out the paint more and more until the final coat was thinned almost 400 percent. The final color coat on any paint job is always the most important; but in paints with high metallic content, it is crucial, because it cannot be sanded smooth. The paint must lay out smoothly on its own, since sanding metallic paint scratches and dulls the ultimate shine.

Throughout the Camaro restoration, research was done to determine the correct way of restoring different parts. The design for the stripes was taken off the Camaro of John Seehuus, who was kind enough to measure the stripes on his original Z28. The 14-inch stripe had a ¼-inch pinstripe, ¼-inch away from the larger stripe. The rear stripes were curved at the top and came ¾-inch away from the base of the rear window. The stripe came over the rear spoiler and ended at the bottom of the trunk lid.

The front stripes were a bit different, curving around the contour of the cowl vents and coming straight off the front edge of the car. The raised cowl induction hood was draped with the stripe, the inside edge lining up with the edge of the cowl vent. A lot of time was spent laying out the tape before a final satisfactory

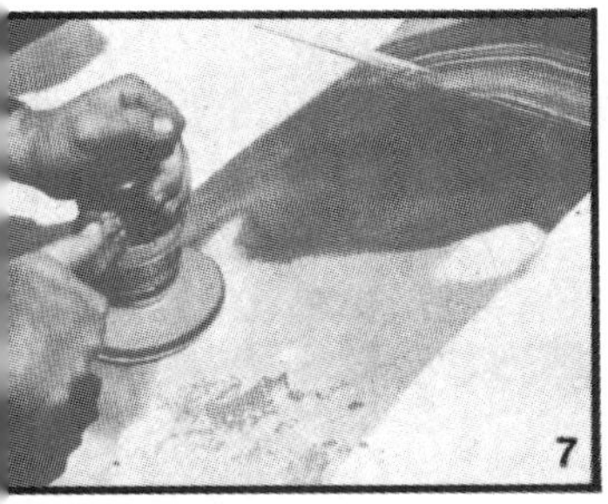
7

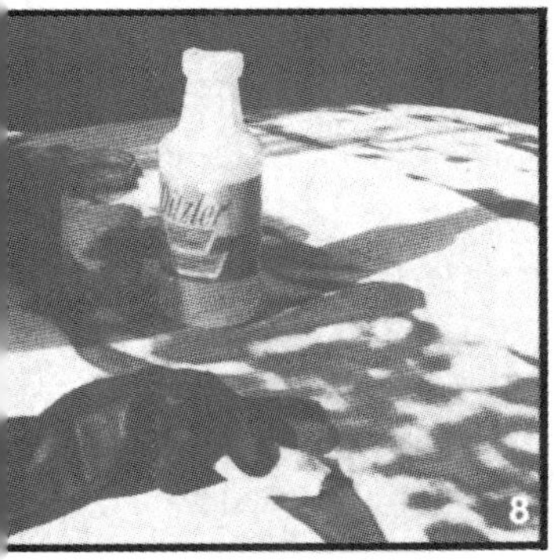
8

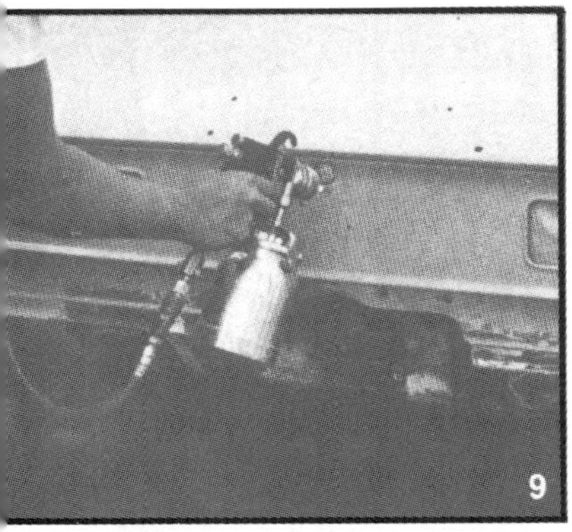
9

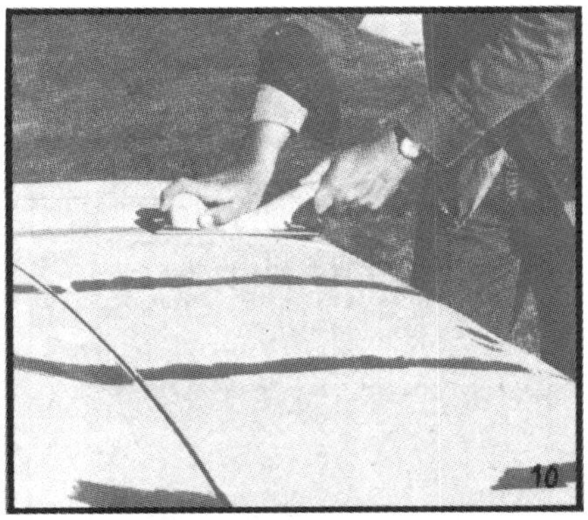
10

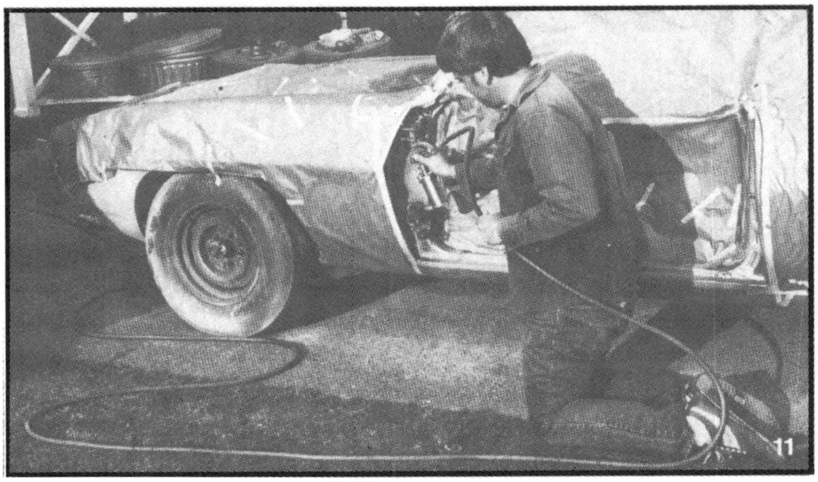
11

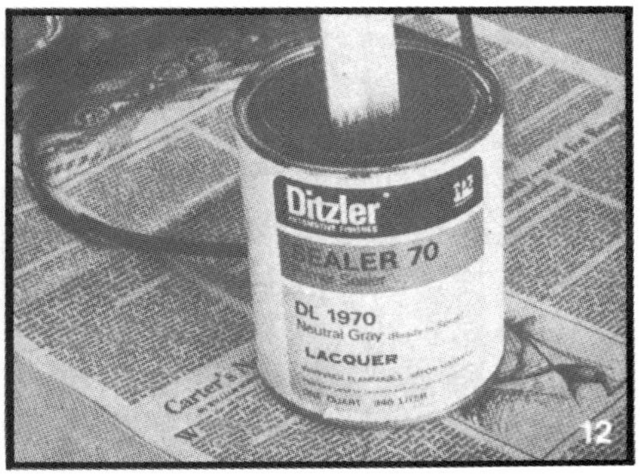

12

(7) The gold-based stripes had cracked and checked badly and were removed, completely exposing the bare metal. A dual-action sander and 150-grit sandpaper were used for this operation.

(8) To avoid rust problems on the exposed metal areas, the surface was treated with Ditzler metal prep, then sprayed with light gray primer.

(9) The primer coats were sprayed on very dry, then sanded smooth with 400-grit sandpaper. The Ditzler gray primer was thinned 150 percent, as per instructions.

(10) Following the first layers of primer, spot putty was used to fill nicks in the car's surface. The edges of the stripes had been cut with a razor blade, leaving scratches that had to be filled and sanded flush.

(11) To assure that the door jambs were painted properly, the doors were removed from their hinges and painted with a Binks No. 15 Touch Up gun.

(12) Because of the poor base paint, the car was covered with Ditzler Primer Sealer before being sprayed blue. The sealer was sprayed unthinned.

design was achieved.

After the stripes were painted, a full gallon and a half of clear lacquer was sprayed over the entire car. Again the clear was thinned experimentally between 200 and 300 percent and then sanded perfectly smooth, using plenty of 600-grit sandpaper.

Russ Johnson, co-owner of Auto Artists (18150 Foothill Blvd., Fontana, CA 92335), was kind enough to buff out the Camaro paint, using an electric buffer and Ditzler rubbing compound. To remove the swirl marks left by the buffer, Ditzler finish glaze was hand-rubbed over the entire car. The result took the Camaro from a dull smooth surface to a shiny, dazzling signboard which drew the attention of half the neighborhood. If we ever tire of the magazine business, we're set for life as backyard painters.

The dark blue interior was in good condition except for a rip in a door panel and a sun-bent dash pad. For trivia

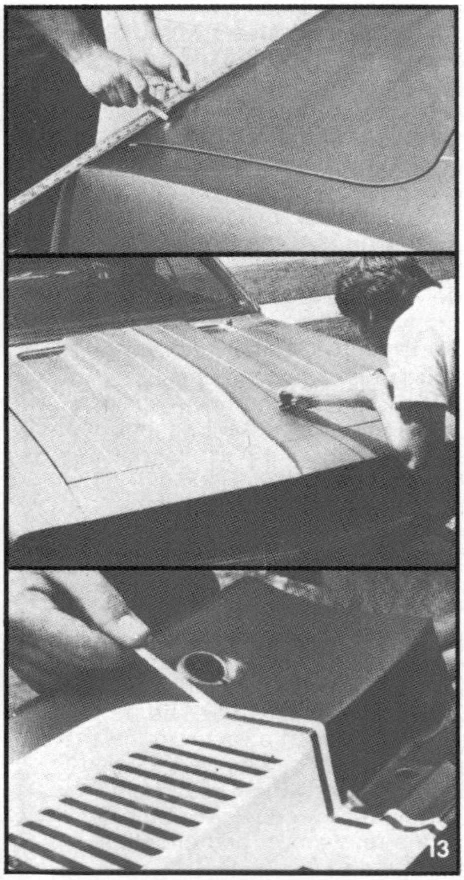

(13) By far the most impressive part of the Camaro paint job was the striping. By using guidelines from an original paint job, the stripes were measured out and marked in chalk on the car's surface; and sighting down the marks helped to position the ¼-inch tape. Next to the original piece of tape, two more ¼-inch rows of tape were laid down with the piece in the middle, then removed to form the guidelines for the pinstripe.

(14) With a Binks No. 62 spraygun, the entire paint job was covered in a gallon and a half of clear lacquer, which was sanded smooth in preparation for final buffing.

(15) Russ Johnson, co-owner of Auto Artists in Fontana, California, buffed the paint to its final luster. Using a cutting pad and Ditzler rubbing compound, he brought out the shine in the paint. To remove the swirl marks left by the buffer, a final finishing rub was applied.

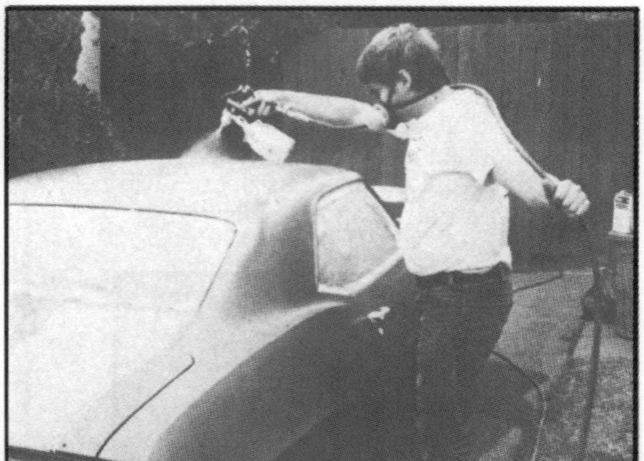

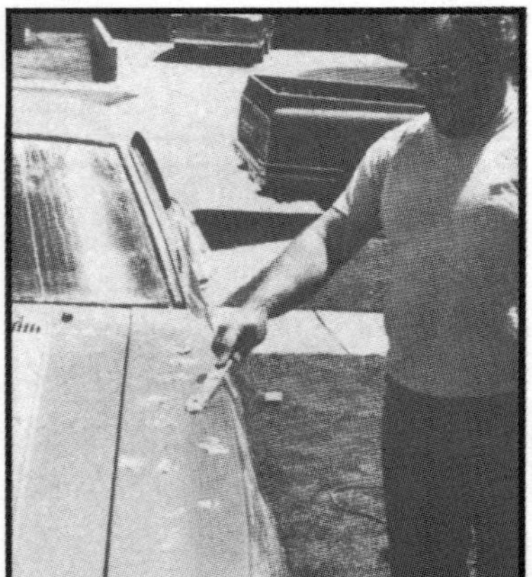

Buildup '69 Z28

experts, the '69 standard door panels had bolt-on armrests, while the custom door panels had molded armrests. This particular car had the custom panels, and the driver's side had a large rip in the center. A custom upholsterer was located, the dark blue material was matched, and the entire panel was recovered in one piece, like the original.

The dash pad was a different story, since although certain outlets still sell the pads, the only available color is black and the cost is over $125. Here I took matters into my own hands. The dash pad was removed and the fiberglass underliner was split at the apex of the warp. The curve was flattened out and glued straight, and the pad was then glued tightly over the straightened fiberglass form. Although the result was not perfect, it was satisfactory—and better than spending the extra money for another part which would need color dyeing from time to time.

As noted ealier, the correct restoration procedures were an integral part of the Camaro restoration. Contacts were made with owners of original Z28s, but for the rare information, a pipeline of experts was set up. The following were responsible for supplying parts and/or information about the original Z-cars.

Dave Longstreth
Arizona Camaro Parts
1708 East McLellan Ave
Phoenix, AZ 85016

Chevyland (Marv)
1748 Fulton Ave.
Sacramento, CA 95825
(916) 481-3906

David Dietz
Louisville, KY
(502) 239-1875

The result of this three-month project was a very quick Camaro with a certain amount of collectable worth. But the greatest prize for a job well done came from the kids who contantly circled the car throughout its preparation. When it was finally completed, the most frequent question was, "Is this *really* the same car?" **HR**

The result of all the hard work—and worth every minute of it!

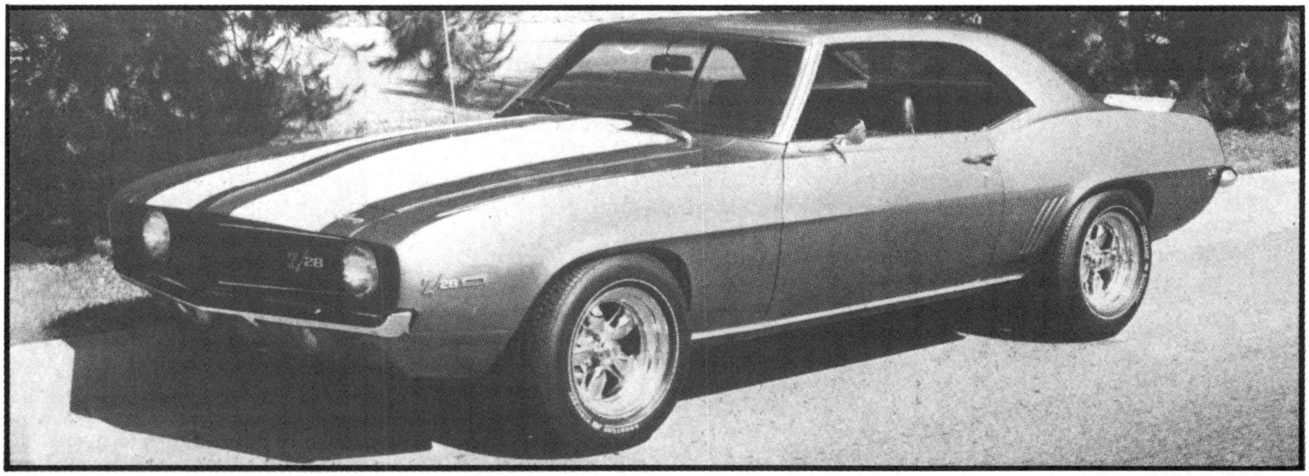

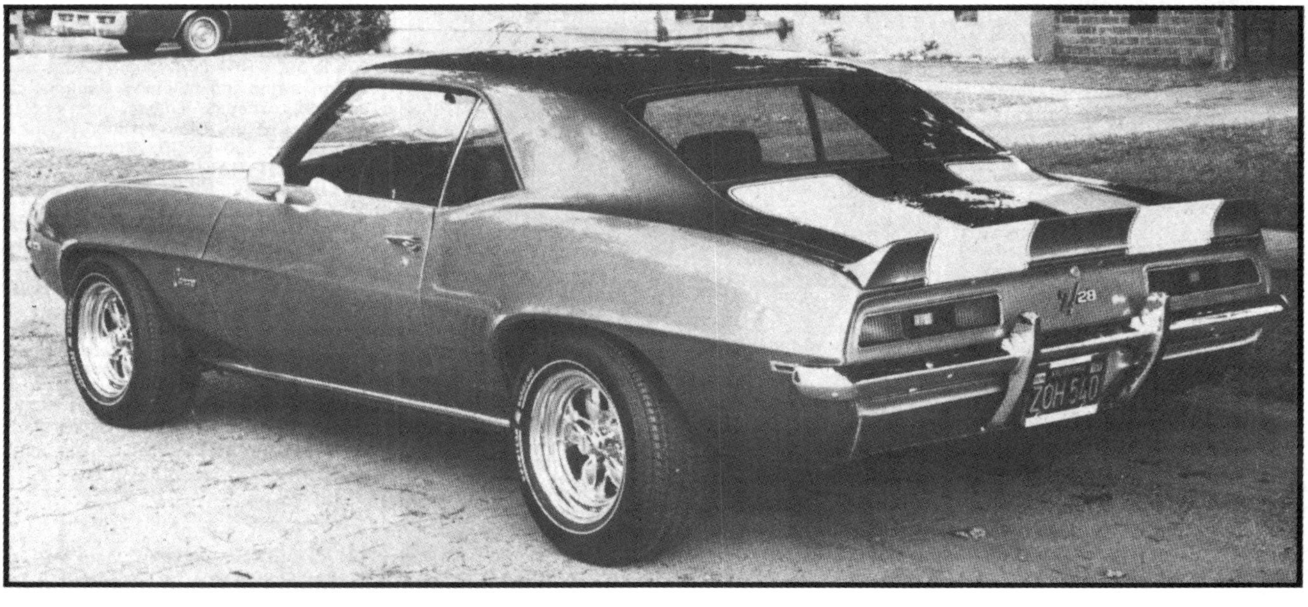

HOW TO MAKE YOUR CAMARO BETTER THAN NEW!

CAMARO RESTORATION GUIDE

One of the persistent automotive myths in America is the story about the secret factory in the Rocky Mountains that produces brand-new 1957 Chevys. Well, if that secret plant really exists, it must be turning out brand-new 1969 Camaros. Go to any street machine meet or car show and you're likely to see row after row of *perfect* Camaros. They have flawless paint, spotless interiors, and pristine engines. In truth, though, these are used (and sometimes abused) cars. They've endured the hardships of highway miles and hail storms. They've survived hit-and-run drivers and half-hearted hop-ups. Every year there are fewer classic Camaros in the world, but the survivors are climbing fast.

The force behind this Camaro revival is the growing interest in restoration. More and more enthusiasts are recognizing the value of older Camaros. "Value" means more than just their worth; it also includes the

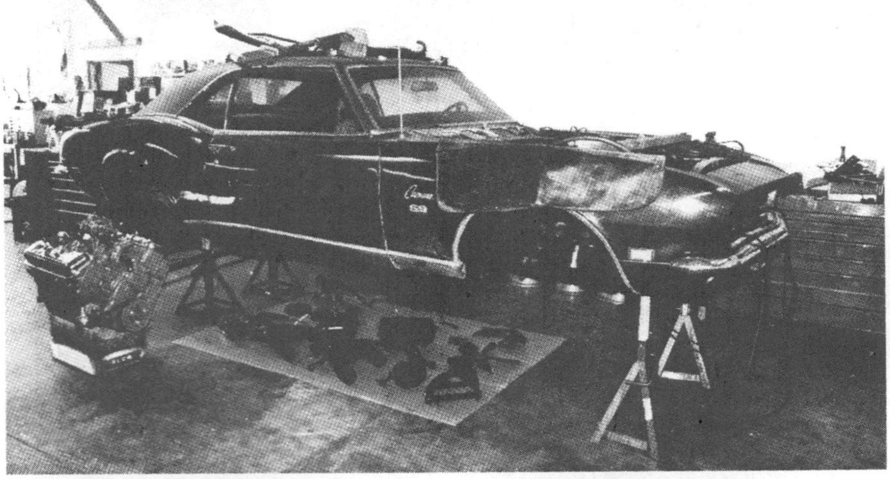

It's been more than 15 years since Chevrolet stopped building classic Camaros, but today many enthusiasts are constructing "better-than-new" F-bodies. As interest in Camaro restoration continues to grow, more specialists are offering hard-to-find parts and supplies.

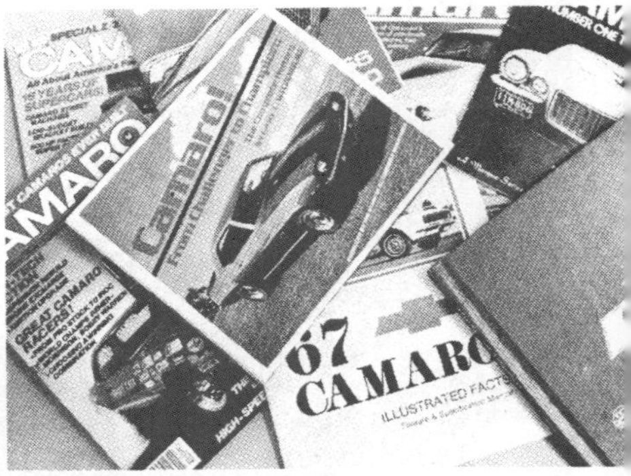

Camaro literature may not make the best-sellers list, but there's an abundance of publications to guide your restoration efforts.

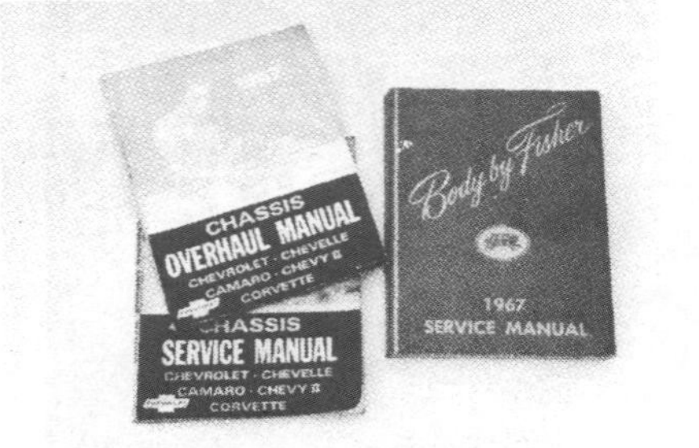

Factory service manuals are an invaluable source of information on overhauling and restoring early-model Camaros. Three volumes cover minor repairs (chassis service), major rebuilds (chassis overhaul), and glass and sheetmetal (Fisher body).

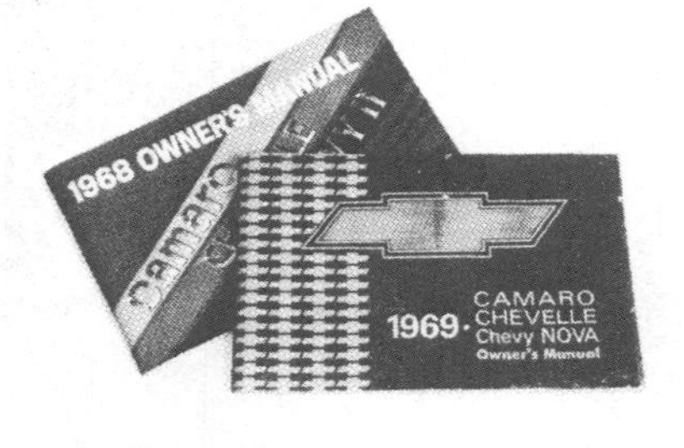

Original and reprinted owners manuals are widely available for classic Camaros. They add a touch of authenticity to a "like-new" F-body.

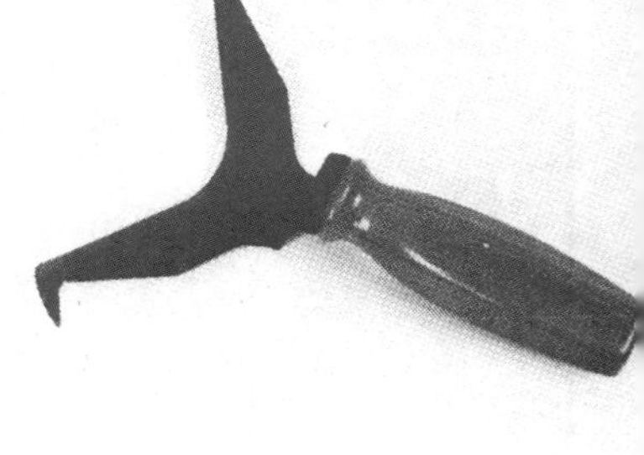

Special tools make the tough jobs easy; this one removes Camaro windshield molding clips.

pleasure of owning and driving a special automobile. Rescuing a neglected Camaro is better therapy than you'll receive from a dozen psychiatrists, and putting a rebuilt Camaro through its paces is the most fun you can have with your clothes on. A bumper-to-bumper rebuild will test your skills as a detective, mechanic, and bodyman. It will try your patience, and put a strain on your monthly budget. But the results are worth every scraped knuckle and late night out in the garage.

RESTORATION BASICS

Every Camaro owner has his own idea of a proper restoration. For some enthusiasts, a restoration should duplicate the car's original condition, right down to the correct hose clamps and battery terminals. Other Camaro owners take a more liberal approach, mixing factory and aftermarket components. After all, radial tires and high-output ignitions were unknown when the first Camaros were produced—why not take advantage of advances in automotive technology to make a good car even better?

The right approach for your Camaro restoration project depends on your car and your ambitions. Not all Camaros were created equal; rare or limited-production models deserve a more rigorous restoration than a basic sport coupe. For example, if one of the 602 Z/28s built in 1967 should come your way, you really owe it to yourself (and to the car) to make it as authentic as possible. On the other hand, a nondescript coupe with a two-barrel V8 and a Powerglide will make a terrific commuter car if it's outfitted with modern tires, an assortment of bolt-on power parts, and a high-tech sound system. There's more than one way to overhaul a Camaro.

The one strict rule of Camaro restoration is this: *Don't do anything you'll regret later.* There are fads in automobiles just like fads in music and hair styles. The difference is that you can always buy new records or grow more hair—but you can't replace a ruined Camaro. Think twice before cutting a hole in your Camaro to install a sunroof or taking a torch to the quarter panels to graft on fender flares. Customized Camaros certainly have a place on the automotive scene—but it doesn't make sense to cut up an original F-body for the sake of a trendy automotive style.

Often the term "restoration" conjures up an image of a bare shell sitting on four jackstands, surrounded by piles of body parts, trim, and upholstery. That's certainly what the antique car buffs mean by "restoration," but few Camaros really require such drastic measures. Of course, there are some enthusiasts who relish the challenge of a frame-up rebuild. But it's certainly possible to achieve outstanding results without disabling the car for more than a weekend at a time. New carpeting, a well-scrubbed engine compartment, and a few days of detailing can often turn a derelict Camaro into a street star.

You can increase the chances of successfully completing your Camaro restoration by setting clear goals at the beginning. Car projects have a way of getting out of hand; the sight of a disassembled Camaro can drain your enthusiasm halfway through the undertaking. Don't try to tackle everything at once; concentrate on one area at a time, and work until it's finished. Don't tear into the engine while you're still doing bodywork—you'll end up with with bondo dust in your cylinder bores, and valve springs in your spray gun. By dividing the project

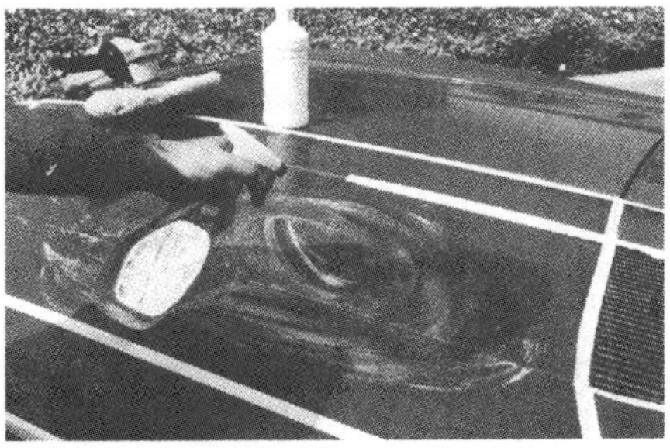

Simple, do-it-yourself detailing can turn a beater into a beauty. If your classic Camaro is in reasonably good condition, a good cleaning will work wonders. Buffing old paint is easier and cheaper than repainting. Use a medium compound and wet the work surface with a spray bottle. Wear goggles, and clean the buffing pad frequently to prevent the compound from gumming up the buffing pad.

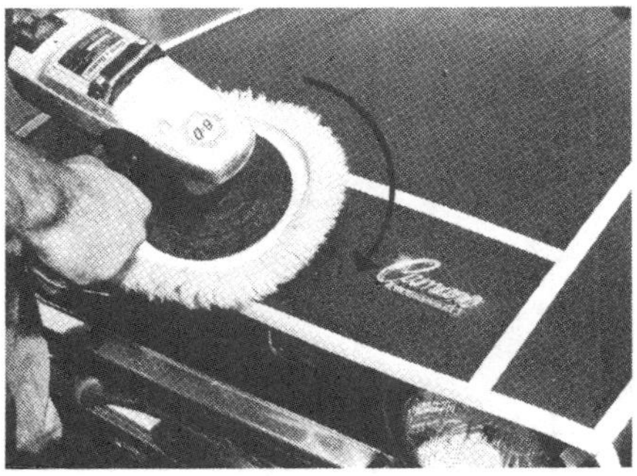

Protect panel edges and body creases with masking tape when buffing old paint to prevent burning through the color coat. Work from the middle of a flat panel out to the edges. Always buff across and off the top of edges, rather than against them. Keep the polisher moving to prevent damaging the paint!

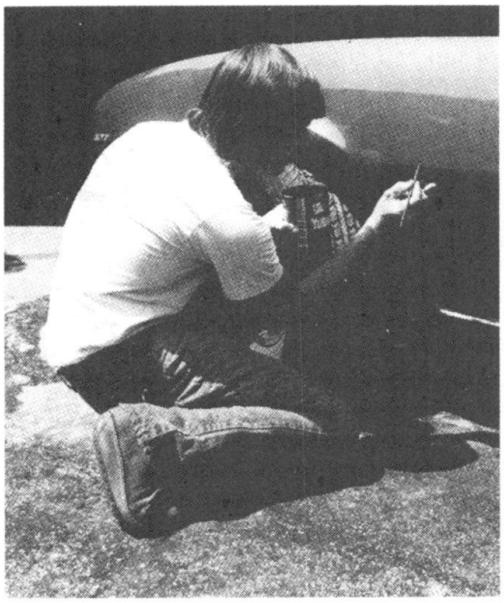

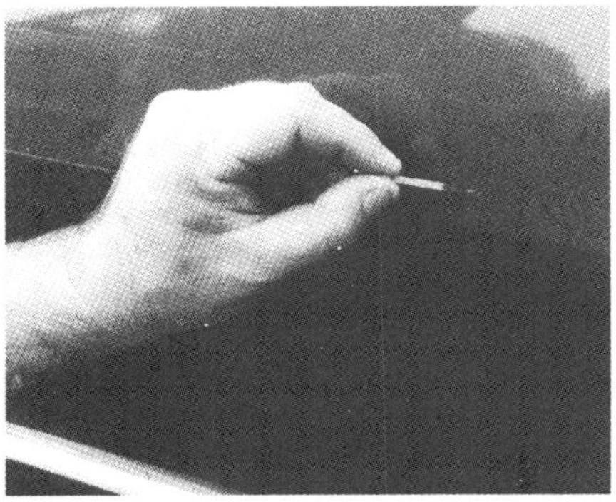

Paint chips and door dings can be repaired with a modeler's brush or a matchstick. Fill the chip by building up layers of paint, then lightly sand and buff the repair. An automotive paint supply can match the original color if you tell them the paint code on your Camaro's cowl tag.

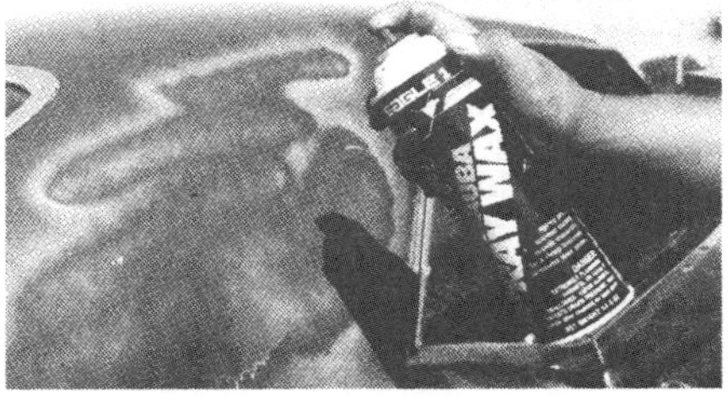

Wax does more than just beautify a Camaro's color coat; it also protects the freshly buffed surface from oxidation and dulling. Don't neglect this important detailing step.

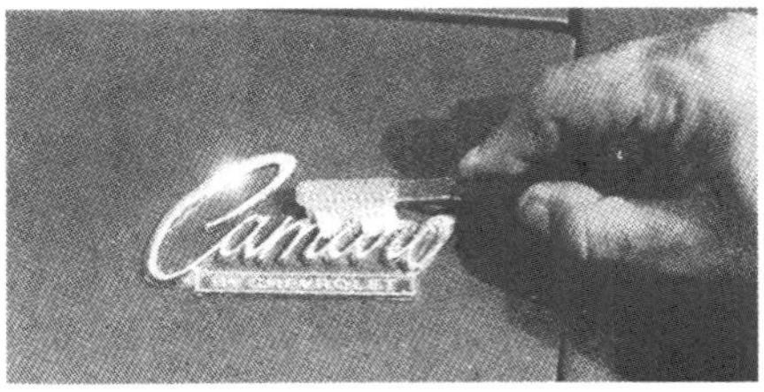

A toothbrush is ideal for removing wax buildup around emblems and trim. Touch up painted medallions with modeler's paint and a small brush.

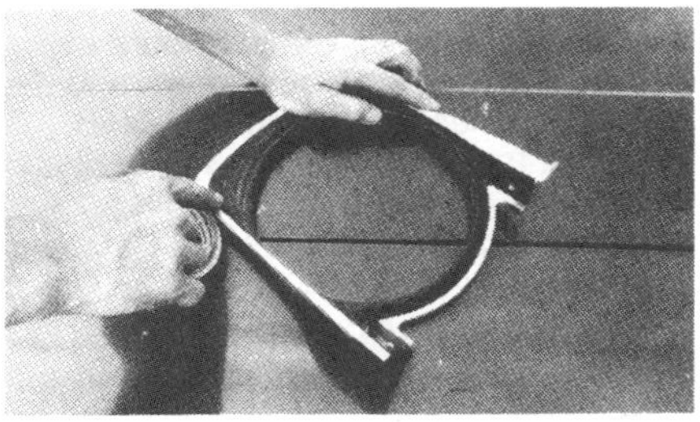

Camaro headlight bezels and grille trim are often pitted and scratched. They can be restored with a little metal polish, sandpaper, and black aerosol paint.

into manageable segments, you'll have the satisfaction of actually finishing one task before starting another.

CASH AND CAMAROS

Don't plan on getting rich rebuilding F-bodies. Some first-time Camaro restorers are motivated by the illusion of fast profits. It *is* possible to make money refurbishing Camaros, but the chances are slim unless you're a professional with the tools and expertise to do the job quickly. Classic Camaros are appreciating, but the days of dirt-cheap F-bodies have passed. As investments, Camaros probably rank somewhere below bonds and slightly above shares in the Brooklyn Bridge. There's not much pleasure, though, in owning a savings passbook or a trust deed—while putting your money in a clean Camaro can provide hours of joy. Dealing in refurbished F-bodies isn't a ticket to Easy Street, but it's reasonable to expect a classic Camaro to provide years of free transportation. Unless you encounter unexpected expenses, you will probably be able to get your money out of a restored F-body when it comes time to sell it—plus a little extra to keep up with inflation.

Money talk always brings up the dark side of Camaro restoration—the problem of counterfeits. There's a considerable difference in value between special models (such as Indy Pace Car replicas, Z/28s, and special COPO cars) and run-of-the-mill Camaros. Unfortunately, some unscrupulous operators are taking advantage of the burgeoning interest in Camaro restoration to pass off imitations as the real thing.

The caveat "buyer beware" should be kept firmly in mind by any prospective Camaro purchaser. If you are paying a premium price for a premium model, you should be certain of what you are buying. Unfortunately, it's virtually impossible to determine the authenticity of many early-model Z/28s unless the original paperwork still accompanies the car. You don't have to rely on just the seller's word, however. It takes just a few hours of research to learn the equipment and options that identify a genuine Z/28. Don't fall for emblems and stripes; take the time to examine the transmission, rearend, and serial numbers before you put down any cold cash.

CAMARO RESEARCH

Social critics have proclaimed this as the Information Age. For a Camaro restorer, reliable information is as important as a set of end wrenches. Fortunately, there are a number of excellent resources available for Camaro owners. The most valuable volume in any restorer's library is a shop manual. The factory has the final word on assembly, installation, and adjustment of critical components. Early Camaro manuals consist of three volumes: chassis service, chassis overhaul, and a Fisher body manual. You'll need all three to guide you through a Camaro revival. Factory assembly manuals have recently become available as well. These books detail the exact sequence used on the Chevrolet assembly line to build Camaros; they are an invaluable guide when you are trying to figure out which way a bracket goes or what size bolt to use.

You can plug into the Camaro information network easily. A subscription to *Hemmings Motor News* (Box 100, Bennington, VT 05201) is your invitation. This amazing periodical carries literally thousands of classified ads every month for cars, parts, and publications. You'll find listings for shop manuals, assembly guides, original factory literature, and Camaro books. Classic Motor Books (P.O. Box 1, Osceola, WI 54020) is another excellent source for information on America's favorite performance car. And don't overlook local and national Camaro clubs; members are always willing to help a fellow Camaro comrade track down an elusive piece of trim or a special carburetor fitting.

DOCUMENTATION AND IDENTIFICATION

If you have ever imagined yourself as an amateur Sherlock Holmes or Sam Spade, documenting your Camaro's history will give you an opportunity to indulge your detective fantasies. Few Camaros remain in showroom stock condition for very long; original wheels, shifters, exhaust manifolds, air cleaners, and radios all disappear over the years. The goal of documentation is to determine how the car was equipped when it rolled off the end of the Chevy assembly line.

There are several clues that will help you track down a Camaro's true identity. If the paperwork from the original invoice has somehow survived, you have hit the jackpot. The window sticker provides a comprehensive list of the car's factory-installed options. The original window sticker is an invaluable aid in verifying the authenticity of an early Z/28. Of course, most window stickers were discarded the day the new owner took delivery.

The "build sheet" rivals the original window sticker as a source of solid information about a Camaro's original equipment. The build sheet is the paperwork that accompanies the car as it travels down the assembly line. It describes the options and powertrain ordered for that particular vehicle. Unfortunately, there is no single place to look for the build sheet; they turn up

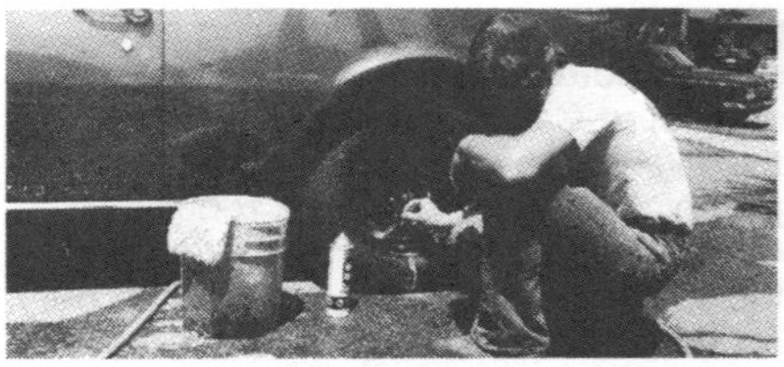

Stock wheels can be renewed with a soap pad and degreaser. Rally wheels look terrific after spraying with Krylon dull aluminum paint and a coat of clear acrylic.

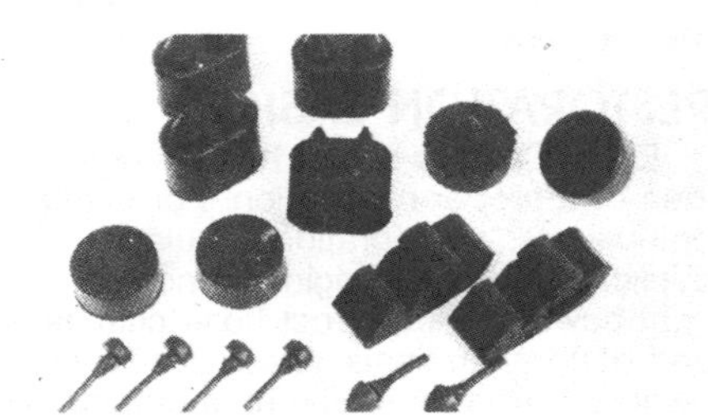

Old rubber body stoppers are often cracked and covered with overspray. Camaro parts suppliers offer complete kits to eliminate squeaks and rattles.

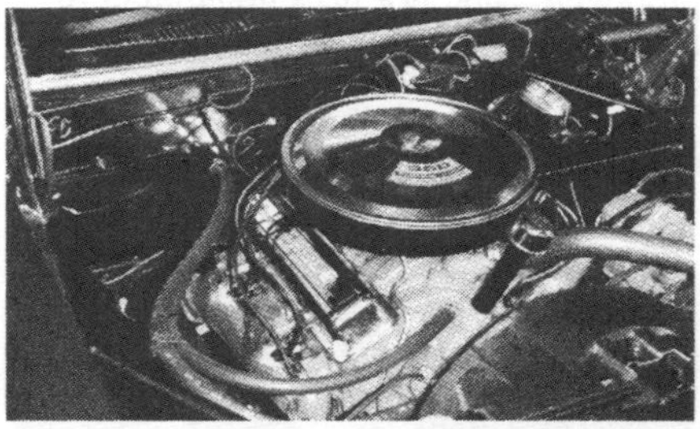

This is how the factory did it: original Camaro engine compartment has bright hardware and fresh semi-gloss black paint.

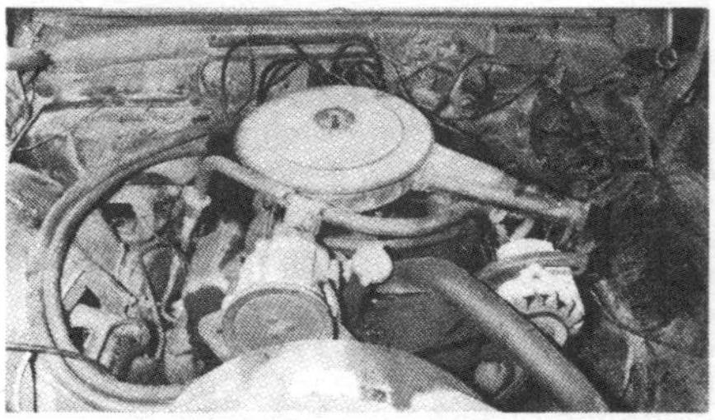

After 15 years of use, most unrestored Camaro engine rooms look more like this. Under that filth and grime is a small-block V8 waiting to be discovered.

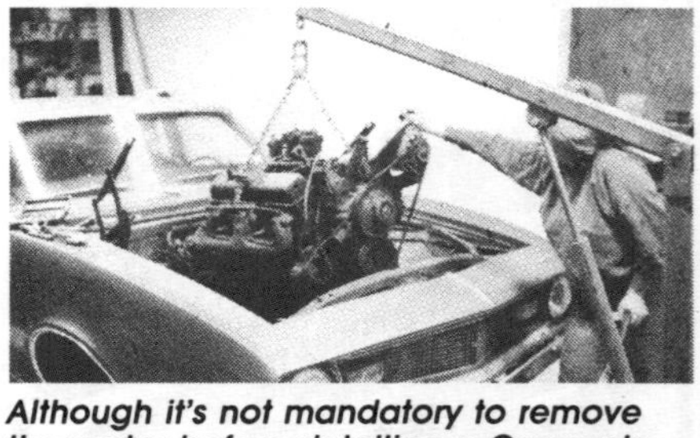

Although it's not mandatory to remove the motor before detailing a Camaro's engine compartment, it's easier to work with all that iron out of the way. You can also inspect and replace the motor mounts, throwout bearing, and other inaccessible components. Camaro motors are relatively simple to remove; rent or borrow a hoist before you start unbolting parts.

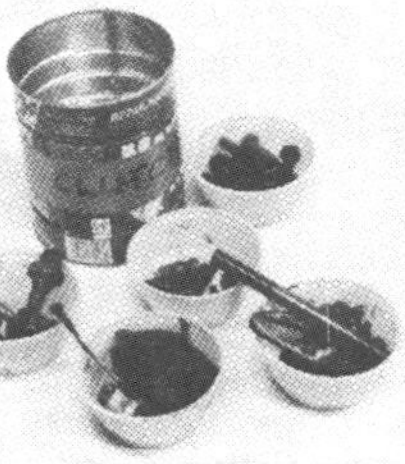

ɔrganize your Camaro's bits and pieces n coffee cans and margarine cups. abel fasteners to refresh your memory vhen it's time to reassemble components.

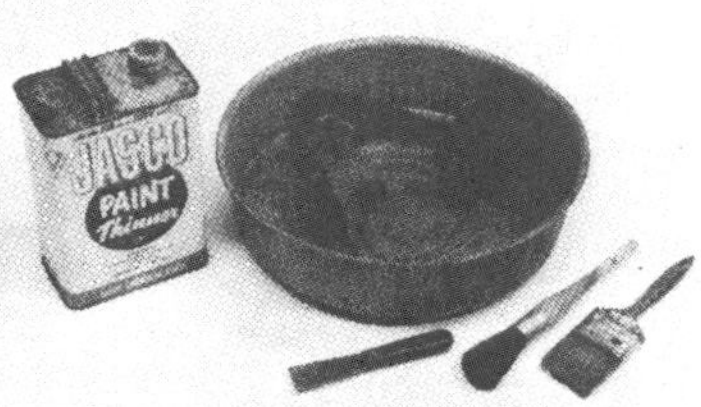

ɹse solvent with a high flash point to :lean greasy parts—never use gasoline! 'utty knives and wire brushes will emove years of grime; use old paint rushes to scrub delicate parts.

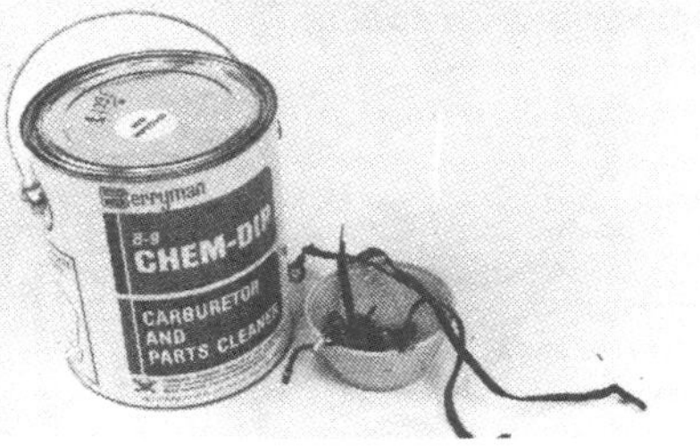

'oak small parts and fasteners in :arburetor cleaner overnight to remove aint and stubborn dirt.

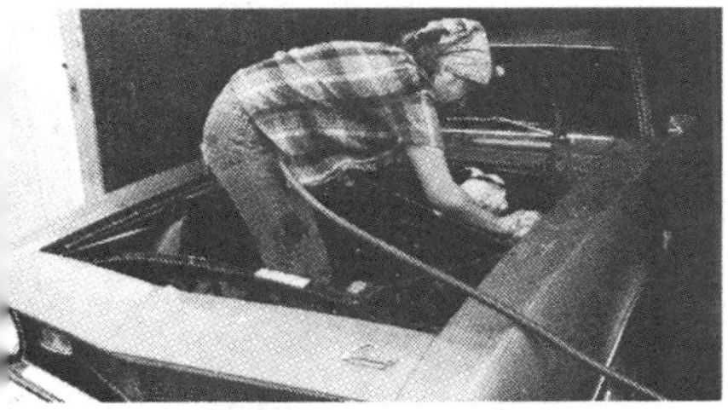

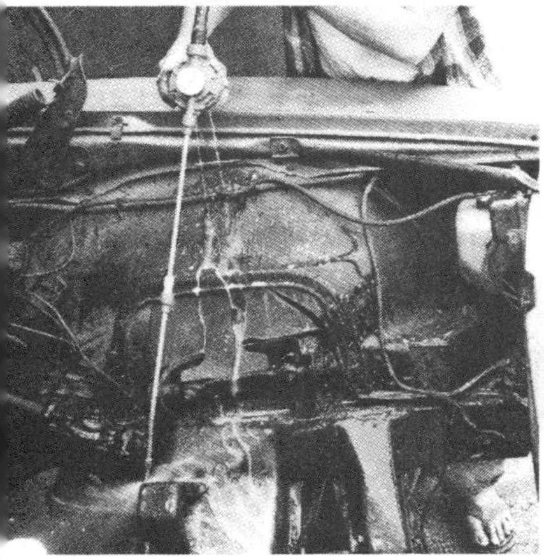

:et into your Camaro cleaning with a ar wash sprayer, hot water, and plenty f detergent. Invest a few dollars at a uarter car wash before you disable our F-body.

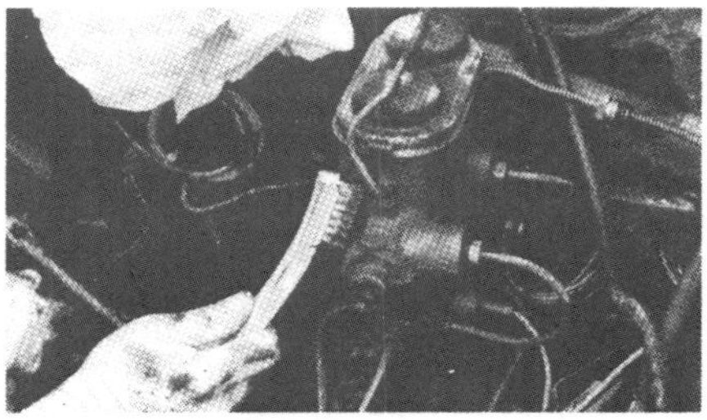

Small, stiff stainless steel brushes will scour grease and corrosion off master cylinders, steering boxes, and other engine compartment items.

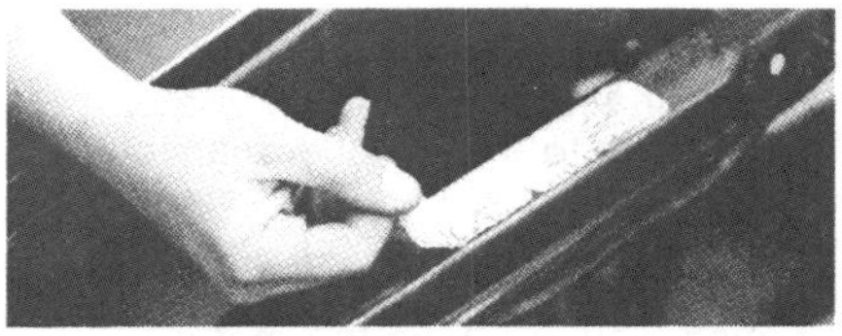

Note location of original emission and cooling system decals; reproduction decals are widely available for classic Camaros.

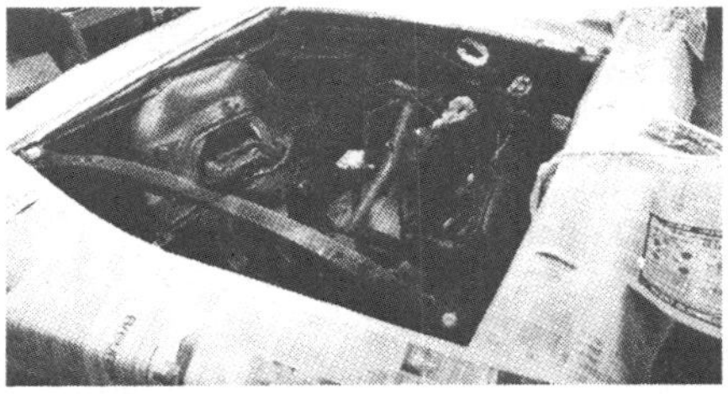

Mask off sheetmetal around engine compartment, then clean firewall and fender wells with degreaser and tack rag before painting.

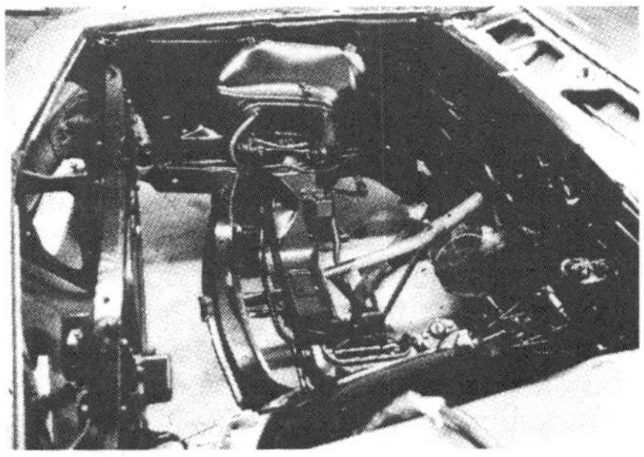

Semi-gloss black engine room paint can be sprayed with aerosol cans or small touch-up paint gun. Epoxy paints are resistant to chemicals and chips. Paint in well-ventilated area, and use respirator mask.

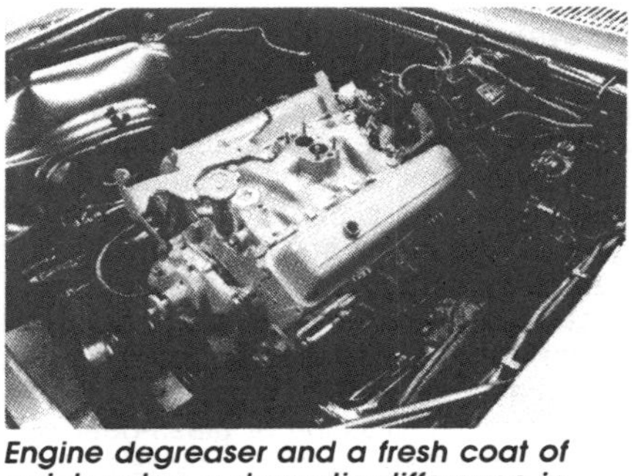

Engine degreaser and a fresh coat of paint makes a dramatic difference in this early-model Camaro's appearance. Early-model Camaro motors should be painted with Chevrolet orange engine enamel for authenticity.

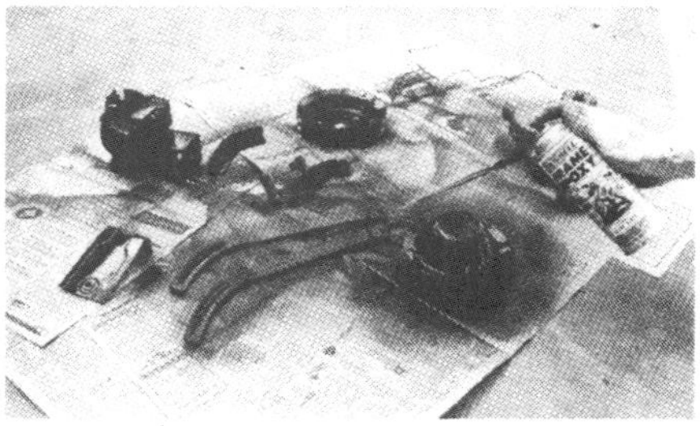

Brackets, pulleys, and miscellaneous hardware can be repainted with spray cans for factory-fresh appearance.

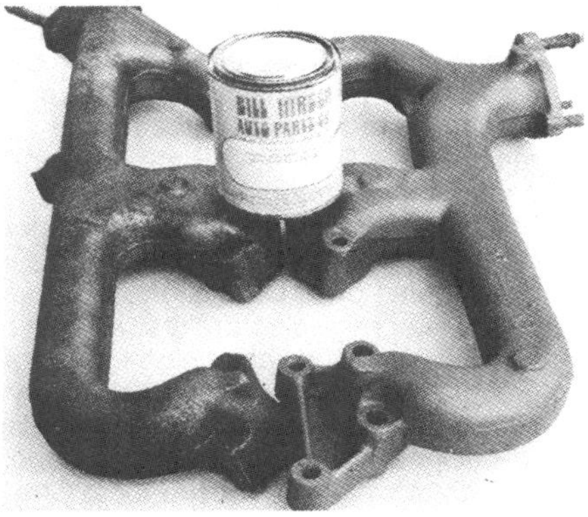

Cast iron Camaro exhaust manifolds are tough to keep looking good. Manifolds should be bead-blasted to remove rust, then coated with heat-resistant paint.

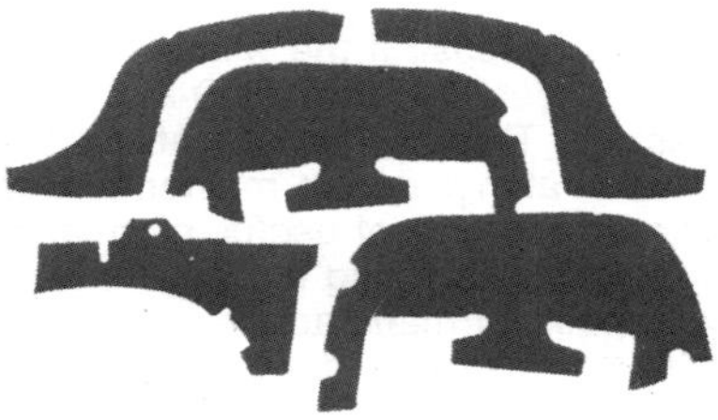

Reproduction fender splash pads replace missing and torn rubber flaps around A-arms. Splash pads keep road dirt out of your clean engine compartment.

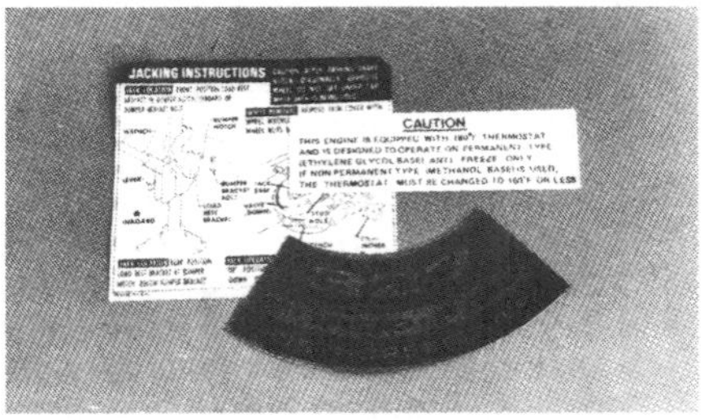

Complete set of reproduction engine and trunk decals costs about $10.

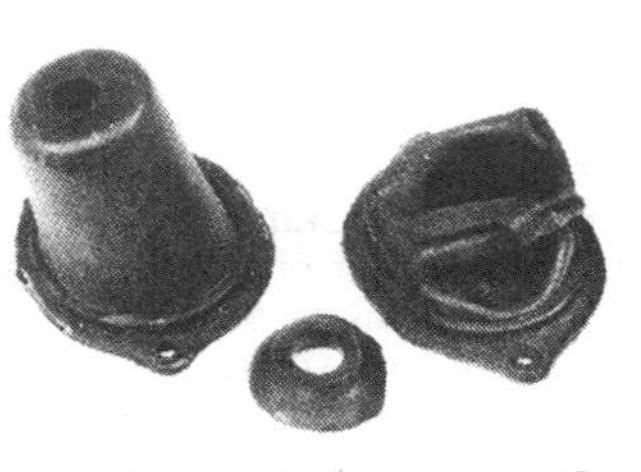

Details make the difference in a Camaro restoration. Replacing deteriorated rubber clutch linkage boot prevents engine fumes and noise from reaching passenger compartment.

CAMARO RESTORATION

under back seats, inside door panels, under carpeting, and on top of gas tanks. If you are planning a total restoration, be alert for the build sheet as you disassemble your Camaro.

Still another way to document a Camaro is through its Protect-O-Plate. The Protect-O-Plate looks like a metal credit card; it was issued to the original purchaser to be used for warranty repairs. The Protect-O-Plate contains a coded list of factory-installed equipment, including the serial numbers for the engine, transmission, and rear axle. It also calls out options such as power steering, disc brakes, air conditioning, and radio. You'll need a General Motors Parts and Accessory catalog for your Camaro's model year to decipher the Protect-O-Plate code. A friendly counterman at the local Chevy dealership can be an invaluable ally in your Camaro restoration project. Ask politely and chances are good that you can obtain a photocopy of the catalog pages you need to unravel the Protect-O-Plate mystery.

Don't despair if the window sticker, build sheet, and Protect-O-Plate are all missing. The cowl tag (located on the firewall alongside the master cylinder) is a storehouse of useful information. In addition to listing the engine type and point of manufacture, the cowl tag also carries codes for body and trim colors. The VIN (Vehicle Identification Number) is also helpful in documenting a Camaro. This number is stamped on a plate riveted to the driver's side door jamb on '67 and '68 Camaros; in 1969, the VIN was moved to the top left side of the dashboard. This number is used by state motor vehicle departments to identify the automobile. By comparing the final eight digits of the VIN with the engine block's identification numbers, you can determine whether the original motor is still in place.

Of course, this emphasis on authenticity assumes that your goal is to restore your Camaro to its original condition. That may not be your aim, however. Some Camaro connoisseurs prefer to mix-and-match components from various years and models, creating a sampler of F-body options. For example, hoods with cold air induction weren't offered by the factory on '67 and '68 Camaros, but a '69 hood with a rear-entry scoop can be retrofitted easily to these models. Similarly, a four-spoke steering wheel from a late-model Z/28 looks perfectly at home in a classic Camaro. Although the purists might object, this is a valid approach to "restoration." Just remember the cardinal rule of the conscientious Camaro builders: *Never throw anything away.* You might not foresee a need for a stock steering wheel, the original fuel line, or a Delco AM radio—but you will probably live to regret having dumped these treasures in a trash can.

DISASSEMBLY

The most valuable tool a Camaro restorer can own is a cheap camera. Take snapshots during each step of the project. These photos will be a useful guide when it comes time to put everything back together—assuming, of course, that your Camaro hasn't already been extensively modified. It's easy to forget how the fuel line was routed, or which way the alternator mount bolts to the exhaust manifold. A stack of photos will jog your memory when you're reassembling your Camaro. And when you're finally finished, you will have the material for a scrapbook of Camaro memories.

Another essential aid during the disassembly phase is an assortment of coffee cans, plastic margarine cups, and paper bags. Assign the fasteners and small hardware for each component to a separate container. This system is much more efficient than trying to sort through a box of bolts to find the fasteners for the bumper brackets or radiator mounts.

CLEANING

Before immobilizing your Camaro, have its engine compartment and undercarriage steam cleaned. This will remove most of the dirt and grease that has accumulated over the years. If you have a stack of quarters, you can spend them at a self-service car wash cleaning your Camaro. Bring along a floor jack and stands so you can clean the underside of your F-body. Remember to protect the alternator, distributor, and carburetor with plastic bags when you are spraying down the engine compartment.

When your Camaro is back home, you can start scrubbing. Cleaning the front subframe and wheelwells is easy when the engine has been removed. You can attack the layers of crud chemically with degreaser and cleaning solvent. Steel wool soap pads, paint scrapers, and wire brushes will loosen thick grease deposits. If you have access to an air compressor, a solvent sprayer will make short work of most road grime.

There are several techniques for cleaning the dozens of brackets, hinges, and miscellaneous small parts that make up a Camaro. A bench grinder with a coarse wire brush is ideal for cleaning bolt threads and nuts. Rusty and corroded parts can be refurbished with a glass-beader or sandblaster. (Use low pressure and fine grit when blasting or glass-beading to prevent changing the surface texture of the component. Blasting can leave a rough finish on parts that should be smooth.) If you are overhauling the suspension during your Camaro

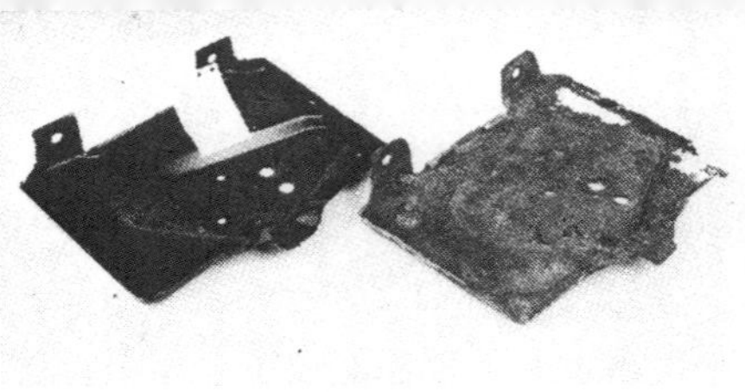

Battery acid quickly destroys Camaro battery trays. Replacements are available from Chevrolet parts departments and Camaro specialists.

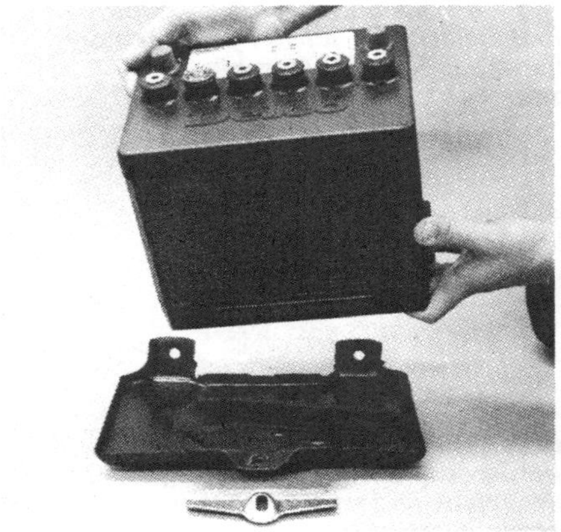

Original NOS (New Old Stock) batteries for classic Camaros are still available from Restoration Battery.

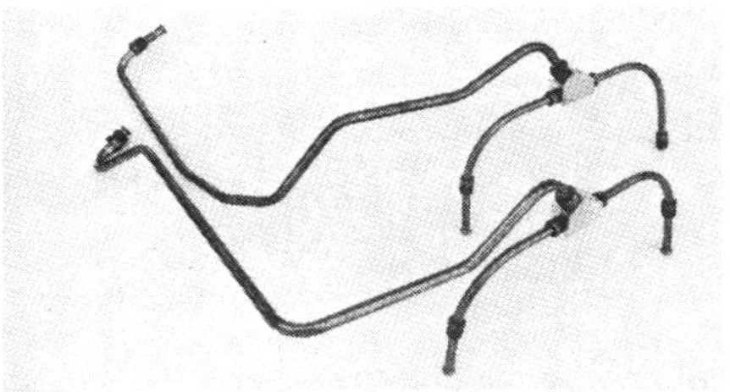

Zip Products manufactures replacement steel fuel lines for early Camaros with correct fittings and tubing sizes.

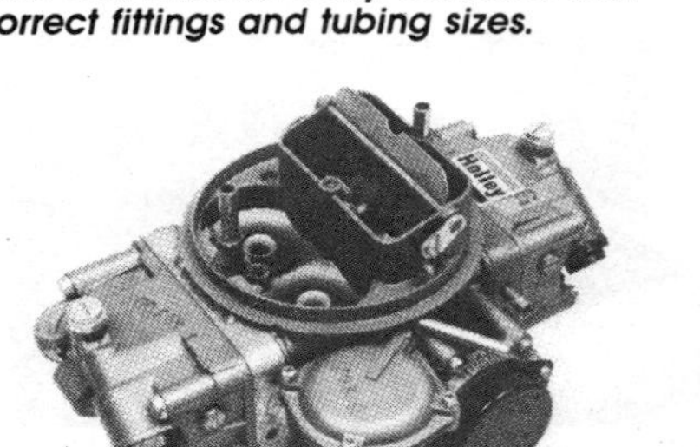

Holley has responded to the Camaro restoration movement by re-releasing original carburetors for popular performance models.

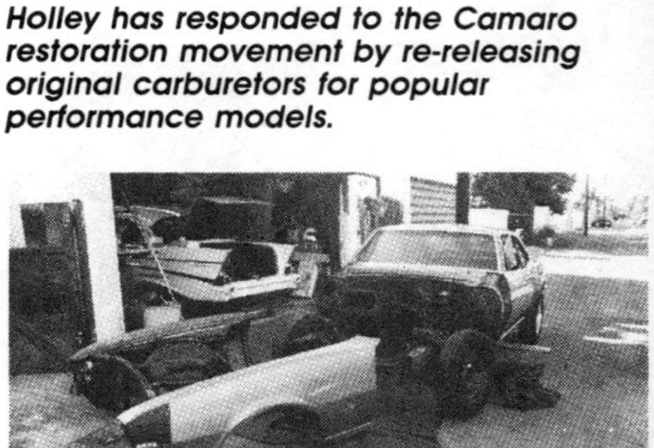

If you're planning an extensive restoration project, remove Camaro fenders and radiator support as a single assembly to allow access to subframe and front suspension.

CAMARO RESTORATION

project, you can leave the greasy A-arms, springs, and steering linkage in a machine shop's hot tank overnight. They'll come out of the tank squeaky clean. Of course, you will have to replace the rubber control arm bushings, ball joints, and steering components afterward—but you already planned on doing that, didn't you?

DETAILING

If you are restoring your Camaro to showroom condition, the undercarriage and engine compartment should be sprayed with semi-gloss black paint. The original engine compartment paint probably had its share of runs, sags, and drips—so an "authentic" paint job wouldn't be flawless. If you want to improve on the factory's paint choice, an epoxy or two-part acrylic paint will maintain its appearance much longer in the harsh environment of the engine room than a "correct" factory finish. A small touch-up spray gun works well in the tight quarters of a Camaro engine room, but equally good results can be obtained with aerosol spray cans.

Some parts require special treatment to restore their original appearance. Because Camaro hood hinges were seldom painted at the factory, they are often marred by rust and corrosion. Such components can be de-rusted with chemical solutions, then sprayed with clear sealer to protect them from future harm. Items such as master cylinder caps and power brake diaphragm were iridited or anodized at the factory. It's often easier to replace these parts with new components than to refinish the old pieces. However, if you're looking for an inexpensive way to dress up your Camaro's engine compartment, consider gold iriditing the underhood fasteners, pulleys, and brackets.

INTERIOR RENOVATION

Refurbishing the interior should be the last item on your Camaro agenda. Finish the bodywork, painting, and wrenching before renovating the passenger compartment. Then you won't have to worry about bondo dust, overspray, and grease stains spoiling your Camaro's new carpet and upholstery After all, the interior is the part of the car you will see most often when your Camaro project finaliy hits the road.

If the upholstery isn't ripped and torn, you can spruce up your Camaro's interior with mild detergent and a bottle of Armor All. Faded vinyl and sun-bleached carpeting can be restored with fabric dye. But if the passenger compartment is too far gone to respond to these cosmetic

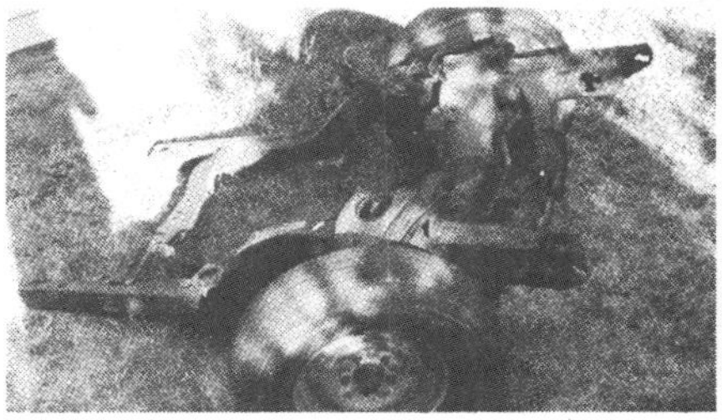

Front subframe can be removed by unscrewing four body bolts. Inspect frame carefully for cracked welds and fatigue around control arm mounts and spring pockets.

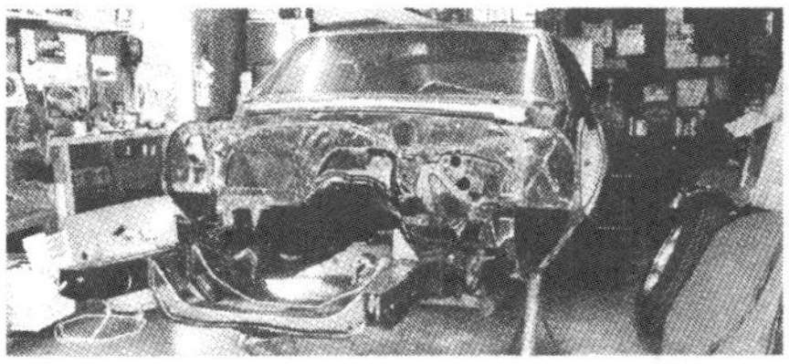

Sandblasted, painted, and reinstalled, this 18-year-old Camaro subframe looks like it just rolled off the assembly line.

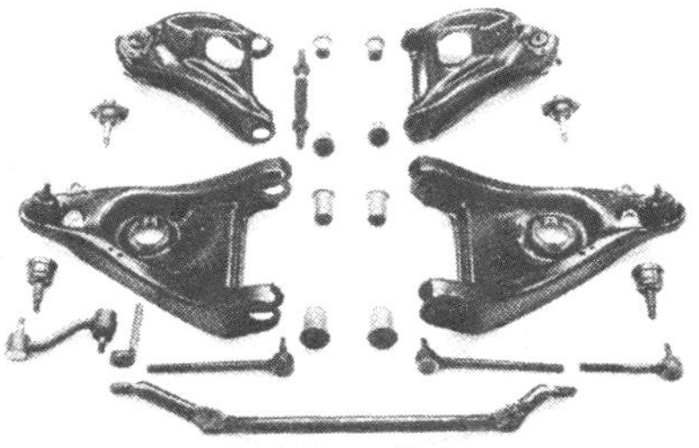

Install new front suspension components to restore precision to early Camaro steering and handling. Upper and lower ball joints, control arm bushings, Pitmann arm, idler arm, tie rod ends, and shock absorbers should be replaced.

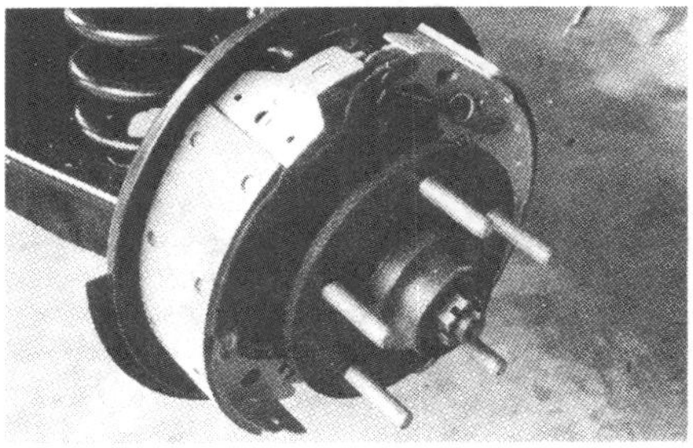

Camaro drum brakes will provide adequate stopping power with new linings, reconditioned drums, and fresh hardware.

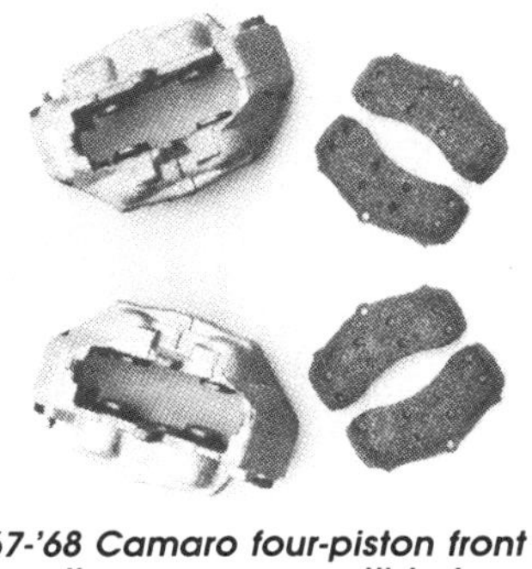

The '67-'68 Camaro four-piston front disc brake calipers are susceptible to corrosion damage. Corvette brake specialists can repair rusted calipers with stainless steel sleeves.

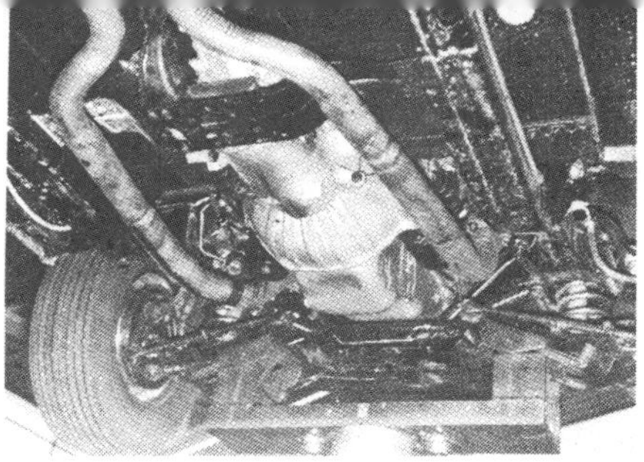

How would you like your Camaro's undercarriage to be this clean? Note spotless bellhousing, transmission, and brake lines.

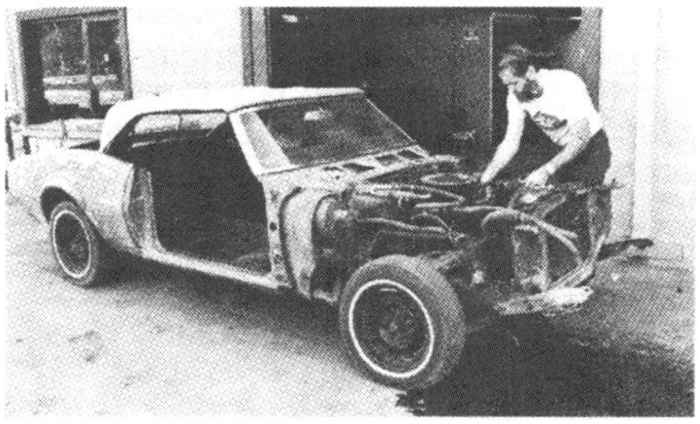

This '67 Rally Sport convertible is stripped to a bare shell for a total restoration. A do-it-yourself restorer can build a "brand-new" classic Camaro for much less than the cost of a new car.

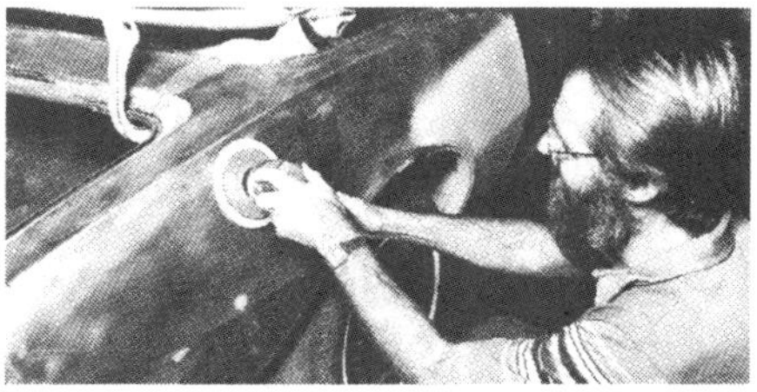

Random orbit pneumatic sander strips off layers of old paint.

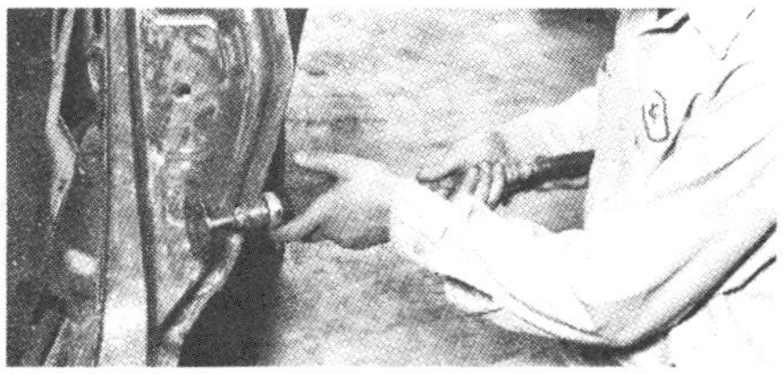

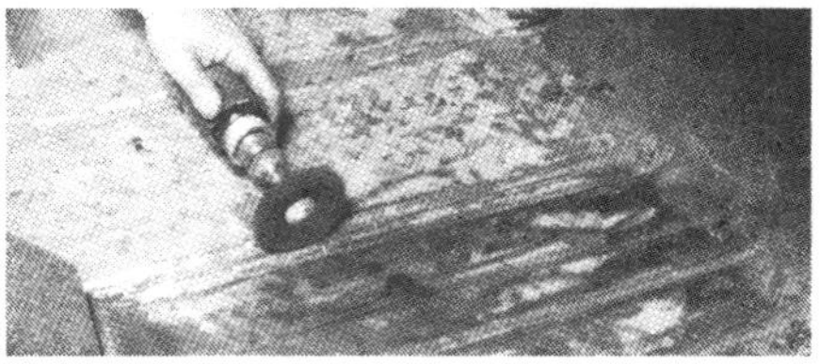

Abrasive 3M foam wheel cleans door jambs and removes rust quickly from trunk floor.

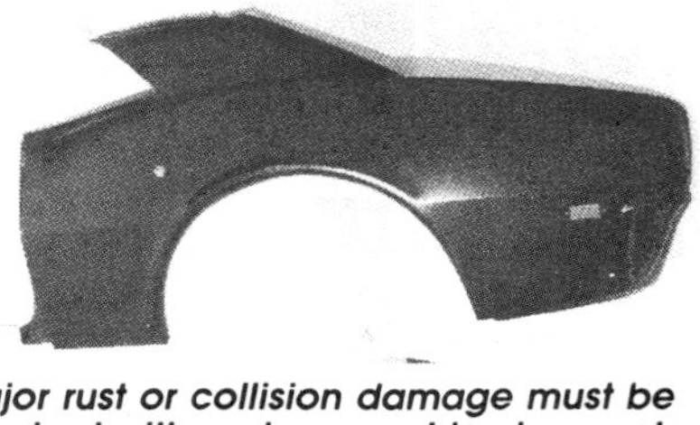

Major rust or collision damage must be repaired with replacement body panels. Smaller "patch panels" are available to fix damage in rust-prone areas such as rear window and wheelwells.

treatments, you can order a complete new interior by mail. Specialty parts suppliers have made it easy to return a classic Camaro to its former glory. Carpeting and upholstery kits are readily available in a wide variety of original colors; even hounds-tooth cloth upholstery and deluxe trim kits are now on the market. Reproduction door panels and dashboard pads can be found at reasonable prices to replace cracked or faded vinyl.

Don't neglect your Camaro's trunk during a thorough restoration. The floor and inner fender wells can be repainted with aerosol cans of Zolotone splatter paint to produce a like-new appearance. A new reproduction trunk mat and spare tire cover are the finishing touches.

THE LAST DETAILS

Details make the difference in a Camaro restoration. Cleaning, painting, and overhauling mechanical systems are expensive and time-consuming. After investing your money and effort in building a first-class Camaro, don't leave the job half-finished by forgetting the subtle touches.

Reproduction decals will give your Camaro a showroom-fresh appearance. There are air cleaner decals that announce your engine's horsepower rating, decals with emission and tune-up information, and even reproduction jacking instructions for the inside of the trunk lid!

Rubber components often deteriorate with age. New rubber pads for the brake, clutch, accelerator and parking brake pedals are inexpensive and easy to install. Old Camaro weatherstripping is usually cracked, torn, and covered with overspray; replacing the weatherstrip around the trunk, doors, and windows will keep both rain and road noise out of the interior. Reproduction rubber bumpers for the doors and hood will make your classic Camaro look like it just rolled off the end of the assembly line. And you can even mail order brand-new rubber splash pads for the front A-arms and new biscuits for the subframe bushings!

It's the dream of every Camaro enthusiast to be able to walk into a Chevrolet dealership and order a new first-generation Camaro. Imagine being able to take delivery on a '67 Z/28 with zero miles on the odometer, or perhaps a big-block '69 Super Sport that's never turned a tire. Even though you can't buy such vehicles, you *can* build them. A restored classic Camaro will provide hours of pleasure in the garage and on the highway. Isn't it time you enlisted in the Camaro restoration movement?

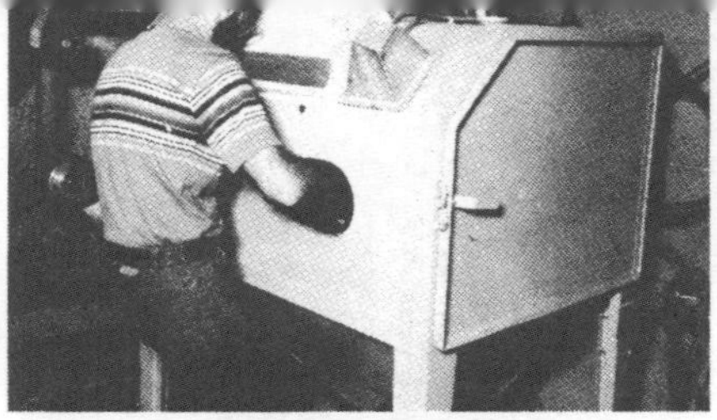

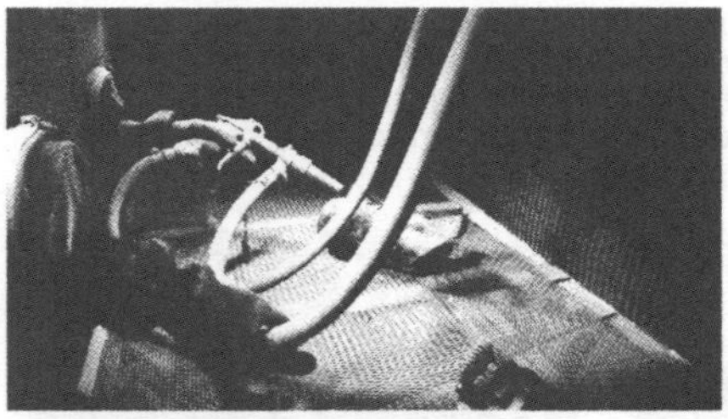

Sandblasting and glass-beading are effective methods of cleaning rusty Camaro components. Use low air pressure and keep the blasting gun moving to prevent damaging the metal surface. Delicate items can be blasted with crushed almond shells or cherry pits, which produce a soft luster.

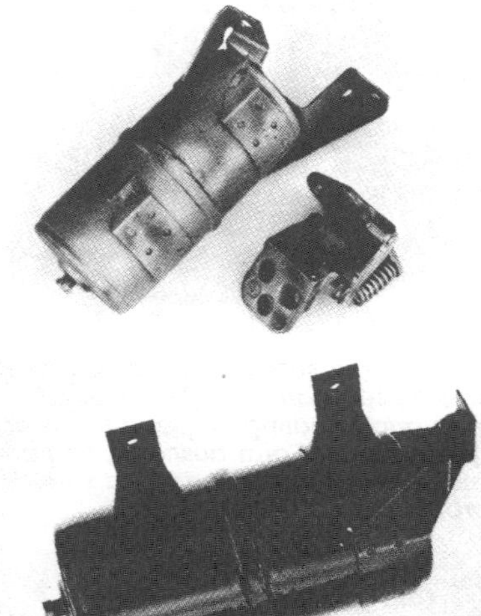

Glass-beading is ideal way to make complex castings and brackets look like new again. Prime and paint parts immediately after cleaning to prevent surface rust from forming.

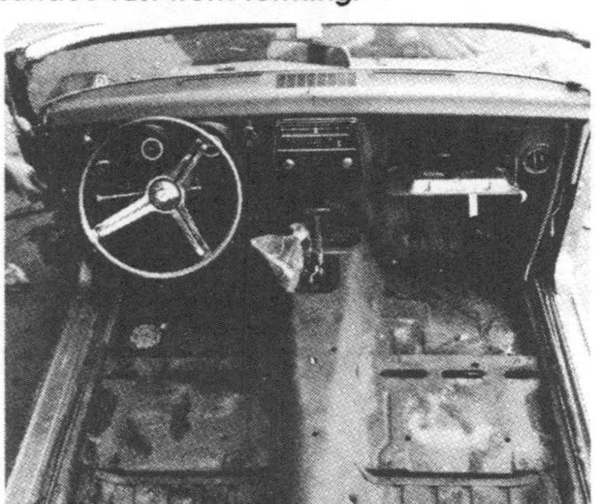

Leaky weatherstripping can allow water to collect in footwells, corroding Camaro floorboards. This stripped Camaro interior will be refurbished with mail order upholstery, carpeting, and door panels.

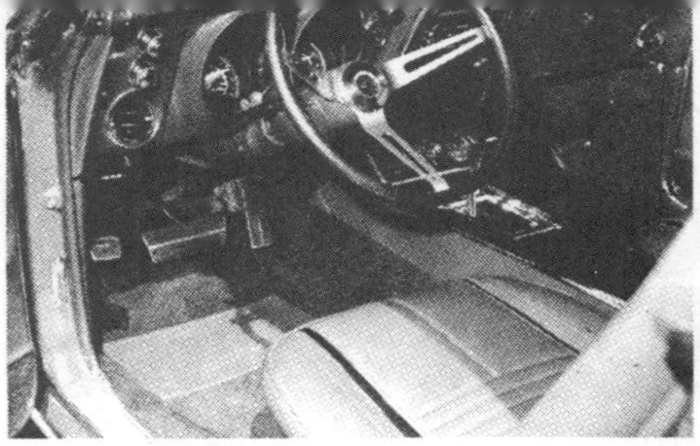

New carpeting, seat covers, and dash pad will make a ratty interior presentable again. Plastic dash vents and switch bezels can be "rechromed" by vacuum plating. Don't forget to replace rubber pads on brake and accelerator pedals. Carpets should include heel pads and good quality underlay.

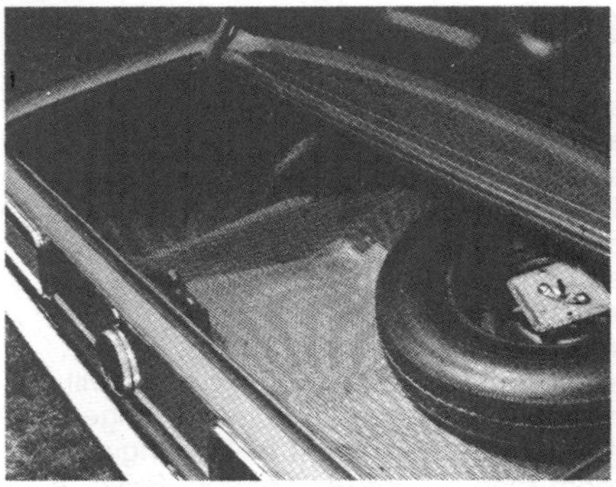

Don't overlook your Camaro's trunk during a thorough restoration. A spray can of splatter paint and a new trunk mat cost very little, but make a big difference in appearance. Early-model Camaro rubber trunk mats and spare tire covers have hounds-tooth pattern.

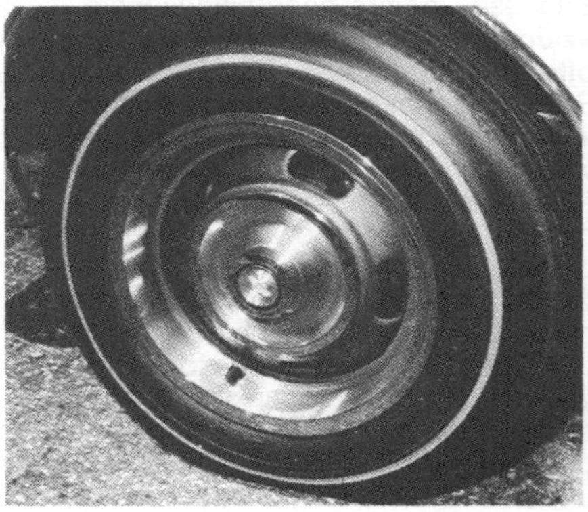

Most Camaro hub caps are still available from Chevrolet. Authentic red-stripe tires can be ordered from specialty tire companies—although modern radials will generally improve handling.

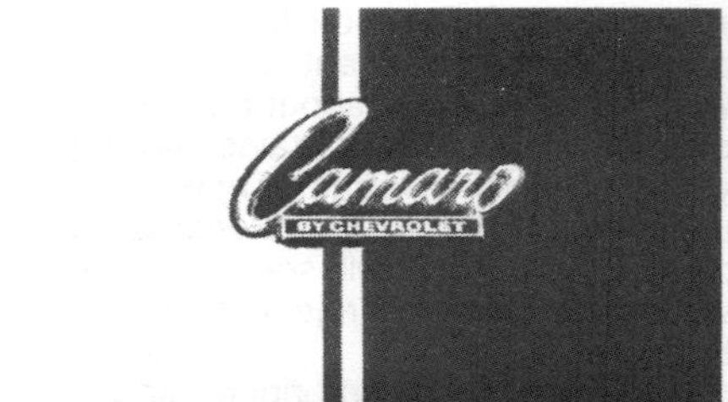

Stencils and decal stripe kits make it easy to duplicate original factory paint schemes.

HOW TO REUPHOLSTER F-BODY BUCKETS

CAMARO COVER-UP

CAMARO RESTORATION

Tijuana once was the world headquarters for car upholstery. Southern California car crazies pulled the door panels off their '55 Chevys, crossed the Mexican border south of San Diego, and sought out back-alley tuck'n'roll specialists. After an afternoon of bull fights, Dos Equis beer, and the Agua Caliente horse races, they'd pick up the reupholstered panels and return to the U.S. of A. with the aroma of huarache sandals, tortilla chips, and fresh Naugahyde filling their camper shells.

Now tuck'n'roll is out—along with crushed velour and mohair. "Driver's seats" are cool, as long as they have at least 47 adjustments for rake, lumbar support, and the phases of the moon. The stock look is also in vogue—although split seams, torn vinyl, and clouds of foam rubber are strictly low-class. Fortunately, Camaro owners no longer have to risk a bout with Montezuma's Revenge to have the best seats in the house. Sitting pretty is as easy as installing a do-it-yourself seat upholstery kit.

New reproductions of original factory

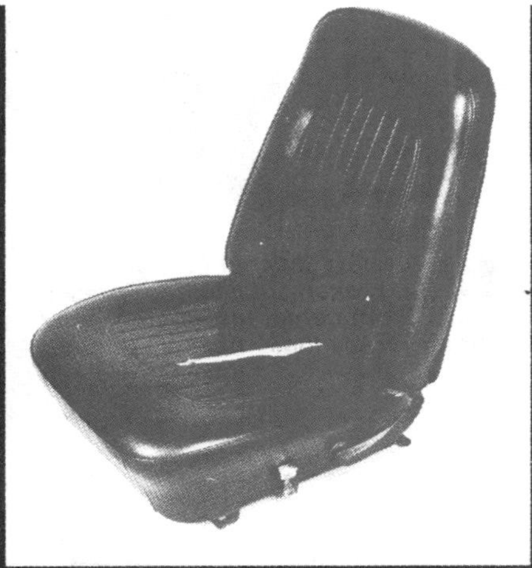

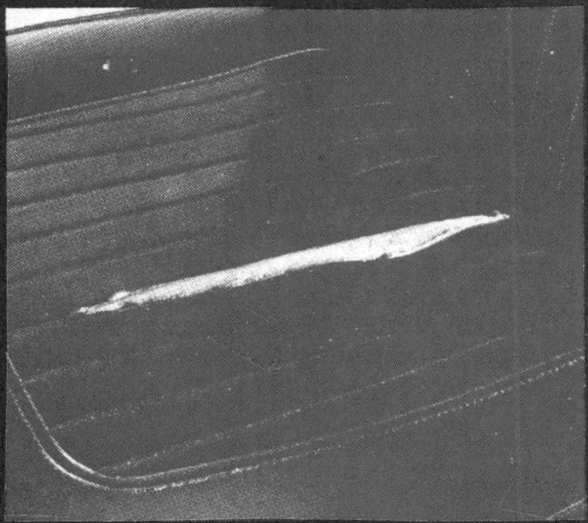

1 *There comes a time in every Camaro's life when duct tape and vinyl glue just won't do the job anymore. Split seams, ripped upholstery, and bare foam rubber don't do anything for a Camaro's appearance—or driver comfort.*

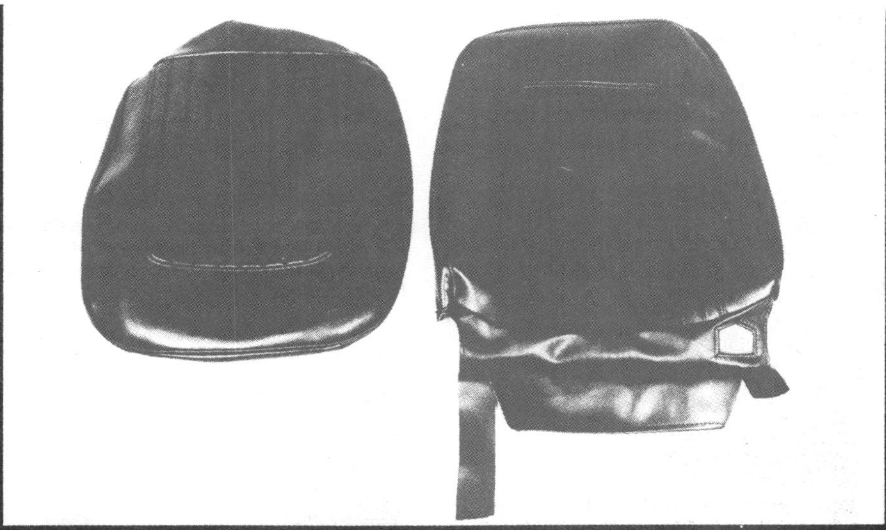

2 *High-quality reproduction upholstery kits are available from a variety of Camaro parts suppliers. These '68 Camaro standard seat covers are dead ringers for the factory upholstery. Kits are also offered for deluxe interiors.*

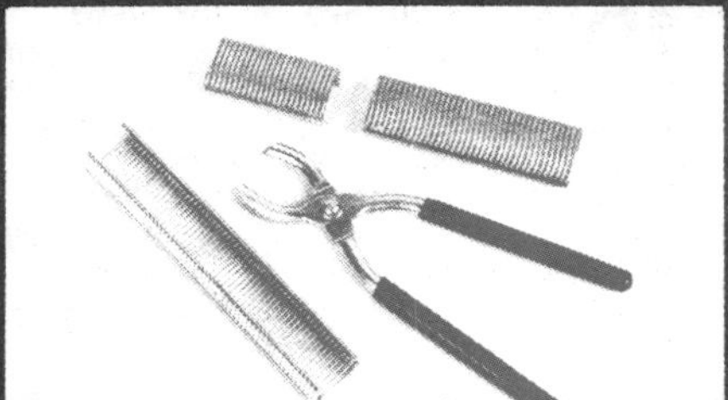

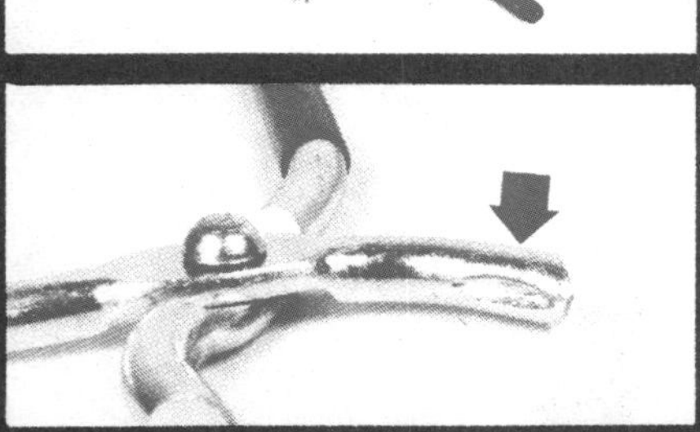

3 *A pair of hog-ring pliers are an essential tool for do-it-yourself upholsterers. If you don't intend to make reupholstering Camaros your life's work, you can purchase a pair of inexpensive pliers and a supply of rings at a well-stocked auto parts house. Conventional pliers can be modified by grinding grooves in the jaws to close the hog rings.*

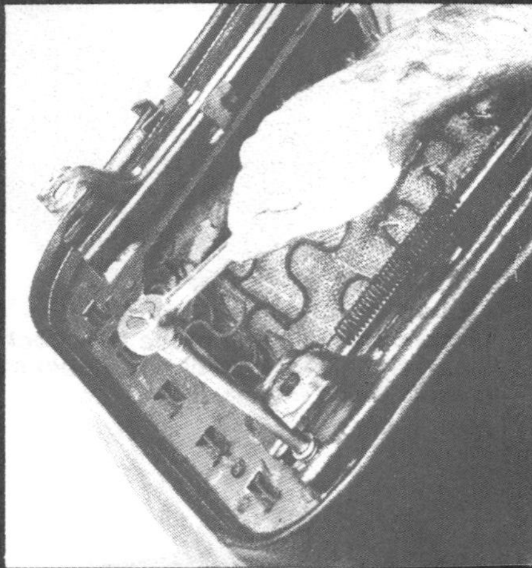

4 *Remove the four bolts which attach the seat tracks to the floorboards, then turn the seat upside down. Unhook the seat track spring and unbolt the tracks. Clean and lubricate the rails, then put them aside for reassembly later.*

COVER-UP

seat covers are easy for a novice upholsterer to install, and they're accurate enough to satisfy any Camaro connoisseur. Standard and deluxe interior kits are currently available from many specialty parts suppliers for most classic Camaros, including seat covers with hounds-tooth inserts for '69 and '70 models.

A pair of hog-ring pliers are the only special tools required to install Camaro seat upholstery. "Hog rings" are semi-circular metal clips which attach the seat covers to the frame. If you're unsure how to install the new upholstery, use your Camaro's second bucket seat as a guide to how the factory does it. Be sure that the new upholstery is centered on the seat form; otherwise your buckets will turn out lopsided. A complete Camaro cover-up can be completed in a weekend—an accomplishment beyond even the reach of Washington politicians!

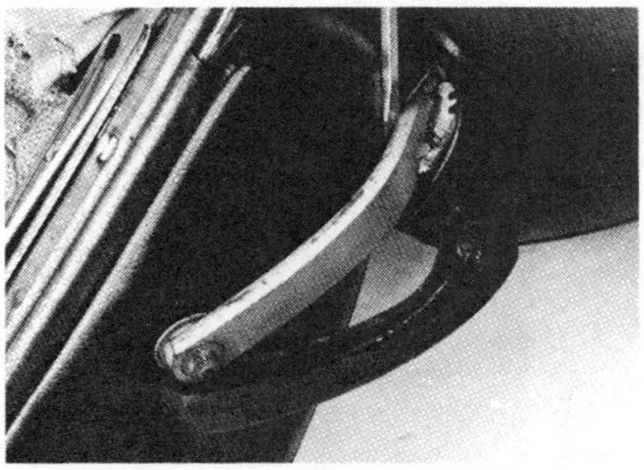

***5** Carefully pry off the plastic hinge covers and remove the seat-back retaining clips. Then spread the hinges and disengage the back from the seat bottom.*

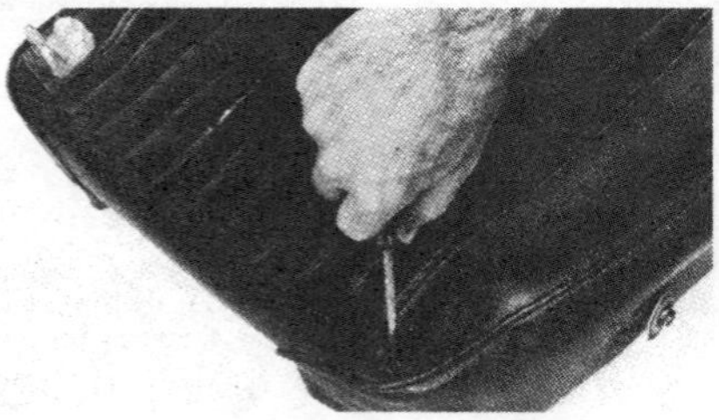

***6** Unscrew the rubber bumper that controls the angle of the seat back. Replacement stoppers are available to fix seats which lean back too far.*

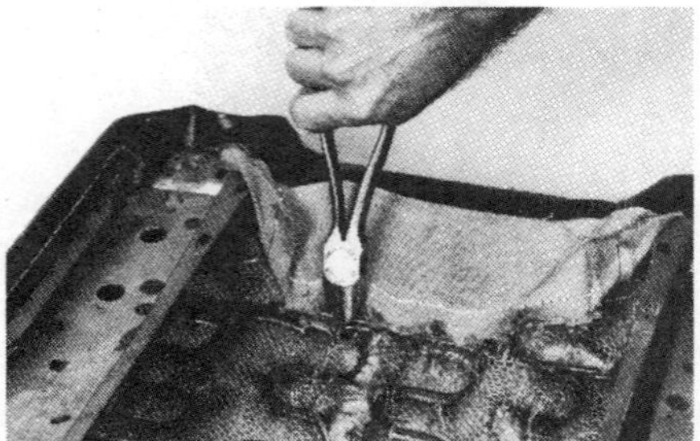

***7** Turn the seat bottom upside down and remove the hog rings at the rear of the seat. Sharp side-cutters make ring removal easy.*

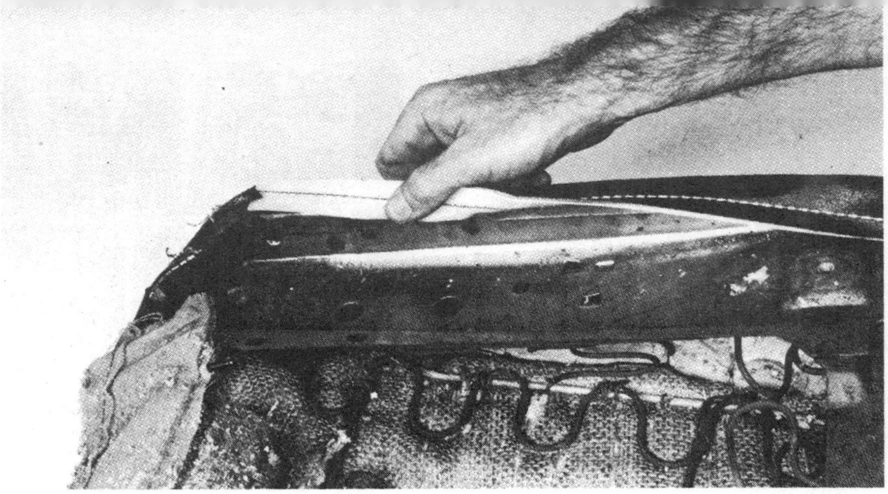

***8** Push down on the seat bottom and pull the upholstery's plastic edging out of its channel in the seat frame.*

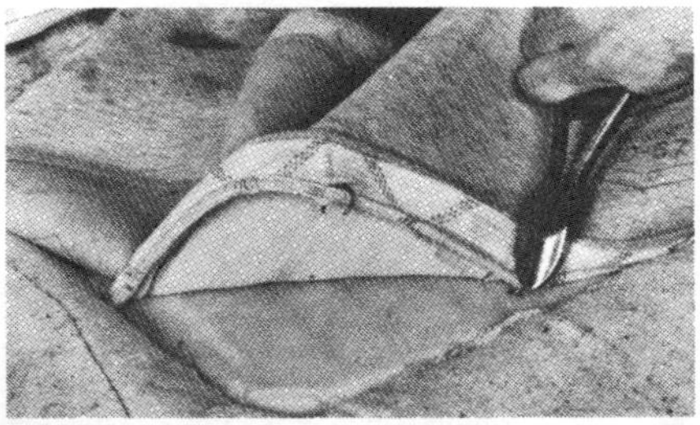

***9** Turn the seat bottom over again and fold back the upholstery. Cut the hog rings that hold the center seat insert to the springs.*

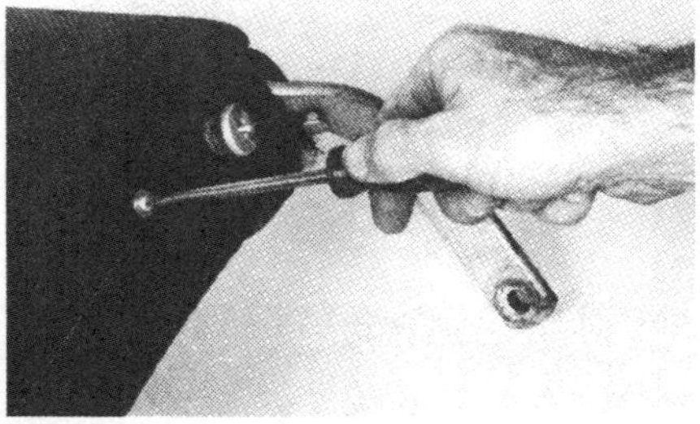

***10** Remove the two Phillips head screws on the bottom of the seat back. Pull the back away from the seat and push up to disengage it from the clips on the seat frame. All 1967 Camaros have stamped steel seat backs; these can be resprayed with satin black paint. The '68 and later plastic seat backs can be restored with Armor All or aerosol vinyl dye.*

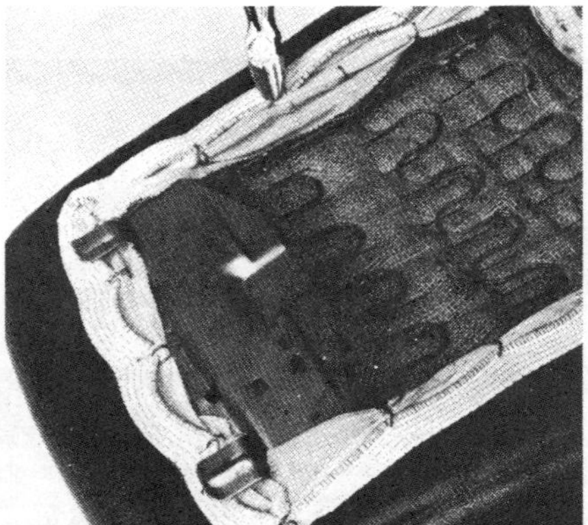

***11** Cut the rings holding the seat-back upholstery, then remove the seat cover by turning the vinyl inside-out over the corners.*

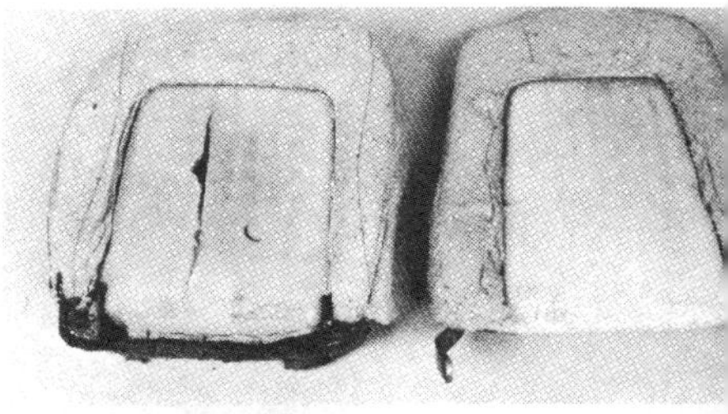

***12** Carefully inspect the seat frames for broken springs and cracked welds. Foam center inserts give early Camaro seats their "bucket" contour.*

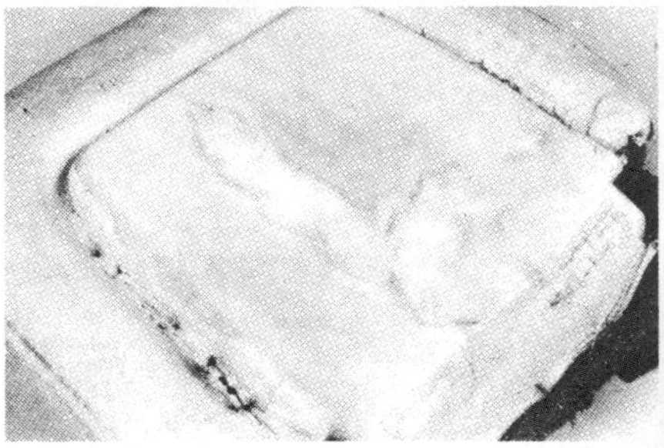

***13** Torn or cracked seat forms can be repaired with polyfill, available at upholstery supplies. Replacement seat forms are available to repair badly deteriorated foam. Torn burlap under the cushion should be replaced with a piece of carpeting to protect the foam from the seat springs.*

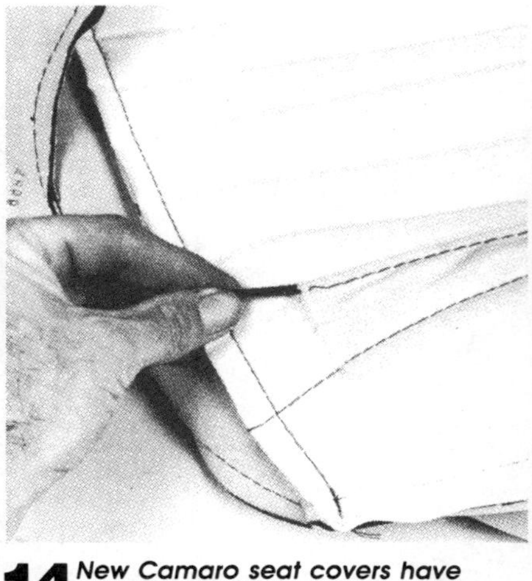

***14** New Camaro seat covers have heavy cord "listing" around the center insert. This listing must be attached to cord under seat to hold center insert in position.*

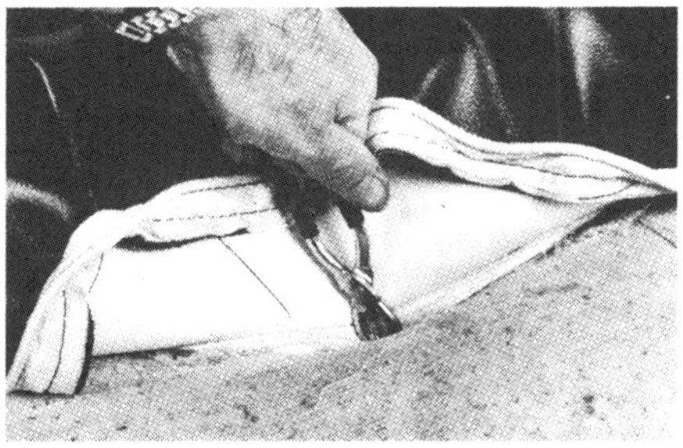

15 Center new seat-bottom upholstery in relation to seat cushion corners. Push down hard with hog-ring pliers and attach listing to cord under seat; rings must penetrate foam and burlap. Start at middle front, then work from side to side. Check that center insert is positioned properly and that upholstery is smooth.

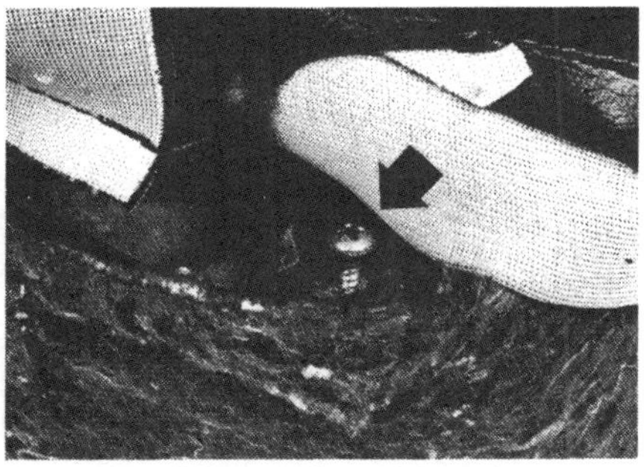

16 Install screws for seat-back bumper and latch before pulling upholstery over frame. You won't have to probe for screw holes later.

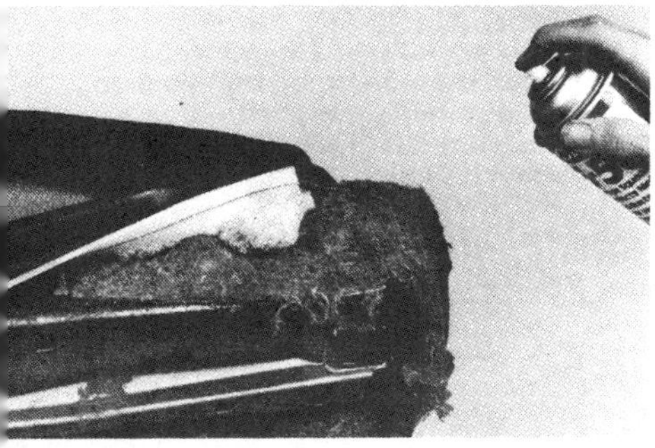

17 Lubricate corners with silicon spray, then stretch upholstery over frame. Vinyl covers can be heated with a hair dryer to make them more pliable.

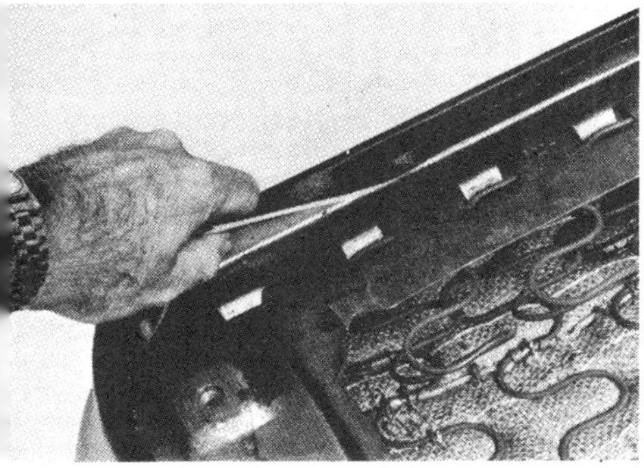

18 Compress seat springs while pushing down on seat cover to engage plastic edging in slot around seat bottom.

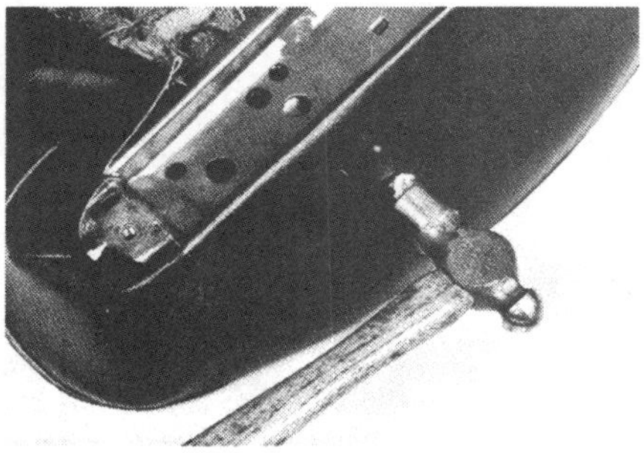

19 Vinyl will be stretched tightly over seat-back pivots. Tap seat cover with hammer to cut holes for pivots.

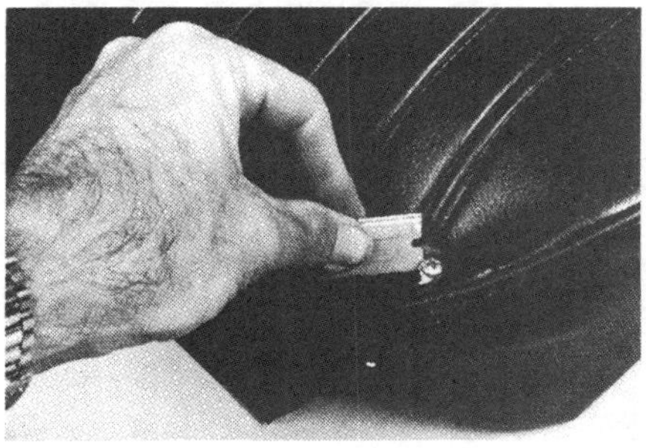

20 Screw heads are easy to find under new seat covers. Cut slits for fasteners with single-edge razor blade.

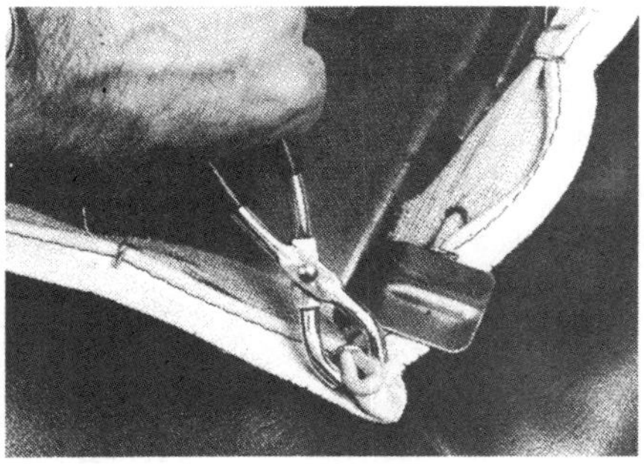

21 Center seat-back cover on foam cushion and attach center insert listing. Stretch upholstery over corners, then gather excess material with hog ring to smooth edges of seat top.

22 Adjust seat-back upholstery to clear latch mechanism and release lever, then pin in place with hog rings.

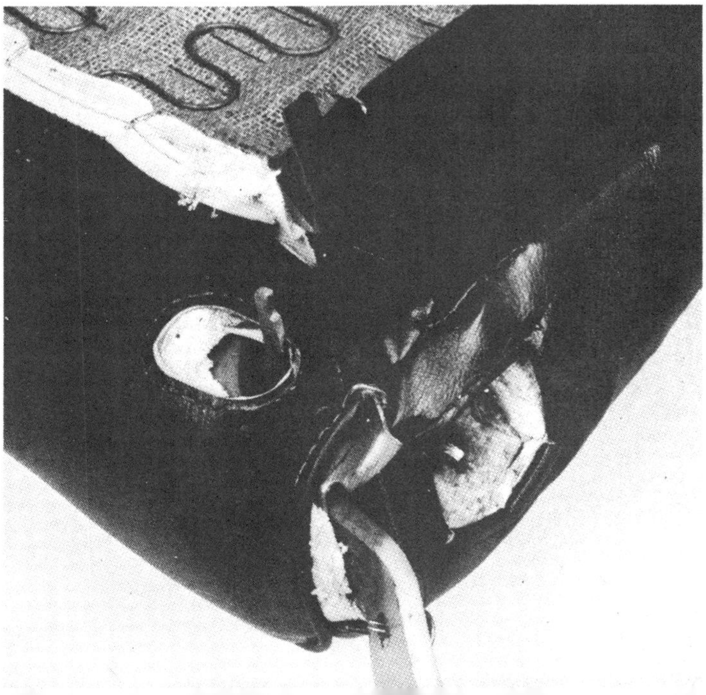

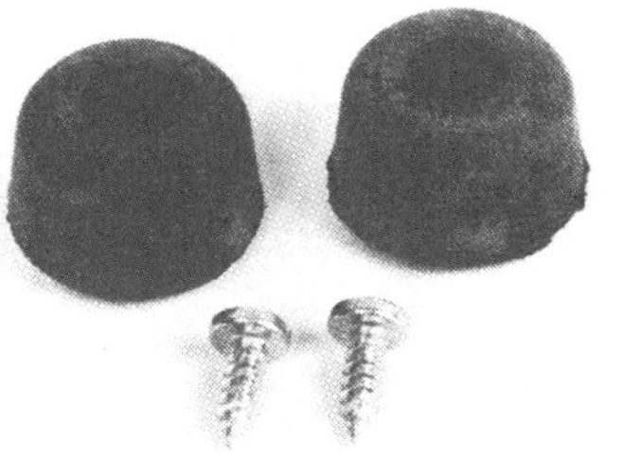

23 Install new rubber seat-back stopper with sheetmetal screw.

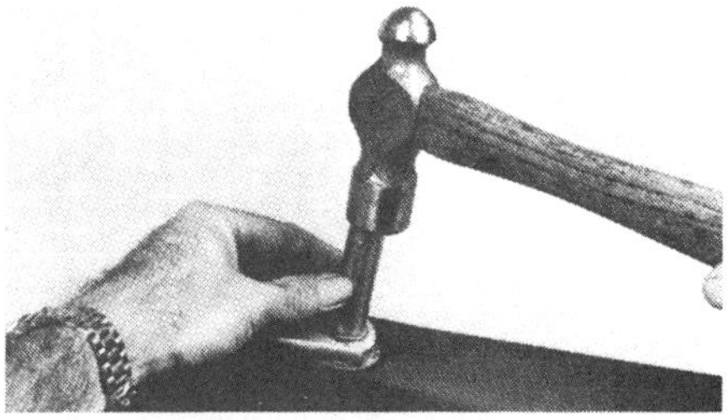

24 Reinstall seat back on base, then drive retaining clips over pivots with a deep socket. Bolt tracks to seat bottom.

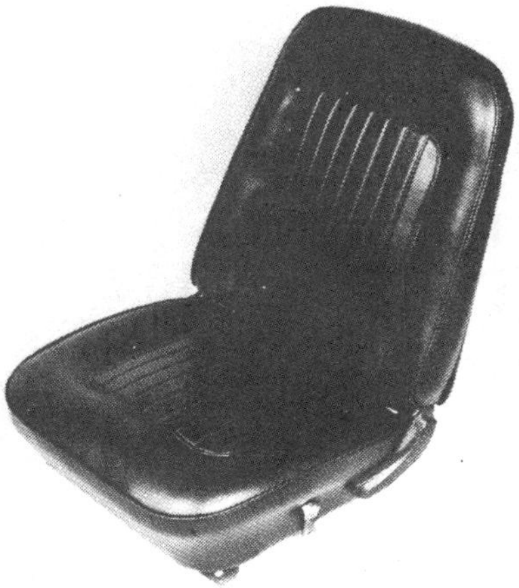

25 Classic Camaro bucket seat has made a dramatic recovery with new upholstery. Cover-up takes only an afternoon of work, and makes a tremendous difference in both appearance and driving comfort.

CAMARO MOTOR REVIVAL

CAMARO RESTORATION

Where were you in '67? Maybe at the corner of Haight and Ashbury in San Francisco, grooving on the Summer of Love? Standing in line at the Army induction center, wondering about some far-away place called 'Nam? Waiting for your turn on the swings at grammer school, or playing with wooden blocks in your parents' nursery? Times have changed since 1967—for both cars and citizens.

The Sixties still survive, if you know where to look. Former hippies are

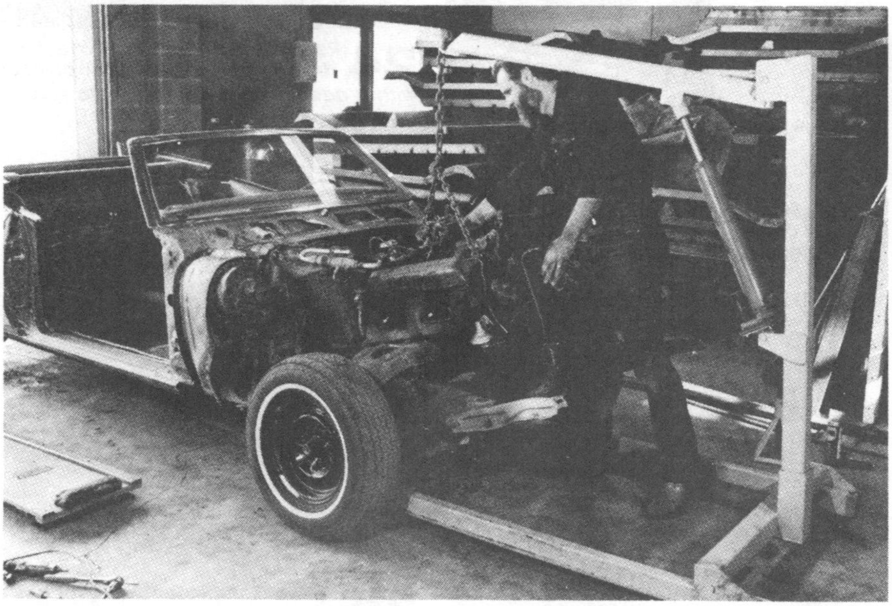

1 Dana Morgan never intended his Camaro project to go this far. What began as a quick paint-and-upholstery clean-up became a full-fledged restoration once Dana started working on his wife's '67 Rally Sport convertible. With the front end sheetmetal already removed, it's a simple task to pull the engine and transmission.

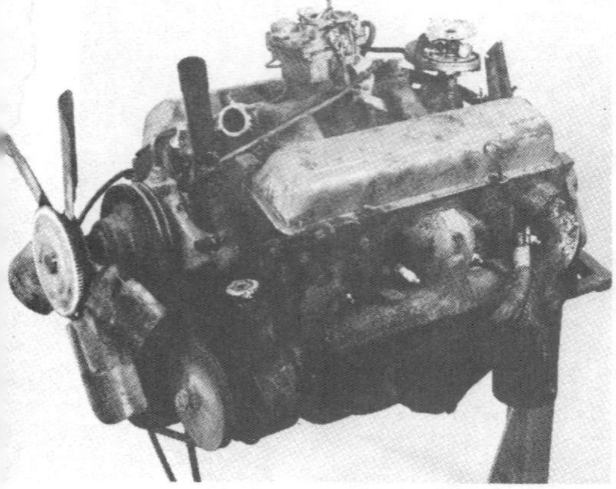

2 Chevrolet rated the 327-cubic-inch small-block in Dana's Camaro at 210 horsepower. Although a 210hp two-barrel isn't something to get too excited about, this engine is extremely responsive and tractable. Late-Sixties small-blocks are gems; this motor has bigger cylinder head ports and stiffer cylinder walls than a new IROC-Z engine! Another bonus: a PCV valve is the only emission-control device required for this 49-state engine. Its 8.75:1 compression ratio is better suited to today's low-octane gas than the 275-horsepower version of the 327ci small-block, which is rated at 10:1. Dana planned to upgrade his Camaro's V8 with a factory four-barrel intake manifold and Quadrajet for stronger street performance.

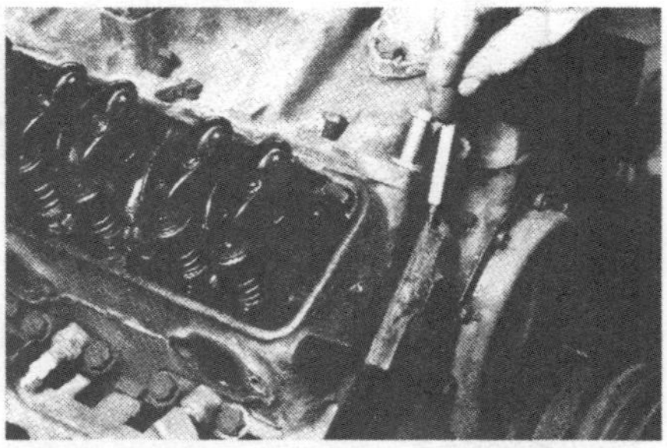

3 Dana verifies the engine's identity by decoding the block numbers. This motor is stamped "V1216ME," which is deciphered as follows: the engine was built at the Flint plant (code "V") on December 16 ("12-16"). The "ME" identifies a 1967-68 327ci V8 with a two-barrel carburetor, no air injection, and a Powerglide transmission. This confirms that the original engine is still in place.

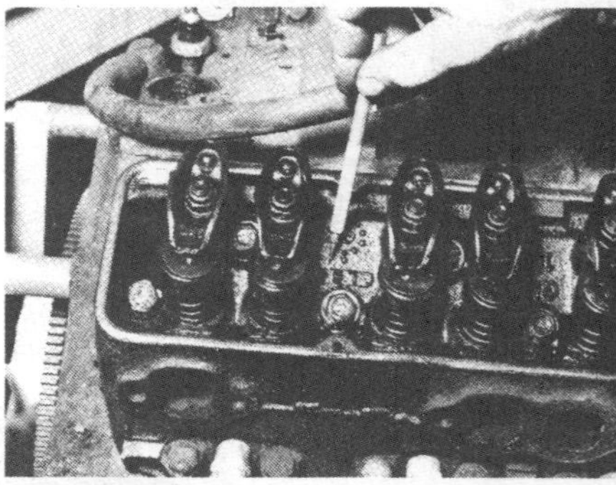

4 Checking the cylinder head casting numbers (3876775) and casting dates (L86) reveals that the heads have not been changed by previous owners. This casting number is correct for a 210hp 327; it indicates a head with 1.72-inch-diameter intake valves, 1.50-inch exhausts, and a minimum combustion chamber volume of 69 cc. The "L86" casting date reveals that the heads were manufactured on December 8, 1966. (Months are coded starting with "A" for January, "B" for February, etc.)

working on Wall Street, Saigon is now Ho Chi Minh City, and Bob Dylan's still singing. Classic Camaros continue to cruise the streets of America, too. And just as flower power has given way to Reaganomics, early-model Camaros have adjusted to life in the Eighties. Gas prices have quadrupled, the national speed limit's taken a tumble, and the EPA has banned lead at the gas pumps. A Sixties-style small-block is as out of place today as love beads and Nehru jackets.

Dana Morgan is a Camaro enthusiast who confronted this harsh reality when he set out to rebuild the engine for his wife's '67 Rally Sport convertible. Dana's a lifelong car nut; his Collector's Choice shop in Scotts Valley, California, is frequently filled with priceless Bugattis and temperamental Ferraris. When Dana decided to try his hand with domestic iron, he wanted to build a small-block that combined dependability, economy, and performance—an ambition shared with countless Camaro owners.

Dana carefully outlined his small-block survival plan before picking up a wrench. Although Dana is not a fanatic on the subject of Camaro restoration, he does recognize the importance of maintaining original equipment and casting numbers whenever possible. That meant rebuilding the car's factory-installed engine (after cross-checking the motor numbers) rather than bolting in a mail-order replacement or a new Targetmaster Chevy. With annual emissions inspections becoming a fact of life, Morgan also resolved to leave the engine's smog equipment and exhaust system untouched.

Early-model Chevrolet engines need a regular fix of lead in their gasoline. Valves and seats are subject to rapid wear without the lubrication that small quantities of lead provide. Although the EPA has assured motorists that reducing the lead content in regular gasoline by 90 percent won't hasten valve wear, Dana wasn't willing to take the government's word for it. After consulting with machinist Larry Hollums at Hollums Racing Engines (Fremont, California), Dana decided to upgrade his small-block with stainless steel valves and silicon-bronze valve guides. Many engine experts are still uncertain

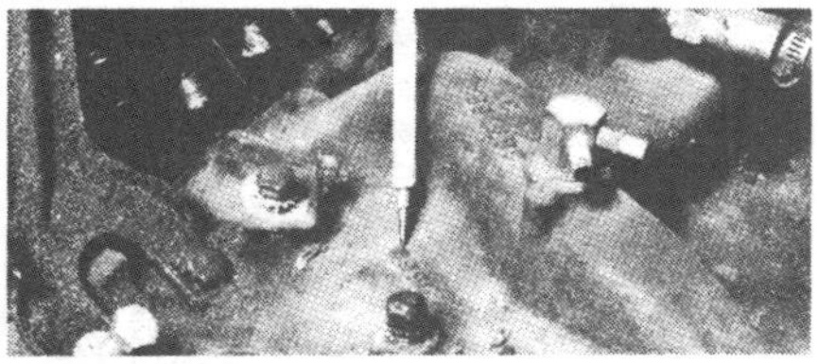

***5** Intake manifold casting numbers are located on the rear runners. The casting date is near the distributor hold-down; this manifold was manufactured September 9, 1966, according to the "196" date stamp.*

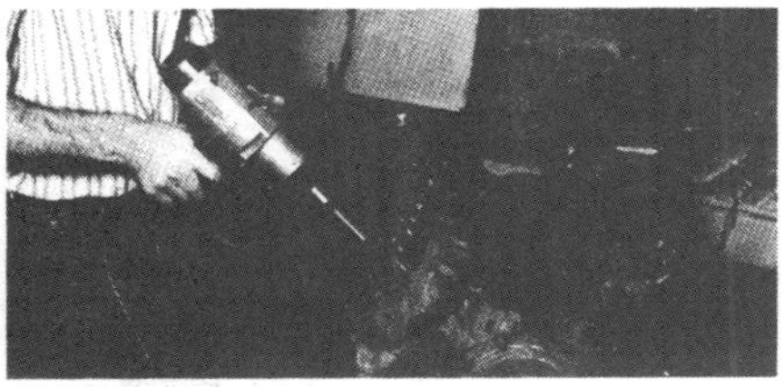

***6** After verifying the engine's authenticity, Dana begins disassembly by removing the cylinder heads. Fasteners and hardware for each component are stored in labeled cardboard boxes.*

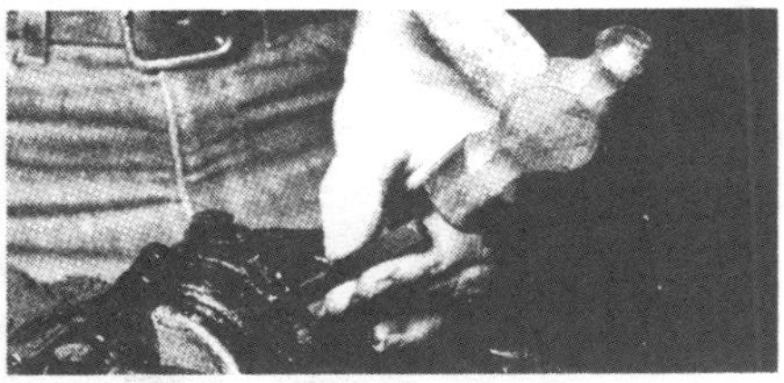

***7** Dana numbers the rod forks and caps with a number punch before removing the pistons. Then he marks the main bearing caps to keep them in order.*

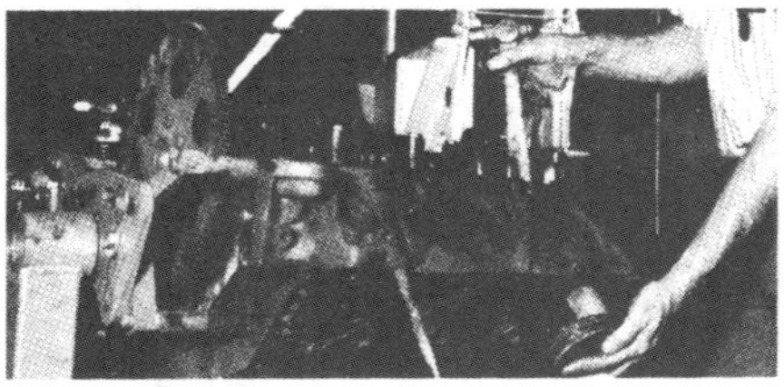

***8** Dana drives the old pistons out of the block with a wooden hammer handle. This prevents scoring or gouging the crankshaft journals.*

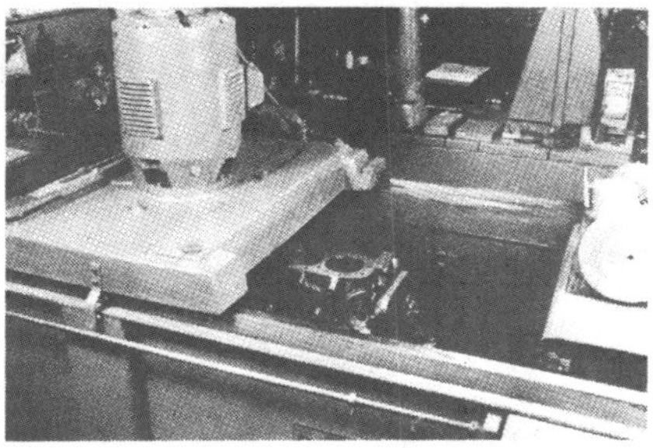

***9** Hollums Racing Engines (Fremont, California) performed the machine work on Morgan's small-block. After align honing and boring the block, Larry Hollums surfaces and squares the decks with a Kwik-Way block grinder. B-H-J "Blok-Tru" fixture ensures that decks are at 90-degree angles. Dana instructed Hollums to remove minimum amount of material from decks to preserve original block stamping numbers.*

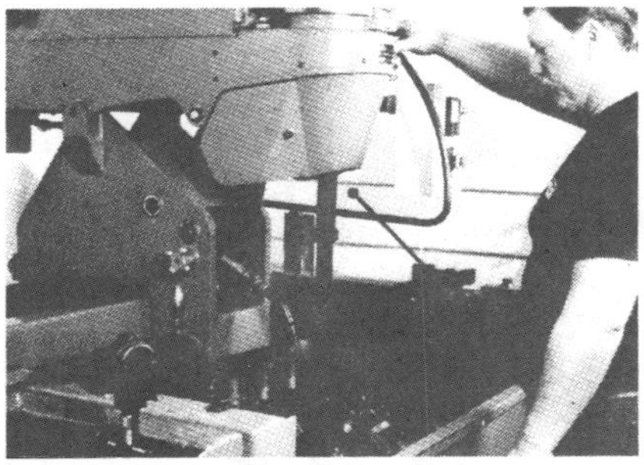

***10** Larry finishes the cylinder bores with fine-grit stones in his Sunnen CK-10 automatic hone. Thick steel deck plate stresses block during honing to simulate distortion caused by torquing head bolts. This ensures that bores will be as round as possible to produce good ring seal.*

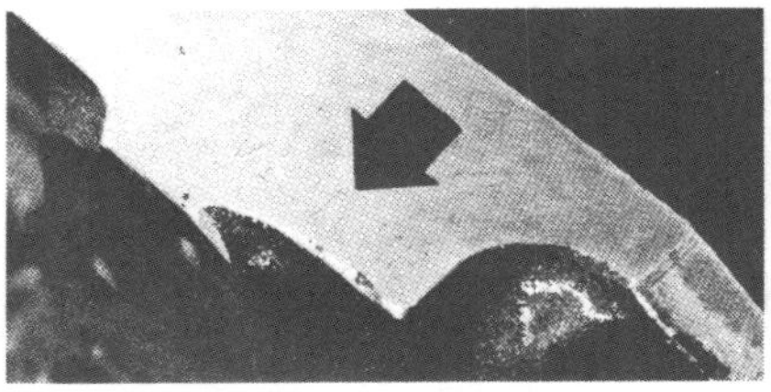

***11** Original block numbers are still visible after decking. Maintaining correct numbers increases car's value to future Camaro collectors.*

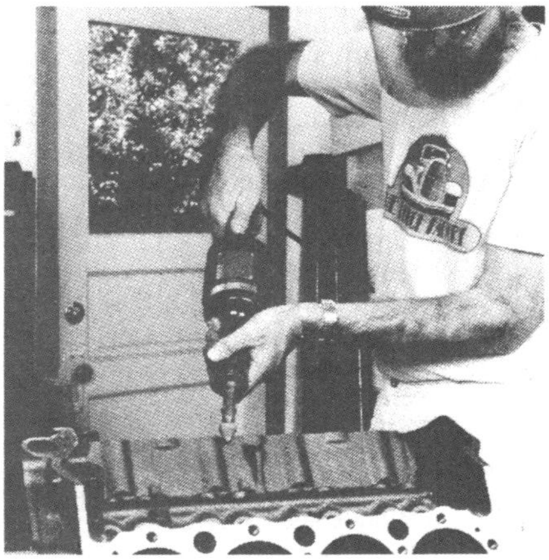

***12** Dana deburrs and details block with high-speed grinder. Sharp edges are smoothed, and oil drain holes carefully radiused to speed lubricant back to pan.*

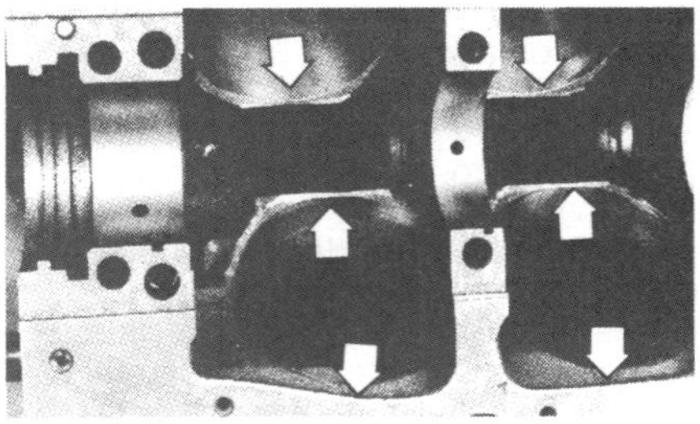

***13** Bottom edges of cylinder bores are ground to eliminate stress risers in thin metal, then dressed with emery cloth. Casting flash in camshaft gallery is removed with long punch.*

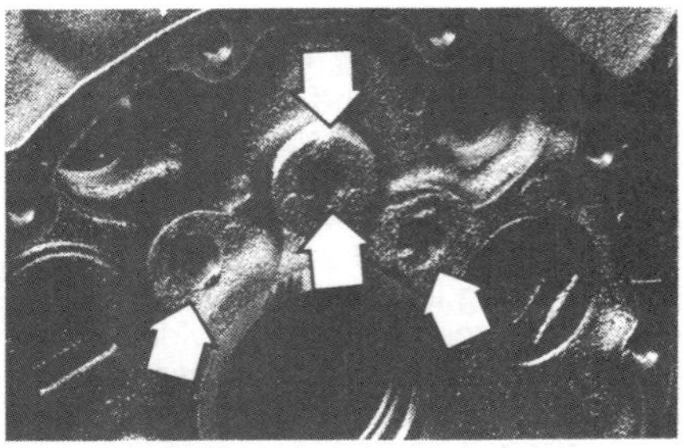

***14** Boss above main oil gallery is ground to provide clearance for wide, high-performance timing chain. Oil galleries are tapped for ¼-inch pipe plugs.*

about the effects of operating early-model engines on unleaded fuel for extended periods. The only permanent solution to rapid valve wear may be to induction-harden the seats or to install hardened inserts. Until the jury comes in on burning unleaded gas in classic Camaros, Dana is relying on high-quality valves and wear-resistant guides to keep his Camaro alive in a lead-free world.

Dana's final Camaro consideration was performance. Judged by the standards of the Eighties, even a base-level V8 Camaro is a fast car—as Dana discovered after test driving a $15,000 Thunderbird! Still, the 327-cubic-inch, 210-horsepower small-block in his Camaro was just too tame for Dana's automotive taste. Substituting a four-barrel manifold and a Quadrajet for the factory-installed two-barrel induction brings the factory power rating up to 250 horsepower—although Chevrolet never actually offered a 250hp/327ci V8 in early-model F-bodies. This engine combination is ideally suited to the Eighties: its small intake runners provide excellent throttle response and fuel economy, and the cylinder heads' large combustion chambers keep the compression ratio under 9:1. Naturally, Dana carefully packed away the original two-barrel setup so he could return the family F-body to showroom condition in the future.

Dana's hot rod heritage became apparent when he decided to blueprint his Camaro's small-block. Not all engines are created equal; correct clearances and careful assembly can make an astounding difference in an engine's performance. There's "free" horsepower to be gained by making sure that the cylinder bores are perfectly round, the deck heights are equal, and the valve seats are concentric. Balancing reduces vibration and extends bearing life. By carefully blueprinting his Camaro's motor, Dana realized solid gains in both performance and reliability—while taking advantage of the low cost and convenience of stock-type replacement parts.

Sixties-vintage Camaro motors are marvelous machines. Compared to the thin-wall castings and pig iron crankshafts found in current engines, Dana's 327 has all the right stuff: substantial cylinder walls, heavy-duty cylinder heads, and a forged steel crank. During its first 18 years of service, Morgan's motor provided 100,000 miles of reliable transportation. And if the Chevrolet small-block gods are smiling, it should log another 100,000 before it's ready for another renewal.

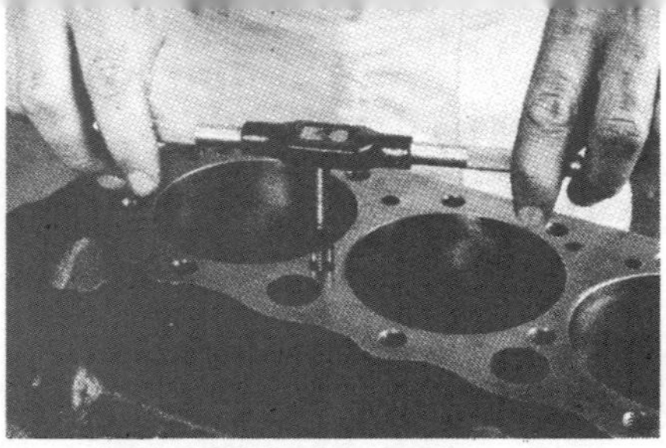

15 *A 7/16 NC bottoming tap cleans old sealer out of head bolt holes and restores threads to ensure accurate torque readings.*

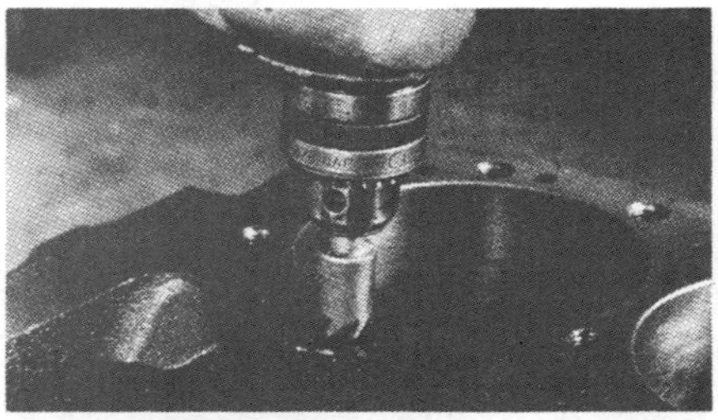

16 *Dana countersinks head bolt holes to prevent fasteners from pulling up top thread.*

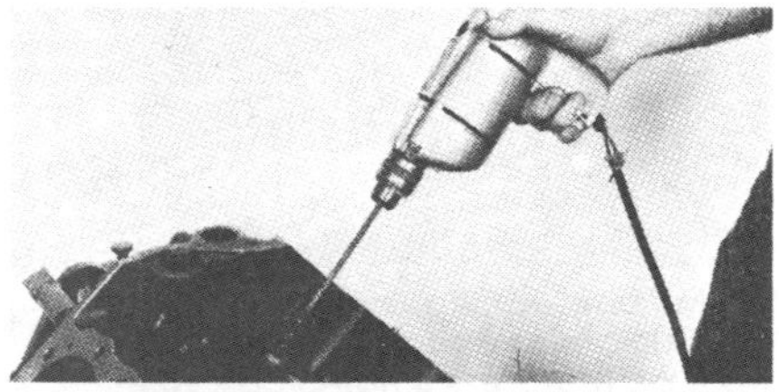

17 *Brake-cylinder-ball hone quickly smoothes and deburrs lifter bores.*

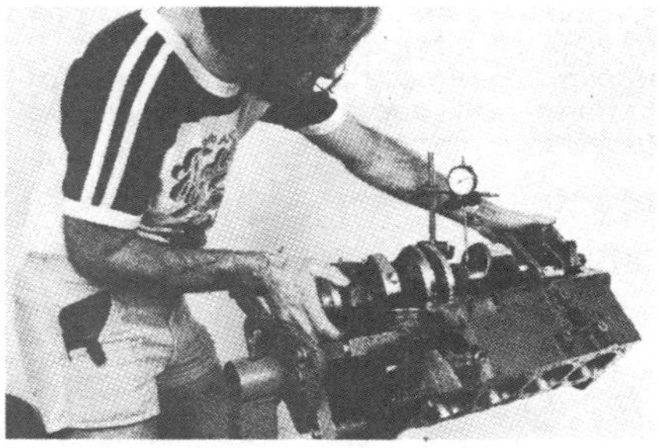

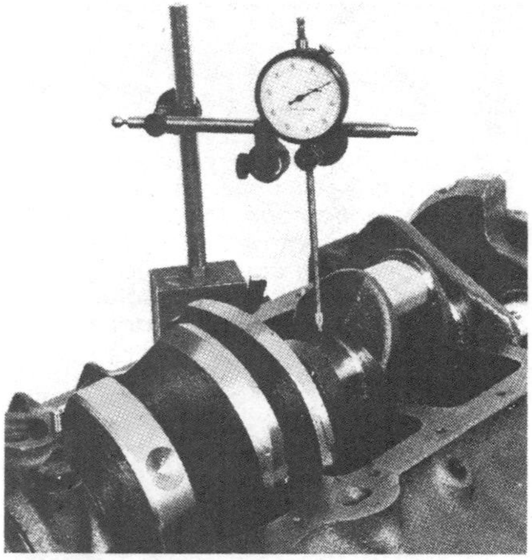

18 *Dana pre-assembles engine to check critical clearances. He checks crankshaft straightness by installing front and rear main bearings, then setting up dial indicator on center journal. Re-ground steel crankshaft has just .0007-inch runout.*

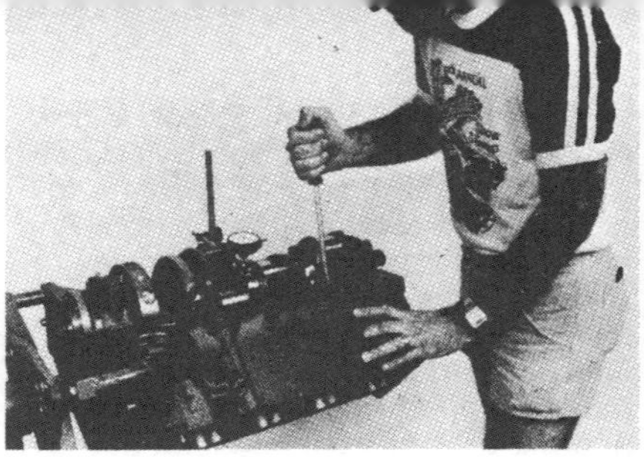

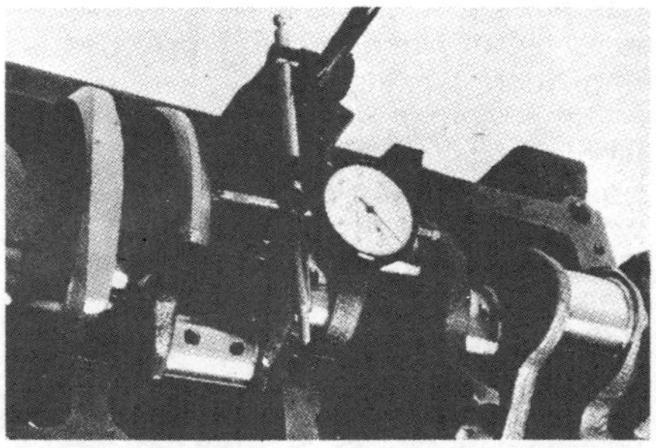

19 *Dana measures crankshaft's thrust clearance (or end play) by prying crank forward and backward in block while noting dial indicator reading. End play is perfect at .005-inch.*

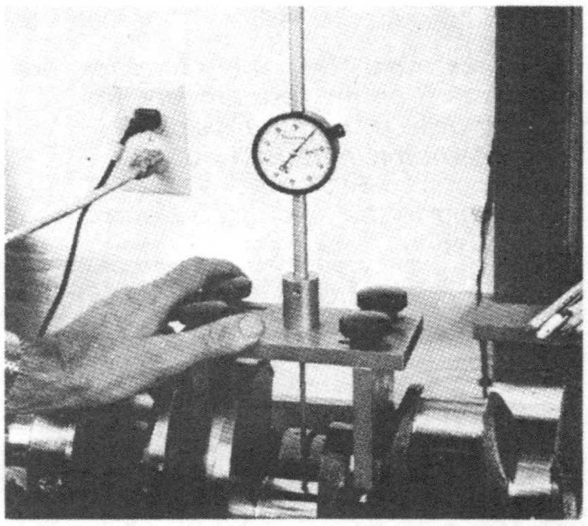

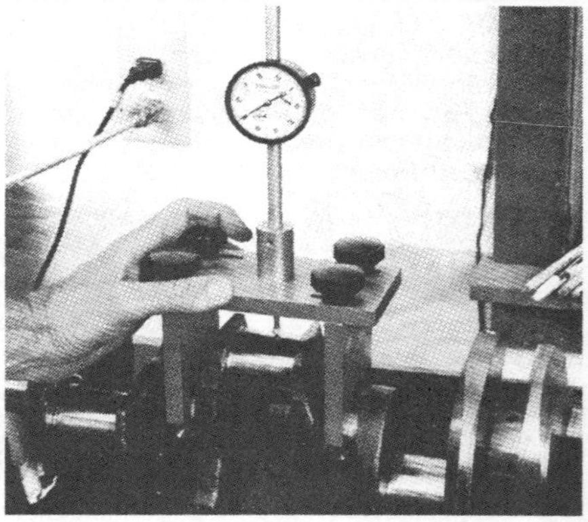

20 *Chrome-A-Shaft crankshaft gauge measures stroke quickly and accurately by reading journal height at top and bottom of travel. Forged steel crankshafts installed in '67 327ci V8s have 3.250-inch stroke.*

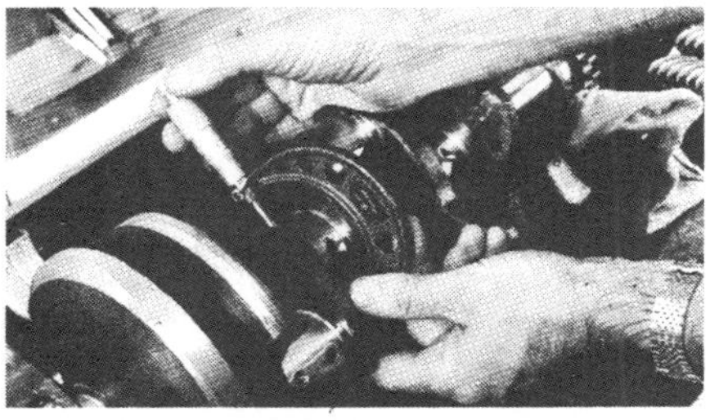

21 *Correct bearing clearances are critical to engine life. Dana measures main journal diameter of re-ground crankshaft with outside micrometer. Journals measure 2.289-inch in diameter.*

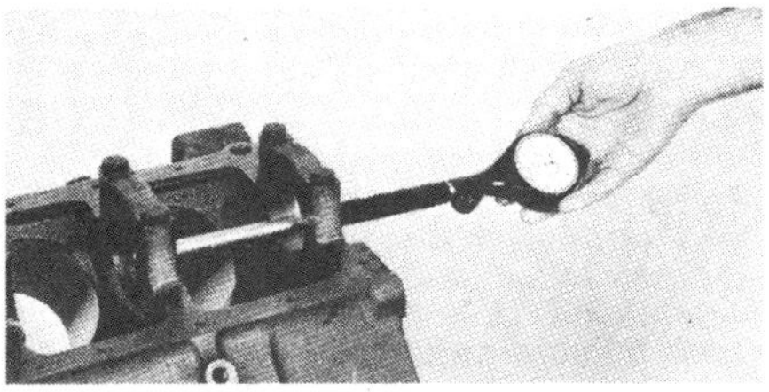

22 *Dial bore gauge measures diameter of main bearing saddles. Hollums align hones housing bores to 2.491-inch diameter.*

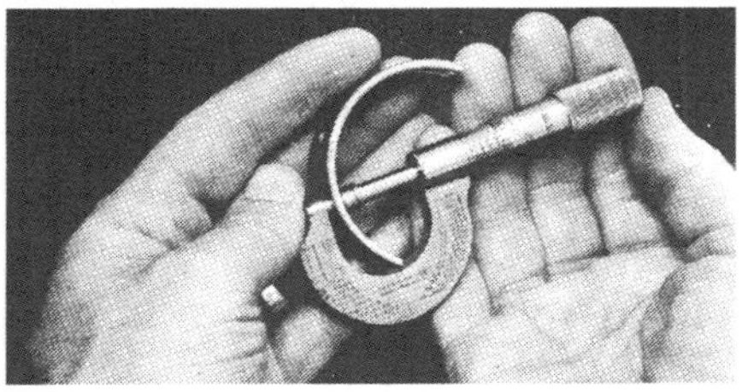

23 *Final step in checking main bearing clearance is to measure thickness of Childs & Albert inserts (part No. 3270-10) with ball micrometer; bearing shells are exactly .100-inch thick. Dana calculates final clearance by subtracting bearing thickness and journal size from housing bore diameter. In this example, 2.491-inch housing bore minus .200-inch bearings and 2.289-inch crankshaft leaves final clearance of .002-inch.*

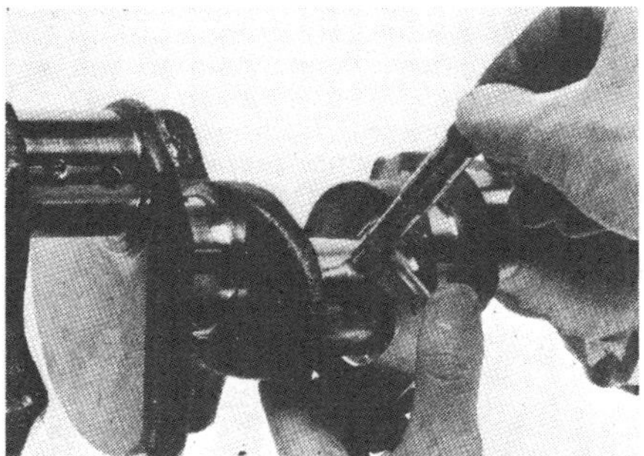

24 *Procedure for measuring rod bearing clearance is similar. Dana finds that all four rod journals measure 1.9895-inch in diameter.*

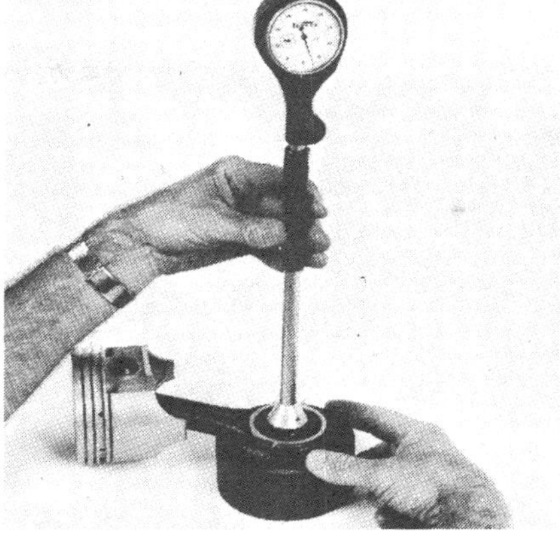

25 *Bearing inserts are installed in rods and rod bolts tightened. Dial bore gauge shows .002-inch vertical clearance with Childs & Albert small-journal bearings (part No. R3270-10).*

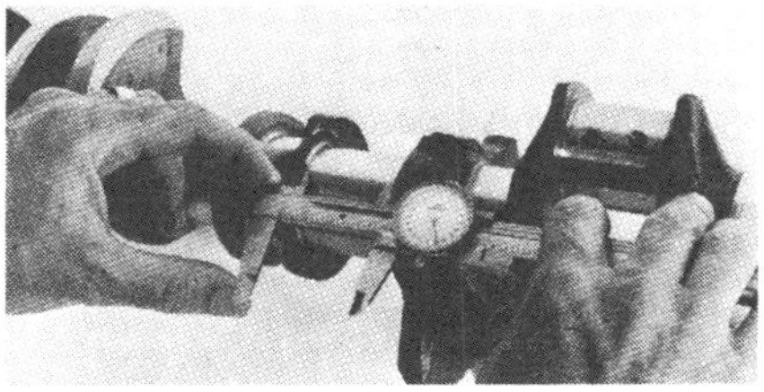

26 *Rod side clearance is important to oil flow across bearing surfaces. Morgan first measures width of rod journal with dial calipers. Distance between crank cheeks is 1.902-inch.*

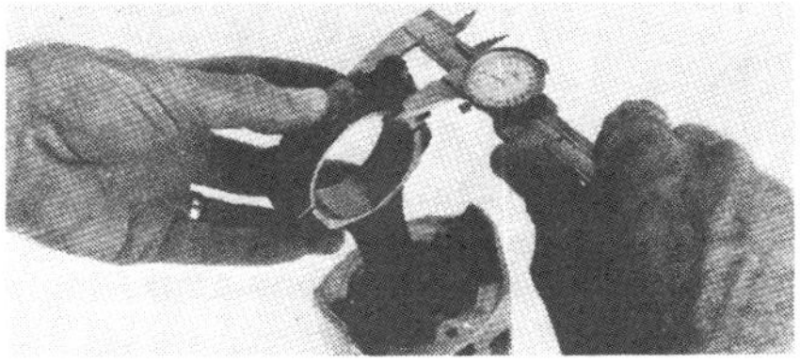

27 *Then he measures width of connecting rods' big ends, which vary from .941 to .943-inch. Subtracting width of two rods from crank journal width reveals side clearance—.016-inch side clearance is ideal.*

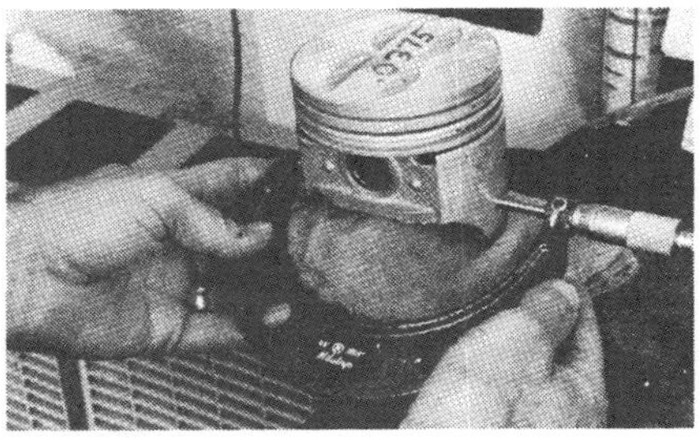

28 *Dana selected cast aluminum pistons for his Camaro's motor. Cast pistons can be installed at tighter clearances than forgings, providing quieter operation and improved ring stability. He measures .030-inch oversize pistons at bottom of skirt tang to determine piston-to-cylinder wall clearance.*

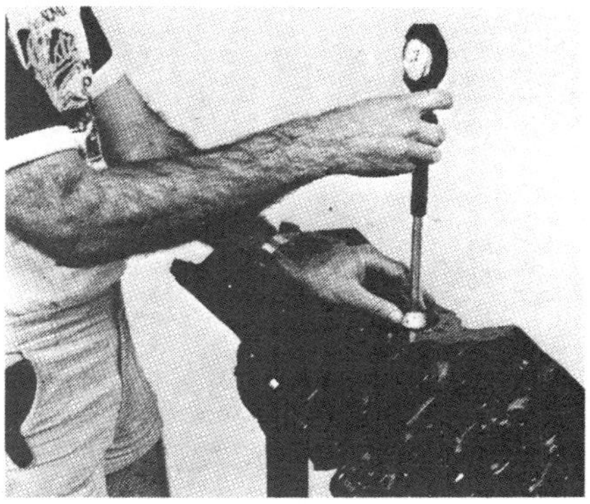

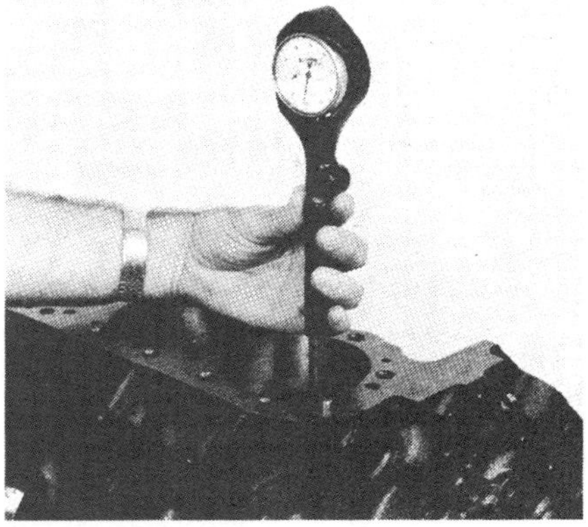

29 *Dial bore gauge is set to piston diameter, then inserted in cylinder. Dial indicator reads out piston clearance; Hollums honed bores to provide .0008-inch skirt clearance. Dana also checks taper by measuring bore diameter at top and bottom of ring travel. Taper is negligible after honing with CK-10 and torque plates.*

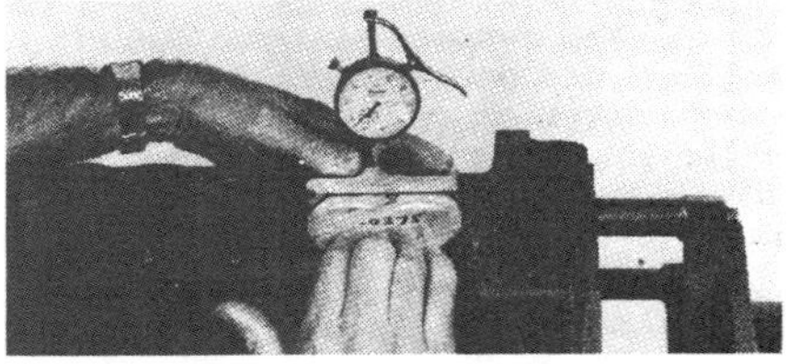

30 *Correct piston deck height is essential to engine efficiency. If piston-to-head clearance is too small, the pistons will hit the heads at high rpm; too much clearance sacrifices compression and invites detonation. Dana measures deck height by bringing piston to TDC (top dead center), then rocking piston from side to side. His pistons are .035-inch below deck surface; this provides total piston-to-head clearance of .050-inch with .015-inch-thick head gasket. Block could be cut another .015-inch to bring deck heights to minimum tolerance, but machining decks would remove original block stamping numbers.*

REVIVAL

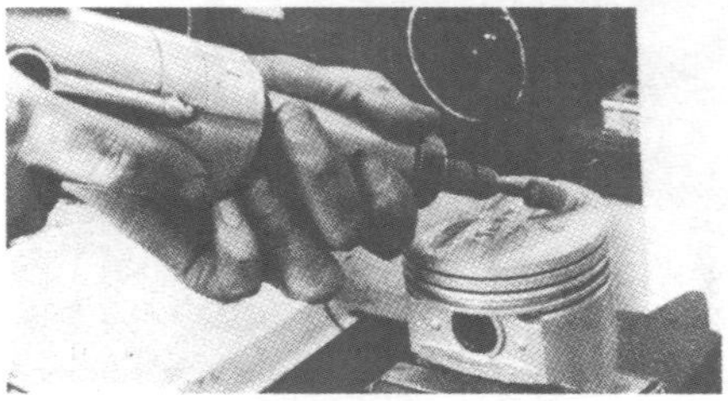

31 *Dana radiuses piston valve reliefs with abrasive cartridge roll to eliminate sharp edges that can foster pre-ignition and detonation.*

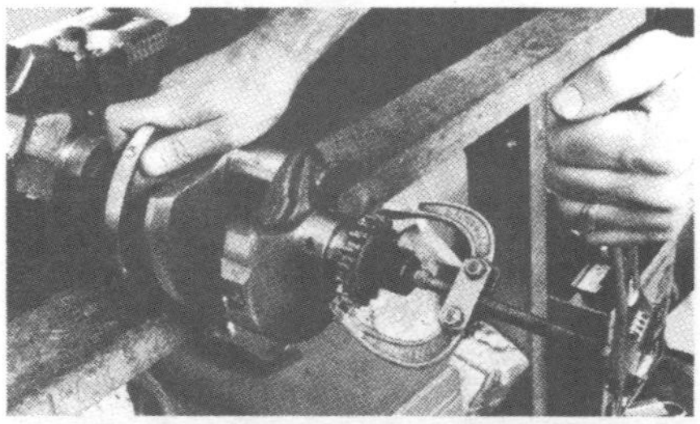

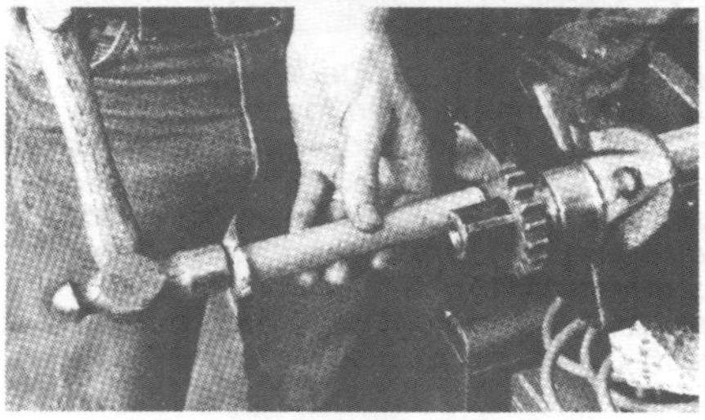

32 *Dana removes worn crankshaft timing gear with puller, then installs new wide gear with soft aluminum punch. Crank gears must be installed carefully to prevent cracks at Woodruff key slot.*

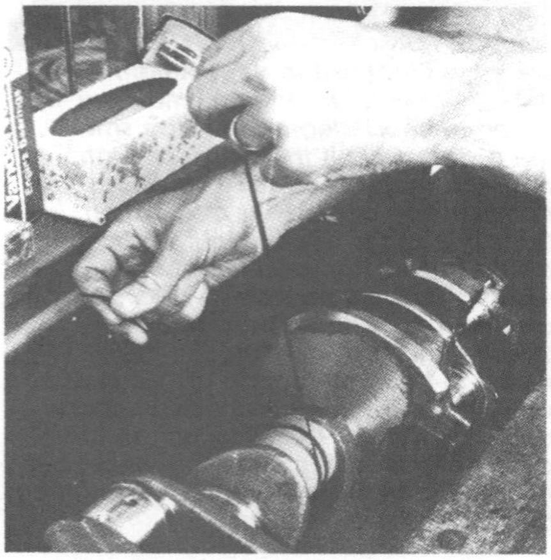

33 *Dana polishes crankshaft journals with 600-grit wet-or-dry sandpaper and a shoelace. Polishing removes tiny nicks and scratches caused during pre-assembly.*

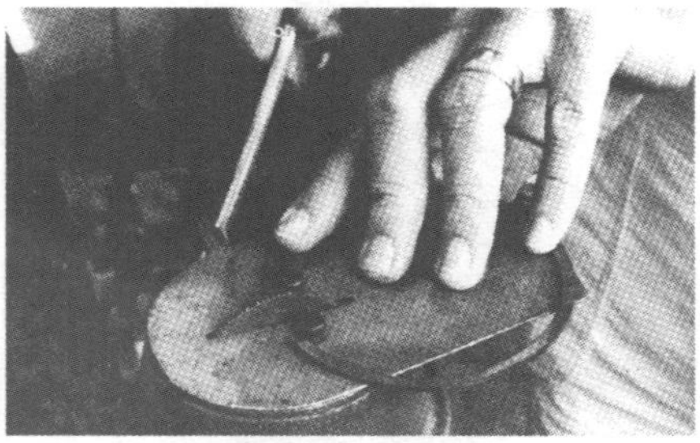

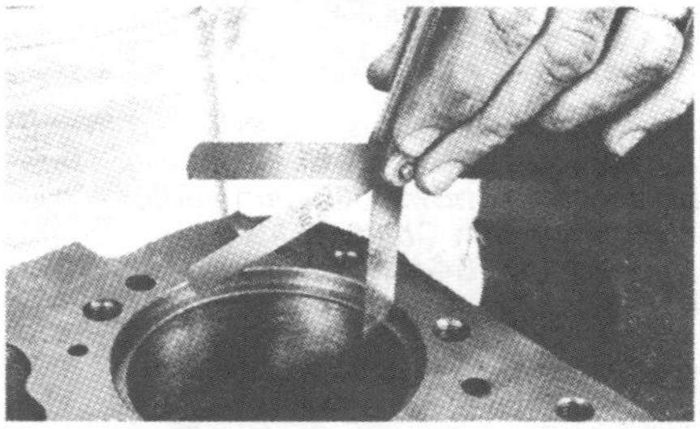

34 *Speed Pro performance piston rings (part No. 9343-035) are .035-inch oversize to allow engine builder to tailor end gaps to his application. Dana files moly-filled ductile iron top compression ring to produce .017-inch end gap. Cast iron second rings are cut to .014-inch end gaps.*

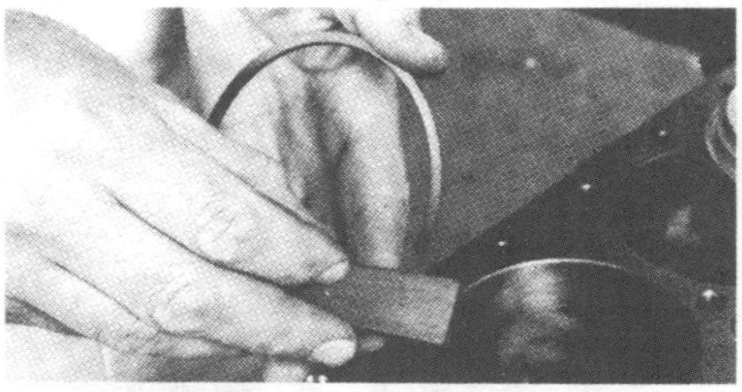

35 *Dana deburrs rings with machinist's whetstone after filing.*

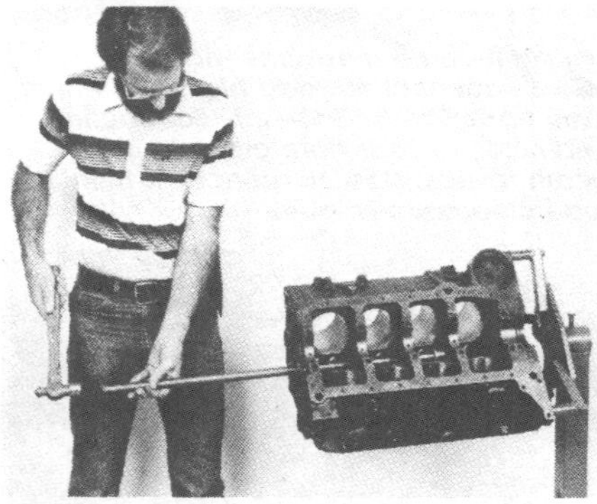

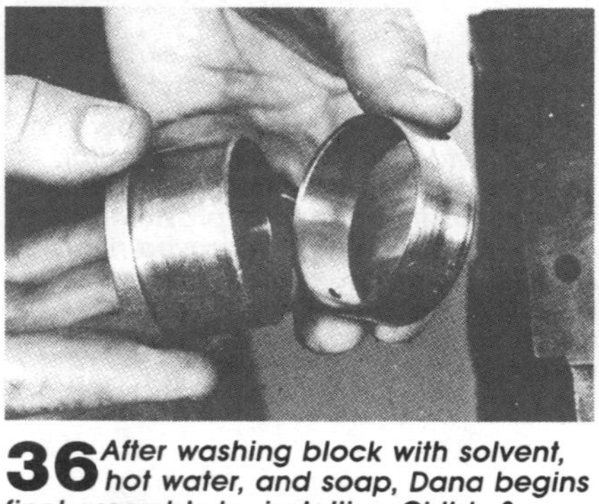

36 *After washing block with solvent, hot water, and soap, Dana begins final assembly by installing Childs & Albert babbitt cam bearings (part No. C3500) with homemade driver.*

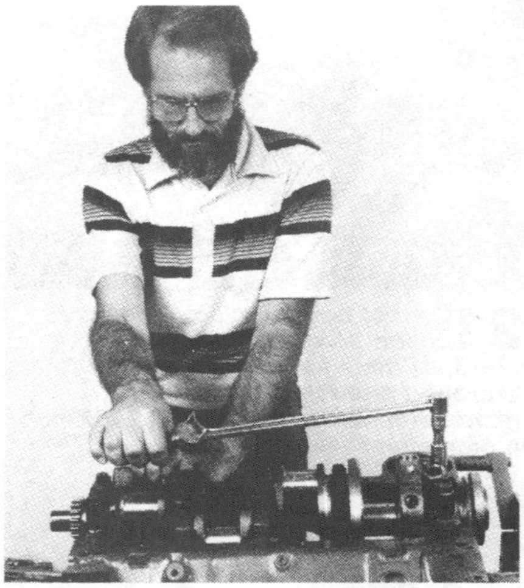

37 *Main bearing shells are washed with acetone, installed in block, and lubricated with 50/50 mix of GM Engine Oil Supplement and 30-weight oil. Dana tightens main bearing cap bolts to 80 ft.-lbs. with beam torque wrench.*

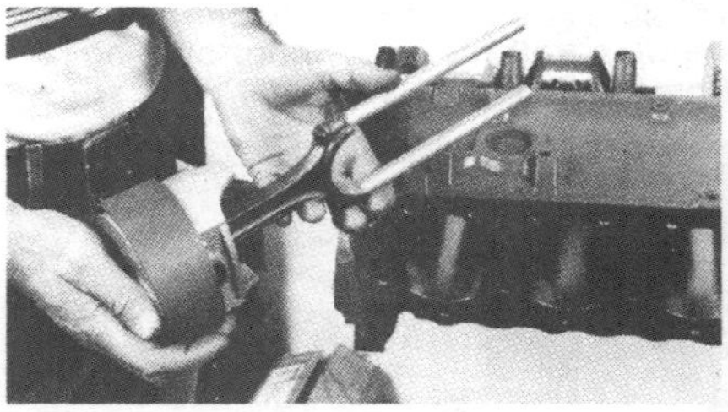

38 *Dana threads aluminum "chopsticks" over rod bolts to protect crankshaft journal from nicks when installing pistons.*

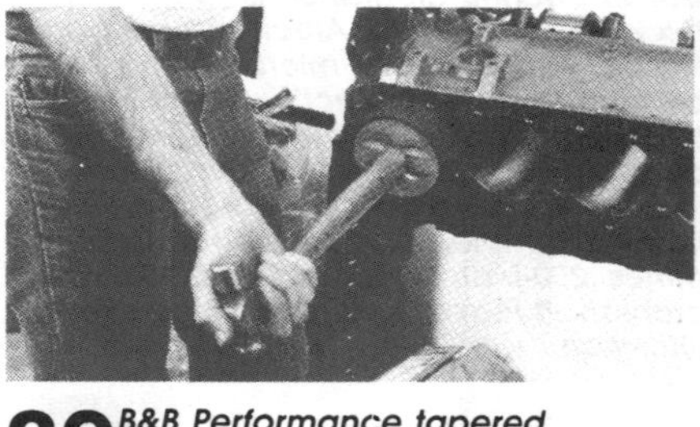

39 *B&B Performance tapered aluminum ring compressor makes piston installation easy. Morgan taps piston into bore with hammer handle while guiding rod onto crankshaft journal. Notches in piston top are installed toward front of block.*

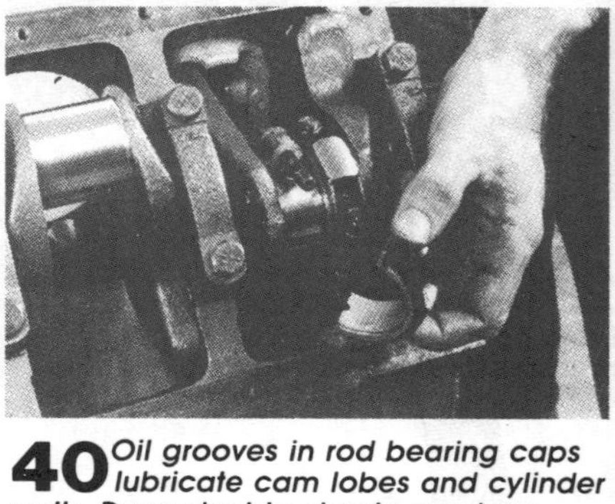

40 *Oil grooves in rod bearing caps lubricate cam lobes and cylinder walls. Dana double-checks numbers stamped on rod caps and forks before tightening nuts.*

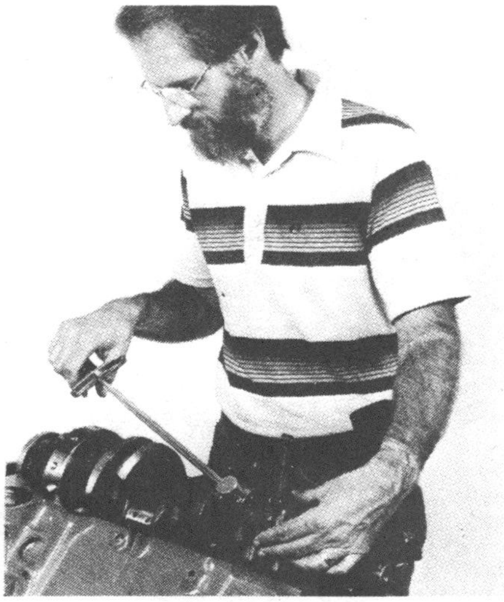

41 One shortcoming of early-model Chevrolet V8s is their 11/32-inch rod bolts; '68 and later small-blocks have stronger 3/8-inch fasteners. Hollums Racing Engines replaced stock bolts with Moroso high-strength fasteners. Dana torques new bolts to 40 ft.-lbs.

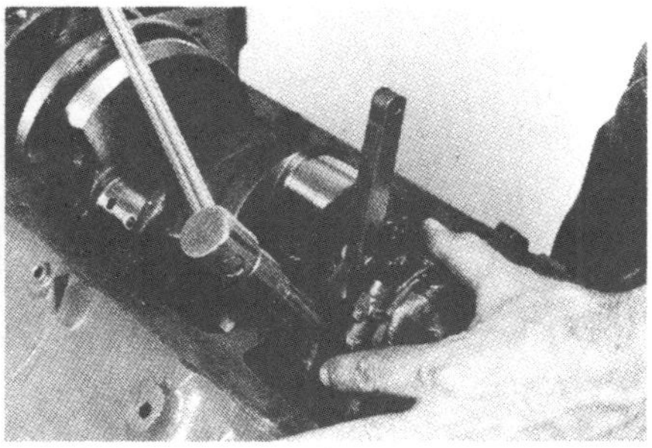

42 Feeler gauge inserted between rods aligns caps and forks while bolts are tightened.

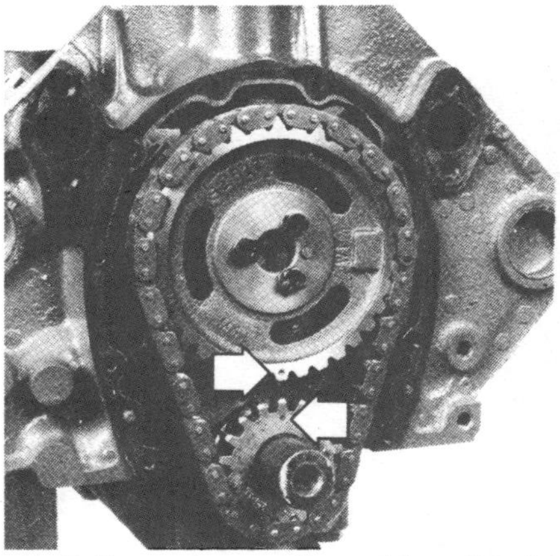

43 Dana chose a Crane blueprinted replacement cam (part No. 968711) for his refurbished small-block. This is a duplicate of Chevrolet part No. 399629 used in 210, 275, 295, and 300-horsepower small-blocks. The Crane profile duplicates the factory's lift and duration (195-degrees intake duration at .050-inch tappet lift, 202-degrees exhaust duration; .390-inch intake lift, .410-inch exhaust lift). Dana begins the cam installation by lining up the timing marks on the camshaft and crankshaft sprockets.

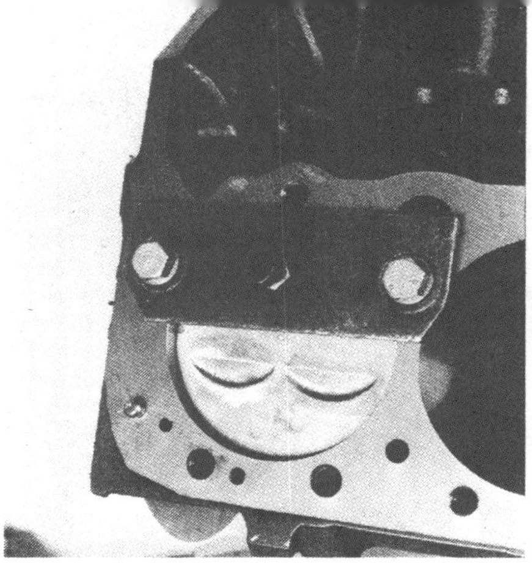

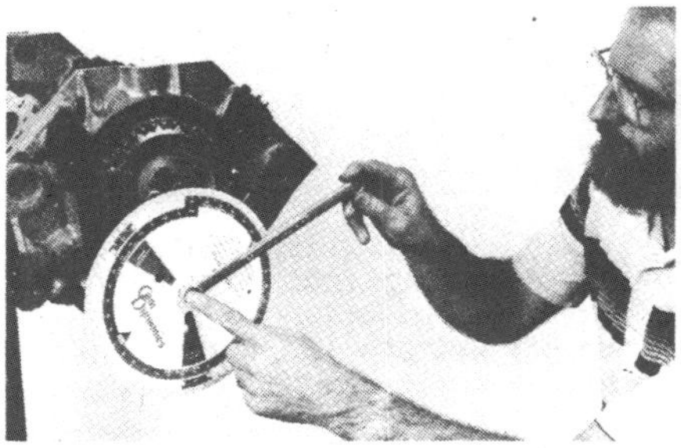

44 Degreeing the cam ensures that it is manufactured and installed correctly. Dana bolts a piston stop to the deck surface, then attaches a degree wheel to the crankshaft snout. Top Dead Center is exactly halfway between the points where the piston contacts the stop when the crank is rotated forward and backward.

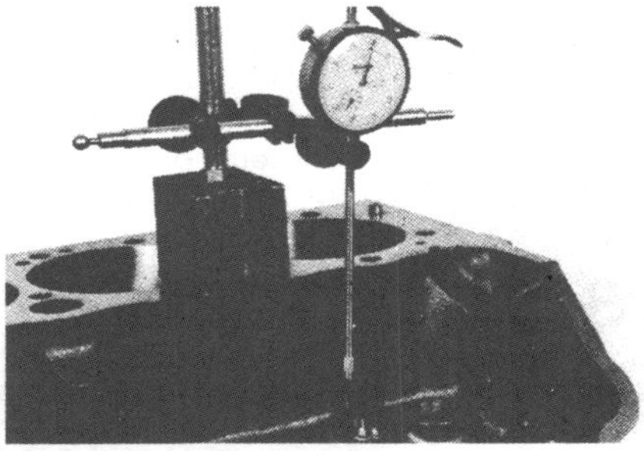

45 Dana positions dial indicator plunger on the edge of the lifter for the No. 1 intake valve. First he checks maximum tappet lift (.260-inch). Then he measures the intake centerline by finding the point where maximum lift occurs—in this instance, 108 degrees after TDC.

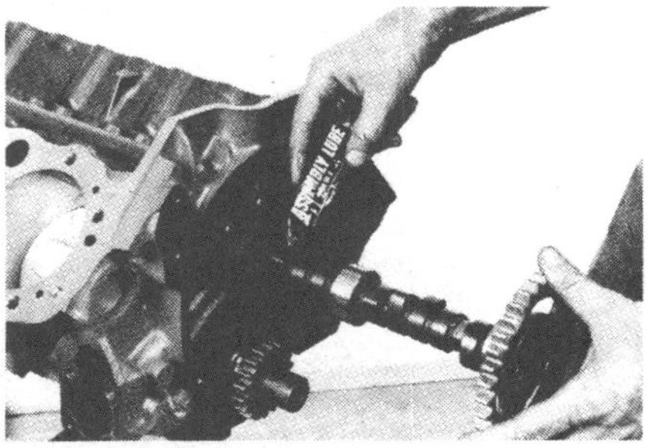

46 Dana removes the cam after checking its timing and coats the lobes with molydisulfide assembly lube. The first few minutes of operation are critical to the life of a flat tappet cam; the engine must not run below 2000 rpm for 20 minutes.

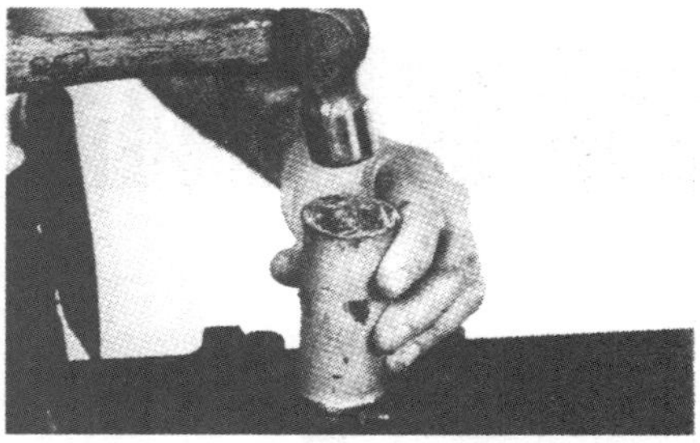

47 Dana installs brass soft plugs to finish the short-block assembly. Plugs are sealed with Gasgacinch adhesive.

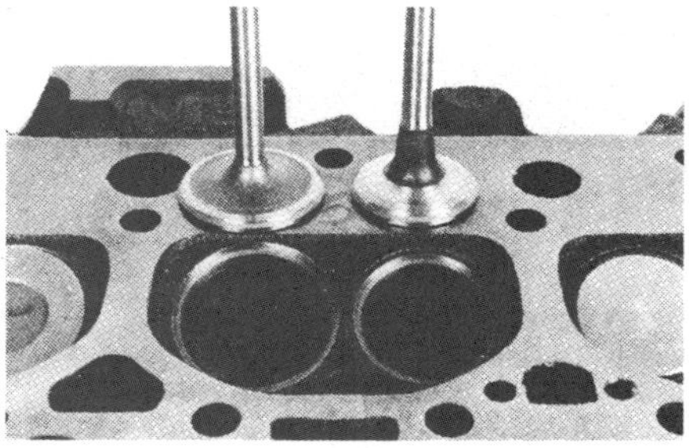

48 Cylinder heads are upgraded with TRW stainless steel valves (1.72-inch-diameter intake, part No. V2694X; 1.5-inch-diameter exhaust, part No. S2323) to retard wear caused by lack of seat lubrication with unleaded gasoline. Larry Hollums back-cut valve heads at a 30-degree angle to improve low-lift flow, then ground three-angle valve seats to promote cylinder filling.

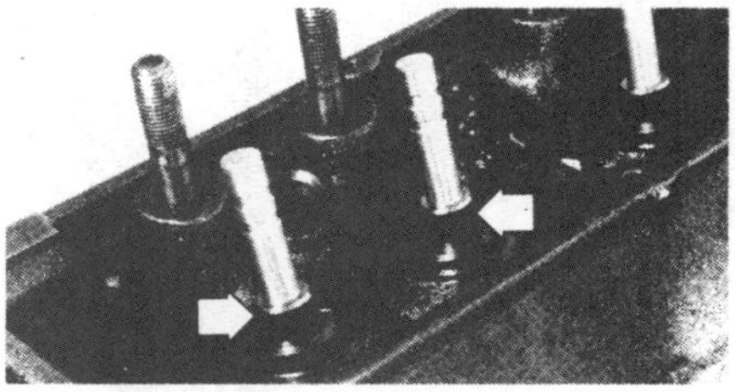

49 Hollums Racing Engines also installed silicon-bronze valve guide inserts to eliminate rapid valvetrain wear caused by lack of stem lubrication.

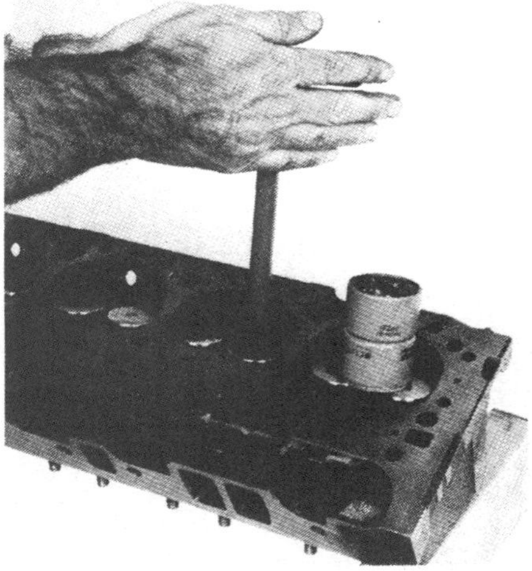

50 Dana laps valves by hand with fine Clover compound. Light gray rings on valve heads and seats highlight sealing surfaces.

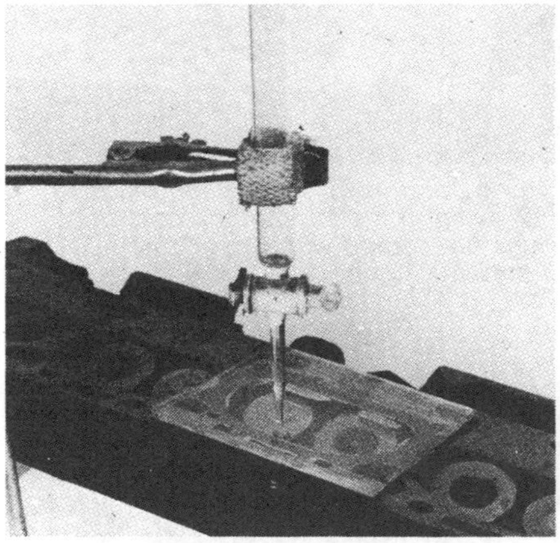

51 Combustion chamber volume measures 73.4 cc after surfacing heads and installing new valves. With flat-top pistons and .015-inch-thick head gasket, true compression ratio is 8.70:1—just right for today's rotten gasoline.

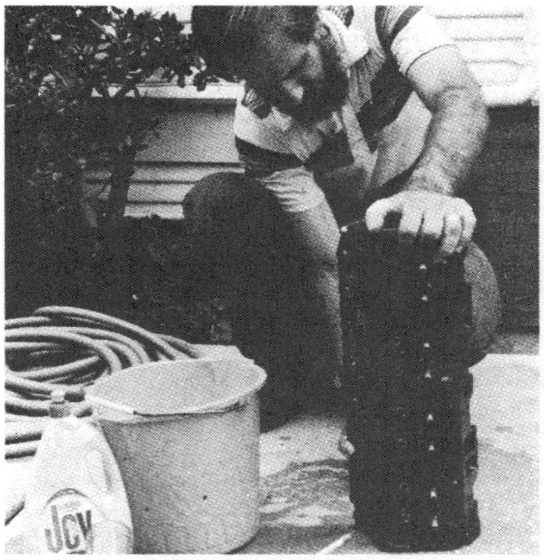

52 Morgan cleans cylinder heads with solvent, soap, and water. Car wash sprayer is an ideal tool for blasting grit from valve guides and head ports.

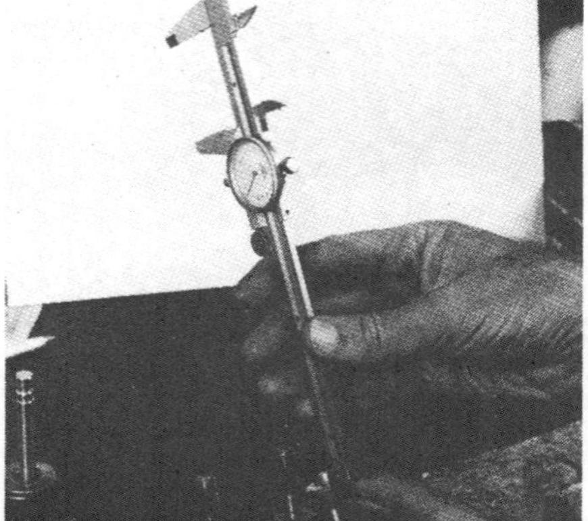

53 Dana begins cylinder head assembly by measuring installed valve spring height with dial calipers. He measures from spring seat to top of retainer.

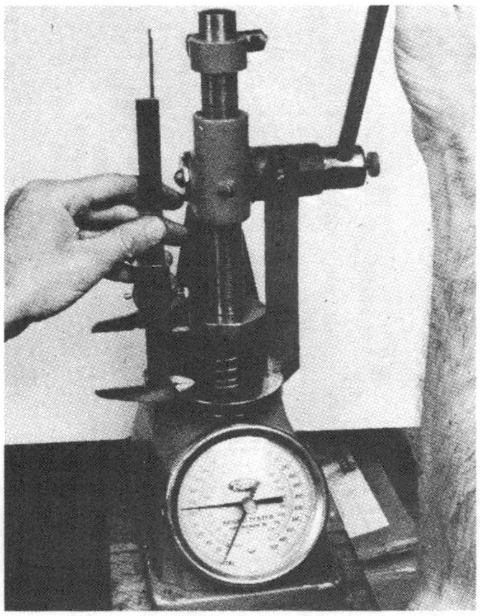

54 Stiff Iskenderian high-performance valve springs (part No. 8005A) are same diameter as factory coils. Dana sets up Isky springs with 110-pounds seat pressure.

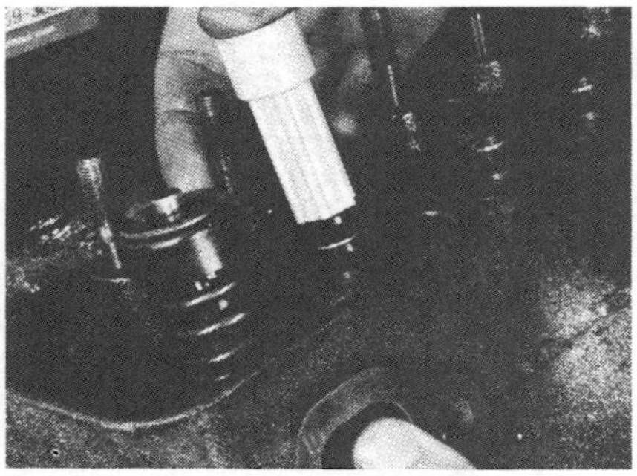

55 Teflon valve stem seals provide better oil control than stock O-ring seals.

56 A lever-action spring compressor makes quick work of assembling the cylinder heads. New seven-degree split locks and Chevrolet steel retainers are stone reliable on the street.

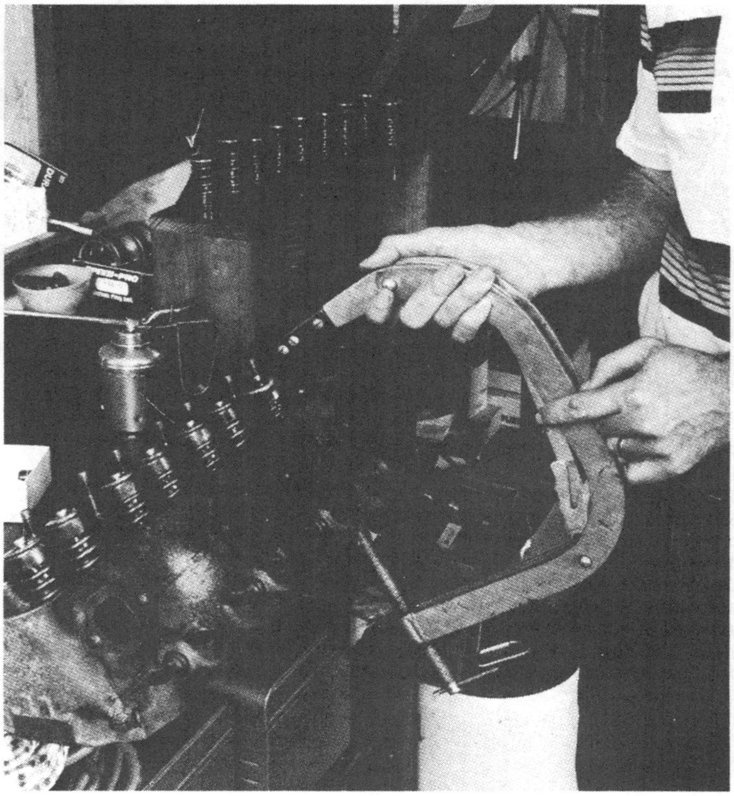

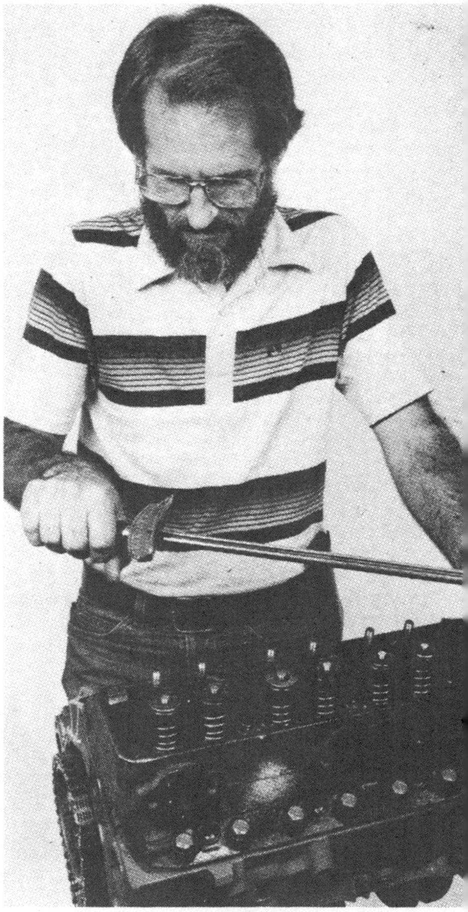

57 Dana torques head bolts to 65 ft.-lbs. in 10-pound increments. Chevrolet stainless steel gaskets (part No. 3916336) and NAPA Hi-Tack gasket compound seals cylinders. Bolt threads and heads are coated with Aviation Form-a-Gasket to prevent coolant leaks.

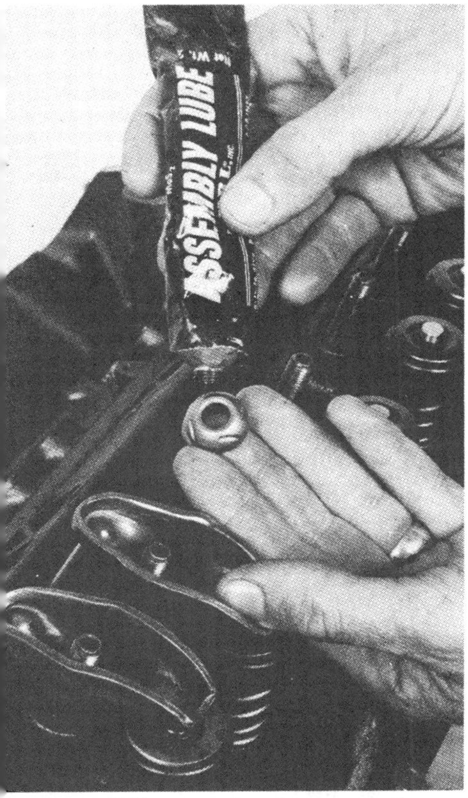

58 *TRW grooved rocker arm balls are less prone to galling than stock pivots. Balls and rockers are coated with assembly lube and Engine Oil Supplement.*

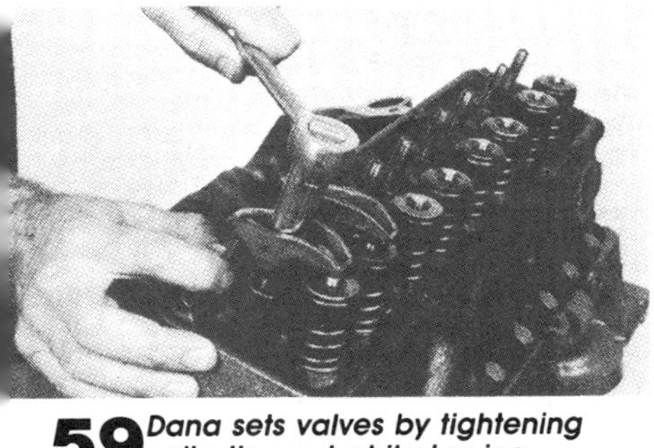

59 *Dana sets valves by tightening adjusting nut while turning pushrod between his fingers. Valve is at zero lash when pushrod no longer turns easily; Dana then tightens nut one full turn to center plunger in hydraulic lifter.*

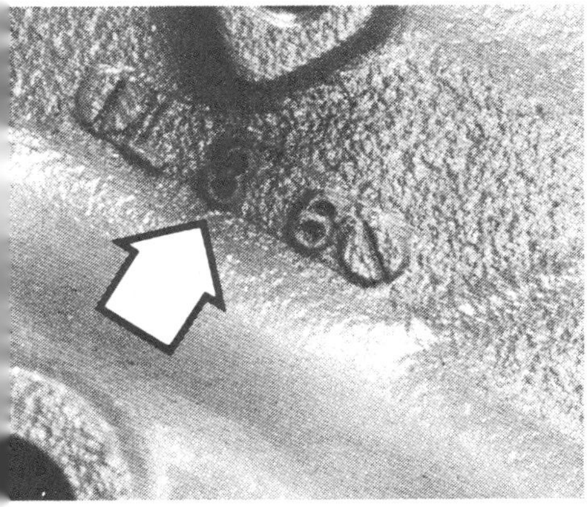

60 *Junkyard four-barrel manifold has December 3, 1966, casting date—just right for a '67 Camaro engine!*

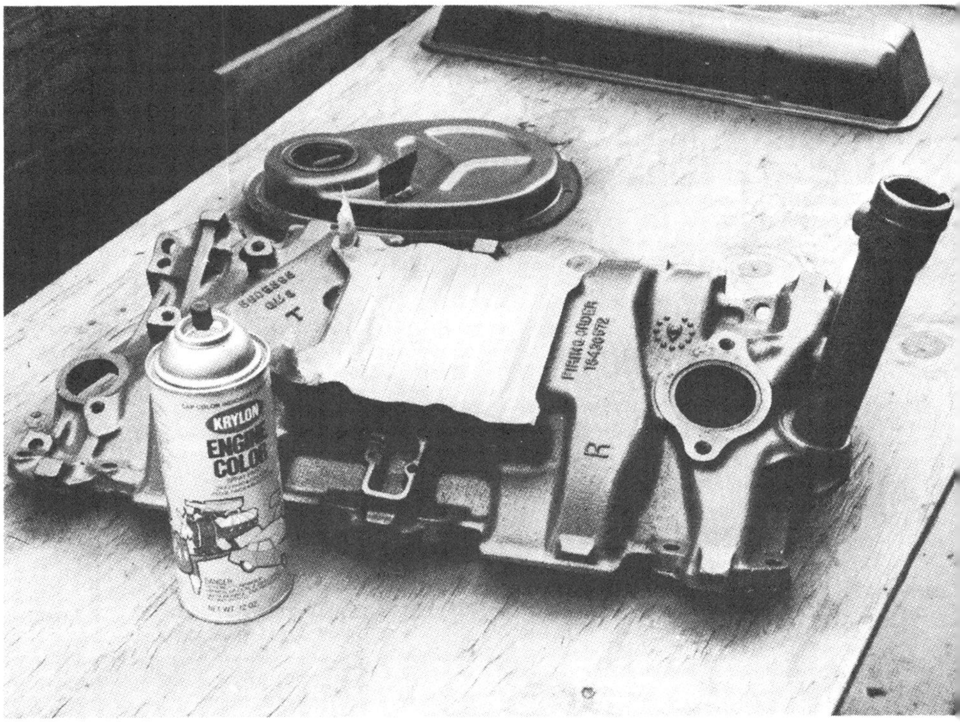

61 *Cast iron Quadrajet intake manifold and original stamped steel valve covers are glass-beaded to remove rust, then painted with Krylon Chevrolet orange aerosol paint.*

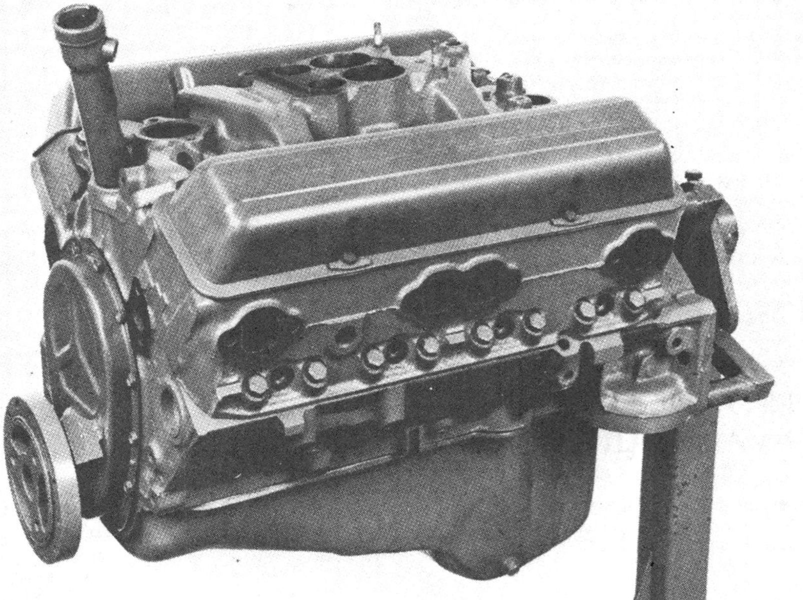

62 *Ready for installation, Dana's blueprinted 327ci small-block is ready to deliver another 100,000 miles of trouble-free service. Taking the time to double-check critical dimensions and clearances pays big dividends in durability and performance. With a better-than-new powerplant, Dana's Rally Sport will soon be ready for some serious driving.*

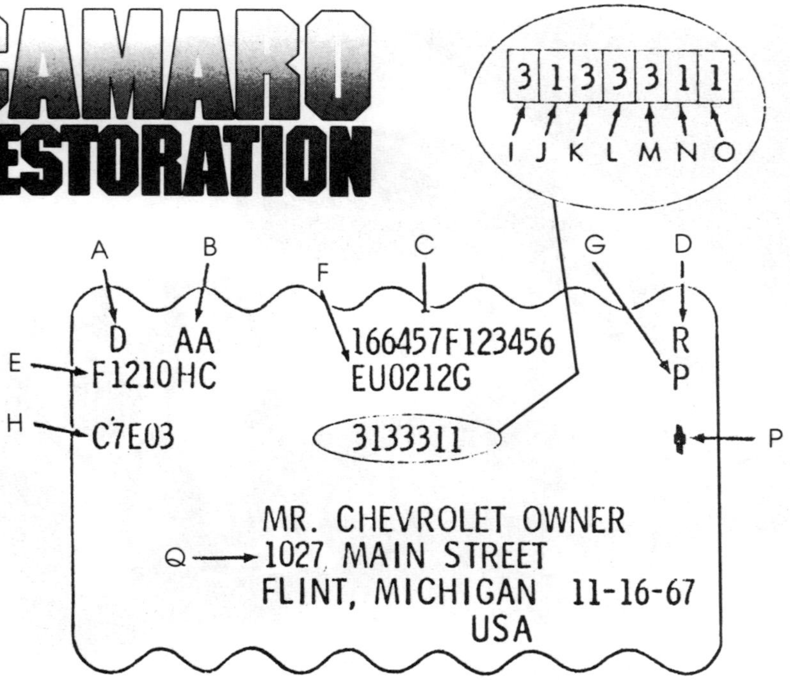

CAMARO CODES

You don't need a Captain Marvel decoder ring to unlock your Camaro's secrets—just a parts book. A company the size of General Motors has millions of parts to keep track of every year. In order to simplify its record-keeping, the General stamps codes on each vehicle's components and its Protect-O-Plate that indicate the particular options installed on that car. You can learn a great deal about your Camaro by studying its Protect-O-Plate, cowl tag, and Vehicle Identification Number.

The following codes apply only to 1967 Chevrolets. Codes for other model years can be obtained from dealer parts books.

PROTECT-O-PLATE AVAILABILITY CHART

LOCATION	DESCRIPTION
A	Interior Trim
B	Body Exterior Paint
C	Vehicle Identification Number
D	Carburetor Source Code
E	Engine Number
F	Rear Axle Number
G	Vehicle Build Month Code
H	Transmission Number
I	Power Steering and/or Power Brakes
J	Radio/Clock Group
K	Radio Code
L	Disc Brake Code
M	Air Conditioning
N	Power Window Code
O	Power Seat Code
P	Chevrolet Trademark
Q	Dealer Supplied Owner Information
R	Carburetor Usage (certain engines only—1966)

LOCATION A—INTERIOR COLORS

CODE	COLOR (not all available for Camaro)
A	Red Vinyl
B	Blue
C	Black Vinyl
D	Red
E	Black
F	Fawn
G	Gold
H	Blue Vinyl
K	Parchment Black
L	Black Vinyl (Opt.)
M	Maroon
N	Black Cloth
P	Plum
R	Bright Blue
S	Blue Cloth
T	Turquoise
U	Blue Vinyl (Opt.)
V	Fawn Vinyl
W	Gold Cloth
Y	Yellow

LOCATION B—EXTERIOR COLORS

All passenger models will have RPO (regular production option) letter designation stamped in this location. On two-tone combinations, the first letter indicates the lower color, second letter the upper color.

LOCATION C—VEHICLE IDENTIFICATION NUMBER

The first two digits denote the vehicle series. The third indicates base engine equipment, and the two remaining numbers identify the body style. When the third digit is odd (1, 3, 5, or 7), the base engine is an inline four-cylinder or six-cylinder. When the third number is even (2, 4, 6, 8), the base engine is a V8. The complete number consists of the model identification, year of manufacture, plant designation, and sequence number.

Example: 124377N100500
12437 = Camaro 2-door coupe with V8
7 = 1967 Model Year
N = Norwood plant
100500 = 500th 1967 Camaro built at Norwood (sequence numbering starts at 100,001 at all plants)

CAMARO MODEL IDENTIFICATION:

6-CYLINDER	V8	DESCRIPTION
12337	12437	Camaro sport coupe
12367	12467	Camaro convertible

ASSEMBLY PLANT DESIGNATIONS:

CODE	PLANT
N	Norwood
L	Los Angeles

LOCATION D—CARBURETOR SOURCE

CODE	PLANT
B	Bay City
C	Carter
H	Holley
R	Rochester

LOCATION E—ENGINE PRODUCTION CODE

Example: F1210ME
F = Source (Flint Motor)
12 = Month (December)
10 = Day (10th)
ME = Engine Suffix (327 2-barrel w/Powerglide)

ENGINE PLANT DESIGNATIONS:

CODE	PLANT
F	Flint Motor
T	Tonawanda Engine Plant
K	McKinnon Industries
V	Flint Engine
S	Saginaw Service

ENGINE SUFFIX DESIGNATIONS (V8 ONLY):

CODE	ENGINE
MO	302
MP	302 with A.I.R.
MA	327/2-barrel w/manual trans
MB	327/2-barrel w/manual trans, A.I.R.
ME	327/2-barrel w/Powerglide
MF	327/2-barrel w/Powerglide, A.I.R.
MK	327/4-barrel w/manual trans
ML	327/4-barrel w/manual trans, A.I.R.
MM	327/4-barrel w/Powerglide
MN	327/4-barrel w/Powerglide, A.I.R.
MS	350 w/manual trans
MT	350 w/manual trans, A.I.R.
MU	350 w/Powerglide
MV	350 w/Powerglide, A.I.R.
MW	396 w/manual and auto trans
MX	396 w/A.I.R.
MY	396 w/3-speed auto trans
MZ	396 w/3-speed auto trans and A.I.R.
EI	396 high-performance
EQ	396 high-performance and 3-speed automatic
EY	396 high-performance and A.I.R.
MQ	396 special high-performance
MR	396 special high-performance and A.I.R.

LOCATION F—
REAR AXLE NUMBER

Example: PA0212G
PA = Camaro 3.08 ratio
02 = Month (February)
12 = Day (12th)
G = Axle Plant (Chevrolet Gear and Axle)

AXLE PLANT DESIGNATIONS:

CODE	PLANT
G	Chevrolet Gear and Axle
W	Warren
B	Buffalo
K	McKinnon Industries

AXLE DESIGNATIONS:

CODE	AXLE
PA	3.08 ratio
PB	3.36 ratio
PC	3.31 ratio
PD	3.07 positraction
PE	3.08 positraction
PF	3.31 positraction
PG	3.36 positraction
PH	3.55 positraction (small ring gear)
PI	3.55 positraction (large ring gear)
PJ	3.73 positraction
PK	3.55 w/4-speed
PL	3.55 w/close-ratio 4-speed
PM	3.73 ratio
PN	3.07 w/metallic brakes
PO	3.07 w/positraction, metallic brakes
PP	3.31 w/metallic brakes
PQ	3.55 (large ring gear) w/metallic brakes
PR	3.31 w/posi, metallic brakes
PS	3.55 (large ring gear) w/posi, metallic brakes
PT	3.73 w/metallic brakes
PU	3.73 w/posi, metallic brakes
PV	3.07 ratio

LOCATION G—
MONTH OF VEHICLE BUILD

CODE	MONTH
V	August 1966
L	September 1966
R	October 1966
K	November 1966
Z	December 1966
S	January 1967
P	February 1967
W	March 1967
N	April 1967
Y	May 1967
T	June 1967
X	July 1967

LOCATION H—
TRANSMISSION NUMBER

Code number identifies the source, model year, month and day produced.
Example: P7EO3
P = Source (4-speed Muncie)
7 = Model year (1967)
E = Month (May)
03 = Day (3rd)

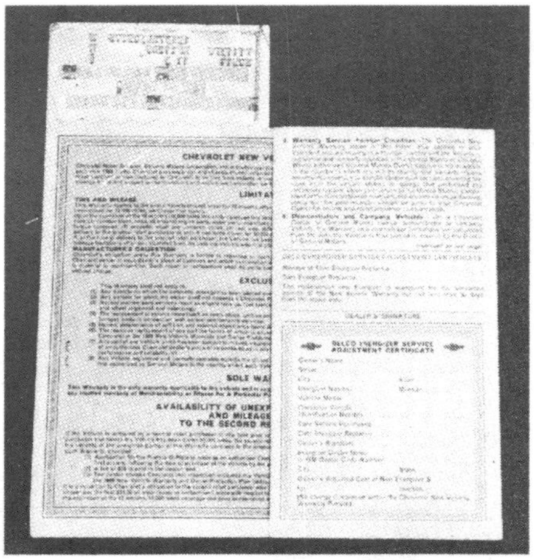

A Camaro's original Protect-O-Plate is a gold mine of information. The numbers stamped on the plate reveal details about the car's original equipment, right down to the date the rear axle was manufactured!

Cowl tags are another source of reliable information about a Camaro. This tag identifies the model year (A—1968), the model identification (B—V8 sport coupe), the plant (C—Los Angeles), the trim code (D—style and color of interior), and the color (E—upper and lower body color).

Vehicle Identification Numbers (VIN) are useful tools for Camaro detectives. VIN numbers for '68 and '69 Camaros are located on top of the dashboard on the driver's side. This VIN tag corresponds to the accompanying cowl tag: the first five digits (12437) indicate a Camaro V8 sport coupe; the next number (8) is the model year, 1968; "L" is the Los Angeles assembly plant; and 317697 is the unit number. The unit number is also stamped on the engine block just above the water pump on '68 and later Camaro motors. If the numbers on the VIN tag and the block match, the engine is original.

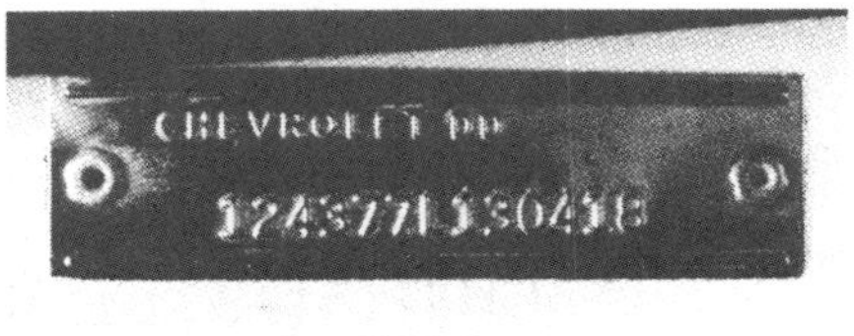

VIN numbers for 1967 Camaros are stamped on a plate riveted to the driver's side door jamb. This tag is for a 1967 sport coupe with a V8 engine which was the 30,418th Camaro assembled in the Los Angeles (Van Nuys) plant.

TRANSMISSION SOURCE DESIGNATIONS:

CODE	SOURCE
D	Saginaw 3-speed w/overdrive
M	Muncie 3-speed
K	McKinnon Industries 3-speed
H	Muncie heavy-duty 3-speed
K	McKinnon Industries 4-speed
R	Saginaw 4-speed
P	Muncie 4-speed
A	Cleveland torque drive
C	Cleveland Powerglide
T	Toledo Powerglide
E	McKinnon Industries Powerglide
X	Cleveland Turbo 350
Y	Toledo Turbo 350

TRANSMISSION CALENDAR MONTH DESIGNATIONS:

CODE	MONTH
A	January
B	February
C	March
D	April
E	May
H	June
K	July
M	August
P	September
R	October
S	November
T	December

LOCATION I—POWER BRAKES AND/OR STEERING

CODE	OPTION
1	Power Steering Only
2	Power Brakes Only
3	Both Power Steering and Power Brakes

LOCATION J—
RADIO CLOCK GROUP

LOCATION K—
RADIO OPTION

CODE	RPO OPTION
3	U63
4	U63/U80
5	U69
6	U69/U80

LOCATION L—
DISC BRAKE OPTION

CODE	RPO OPTION
3	J52 (front discs)
5	J41/J52
6	J56 (heavy-duty brake)
8	J41/J56
9	J52/J56

LOCATION M—
AIR CONDITIONING

CODE	RPO OPTION
1	C60 (all-weather)
4	C48 (less heater)

LOCATION N—
POWER WINDOW EQUIPMENT

CODE	RP OPTION
1	A31 (electric window control)
5	A31/B93 (electric window, door edge guards)

INTERIOR RECOVERY OF THE CUSHY KIND

By Rob Kinnan

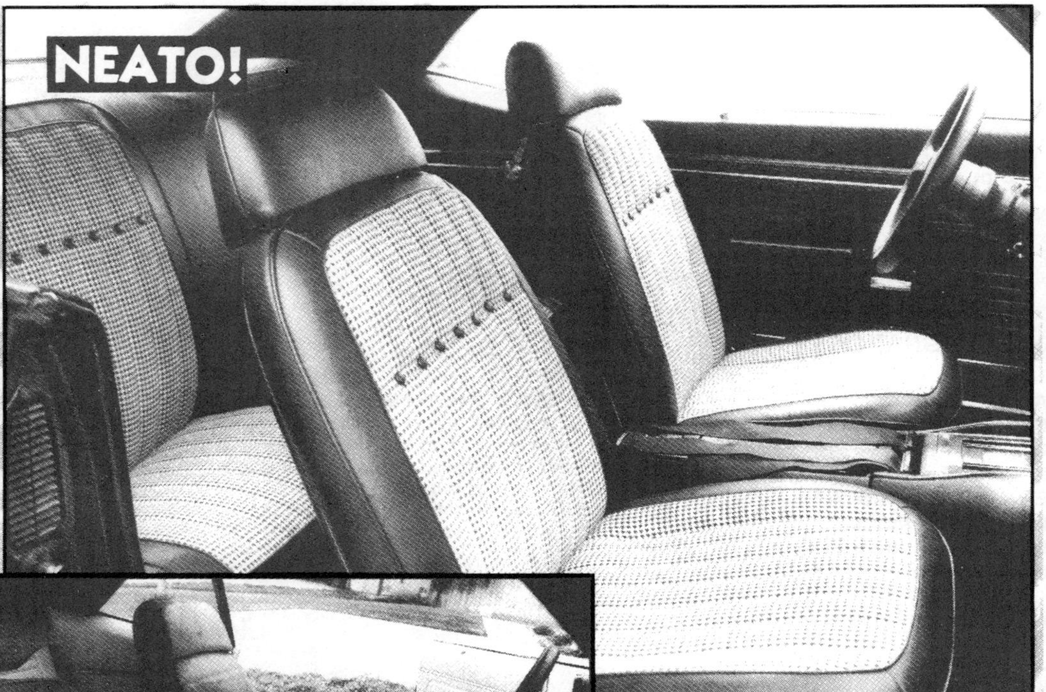

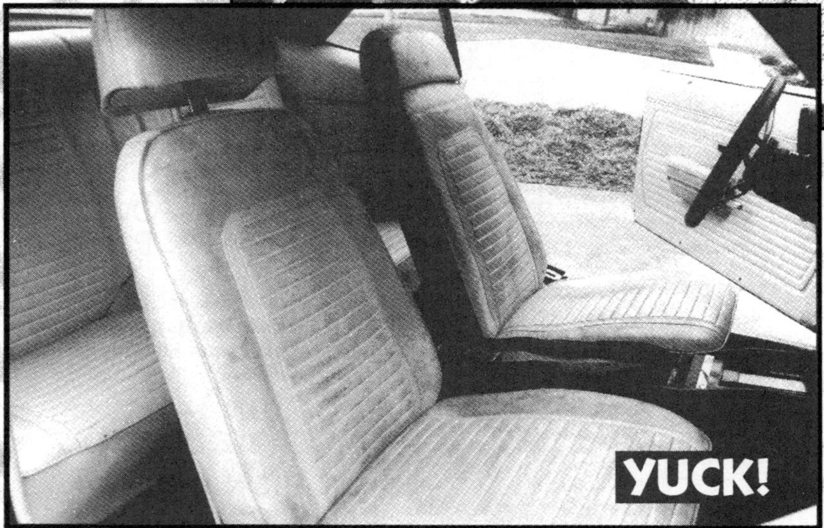

Here's the starting point: one well-used '69 Camaro white-vinyl interior. As you can see, the seats clearly can't be saved. The carpet was worn badly in the driver's footwell and the headliner had a couple of unsightly tears.

Father Time sure can be a nasty old bugger, can't he? Among wine connoisseurs, he's a hero, but members of the musclecar flock often have nightmares of the old man standing on their hoods with a sledgehammer and a smile. The sledgehammer, of course, is symbolic of the effect he has on an aging musclecar. Time spent outdoors not only causes that beautiful megabuck custom paint job to fade and oxidize, but it also plays havoc on the interior. Imagine your painted body panels in a greenhouse, where the sun's rays are magnified several times, and you can understand what your poor interior goes through on a summer day. Added to that is the wear and tear that's caused every time you get in and out of the car. Needless to say, unless you never drive your car and religiously protect it with a garage and car cover, your threads are probably in need of replacement.

Once you've determined that your car's interior does indeed need freshening up, you basically have two ways to go. A custom-sewn interior looks great, but it can cost a whole bunch of money. The other avenue is to look to one of the aftermarket reproduction suppliers for a stock pattern. This is definitely the way to go if you want to retain the flavor of your musclecar as it was meant to be. Besides, in most cases, you can't beat some of those original factory patterns for durability and appearance. This is the route we chose for the interior make-over on a '69 Camaro.

The Camaro began life with white-vinyl seats and door panels, while the dash, headliner, and carpet were black. If you've ever owned a car with a white interior, you know how difficult it is to keep clean. Worse yet, the previous owner didn't take especially good care of the upholstery, and hours of elbow grease and a gallon of Simple Green only resulted in what can best be described as a mediocre cleanup. The dirt and grime simply didn't want to come out. Not only was the vinyl hard to clean, but it was also rather hard and very flat. Every time we turned a corner, it was a battle just to hold on to the wheel to keep from sliding out of the seat.

With our repro ideas in mind, we

Camaro Interior Make-over

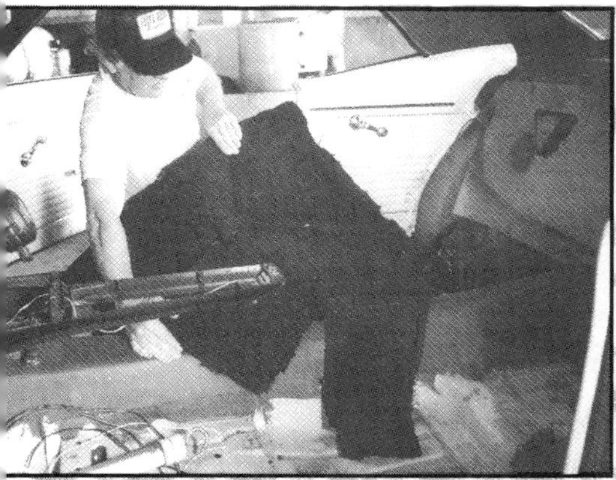

1 *The first step is to remove the front and rear seats. The rear-seat bottom is removed by pushing in on the bottom edge and then lifting up. The back cushion is secured by two bolts at the bottom. The door scuff-plates and the center console should now be removed, allowing you to pull out the old carpet. Under the carpet, you'll find the factory soundproofing mats. These were replaced with new rubber pieces from USA-1. The mats are rather heavy, so if the weight savings is important, you may not want to use them. If your kick panels are showing their age, now's the best time to remove them for painting or replacing.*

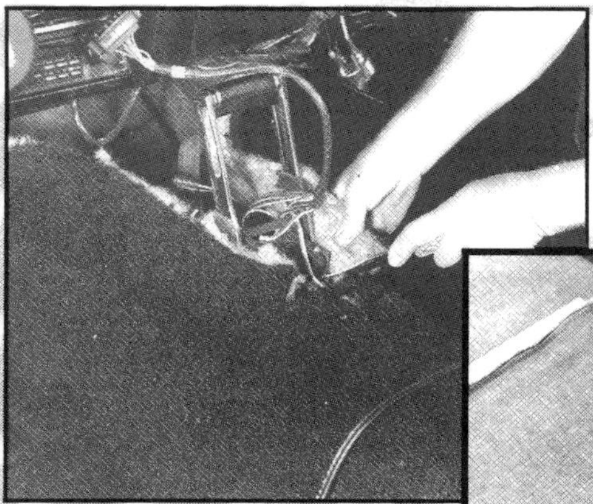

2 *The carpet kit comes in two pieces, one for the front and one for the back. The back is pretty straightforward—it falls into place and the only areas you really need to trim are around the bottom seat-cushion anchors. We used a pocket knife, but using a sharp carpet knife will make life much easier. For the front, set the carpet in place, lining up the molded areas of the carpet with the footwells. If you have a center console, the carpet won't just fall into place. You'll have to cut an opening for the console. Cut a slit from the front of the carpet back toward the shifter, stopping just after the shifter. Carefully trim the carpet to conform to the edges of the shifter base. Don't cut too much or the console won't hide it. All that's left to trim now is any excess carpet behind the kick panels and under the scuff-plates. Leave at least an inch or more of overhang so that when the parts are re-installed, the carpet tucks underneath.*

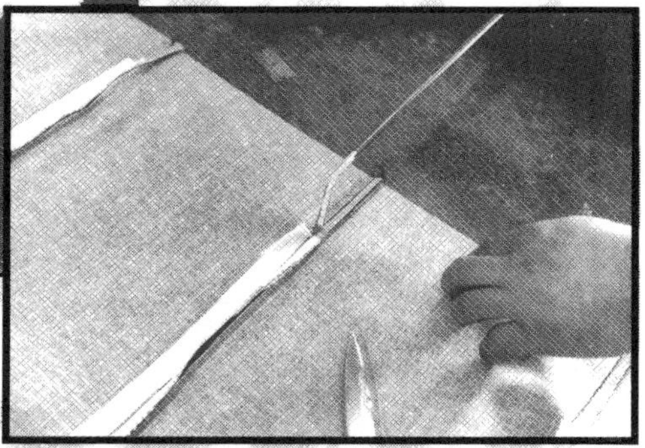

3 *With the seats out of the way, now is the best time to replace the headliner. All of the hardware on the ceiling (mirror, visors, shoulder belts, and those little clips for clothes hangers) must be removed first, then you can remove the old headliner. Deadlines being what they are, we had the pros at Precision Auto Upholstery in Los Angeles install our headliner and seat covers. Here, Chris removes the bows from the old headliner and inserts them into the new one. The ends of the fabric loop are sewed shut—so center the bow, then cut openings for the ends. After all the bows are put in the new headliner, Chris sprays the edges of both the headliner and the inside of the roof with Weldwood Contact Adhesive.*

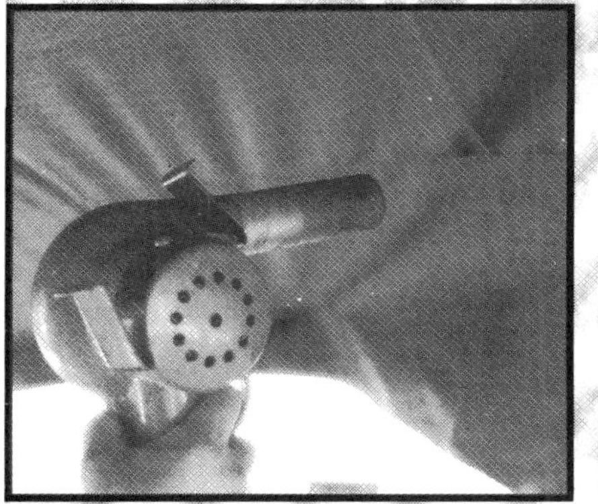

4 *Insert the bows into their holes and make sure the fabric is centered in the car. Measure if you have to. Once you're sure everything's centered, snap the bows into the plastic clips that support them in the middle. With the center held in place, Chris stretched the edges out and stuck them to the glue around the edges of the roof. He used this heavy-duty hair drier to soften up the vinyl and make it more pliable. After the edges are set, trim off any excess, push on the molding, and you're done.*

called USA-1 Interiors and ordered custom hound's-tooth check seat covers, black door panels and dash, and many other little parts that make the difference between a good job and a great one. The seat covers were an option on the early F-bodies. They are vinyl with black-and-white, checkered-cloth inserts. They're very comfortable, look fantastic, and, since they're cloth, they're easier on the posterior when the summer rays make things nearly unbearable in a car. The cloth also provides a little more "grab" on the curves, which is a welcome addition. To replace the well-worn carpet, we used an Auto Custom Carpets set to complete the make-over. So follow along on these pages and watch as we turn our sad-sack interior into the cushy cabin it always wanted to be.

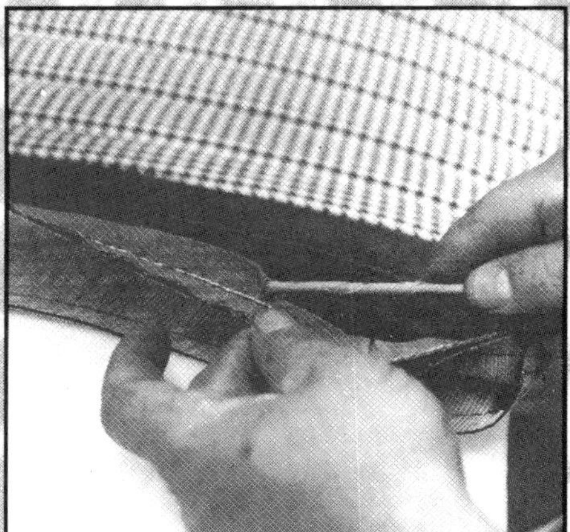

5 *The seat upholstery is held to the wire backing by hog rings. To remove the covers, cut these rings with side cutters and peel off the covers. There are stiff wires on each edge of the cover that keep the edges straight and taut. These must be removed and inserted into the new cover, as Jaimie does here.*

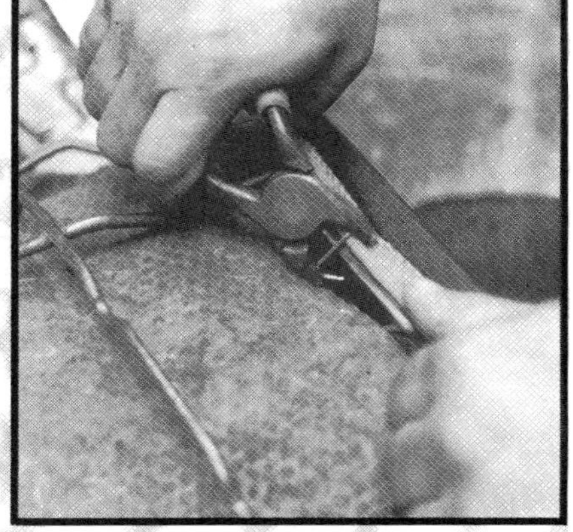

6 *Jaimie did the back seat first, reusing the original foam padding but supplementing it with a thin layer of ½-inch foam. Contact adhesive holds the layers together. The cover slips over the padding and the top edge is fastened with these hog rings. They poke through the fabric and loop around the wire backing.*

F-STOPPING

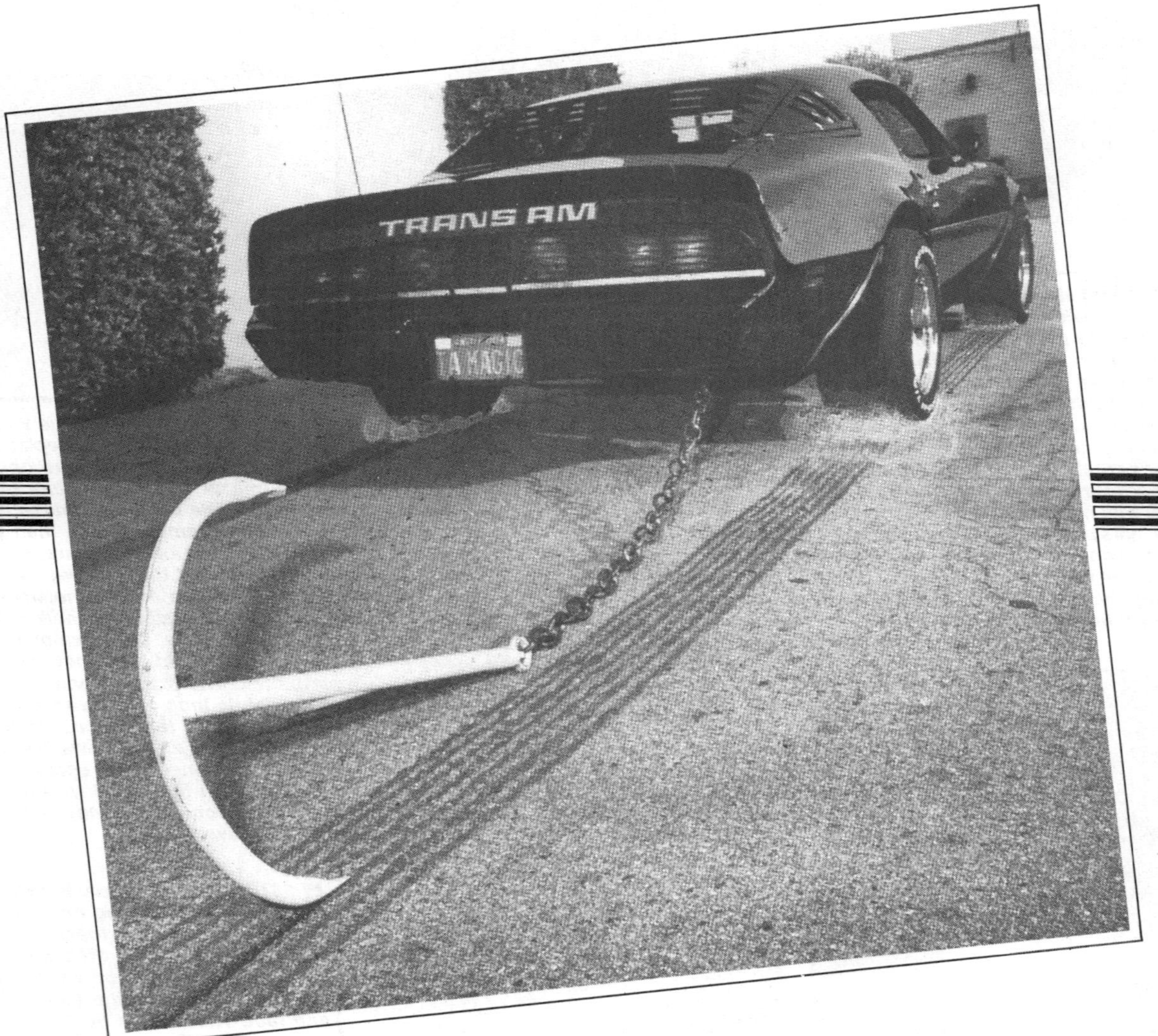

The old adage about all good things coming to an end has no better application than when you bring your street machine to a rotor-smoldering stop after a high-speed pass down your favorite stretch of open road. After years of arduous effort, you've finally built your Camaro or Firebird into a ground-blistering bullet, capable of reaching Mach 2 in ten seconds or less. But, you say, it still won't stop when you want it to?

So, you've already tried the old friction-block-against-the-tire routine, huh? And so far you've had your Guccis resoled six times after throwing your door open in those "emergency" situations? You say that your girlfriend has left your warm embrace for some aristocratic, leather-gloved stockbroker driving one of those flashy "furrin" cars?

Well, fear not, asbestos breath. We at CAR CRAFT have come up with an ingenious plan that should help keep your brake pedal from turning to sponge (and

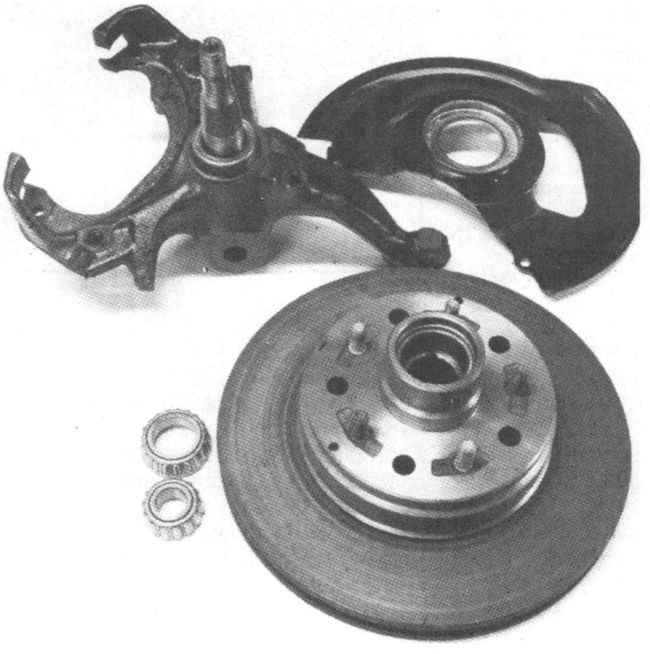

These are the parts required to upgrade the brakes on your '70½ through '81 Camaro or Firebird. Of course, you will need a pair of each of the components to complete the swap.

Bigger Brakes for Your Camaro or Firebird

By Chuck Hanson

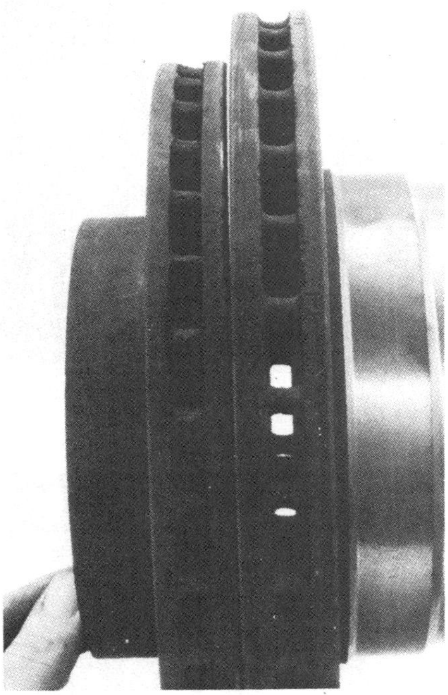

This side-by-side comparison illustrates the difference between the 11- and 12-inch rotors. Substituting the 12-inch rotors increases the overall brake-swept area by 13 percent, adding more than a modicum of increased stopping ability.

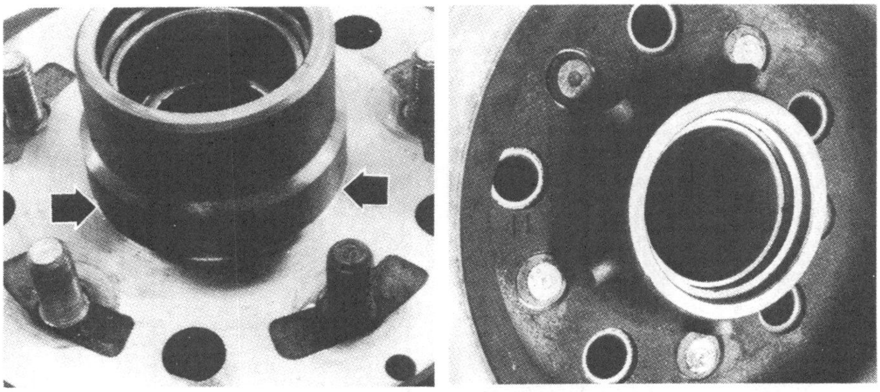

The only machine work required entails reducing the hub diameter (arrow) from 3.060 to 2.780 inches in order to properly locate a set of stock Trans Am wheels. If you use aftermarket wheels, this step may not be necessary, but it is definitely worth checking. We also had our hubs re-drilled to accept 7/16-inch studs in the required 4¾-inch bolt circle. Re-drilling the hubs required removal of material from the ribs found on the backside of the rotor and counter-boring the stud holes for stud head clearance.

your legs from turning to rubber) the next time you need to bring your land-locked missile to a tire-screeching halt. And while it will cut deeply into the annual income of your favorite shoemaker, we can't guarantee it will bring your girlfriend back.

Using pieces that can be bought in the junkyard for reasonable prices and are available in almost limitless quantities, the accompanying mods will increase the effective brake-swept area of your favorite '70½ through '81 Camaro or Firebird by a whopping 13 percent! Actually, credit for this stop-swapping endeavor should go to our GM parts guru and good friend, Jim Losee (The Factory Pipeline, 201 West "D" Street, Wilmington, CA 90744-5694). He is responsible for conducting all the research as well as making the final determination on spindle, ball joint, and steering arm compatability. As a result, anyone owning a second-generation F-body is the beneficiary of his efforts.

There are a couple of ways to go about fitting your F-body with bigger binders. The first is expensive and entails purchasing all the required pieces from the dealer. The second is considerably cheaper and calls for a trip to the local auto dismantling facility (we call 'em junkyards in Tennessee). If you decide to go the junkyard route, try to find one that will let you remove the parts yourself. Not only does this save you from having to pay their labor costs, but it allows you to inspect and choose all the components before laying down your long green.

The brake assemblies we wanted used 12-inch rotors, so we began our search by looking under the front-ends of all '78 and up, full-sized GM cars (especially station wagons). Unless you have a calibrated eye, it's a good idea to bring along a tape measure to ensure you're getting the 12-inch setup. Since the calipers interchange between the 11- and 12-inch versions, we elected to purchase only the spindles, dust shields, and rotors. Fifteen-inch wheels are also necessary to clear the larger diameter of the 12-inch brakes, so while you are at the junkyard you may want to look around for a set of Corvette or WS-6 wheels. Both come in 15x8-inch sizes and are relatively easy to find. Most aftermarket wheels will work too, as long as they are of the 15-inch variety and have no unusual machining contours. We recommend checking the wheel-to-brake clearance very carefully before taking off down the street.

A slight amount of machining is required on the rotors to make them work, while the spindle literally falls in place, bolting directly to the existing ball joints and steering rods. We recommend that before removal of your old brake assemblies you check the ball joints for excessive wear. If they are questionable, re-

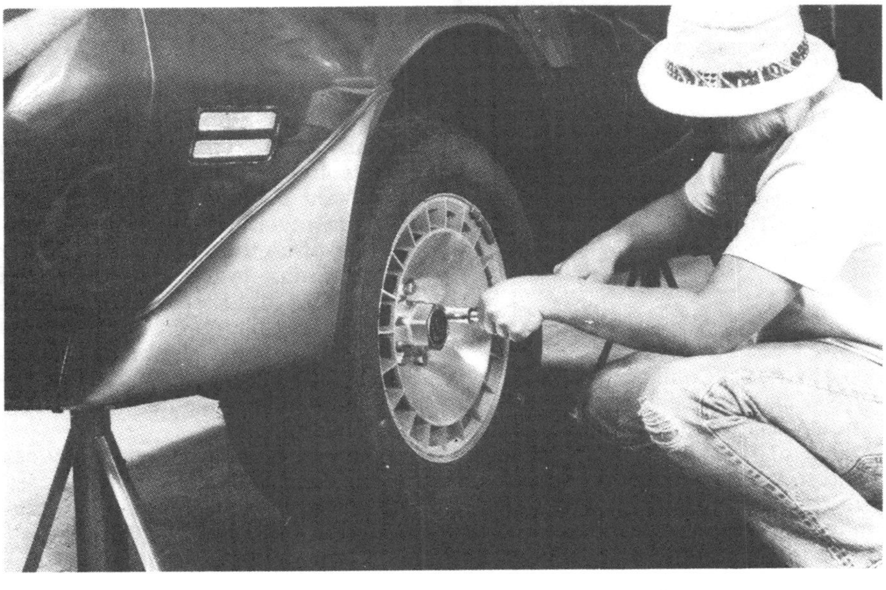

Naturally, you need to remove the front wheels to get to the heart of the situation. Be sure to support the car firmly on jackstands. Failure to do so could prove hazardous to your health.

F-STOPPING

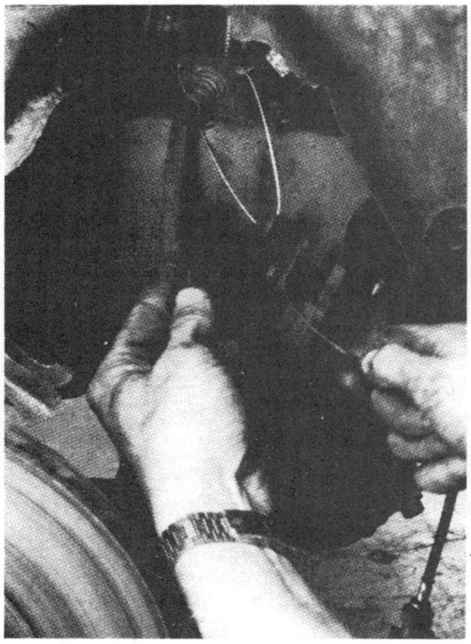

Two Allen-head bolts secure the caliper to its mount. After their removal, the caliper should be suspended by wire from the brake line mounting bracket. Don't allow the caliper to hang by the hose.

place them, along with any other worn items such as wheel bearings and brake pads.

Separating the spindle from the ball joint usually requires the use of a "pickle fork" and a big hammer. If you don't have one, they can usually be rented from an auto parts store for a nominal fee. Unfortunately, the pickle fork will most likely render your ball joints useless; don't use it unless you plan to replace your ball joints or if there is no other way to break the spindle loose. Our alternate method involved backing the ball joint nuts off a few turns, then rapping the spindle pads with a hammer near the ball joint to break them loose. The watchword here is caution. Installing the new brake units involves a reversal of the disassembly process. However, be sure that proper assembly techniques and torque values are used.

The final phase of our brake swap includes a little homespun alignment exercise (see photo). While it is not recommended that this alignment be retained for everyday driving, it will allow you to get down to your local alignment shop for the "real" thing.

All that's left now is for you to round up all the proper tools and equipment, check out the accompanying photos, and spend a couple of hours completing the swap. The end result will be felt the first time some clown darts out in front of you from a side street and you find that your car can stop on the proverbial dime. What's even more pleasing is that you'll probably even get back nine cents change!

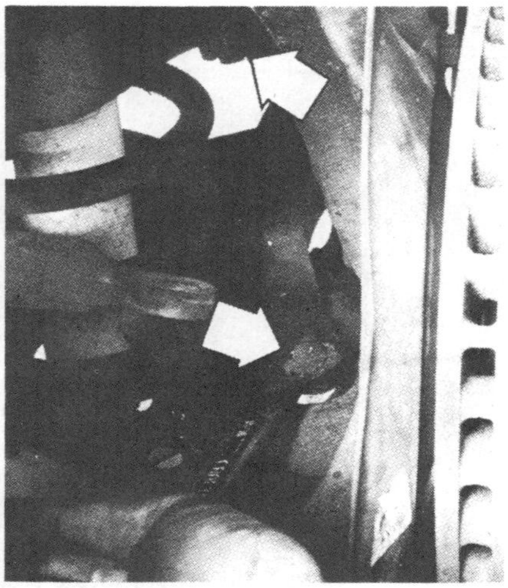

Support the lower control arm with a floor jack while backing off the upper and lower ball joint nuts a few turns (arrow). Lower the floor jack, and with a few good raps to the spindle from a sizeable hammer, dislodge it from the ball joint studs. After disconnecting the tie rod end from the steering arm you can raise the floor jack and remove the ball joint nuts.

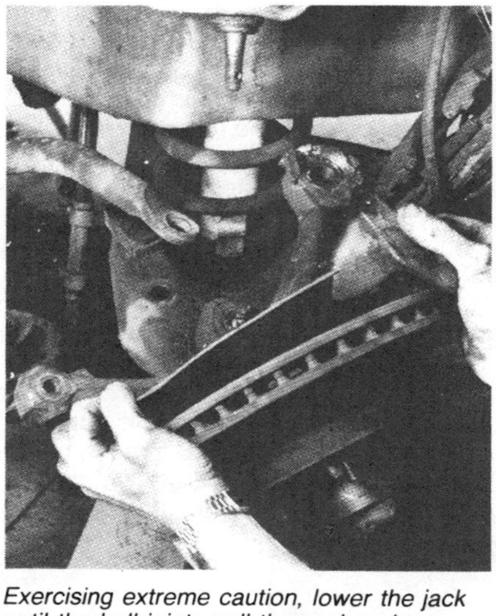

Exercising extreme caution, lower the jack until the ball joints pull themselves loose from the spindle. This allows you to remove the spindle and rotor as an assembly. Your shocks and stabilizer bar should still be connected (although our photo shows the stabilizer bar loose), and will secure the lower control arm and coil spring in their proper locations.

STOP SHOPPING

Late-model Camaros aren't the only GM cars that can receive a brake update using stock components. Because of the "General's" foresight (and to keep production costs down, we're sure), many of the various makes and models manufactured under the GM banner enjoy a high degree of brake component interchangeability. Given the incredible number of these cars in boneyards across the country, it is extremely easy (and cheap) to convert your standard drum brake setup to the more desirable front-disc variation.

Front discs didn't become an option on GM cars until 1967, with drum brakes being offered as standard equipment on most models through 1972. However, the disc brake combinations offered from 1969 and later are preferable to the earlier versions because of fluid leakage problems with the four-piston calipers. The '69 and later versions are of the single-piston, floating caliper style, and offer a marked improvement from a reliability standpoint.

True bolt-on interchangeability can be achieved with the following cars: '68-74 X-bodies (Novas, Omegas, Venturas, and Apollos); '67-69 F-bodies (Camaros and Firebirds); and '64-

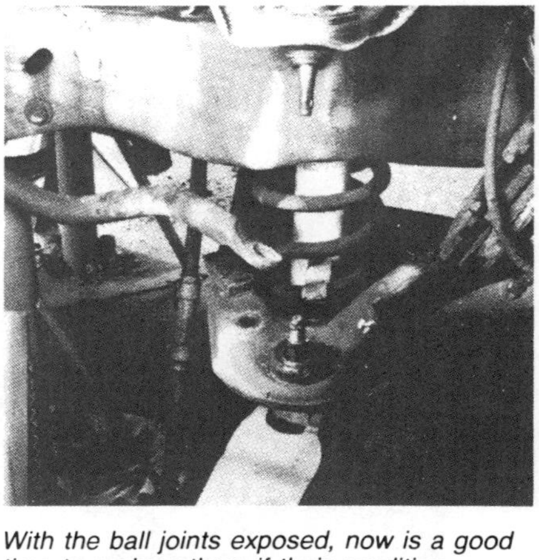

With the ball joints exposed, now is a good time to replace them if their condition warrants it.

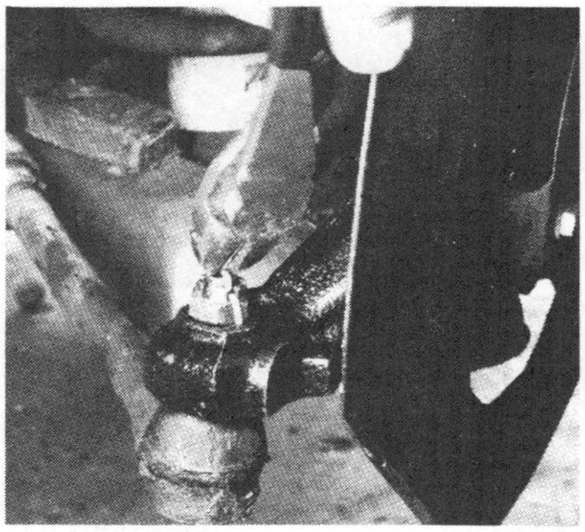

Installation of the new pieces is basically a reversal of the foregoing procedure. Torque the upper ball joint nut to 50 ft./lbs., the lower to 65 ft./lbs., and the steering arm-to-tie rod connection to 35 ft./lbs. We also installed new castellated nuts and cotter pins at all attaching points, giving us an added sense of security.

F-STOPPING

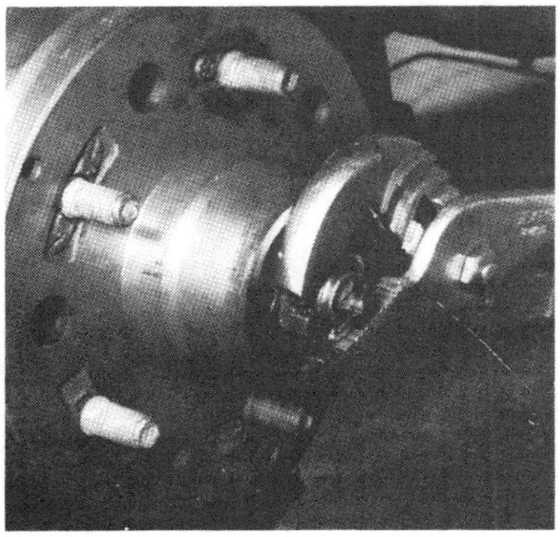

Wheel bearings are adjusted by tightening the spindle nut while at the same time rotating the rotor. After all the "play" is removed from the wheel bearing, back the nut off to the next castellation using either the vertical or horizontal spindle hole to minimize wheel bearing clearance.

72 A-bodies (Chevelle, Monte Carlo, Cutlass, LeMans, and Skylark). When interchanging components between body styles, consideration should be given to the steering arm location. It is bolted to the spindle, and the stock piece should remain with the vehicle you are putting the discs on, since steering arm location varies from model to model.

When converting from drum to disc brakes, the larger capacity master cylinder from the disc brake car is recommended along with the distribution valve. This valve is a very important part of the overall package since it controls the amount of pressure distributed to the front (disc) and rear (drum) systems, and each system has different pressure requirements. The valve can be found either below the master cylinder or on the inside of the driver's side frame rail.

Pre-'67 models were all equipped with drum brakes and a single reservoir master cylinder. Therefore, when converting to disc brakes on any of these cars, it is imperative to use the dual-reservoir master cylinder and distribution valve from the disc brake car. Removal of all the necessary brake lines will also facilitate plumbing the separate fluid circuits of the dual-reservoir braking system. Also, since each model has its own distribution valve according to the weight considerations of that particular vehicle, try to find one from the same "bodied" car. For example, for a '70 Chevelle, use components from any '68-72 A-body car.

Those of you who drive "early" H-bodies (pre-'76) can also benefit from the wide array of interchangeable brake pieces that originate in the GM parts bins. The H-body line includes the venerable Vega, along with the Monza, Astre, Starfire, Sunbird, and Skyhawk models. The most notable of stop-swap technology applied to these cars is the use of '77 and later Monza brakes. The gain here is due to a ventilated rotor which dissipates the heat more quickly than the solid hub rotor variety. Again, the swap is a direct bolt-in, and in this instance won't even require the use of a different master cylinder or distribution valve.

So, if your stock brakes just aren't getting the job done anymore, and you're getting tired of throwing out the anchor every time you want to stop or slow down, take a tip from us. High-caliper braking is where it's at for high-caliber cars.

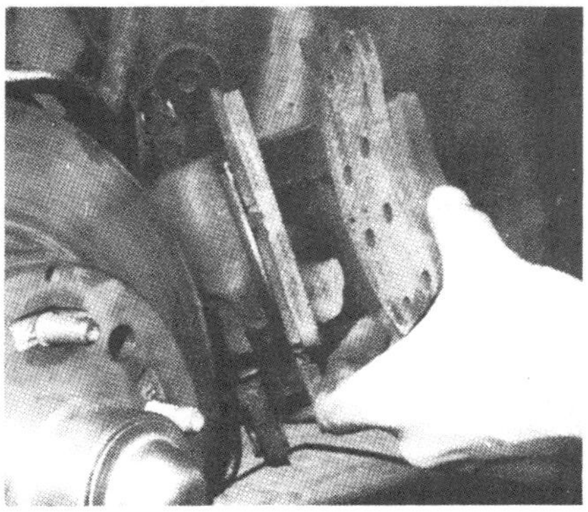

Before we installed the calipers, we outfitted them with new WS-6 brake pads. These pads have a higher metallic content and are listed under part number 12300192.

KITS & PIECES

A partial conversion kit consisting of new OEM rotors, bearings, seals, and studs is available from The Factory Pipeline. Priced at $237, this kit leaves you the task of locating the required spindles and dust shields, as well as having the rotors machined to the recommended specifications. For further information, contact The Factory Pipeline direct.

However, if you prefer to buy your parts from the local Chevrolet dealer rather than scrounge the junkyards or buy the kit, the necessary part numbers are:

Rotor	14008641 both	
Spindle	14012589 LH	14012590 RH
Shield, dust	344023 LH	344024 RH
Caliper	18005264 LH	18005265 RH
Bearings	7450630 inner	7750627 outer
Seal	3965092 inner	

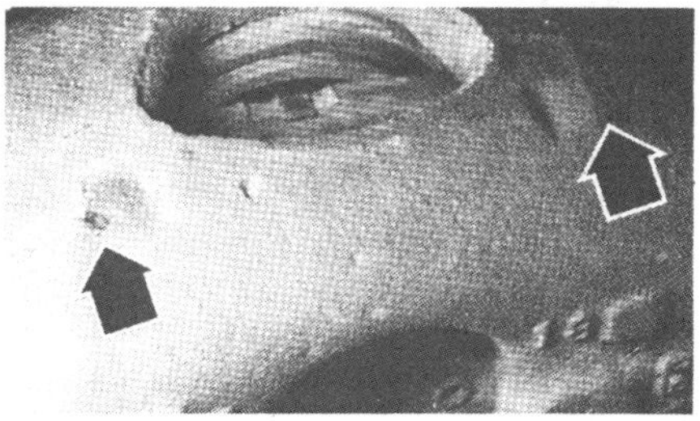

Due to the increased diameter of the new brakes we encountered a slight interference in the areas indicated. A couple minutes' worth of filing solved our problem.

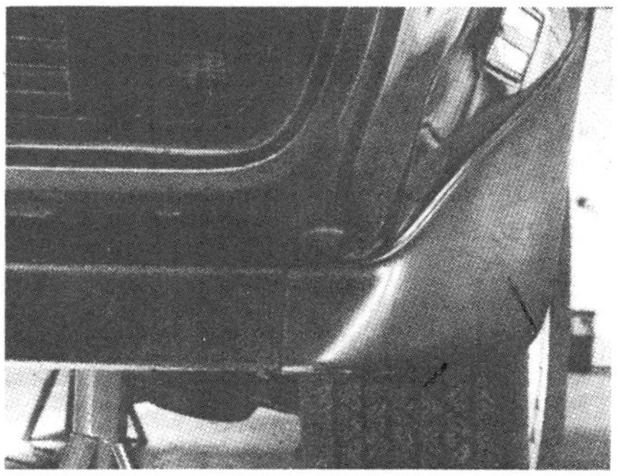

Once everything was back together we encountered a slight toe-in problem. In order to drive the Trans Am to an alignment shop, we had to perform a "quickie" alignment.

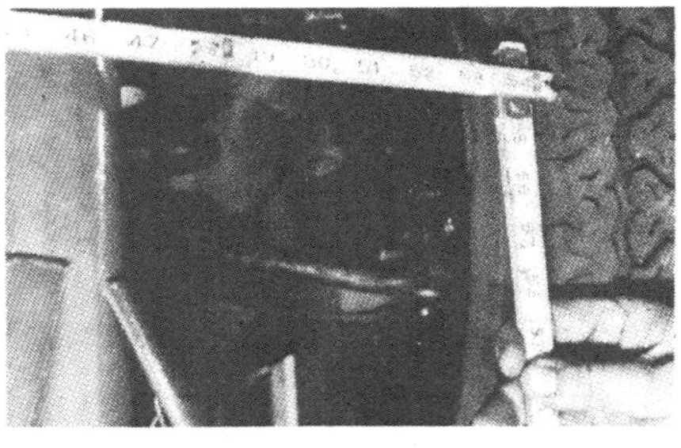

We measured the distance between the tires in front of and behind the control arm to see how far out we really were. After determining the amount, we lengthened the tie rods by like amounts until the measurements inside the tire were approximately the same. Remember, this is only a ballpark setting and should only be used to get the car to an alignment shop. After driving the car for 25 to 30 miles, check and retighten all fasteners as necessary, especially the lug nuts.

Listed below are spindle and brake assemblies that will interchange freely. Special attention should be given, however, to the steering arm location, since it varies from body style to body style. If it is not the same, merely unbolt it from the old spindle and re-attach it to the new one.

'68-74 X-bodies (Nova, Omega, Ventura, Apollo)
'67-69 F-bodies (Camaro, Firebird)
'64-72 A-bodies (Chevelle, Monte Carlo, Cutlass, Lemans, and Skylark)

BEATING THE DRUM

HOW TO REBUILD CAMARO DRUM BRAKES

Disc brakes may get all the good press notices, but drum brakes have the advantage of numbers. With the exception of a very few Z28s blessed with four-wheel discs, a set of drums can be found on at least one end of every Camaro. In the early days of GM's F-body, it was possible to order a Camaro with drum brakes behind both the front and rear wheels. When the second generation of Camaros arrived in mid-1970, front discs became standard equipment, while drums were retained for the rear axle. This disc/drum combination has remained essentially unchanged, even with the debut of the third generation of Camaros in 1982. Despite minor changes in hardware, drum brakes have always been a part of most Camaros' underpinnings.

Brakes generally hold a low standing in a Camaro enthusiast's consciousness until that frightening moment when he stands on the brake pedal and nothing happens. All Camaros are equipped with independent front and rear brake systems, so a total loss of stopping power is thankfully rare. In this failsafe dual system, the master cylinder is divided into two chambers; the forward half operates the front brakes, while the rear reservoir serves the back brakes. If one part of the system loses hydraulic pressure, the piston in that chamber must bottom out before pressure builds in the other half. This means that the brakes won't apply until the pedal is almost on the floor. Avoiding these adrenalin-producing incidents with a regular program of brake inspection and maintenance will do wonders for your coronary system—not to mention your Camaro's sheetmetal.

Brake maintenance is more than just checking the fluid level in the master cylinder every six months. Don't wait to inspect the linings until the brakes start squealing; at that point, it's already too late. Once the lining rivets (or worse yet, the brake shoes) contact the drums, the metal-to-metal friction will machine deep grooves in short order. In fact, the rivets can cut halfway through the drums without making a sound. The only reliable way to find out how much of the lining remains is to pull the drums and look *before* the brakes announce their retirement. Pay heed to warning signs, such as a parking brake that won't hold the car on a grade or a low brake pedal.

The advantage of a do-it-yourself drum brake overhaul is that you can take the time to do the job right. Sunday newspapers are full of ads for bargain brake jobs, but the fine print reveals that the low prices are usually for linings only. If your Camaro has accumulated more than a few miles, it's likely that new linings are not the only components ready for replacement. Drums, wheel cylinders, return springs and flexible lines may all be in need of repair. If you've recently joined the Camaro restoration movement, the occasion of a comprehensive brake rebuild is a perfect opportunity to degrease and refurbish the suspension. If your automotive tastes run more to autocrossing or grand touring, then now may be the time to install the semi-metallic brake linings and cooling ducts you've been considering.

Or perhaps you're a drag racer who's searching for ways to shave a few more hundredths off your Camaro's elapsed time. Installing super-soft organic linings will make it easier to stage the car on the starting line, and discarding the self-adjusting linkage in favor of manual adjusters will eliminate all traces of brake drag. Even if your motivation is only a desire to make sure the job is done right (and perhaps save a few dollars in the process), a Camaro drum brake overhaul is an easy weekend project.

Make sure you remain in good health to enjoy the results of your labors by observing some common-sense safety precautions. Camaros are heavy, so be certain the chassis is well-supported by a pair of sturdy jackstands before diving underneath it with a handful of wrenches. Asbestos dust has been associated with all sorts of distasteful health problems which may not appear until years from now. Never blow off a dusty drum or backing plate with an air hose; instead, wash the lining material away with soap and water or an aerosol brake cleaner. Finally, put away your solvent can when cleaning brake components. When rubber brake seals react with mineral spirits, they swell up bigger than your toe the last time you dropped a four-speed on it. Use only alcohol or hydraulic system cleaning fluid. And keep your greasy fingers off the linings at all times.

Although rebuilding Camaro drum brakes is a straightforward procedure for most weekend mechanics, there are two specialized tools that will simplify the work. First is a pair of brake pliers. These resemble water pump pliers, but their long handles are designed to

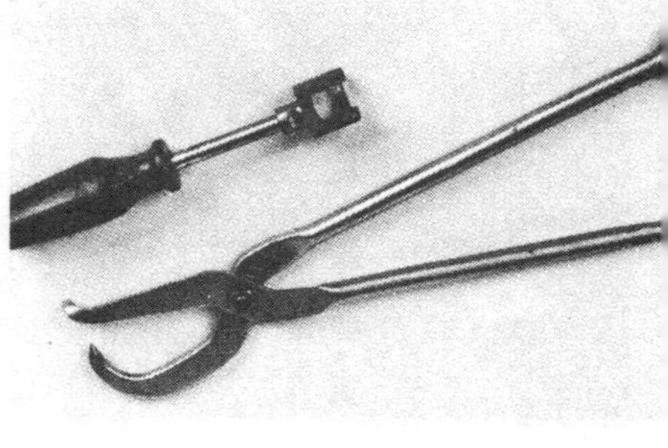

install and remove stiff brake return springs. Second on the list is a brake shoe tool; with a handle like a screwdriver and a slotted ring on its business end, this device quickly compresses the springs and twists the metal cups which anchor the linings to the brake backing plates. Both tools are reasonably priced, and they will make the job much easier.

Planning ahead can reduce your trips down to the auto parts store. It's almost certain that used drums will need to be machined to true their inside diameters; before hauling them into the shop, take an inventory of the other items you'll need to put the brakes back into service. Are the hoses cracked? Do you have enough brake fluid? Are the wheel cylinders leaking? Camaro brake parts are usually in stock at larger parts houses, but occasionally a hose or wheel cylinder will require a special order.

With these preliminaries out of the way, follow along as we demonstrate how to give your Camaro a brake.

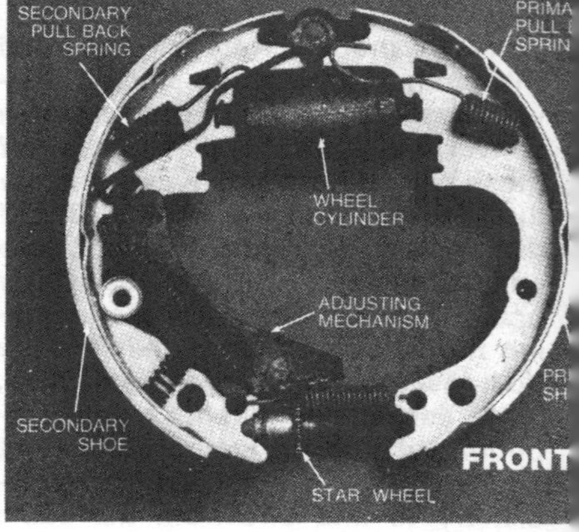

1. This guide will help you talk brakes with your Camaro. Note that the *primary* brake shoe is at the front of the car, while the *secondary* shoe is always to the rear. The secondary shoe is easy to identify, since its lining is longer than the primary lining. Camaro drum brakes are self-energizing; this means that when the shoes are applied, the rotation of the drum forces the lining against its friction surface. This feature decreases the pedal effort required to bring the car to a halt. This self-energizing effect works whether the car is moving forward or in reverse.

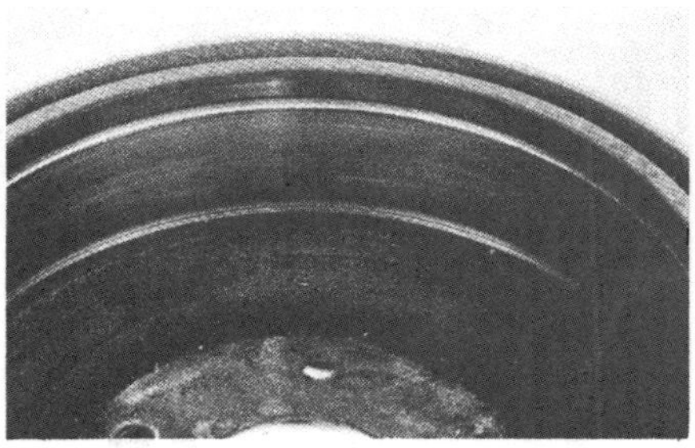

2. Camaro front and rear drums are easily removable. If the drums won't slide off the front hub or rear axle flange after removing the wheels, make sure that you have not overlooked any retaining clips on the wheel studs. A badly worn drum may hang up on the linings; in this case, the shoes must be retracted before the drum will slide over the linings. To retract the brake shoes with the drum in place, knock out the access plug in the front of the drum. Use a small screwdriver handle to retract the shoes. Pull the drum straight off the wheel studs and examine its inside surface. Score marks, grooves and hot spots are evidence the drums should be trued. If a drum is more than 0.008-inch out of round, it should be reconditioned.

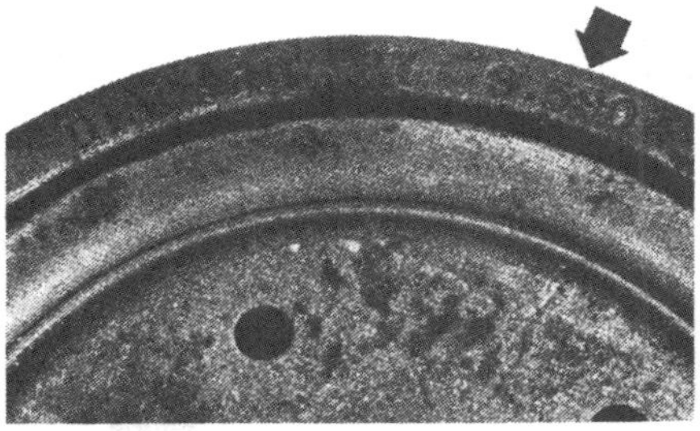

3. There is a definite limit to how far a Camaro brake drum can be turned oversize. The maximum inside diameter is stamped directly on the drum—in this instance, 9.590 inches. It's always unsafe (and usually illegal) to machine the drum past this dimension, since the drum's ability to dissipate heat and to retain its shape are affected. If a drum must be enlarged past this figure before it will clean up, then the drum should be replaced. Figure on spending approximately $50 for a replacement.

4. Once the drums are off, don't depress the brake pedal under any circumstances because the pressure will launch the pistons out of the wheel

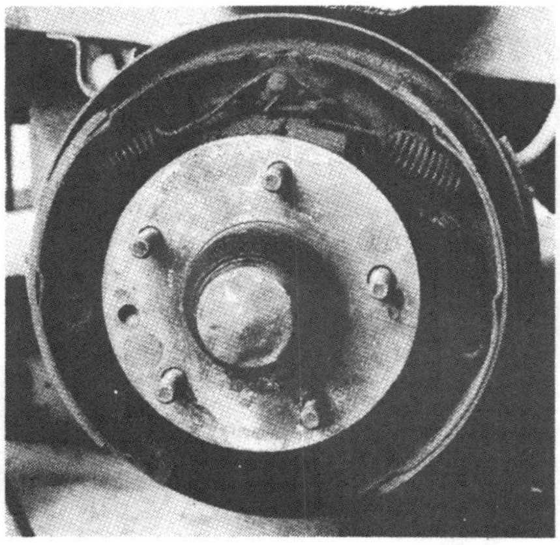

cylinders. Carefully inspect the brake hardware and look for broken springs and anchor pins. Work on one brake at a time; you can use the brake on the other side of the car as a model to refresh your memory about where all the parts go.

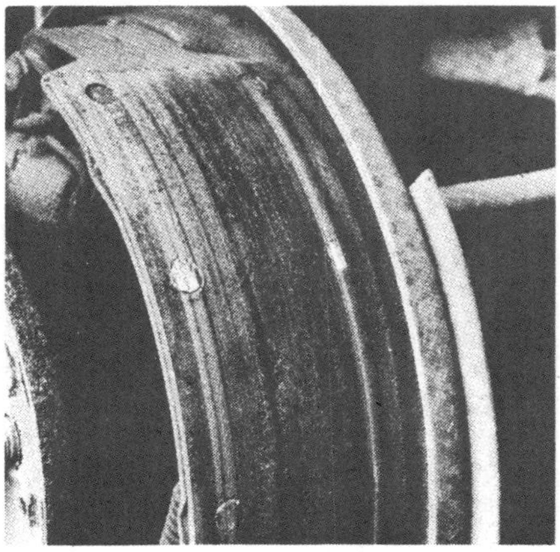

5. These linings are long overdue for replacement. Although the rivets were cutting into the drums for a considerable time, they never made a sound—a strong argument for regular brake inspections.

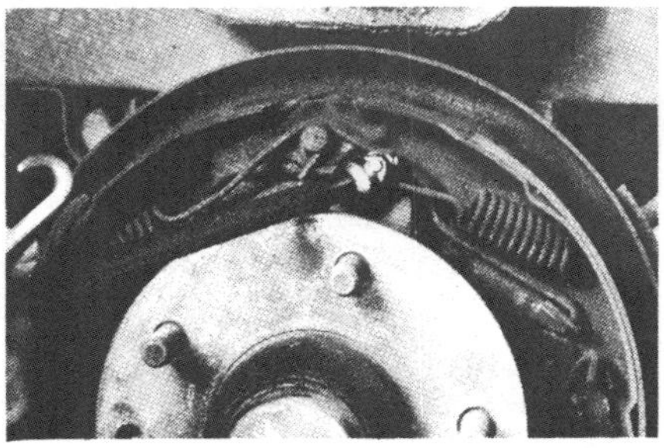

6. Use your brake pliers to disengage the pull-back spring for the secondary shoe from the wire link on the adjusting mechanism. Insert the wire link in the hollow plier handle, then lever the spring off the link by turning the pliers. Next, remove the wire link from the anchor pin above the wheel cylinder.

7. Now you can remove the primary pull-back spring. Again, use the hollow

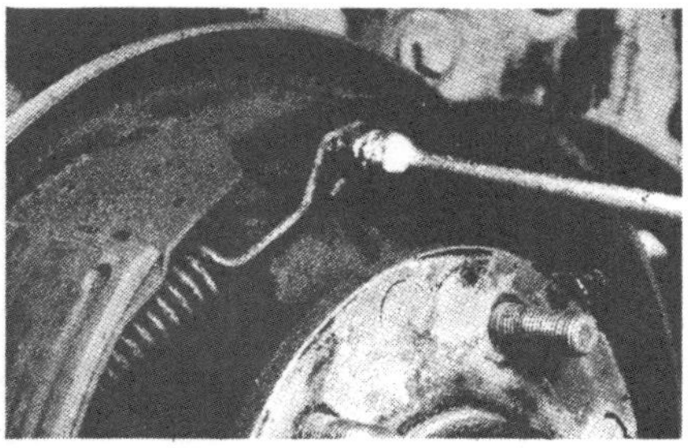

handle of the brake pliers to walk the spring off its anchor pin.

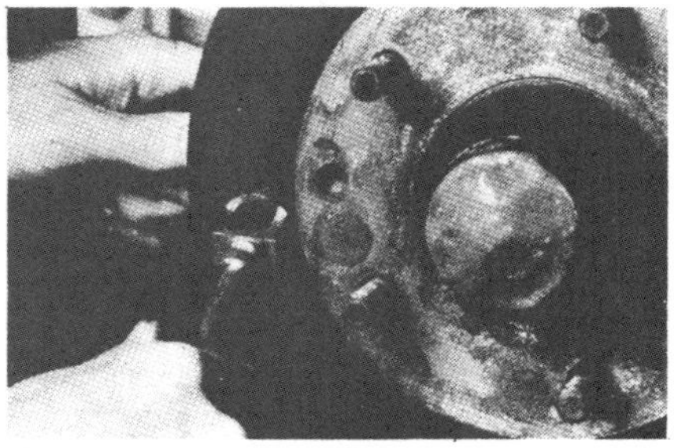

8. Now it's time to remove the hold-down springs. Compress the springs with the brake shoe tool while simultaneously turning the metal-retaining cups. When the slots in the retaining cups line up with the flat end of the hold-down pins, the springs will pop off their pins. Note that the cup on the secondary shoe has a deeper dish than the cup on the primary side because of the extra thickness of the self-adjusting linkage.

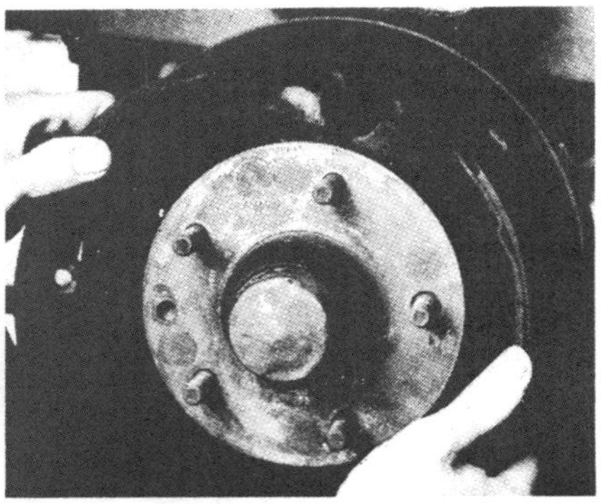

9. Spread the old linings and pull them over the wheelhub or axle flange. Disengage the parking brake linkage on the rear brake shoes. Then disassemble the star wheel and spring from the bottom of the linings.

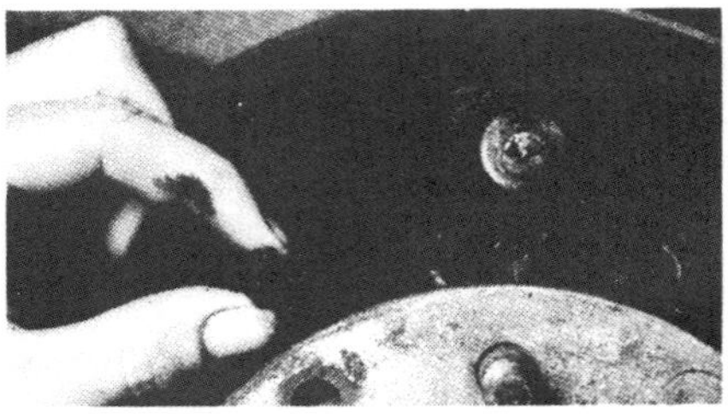

10. Inspect the wheel cylinders by pulling out the small pushrods. Look for cracks and tears in the end seals. Carefully pry back one edge of the seal with a small screwdriver. A small amount of fluid behind the seal is normal, since this lubricates the pistons. If brake fluid runs out when you pry back the seals, it's a sure sign there's excessive leakage past the pistons.

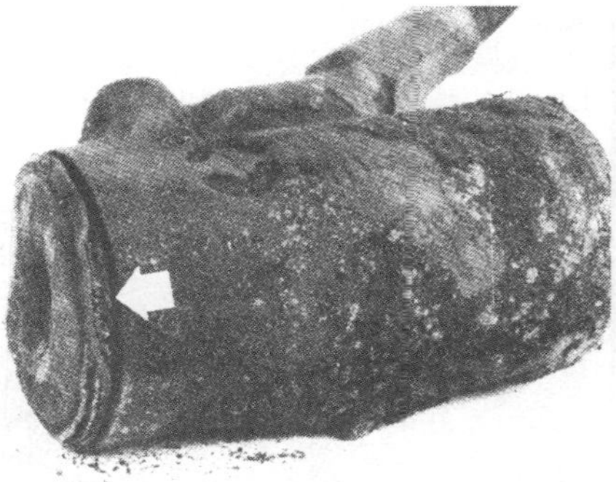

11. This deteriorated wheel cylinder seal calls for immediate replacement. You can either rebuild the cylinders with new pistons and seals, or simply replace the entire assembly. Moisture is a brake system's mortal enemy, since it fosters corrosion on the sealing surfaces of the master cylinder and wheel cylinder bores. Before rebuilding a wheel cylinder, hone its bore to remove all stains and discoloration. Use hydraulic cleaning fluid to clean the wheel cylinder (never solvent or gasoline). Don't use a rag to dry the cylinder, since the lint will contaminate the hydraulic system. If the inside diameter of the wheel cylinder is rough or pitted, replace it with a new one. If a wheel cylinder fails and leaks fluid on your new linings, all your work will have been wasted. Why take that chance? Go ahead and put in new parts.

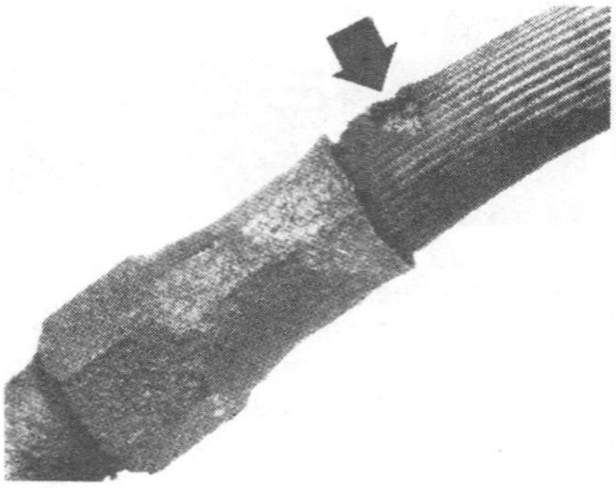

12. Years of use and the elements have ganged up to destroy this flexible brakeline. Even though there may be no sign of fluid leakage yet, the line should be replaced as a precaution against future failures. Small flaws such as this are often overlooked, so check every part of the brake system carefully.

13. It's very easy to distort or round the edges of a small brakeline nut. Always spray the fitting with penetrating

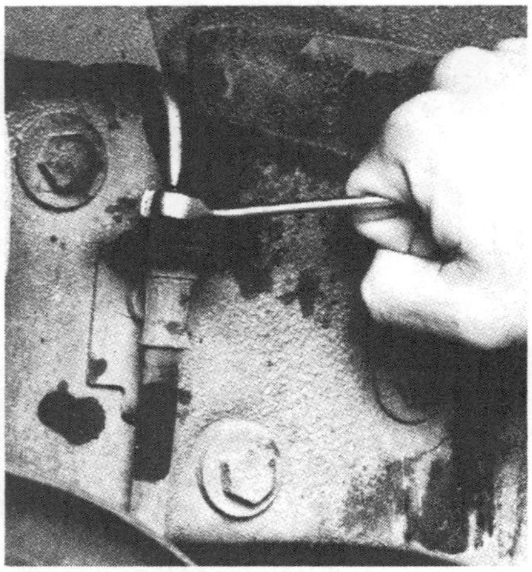

oil and use a line wrench when replacing flexible hoses on your Camaro.

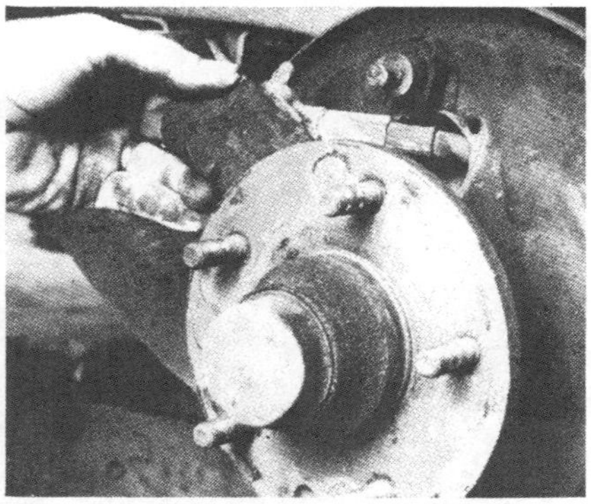

14. A pair of ¼-inch bolts fastens the wheel cylinder to the backing plate. Loosen these two bolts and pull the wheel cylinder and its hose through the backing plate.

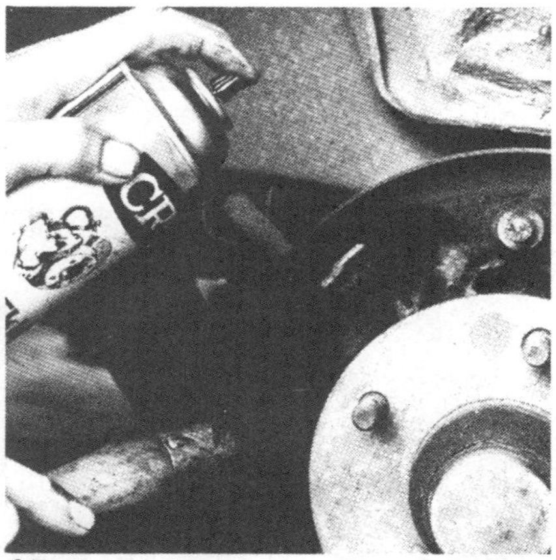

15. Scrub the backing plates with non-flammable cleaner and a stiff brush to remove the brake lining residue. Take care to avoid inhaling the dust. Check the tightness of all the backing plate and steering linkage bolts.

16. With coarse emery cloth, clean the raised pads on the backing plates which contact the edge of the brake linings. Then apply a thin coating of high-temperature grease to ensure the

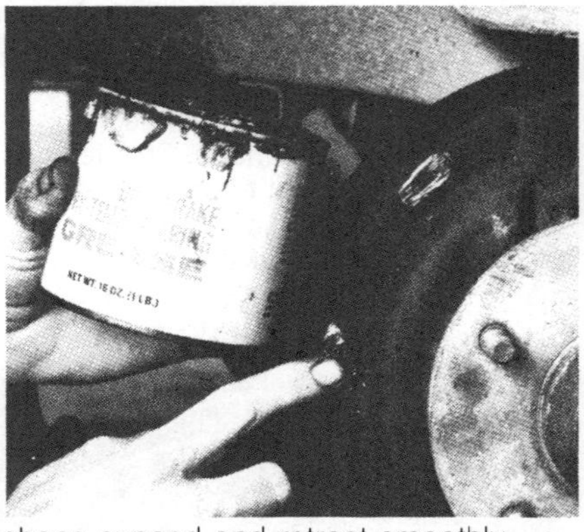

shoes expand and retract smoothly. Install and tighten the wheel cylinders.

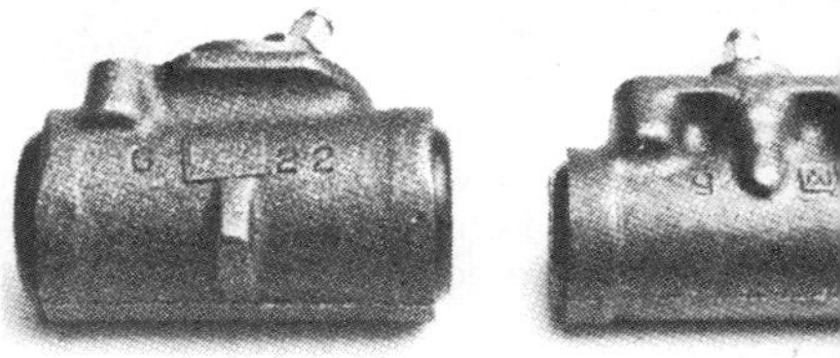

17. The difference in size between early-model Camaro front and rear wheel cylinders is striking. A front cylinder (left) has a 1⅛-inch bore, while the smaller rear cylinder measures only ⅞ inch in diameter. Anyone who has tried to stop a bicycle or motorcycle quickly knows that it's the front wheel that supplies most of the stopping force.

Although the effect isn't as obvious when you're braking a Camaro, the front wheels are still doing most of the work. Chevrolet engineers selected these wheel cylinder diameters to provide the proper front-to-rear braking balance. Some Camaros are also fitted with proportioning valves, which control the hydraulic pressure reaching the rear brakes. This feature further refines the brake balance.

Front wheel cylinders are not interchangeable from side to side, since the brake hose enters the cylinders at an angle. Rear wheel cylinders, however, can be installed on either the right- or left-hand brake.

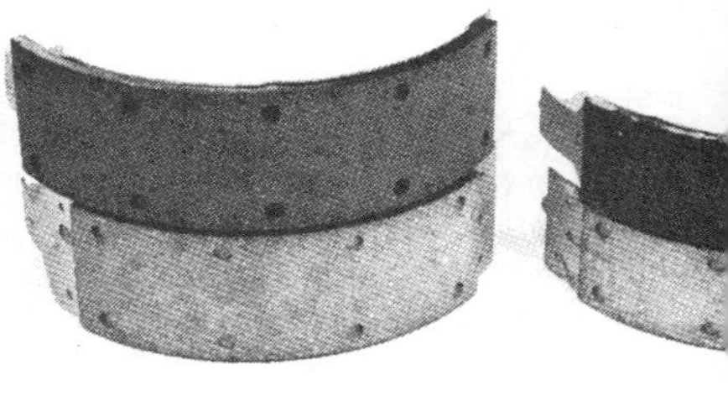

18. The '67- '69 Camaro front and rear drum brake linings also differ in size. Front shoes measure 2½ inches across, while rears are only two inches in width. (The front and rear drums are both 9½ inches in diameter.) As with wheel cylinder sizes, the difference in size is due to the higher loads created at the front wheels when the car's weight transfers forward during braking.

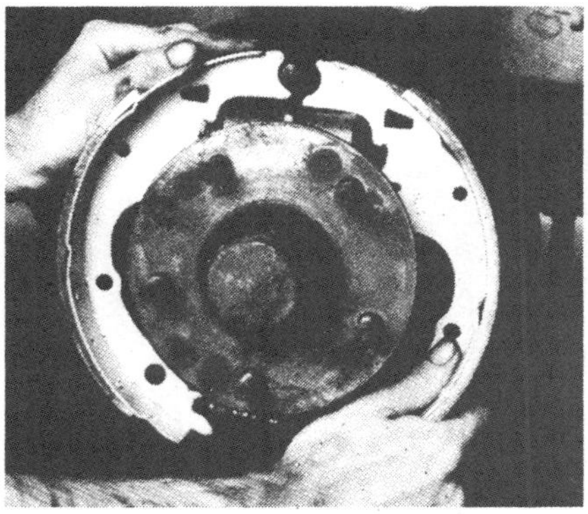

19. Before reinstalling the linings, disassamble, clean and lubricate the brake-adjusting assembly. Pay particular attention to the threads on the star wheel, making sure the wheel turns freely. Hook together the primary and secondary shoes with the adjusting spring, then insert the star wheel assembly between the two shoes, positioning the wheel alongside the secondary shoe. Make sure the adjusting spring doesn't interfere with the star wheel. Spread the shoes and slip them over the wheelhub or axle flange, then align the wheel cylinder pushrods in their slots in the linings.

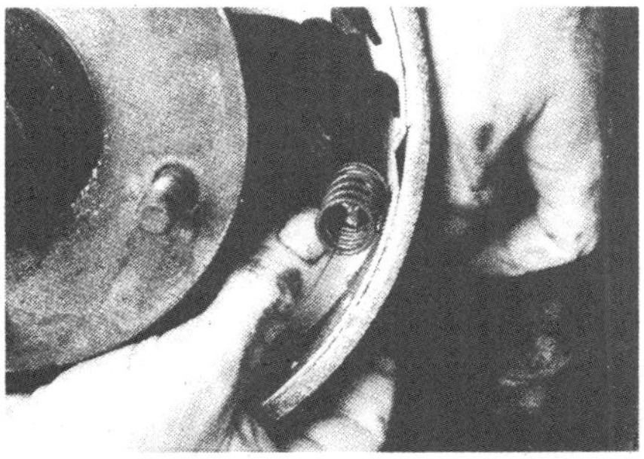

20. A metal cup under the hold-down spring locates the adjusting linkage on the secondary brake shoe On the rear drums, insert the parking brake linkage between the primary and second shoes, and check for interference between the axle flange and the linkage.

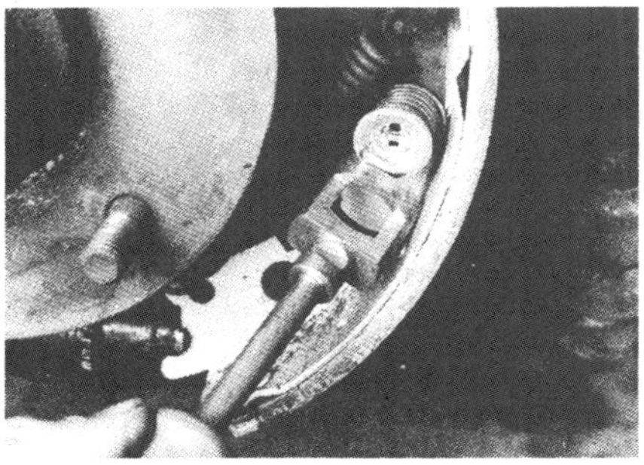

21. Compress the hold-down spring and anchor the secondary shoe on its retaining pin with a slotted cup. Make sure to install the deep-dish cup on the secondary side; then install the flatter cup on the primary side.

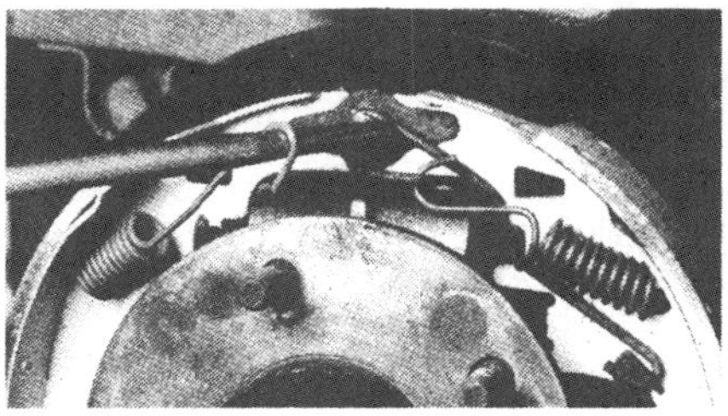

22. Using the tapered handle of the brake pliers, stretch the pull-back springs and slide the hooked end over the anchor pin. Check that the linings are seated properly on the raised pads on the backing plates, and make sure the wheel cylinder pushrods are still properly aligned in their slots.

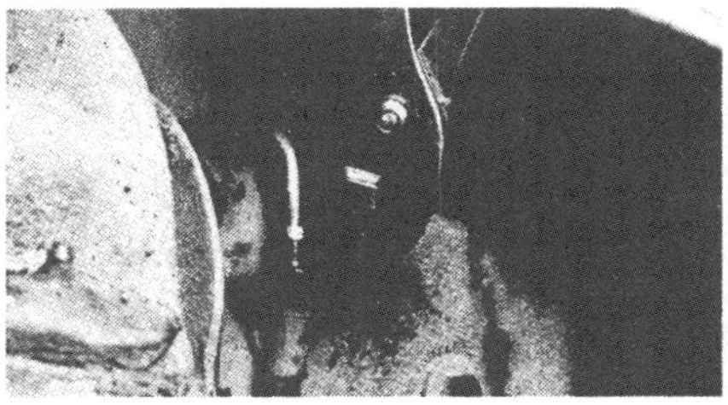

23. If you've replaced or rebuilt the wheel cylinders, reconnect the flexible hose. Insert a soft copper washer between the wheel cylinder and hose end, then tighten the connection. Insert the free end of the hose in the 12-point bracket on the frame; do not twist the hose. The hose must curve inward to prevent it from contacting the wheel and tire. Attach the hose to its support bracket with the U-shaped spring clip, then reconnect the steel brakeline and tighten its nut with a line wrench.

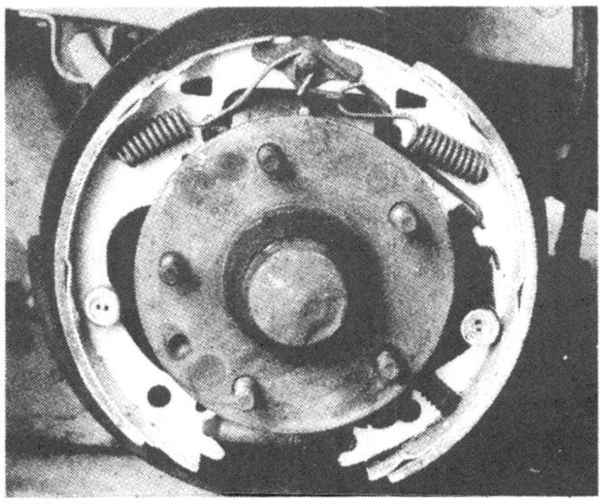

24. Double-check your work by comparing the position of all the components with the brake on the other side of the car. You can make a preliminary brake adjustment by turning the star wheel until the drum will just slide over the linings; then back off the star wheel one full turn. Reinstall the drum.

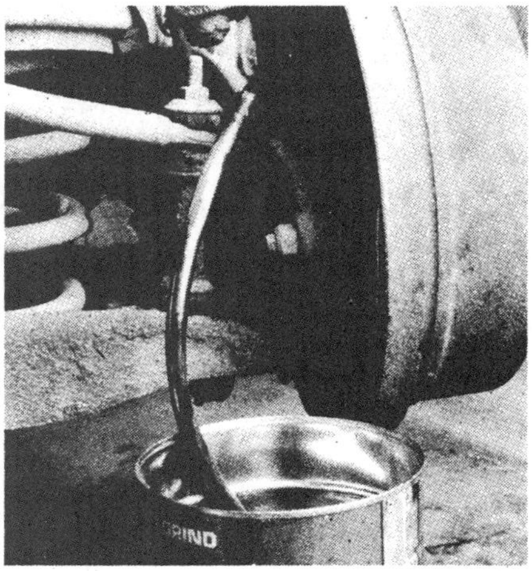

25. The final step in your Camaro brake overhaul is the most important. The brakelines and wheel cylinders must be bled to remove air from the lines. Bubbles remaining in the hydraulic system will cause the pedal to feel spongy, and the brake performance will suffer. It's helpful to have a partner when bleeding the brakes. Always start with the wheel which is farthest from the master cylinder. If you've overhauled all four brakes, begin with the left rear, then go on to the right rear, right front and finish at the left front. Of course, if you've only serviced the front or rear brakes, the other half of the system will not require bleeding.

To bleed a Camaro's brakes, pour some clean brake fluid into a jar or can. Attach a length of vacuum hose to the bleeder valve on the back side of the wheel cylinder, and submerge the other end of the hose in the brake fluid. Fill the master cylinder reservoir to within ¼ inch of the top, then replace the top. Open the bleeder valve ¼-½ turn, then slowly release the brake pedal. Refill the master cylinder reservoir and then repeat this procedure until no bubbles appear in the brake fluid. It's important to keep the end of the bleeder hose submerged in the fluid to prevent air from re-entering the hydraulic system. When no more bubbles appear, tighten the bleeder valve and go on to the next brake in the sequence. When you've finished, make sure the fluid level is ¼ inch from the top of the master cylinder reservoirs and test the pedal for firmness. If the pedal doesn't feel hard, repeat the bleeding process until you have purged all the air from the system.

The final brake adjustment is done automatically by making a series of stops in forward and reverse. Go slowly at first to make sure the brakes are functioning properly and to seat the linings against the drums. Then recheck the fluid level and the tightness of the nuts after a short road test. Make a few safe, straight stops, and you'll have to agree that it's hard to beat a Camaro drum brake system.

100 HOT TIPS FOR A COOL CAMARO

HOW TO BUILD A BETTER CAR BY THE NUMBERS

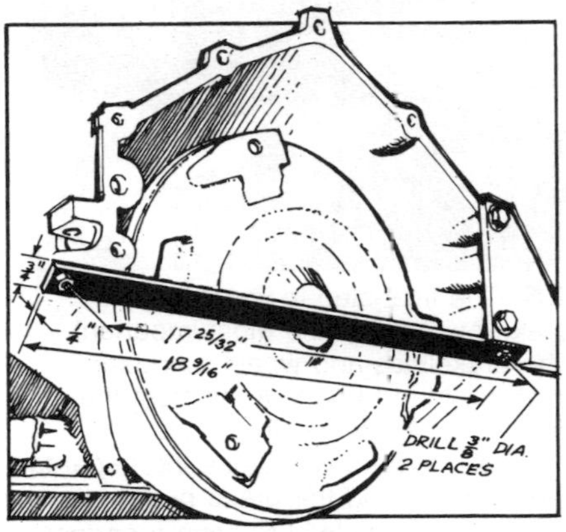

1. Whether you're restoring, rebuilding or racing your Camaro, one of the chores that's probably confronting you is replacing the torque converter. Wrestling an automatic transmission into place is never fun, but it can be made easier by making a converter-retaining strap from a scrap of aluminum or steel. Install the converter on the transmission input shaft, making sure that you fully engage the pump. Then bolt the retaining strap across the bottom of the bellhousing as shown, securing it to the holes for the converter dust cover. Now you can either jack or bench press the trans/converter combination into place without worrying about the converter slipping off the input shaft. When the trans is bolted in place, remove the strap and replace the dust cover.

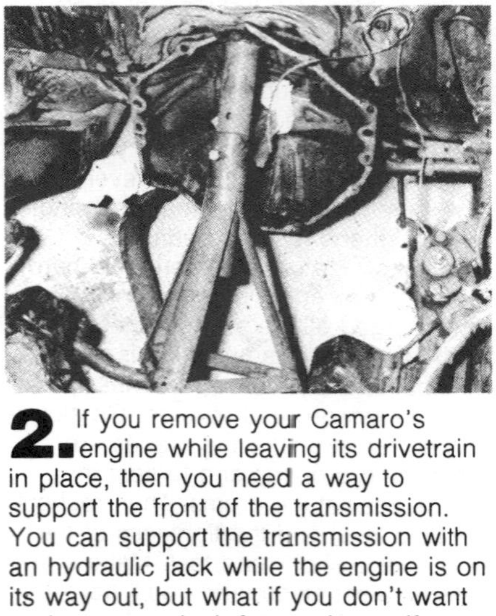

2. If you remove your Camaro's engine while leaving its drivetrain in place, then you need a way to support the front of the transmission. You can support the transmission with an hydraulic jack while the engine is on its way out, but what if you don't want to tie up your jack for weeks, or if you need to move the body? Simple solution: Just put a jackstand between the bellhousing and front crossmember as shown. Adjust the jackstand to prevent the front of the transmission from drooping, which can damage the transmission mount.

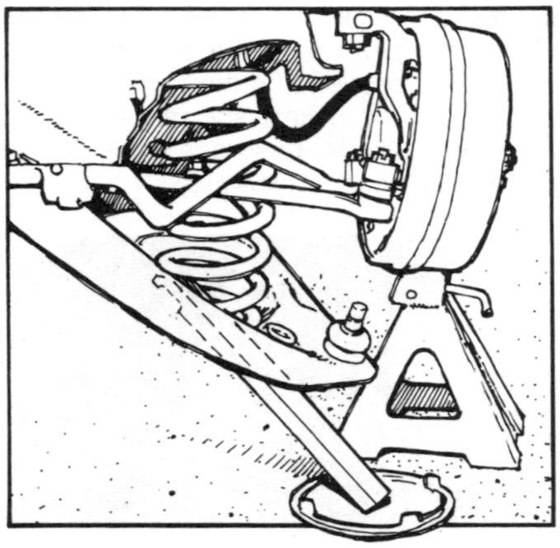

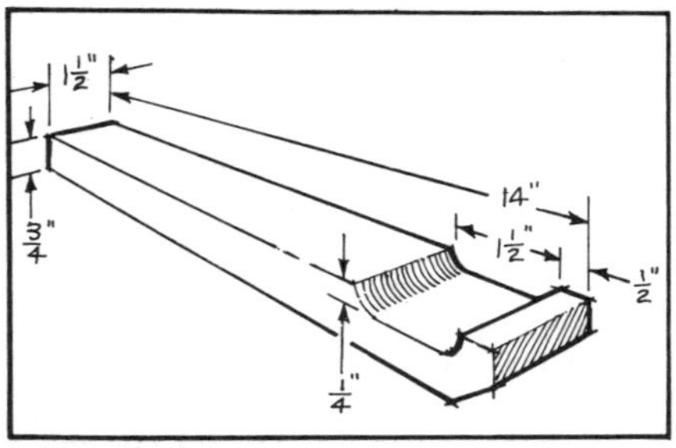

3. Removing or replacing a Camaro's front coil springs is always a battle. It seems as though a floor jack under the A-arm never has quite enough leverage to compress the spring far enough to allow you to bolt up the ball joints. You can give your jack an advantage by fabricating a simple extension like this. This tool also holds the spring in its seat, reducing the likelihood of the coil taking flight across your garage.

4. Part of the styling appeal of any Camaro is its wide expanse of side window glass. Camaro owners know that all that glass can make a terrible racket if the window starts rattling. You can restore peace and quiet to your Camaro's interior if its doors are equipped with anti-rattle adjusters. The soft-faced brackets shown on this '69 Camaro's doors should be adjusted to just touch the window; this steadies the glass in its

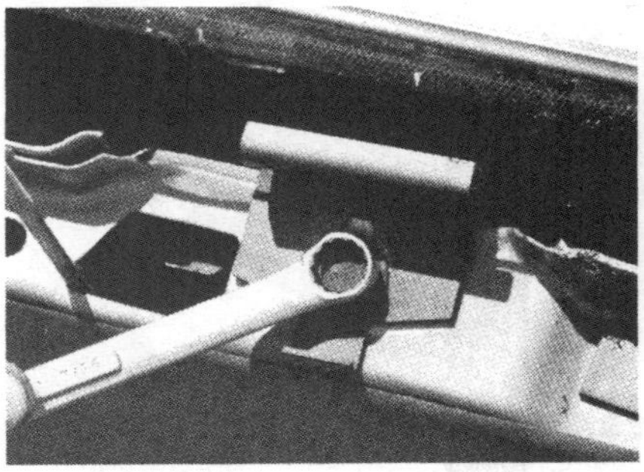

track, which helps curtail wind noise. You can help these devices hold their adjustment by always closing your doors by their handles, not by the windows.

5. Keeping a Camaro engine compartment spotless is hard enough without oil leaks contributing to the grime. If your Camaro has traveled many miles, chances are good that the timing chain cover seal is seeping oil. Replacing the seal is a good start, but usually it's a groove worn in the harmonic balancer that's the real source of the gusher. You don't have to buy an expensive new balancer to stop the oil flow, though. Dorman makes a thin, press-on balancer sleeve to repair the worn balancer snout. The sleeve is chromeplated, so its super-hard surface will endure many miles of use.

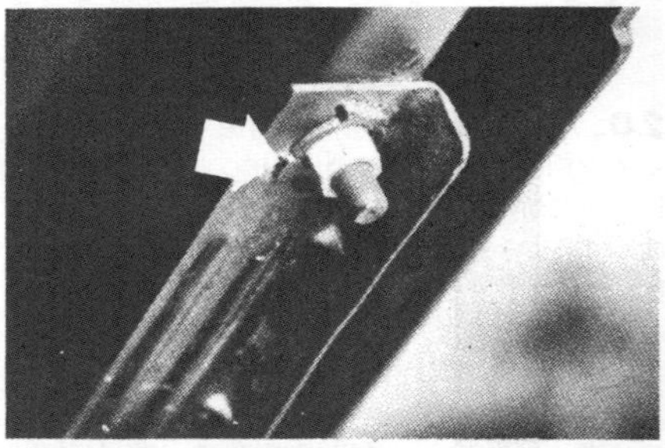

6. Working under a Camaro's long hood is often awkward, and always dark. Removing the hood just doesn't seem worth the effort for most jobs, though, since starting the bolts for the hood hinges is guaranteed to ruin a good mood. So why not replace the bolts with Allen-head studs like this sharp racer? The steel hood can be removed or replaced in seconds with this setup. And if you scribe a mark around the hinges, the hood will be a snap to realign.

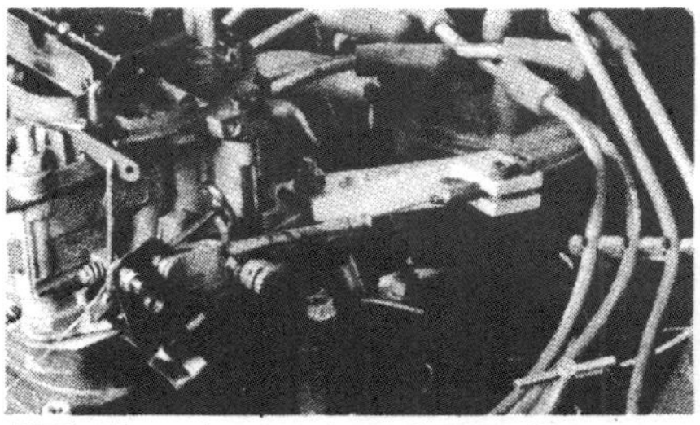

7. If you've equipped your Camaro with a high-rise manifold or an aftermarket carburetor, the throttle linkage may be giving you trouble. One Camaro owner solved this problem by fabricating the combination carb spacer/linkage mount. After replacing his early-model Camaro's mechanical linkage with a cable setup out of a late Chevy II, the throttle always works smoothly, regardless of which manifold or carburetor he happens to be testing.

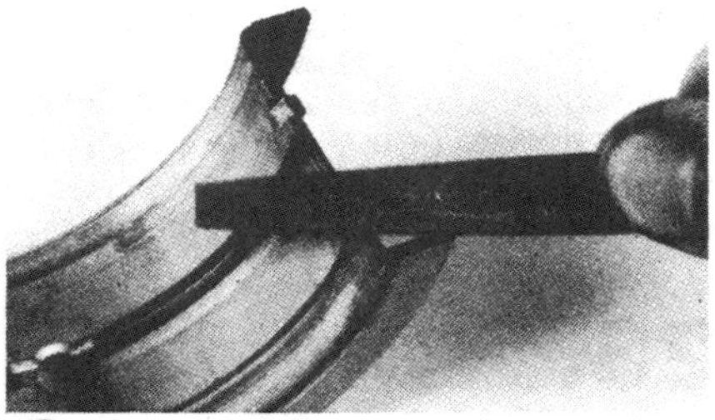

8. A hot rod Camaro puts a great deal of stress on its engine's thrust bearing. A stiff pressure plate forces the crankshaft against the thrust bearing every time you step on the clutch pedal. Replacing the thrust bearing is definitely a chore to be avoided. You can extend its life expectancy by carefully filing the chamfer at the bearing's parting lines to increase the oil supply. In extreme cases, you may want to cut a groove across the bearing with the edge of a file leading directly from the bearing's oil hole to the thrust face. Deburr and wash the bearing insert carefully after these modifications.

9. One of the most difficult chores of rebuilding or modifying a Camaro's front suspension is dislodging the spindle from the tapered ball joint studs. If you plan to replace the ball joints, a "pickle fork" and a little applied brute force will separate the spindle. But if you don't want to destroy the seals, a different approach is required. Loosen the ball joint nut and back it off two turns—do not remove it. With jackstands supporting the car securely, strike the spindle simultaneously with two heavy hammers. There are small pads on the spindle around the ball joint hole for this purpose. As soon as the spindle relaxes its grip on the studs, the pressure of the front coil spring will pop the spindle loose.

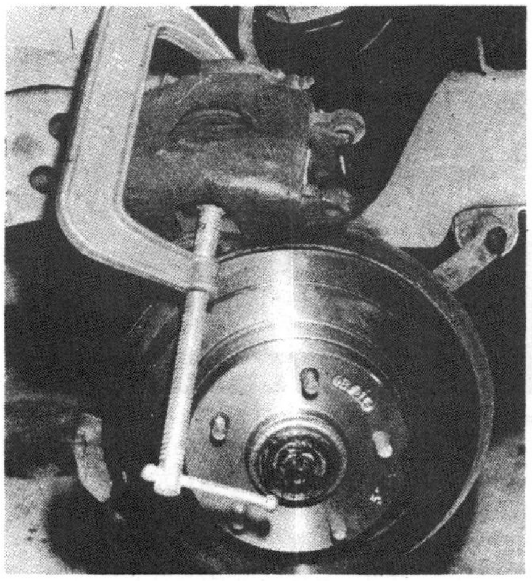

10. Before installing new front disc brake pads in your Camaro, first pull back the pistons in the caliper. A C-clamp and a block of wood are all it takes. Sandwich the wood between the used pads, then position the clamp as shown. When you tighten the clamp, it will quickly push the piston back into the caliper.

11. Keeping rust at bay is a constant battle for anyone who cares about his Camaro. If you're restoring or refurbishing a Camaro, chances are you've found spots where rust has already started its dirty work. You can stop corrosion in its tracks by spraying the affected metal with zinc phosphate. The only preparation required is sanding or wire brushing the surface to remove flakes and scale. Then spray on the primer; it penetrates the metal and neutralizes the rust. This is an ideal treatment for subframes, inner fender panels and chassis components that don't require a smooth-as-glass final finish.

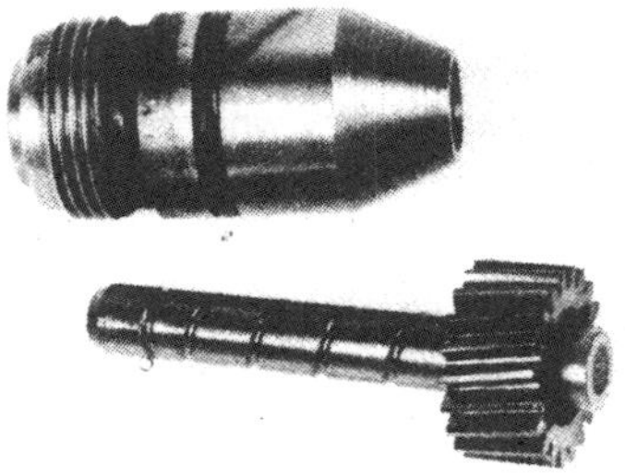

12. If you've changed your Camaro's rear tires or rearend gear ratio, the speedometer reading may be wildly inaccurate. This can cost you some serious speeding fines, and make it impossible to obtain accurate mileage readings. There is an easy way to recalibrate your speedometer, however. Just remember that adding or subtracting a tooth from the speedometer drive gear changes the reading approximately five mph when you're traveling at 60 mph. For example, if your speedo reads 10 mph too slow when you're driving down the highway, the speedometer drive gear should have two *less* teeth. If the speedo is 10 mph too fast, then install a gear with two *more* teeth than the one in the transmission.

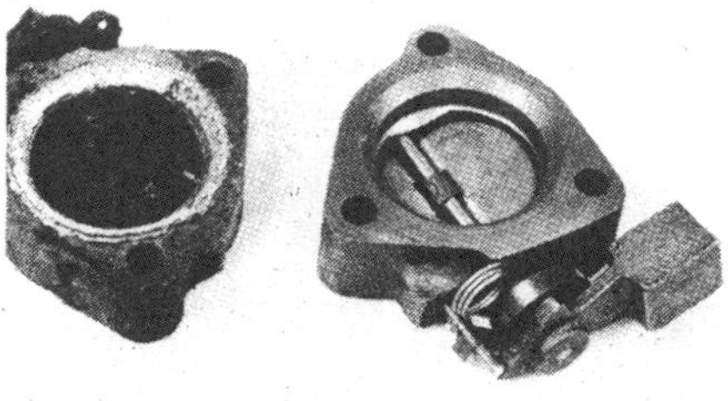

13. If you like the idea of quiet cast-iron exhaust manifolds on your Camaro but don't care for restricting the flow of exhaust gases, you may want to rework the heat-riser valve. By eliminating the butterfly and plugging the pivot holes with set screws, you can free the exhaust flow, and also eliminate problems with noisy and stuck valves. This modification allows exhaust gas to pass down both sides of a dual exhaust system, which reduces corrosion in one pipe when a Camaro is used for only short trips. Warm-ups on cold mornings will take longer, though, since hot exhaust gases are not forced through the intake manifold's heat-riser passage.

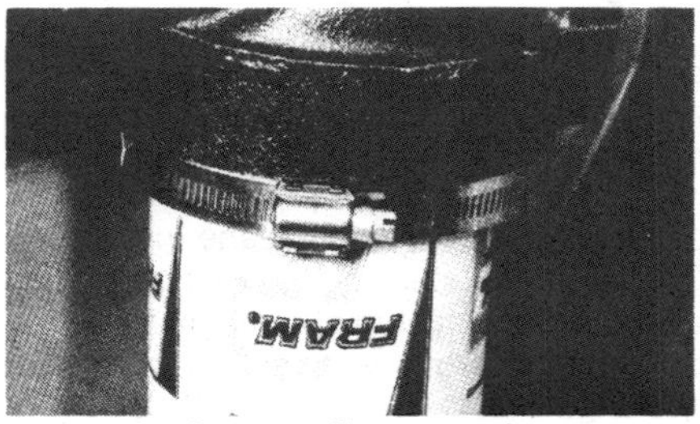

14. Spin-on filters can have a disturbing habit of spinning themselves off the engine block. If you overtighten your Camaro's filter, the gasket can be squeezed out, which only makes the problem worse. The solution? Tighten a hose clamp just under the filter lip. This prevents the filter housing from ballooning if the oil pressure goes sky-high when you rev the engine before the lubricant has had a chance to warm. Then loop a piece of safety wire from the block to the clamp to prevent the filter from unscrewing itself.

15. Although a loud exhaust might sound tough to some performance enthusiasts, a leaky exhaust system just sounds bad to knowledgeable Camaro drivers. Cinching up the tailpipes to the exhaust manifolds is a risky job, though, since the factory-installed nuts have a tendency to weld themselves to the manifold studs. Tighten them too much and the studs may snap off without warning. If you substitute corrosion-proof brass nuts for the steel stockers, you can tighten up the manifold connections without fear.

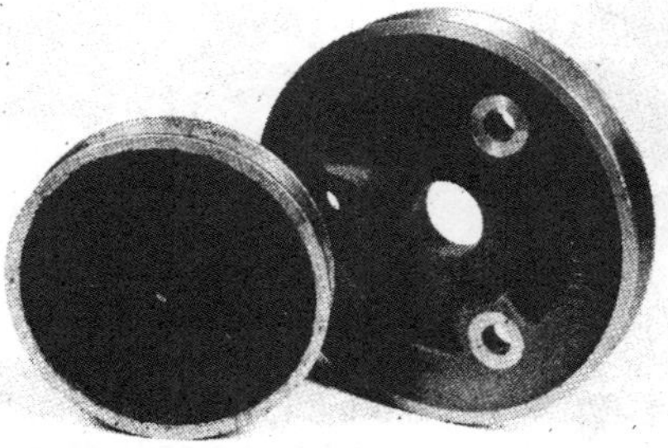

16. Free horsepower is the very best kind. It takes power to spin the alternator, water pump and other accessories on the front of a Camaro motor. One way to minimize these losses is to install an underdrive crankshaft pulley. A small-diameter pulley will slow down the belt-driven pumps, reducing power losses. In fact, a high-speed engine may actually run cooler with an underdrive pulley, since the coolant spends more time in the radiator and the pump is less prone to cavitation—but test this before heading out across the Mojave Desert in your hot street Camaro.

17. Carburetor spacers allow a Camaro owner to tune his car's induction system inexpensively and effectively. If you have a dual-plane manifold, an open spacer can give your motor the top-end power usually associated with a single-plane intake. Conversely, a four-hole spacer bolted to a single-plane just might supply the bottom end torque you've been seeking. Aluminum and plastic carb spacers are available in a variety of thicknesses to accommodate your wildest tuning impulses.

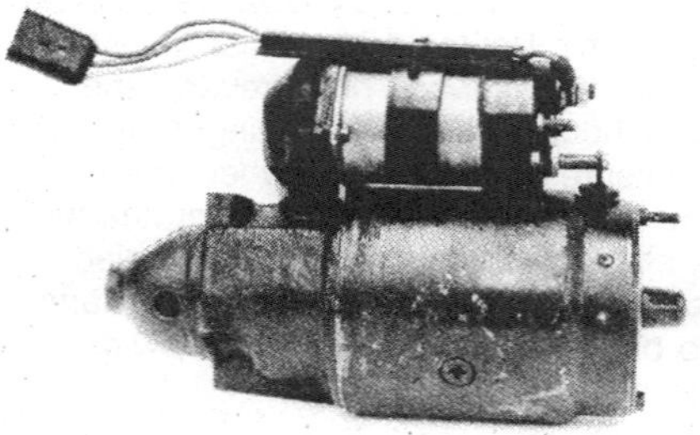

18. When it's time to remove or replace an engine, it seems as if there are thousands of wire connections. Getting electricity to and from the starter solenoid is much simpler if you install a harness from Mr. Gasket, Moroso or B&B Performance. These plug-in harnesses eliminate struggles with the tiny nuts on Delco starters, and prevent you from getting your Camaro's wires crossed.

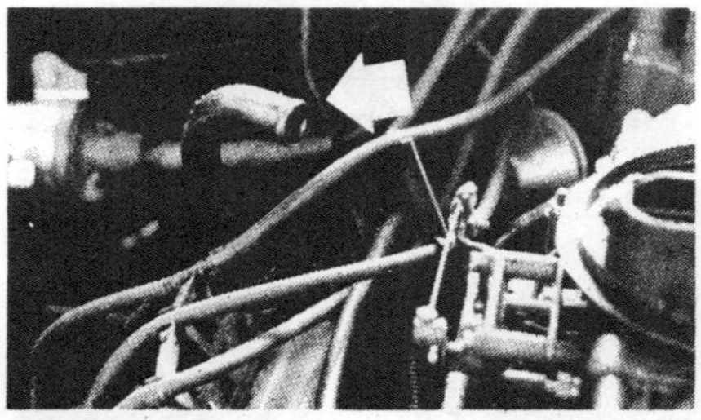

19. No one doubts the cornering power of a Camaro. In fact, high-speed maneuvers can occasionally overcome the engine's oil control, allowing oil to enter the intake manifold from the valve covers through the breather lines. A small piece of Scotch Brite inserted in the breather hose works well as an oil separator, and will keep your Camaro from being black-flagged on the autocross course for conspicuous oil consumption.

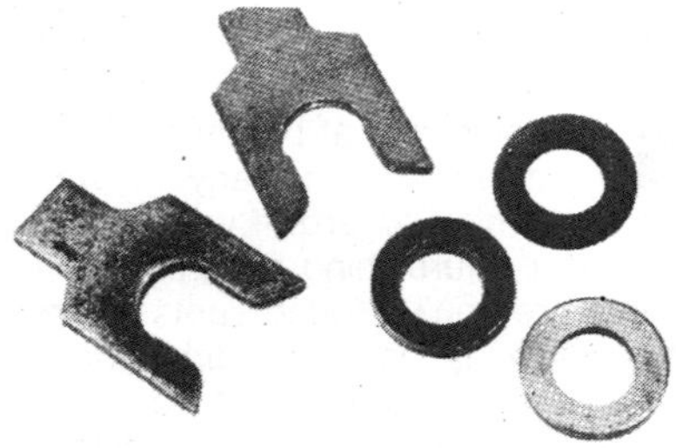

20. U-shaped front end alignment shims slide easily between a Camaro's A-arms and subframe. Unfortunately, they can just as easily slide out again. That's why drivers who use their cars for serious handling contests always substitute round washers for these shims. After the alignment settings are finalized with conventional shims, the upper A-arms are removed and an equal thickness of hardened washers is inserted. This eliminates the radical changes in suspension geometry which can occur if the shims suddenly decide to depart.

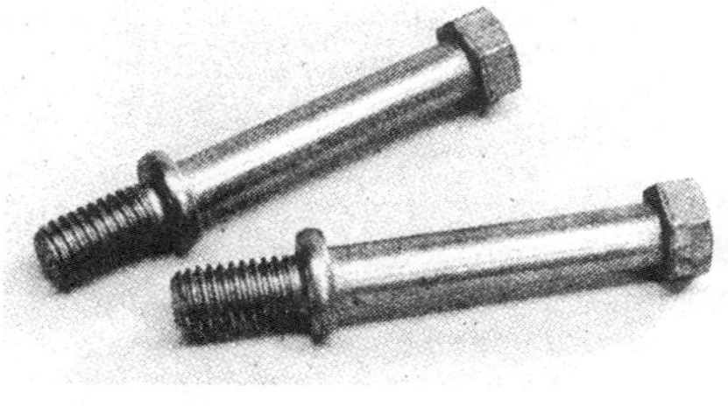

21. The chore of changing a Camaro transmission can be minimized with extended crossmember bolts. Whether you make your own or buy them, the operating principle is the same. Three-inch-long shouldered bolts like these will extend all the way through the subframe, making it easier to align the transmission crossmember with its mounting holes. Also, the extended bolts are much easier to wrench if frequent transmission swaps are part of your Camaro performance plan.

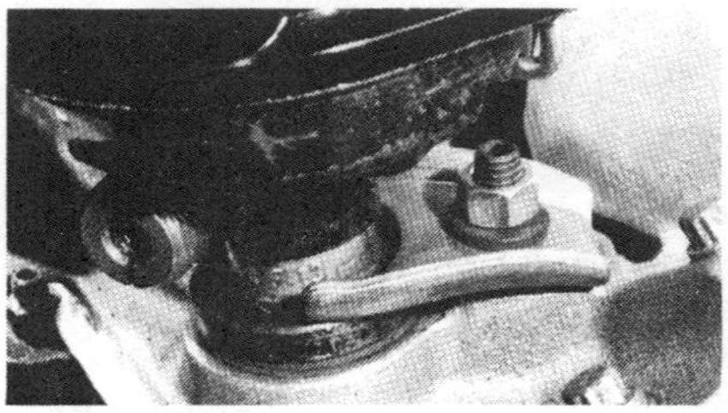

22. An accountant at Chevrolet figured out how to save a few pennies on every engine by using wire distributor clamps on some Camaro engines. The problem with this approach is that these cheap clamps allow the distributor to turn against the resistance of a high-volume oil pump, and heaven forbid you should ever try to hold down a magneto with one of these clamps. The solution: Pirate an old-fashioned stamped steel clamp from another engine, or buy one of the heavy-duty clamps offered by the aftermarket.

23. A Camaro's brake system turns kinetic energy into a lot of heat. Under the kind of heavy use that's typical on an autocross course or a winding mountain road, this excess heat can cause severe brake fade. You can help your Camaro's brakes keep their cool by ducting air from the high-pressure area at the front of the car to the rotors. A spoiler is an ideal pick-up point. Use flexible tubing at least two inches in diameter, and secure it to the subframe with nylon tie wraps on sheetmetal straps.

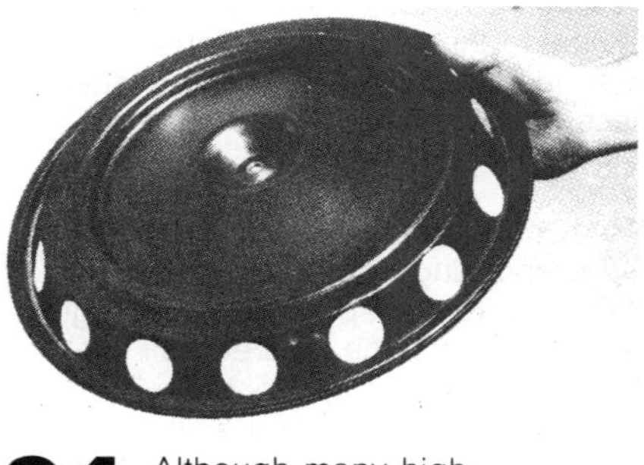

24. Although many high-performance Camaro engines were originally outfitted with open-element air cleaners, noise and emissions laws have forced the engineers to use restrictive single-snorkel air cleaners on all late-model motors. If you still want the protection a filter element offers along with an unrestricted air supply, then an afternoon with a hole saw will soon have your Camaro motor breathing freely. Perforate the air cleaner lid as shown, then add or open up its hood scoop. Your off-road Camaro will be breathing freely soon afterward.

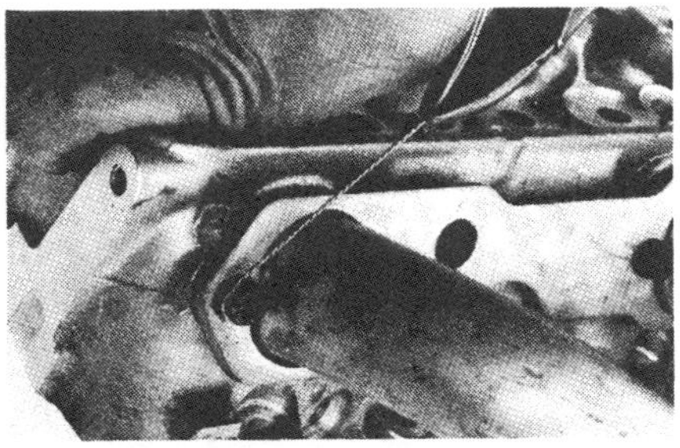

25. Headers on a Camaro should be seen and not heard. If you're tired of retightening the header bolts every week, why not take a tip from airplane and race car mechanics? A strand of safety wire looped between each pair of bolts will prevent the bolts from backing off as the headers expand and contract. Special safety wire pliers are available through race car and aircraft supply houses, but for occasional safety wiring jobs, a pair of needlenose pliers will work just fine. Always loop the wire so that it tightens the fasteners.

26. A self-adhesive timing tape is a low-cost alternative to an expensive degreed balancer. These tapes make setting the ignition timing a snap, and you can use them to determine your distributor's centrifugal advance curve by noting the ignition timing at various rpm points. Timing tapes are available for all Chevrolet harmonic balancers, including six-, seven-, and eight-inch-diameter versions.

27. Header tubes and hot exhaust manifolds can melt the insulation on the battery cable in your Camaro's engine compartment. Clearance can be particularly tight if you've relocated the cable to bring current from a trunk-mounted battery. To prevent a meltdown, protect the cable with an insulating sleeve of high-temperature asbestos. Companies which specialize in race car and aircraft plumbing are the best sources for this material.

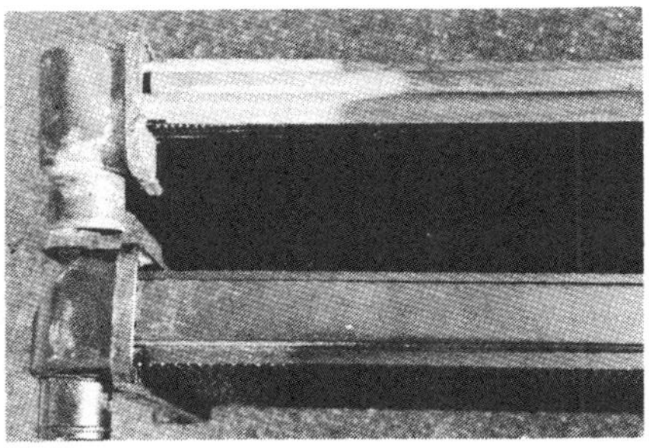

28. Almost any modification you perform on your Camaro's engine will increase its cooling needs. If the high-performance small-block in your early Camaro is feeling the heat, consider installing a big-block radiator core. These rat motor radiators have thick cores and measure three inches wider than the small-block coolers. And best of all, the holes are already drilled in the radiator brace to make this a bolt-in swap.

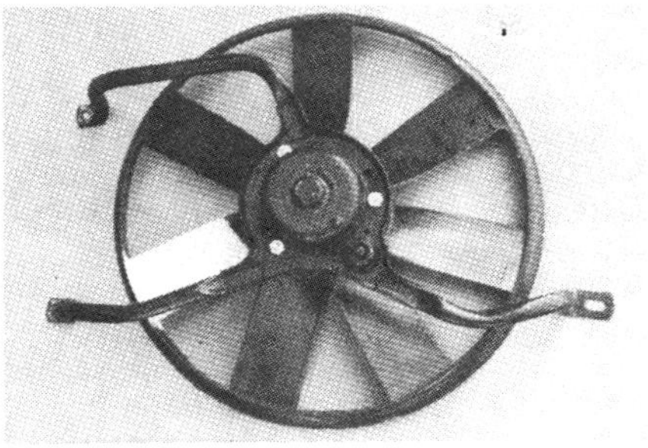

29. How much power does it take to spin a cooling fan? Enough that Chevrolet engineers decided to install an electric fan in the new High-Output L-69 Z28s. This same fan can be adapted to earlier Camaros to free some horsepower that can be better applied to turning the rear wheels.

30. Camaros have never been famous for their cavernous trunks. There's barely enough room for a toolbox and an overnight case in the trunks of '70-'81 Camaros. Of course, you've probably already scrounged a space-saver spare tire to conserve room. But did you know there are *two* spare tire hold-down brackets lurking on the trunk floor? It may be a bit of a stretch, but if you attach the spare to the bracket behind the back seat, you'll gain enough trunk capacity to carry along a shaving kit.

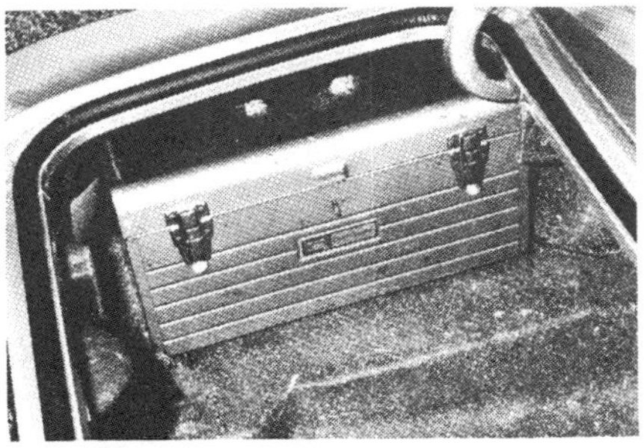

31. No one who drives a hot rod Camaro ventures very far from home without a roadside survival kit. An assortment of hand tools and a few carefully chosen spare parts can usually solve a minor breakdown. The first trick is to find a toolbox that will fit in the nooks of a Camaro trunk. This Sears No. 9-6501 box measures 18x8x9 inches, and is a perfect fit behind the wheelwell of an early-model Camaro. Strap it in place with an aluminum strip or a bungee cord to prevent it from denting the fender when you rocket around freeway on-ramps.

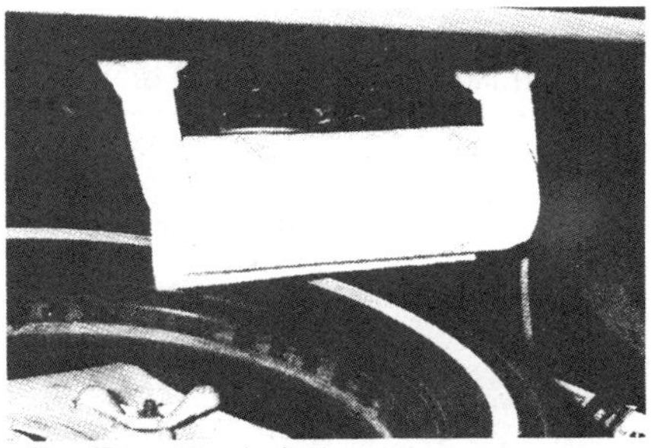

32. Why is it that you can never find a paper towel when you need one? Checking the oil or changing a tire on your Camaro will have you looking for someplace to wipe your hands. If you install a paper towel holder under the package tray, your search will end quickly.

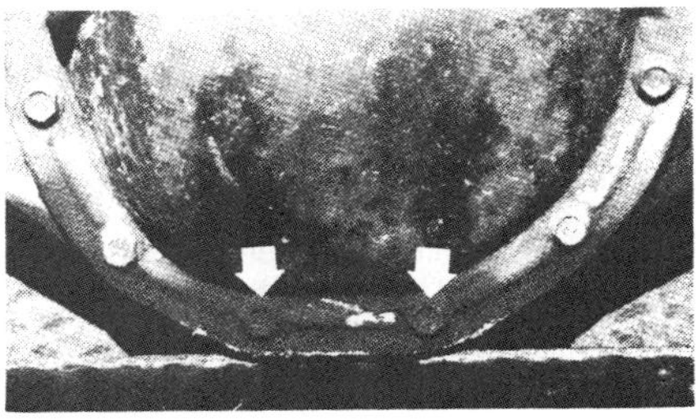

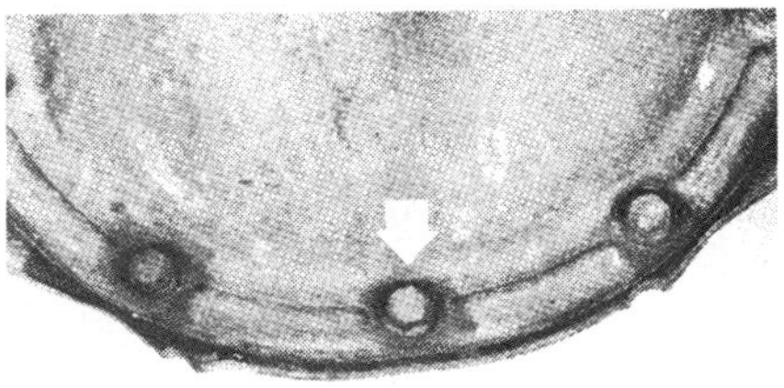

33. Whether you are stalking the salvage yards or sizing up the competition at a stoplight, it's a great advantage to be able to recognize a heavy-duty 12-bolt Camaro rearend instantly. Sure, you can count the bolts on the rearend cover, but there is a quicker method: 12-bolts have *two* bolts in a line at the bottom of the cover; 10-bolts have *one* fastener on the bottom. With good eyesight, you'll be able to spot a 12-bolt from 50 feet away.

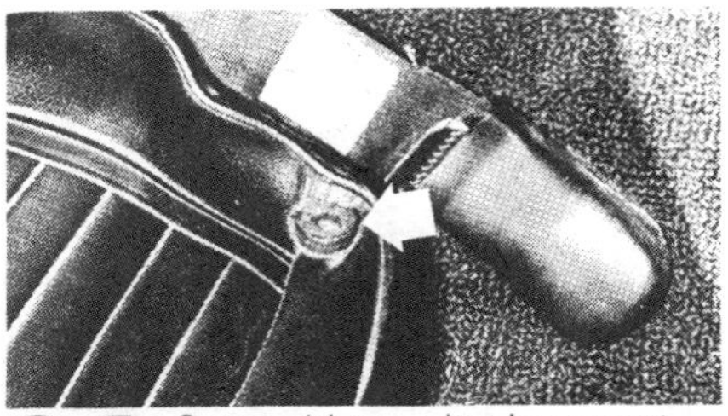

34. Some drivers simply cannot get comfortable in a Camaro seat. If you fall into this category, you might try adjusting the angle of the seat back by shimming or trimming the rubber bumper at the base of the seat back. Removing a little rubber reclines the seat back, creating more headroom and permitting an "arms out" driving style. Adding a washer under the bumper brings the seat back closer to the steering wheel.

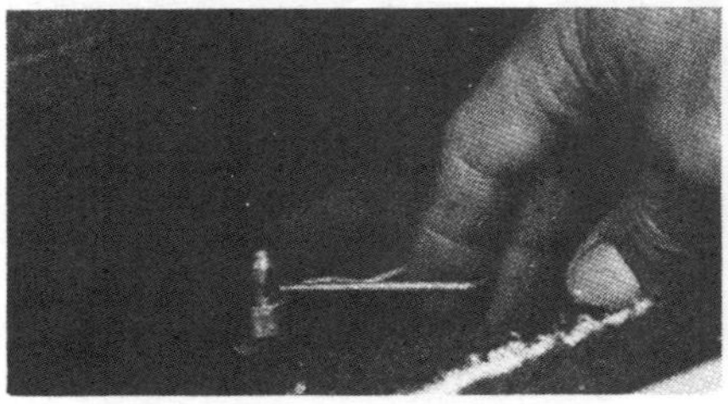

35. Setting up a Camaro for competition means making the car easy to service. For most maintenance chores under the dashboard and around the shifter, the seats are in the way. You can make "quick change" mounts for stock and aftermarket seats similar to hood pins. Insert studs in the four seat-bolt holes, then mark and drill them for hitch pins. A spacer or shim between the pin and the seat mount will hold the seat tightly against the floorboards.

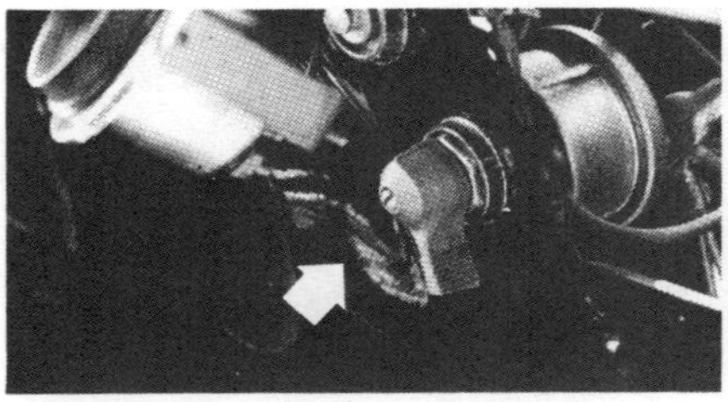

36. Most Camaro race cars are equipped with master electrical switches because the racing rulebooks require them. Street cars can benefit from these devices as well. Imagine the convenience of being able to shut down the electrical system during routine maintenance—and the safety of being able to turn off the current in case of short circuit or electrical fire. This Camaro owner replaced the ignition key with super-reliable aircraft switches, then located a master switch in the vacant dashboard hole. All electrical current except the main starter circuit is routed through this one switch.

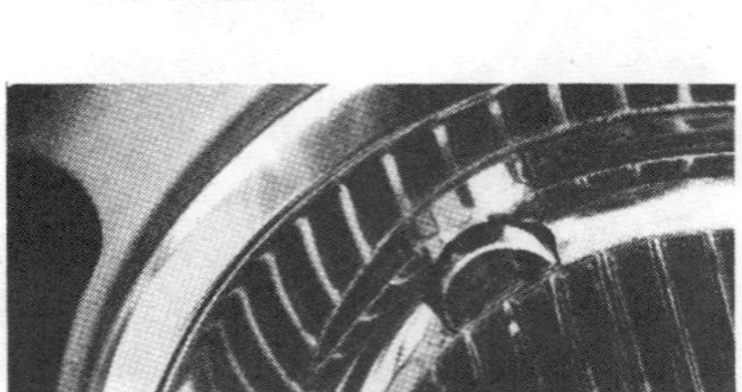

37. Camaro owners used to have trick mag wheels with special locks. Now, even original Rally wheels have to be guarded against thieves. One Camaro owner carefully notched his Rally hubcaps to clear the mag wheel locks he had installed. Do the job carefully and no one will notice—except the thief trying to make off with your 15x7-inch rims.

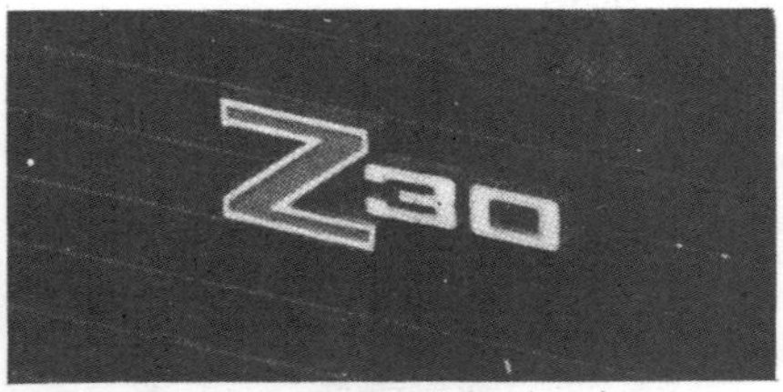

38. The Z28 designation has taken on a special significance for Camaro enthusiasts. This particular combination simply reflects the next RPO code (regular production option) which was available when the 302 engine package was introduced. Some Camaro owners express their individuality by making their own insignia, like our example. Almost any combination is possible by rearranging the numerals found on engine medallions. What's a Z30? How about a 454-cubic-inch late-model Camaro!

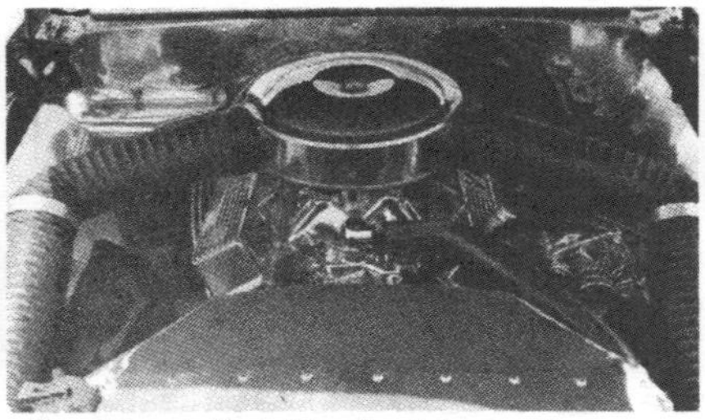

39. A Pro Stock-style engine compartment may be as close as the nearest hardware store. This enterprising Camaro owner gave his street machine a new look by covering the inner fender panels and firewall with sheet aluminum. The flat, angular sheets eliminate tricky bends and complex metal work. The cold-air induction system is also racetrack inspired, with a pair of flexible inlets feeding an oversized air cleaner.

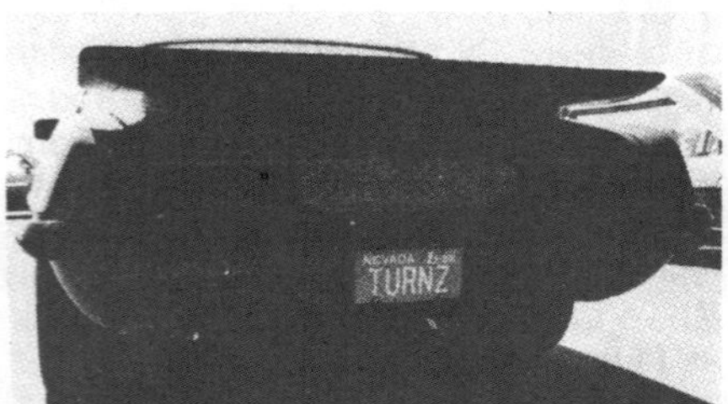

40. Chromeplating is now "out" with the automotive stylists. Today's Eurostyle-look insists on a flat black and clean lines. This '69 Camaro has been updated in the best cafe racer tradition with a painted bumper, black trunk lock, one-piece taillights and a whale tail rear spoiler.

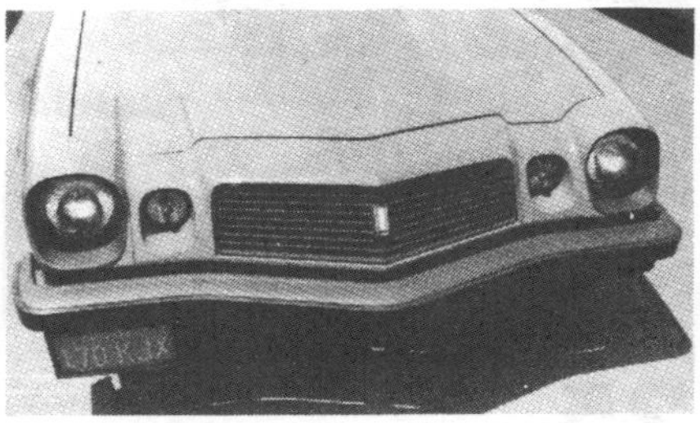

41. Second generation Camaros produced between 1974-'78 are notable chiefly for their large aluminum "battering ram" bumpers. These styling liabilities can be converted into assets with the careful application of some body-colored paint, as this Camaro demonstrates. The black rubber strip is quite an accent against the painted bumper, or consider black anodizing the aluminum as a high-tech alternative.

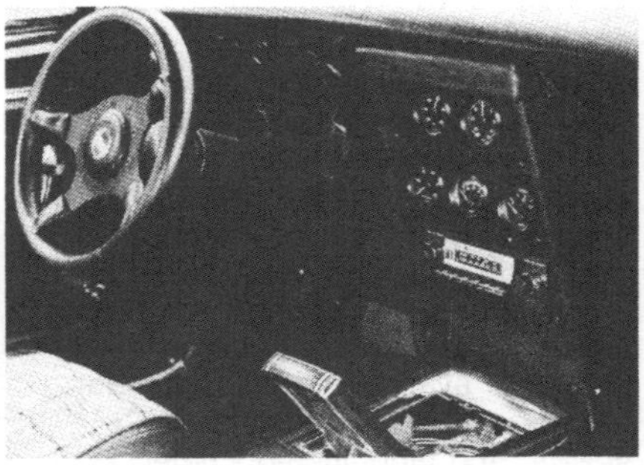

42. Finding a good location for accessory gauges is not easy in an early-model Camaro. The factory's preferred placement on the console puts them well out of the driver's line of sight. This '67 Camaro demonstrates a different solution, with a neat console extension that looks as if it were designed in Detroit. The businesslike gauge arrangement is easy to see, and the console accommodates a standard dash-mounted stereo. Upholstery and a mat black color lend the finishing touches.

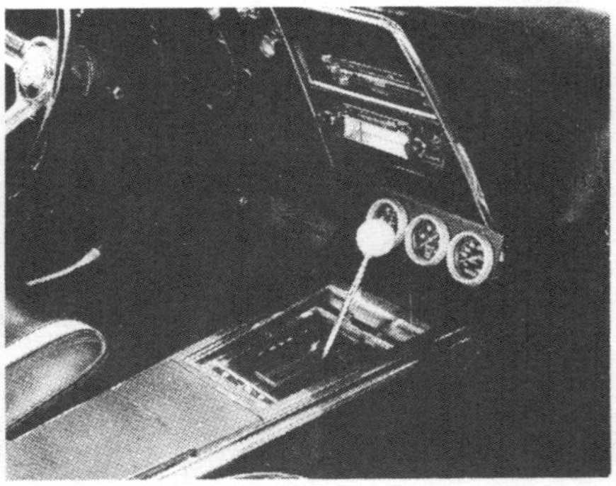

43. You have to look twice to appreciate this interior

treatment. At first glance, it all looks original. Then you notice the Hurst shifter handle protruding through the console in place of the OEM model. Soon you spot the three large-diameter Autometer gauges (indicating oil pressure, water temperature and amps) where the tiny, uninformative factory gauges formerly resided. The effect is a definite improvement over Chevrolet's efforts, while retaining an original "look."

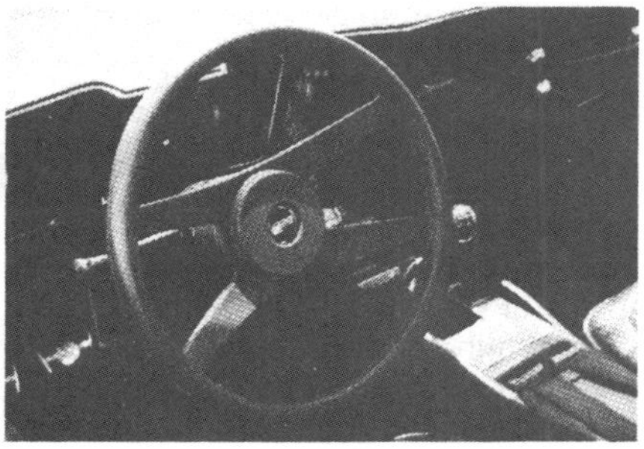

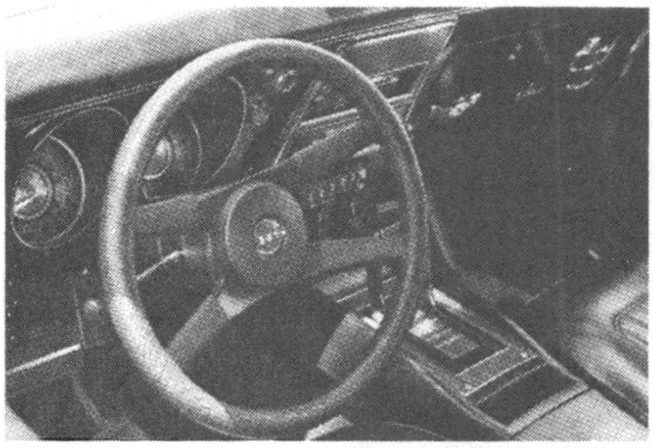

44. Here are two examples of how Camaros can be customized using factory parts. The owner of one '69 updated his car's interior with a four-spoke steering wheel from a second generation Z28, while the '68 hardtop is upgraded with a wheel from a '78 model, featuring a simulated string-wrapped rim. Both wheels are a considerable improvement over the original skinny plastic rims, yet they retain the appearance of a factory installation.

45. Since every Camaro has a minimum of four wheels, coming up with a wheel treatment takes some thinking. These two approaches are simple and attractive. The '67 model has been outfitted with a set of painted "mag style" steel rims from a

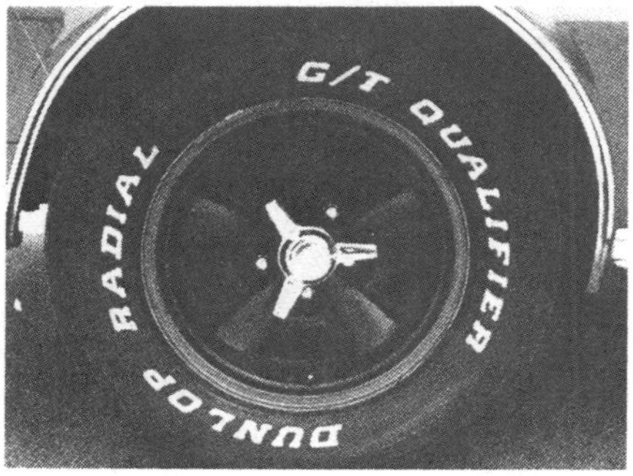

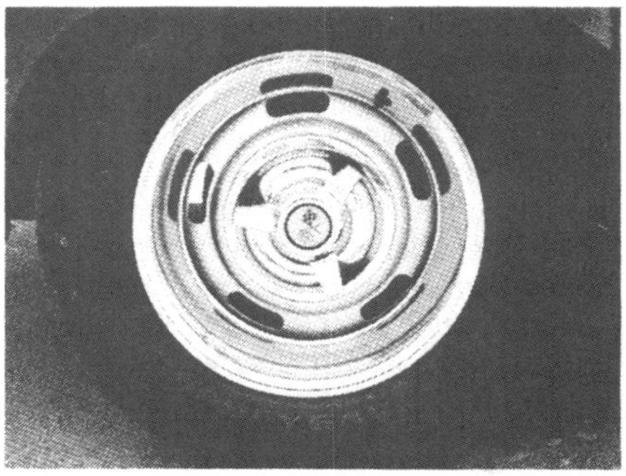

second generation Camaro; they're highlighted with knock-off centers and chrome lug nuts. The '69 Camaro slotted Rally wheels are updated with Corvette-style knock-offs. By mixing and matching parts from various models and years, the owners of these Camaros have given their rides an individual identity.

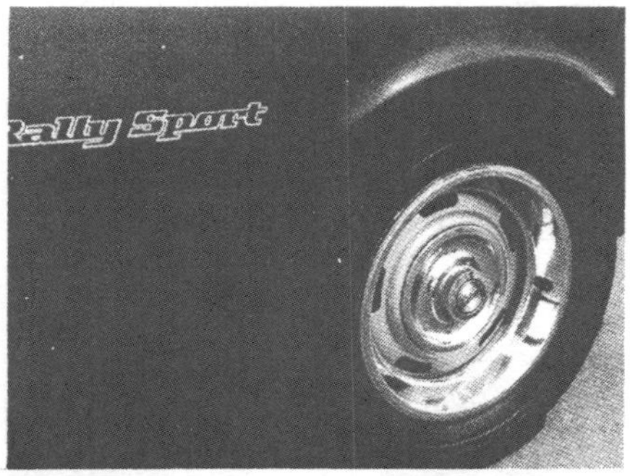

46. This clean '67 Camaro Rally Sport was dechromed to accentuate its smooth black paint. Then the owner added Rally Sport graphics from a '75 Camaro to herald his car's model identification.

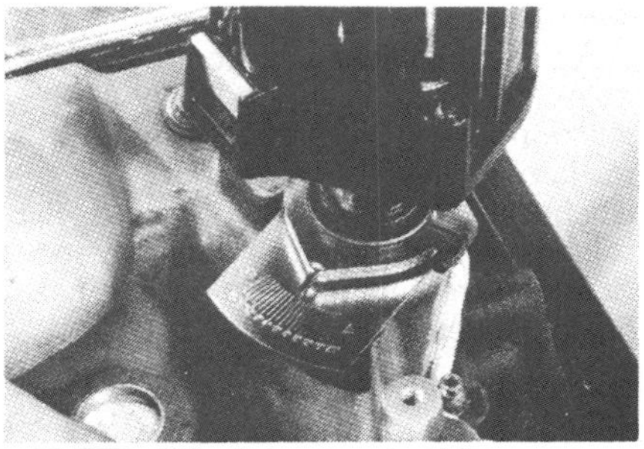

47. Whether your goal is maximum performance or maximum mileage, proper ignition timing is critical. If you enjoy tinkering with your Camaro's spark advance but don't want to hassle with a timing light each time you make an adjustment, consider adding a distributor timing indicator like this neat piece from Moroso Performance. After the degree plate has been attached to the intake manifold and the pointer clamped to your distributor body, accurate changes in spark timing can be made without a strobe light. With a constant reference point, you can easily vary your street machine's spark advance to cure pings and knocks caused by occasional tanks of bad gas.

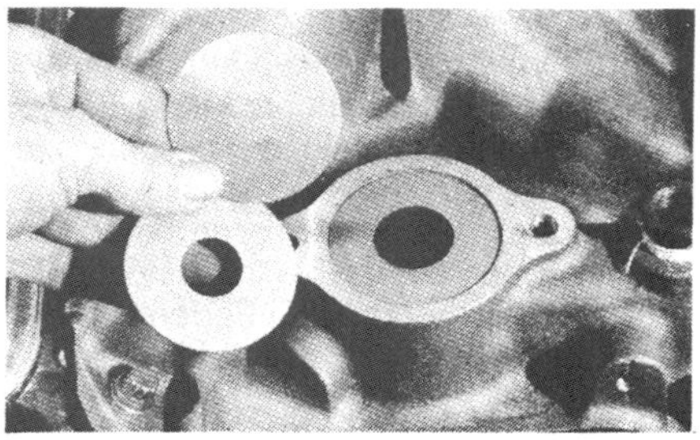

48. The worst thing you can do to cure an overheating Camaro is remove the engine's thermostat. The thermostat body restricts the flow of water through the engine, which allows the coolant to absorb and radiate heat. Since thermostats have an annoying tendency to stick at the most inopportune moments, you may want to remove the spring and plunger from the thermostat for a modified Chevy. Moroso offers an assortment of water-outlet restrictors with various size holes which allow you to tailor the flow of coolant to the engine's cooling requirements.

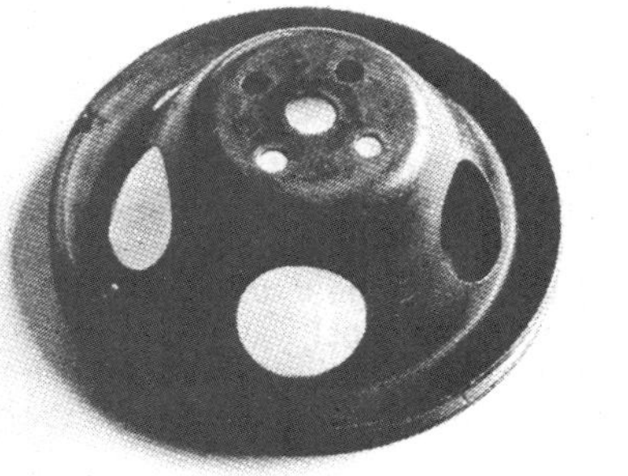

49. Today a Camaro owner can purchase engine accessories inspired by race motors, providing he has an unlimited supply of money. You can also dress up a Camaro engine compartment inexpensively by drilling the various pulleys with a hole saw, as shown. Although the weight savings are negligible, the eye appeal is irresistible. Follow up with a coat of gray or silver paint and you'll have all the elements of a high-tech engine without the high price.

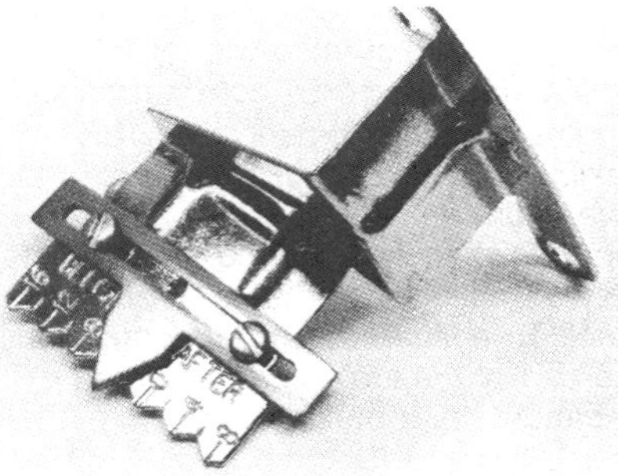

50. The timing mark on a Chevrolet's harmonic balancer

is an all-important reference point for setting spark timing. Unfortunately, very few timing marks precisely indicate Top Dead Center. To further add to the confusion, there are several different timing tabs and balancers in circulation. If the tab and balancer are mismatched, the timing marks can be grossly inaccurate. This problem can be solved by bolting an adjustable TDC indicator from Mr. Gasket or B&B Performance onto the front of your Camaro's V8.

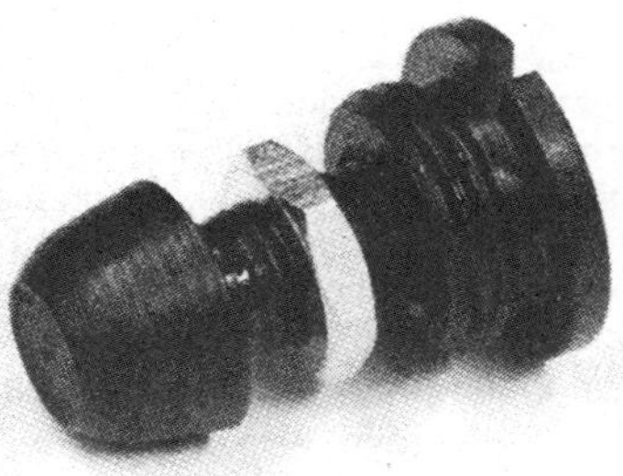

51. Camaros owners who prefer to shift for themselves eventually have to confront the performance deficiencies of a factory-installed diaphragm clutch. The torque output of a healthy small-block or rat motor can overwhelm these standard clutches. Substituting a Borg & Beck or Long-style clutch usually requires adjusting the clutch pivot ball to prevent the throwout fork from hitting the clutch cover. Similarly, most hydroformed scattershields are slightly longer than a stock aluminum bellhousing, which also requires a change in the pivot ball height. The neatest way to put your Camaro's pivot where you want it is to install an adjustable ball from Lakewood or McLeod.

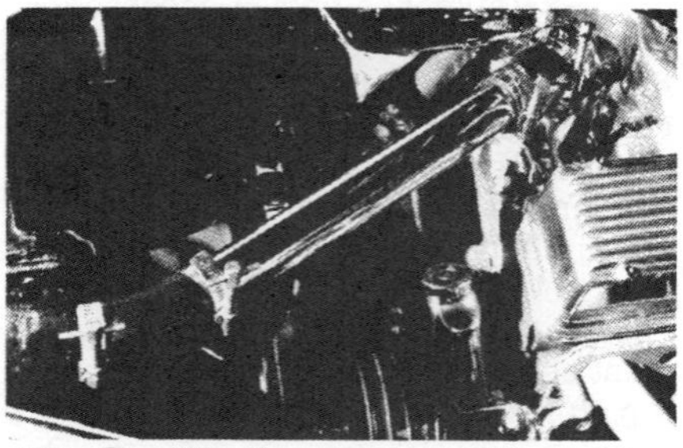

52. What's stronger, lighter and less prone to leakage than radiator hose? Muffler tubing! The next time you're confronted with plumbing your Camaro's motor, consider using lengths of muffler pipe in place of flexible hose. It can also be painted to give your competition-style Camaro a more finished appearance, and it costs only a fraction of the price of braided steel. To ensure leak-proof connections, the end of the tubing can be flared or expanded by a muffler shop.

53. If you're looking for a way to take some weight out of your drag racing Camaro, take a cue from Chevrolet's weight-conscious engineers. The aluminum brake drums found on late-model Malibus will fit early Camaro front spindles with only minor modifications. These lightweight drums have steel liners, so standard brake shoes can be used. The savings is several pounds per side, so the effort and expense is definitely worthwhile if you want to pare down your Camaro.

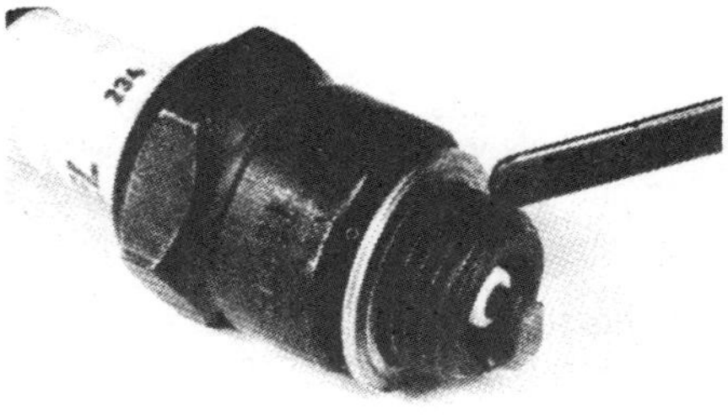

54. When changing the spark plugs in your Camaro, it's always a good practice to lightly lubricate the plug threads with a drop of oil. Fortunately, every Chevy engine has a convenient source of oil—its dipstick! Just touch the side of the dipstick to the plug before threading it into the head.

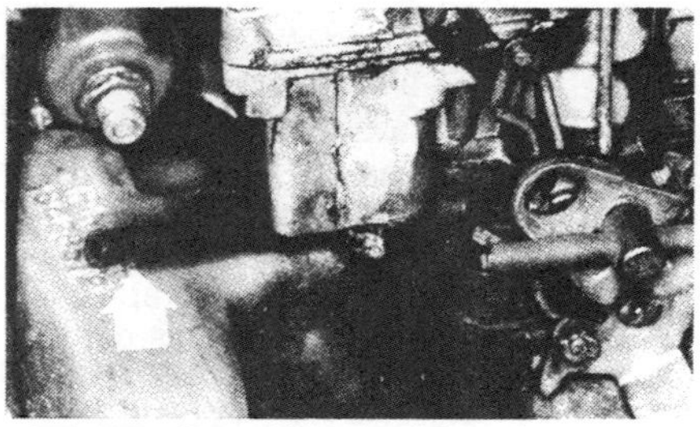

55. Reaching the idle-mixture screws on a Camaro carburetor requires a contortionist. The collection of hardware on top of the intake manifold limits access to the mixture screws on the front of Rochester 2V and Quadrajet carbs. Slip a length of vacuum hose over the end of each screw to make an easy-to-use flexible extension.

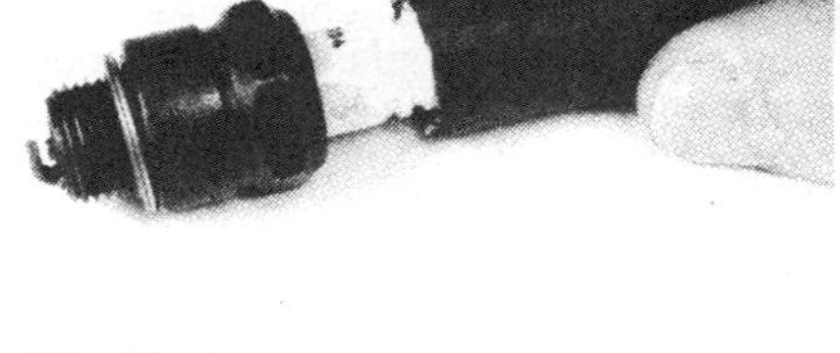

56. Even though spark plug access is generally quite good on Camaros, it's still difficult to work around header tubes and exhaust manifold heat shields when you're changing plugs. A piece of rubber fuel line makes a perfect spark plug installation tool; just slide it over the plug insulator. It's easy to get a grip on the hose when you're tightening or loosening the plugs. Of course, you should finish the job with a plug wrench and socket.

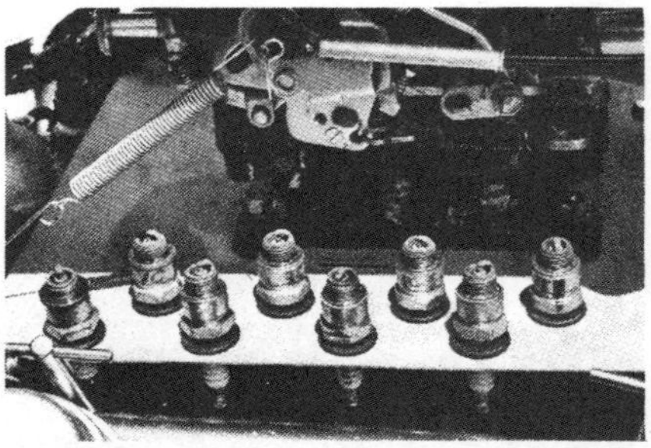

57. The benefits of a carburetor heat shield are well-known. A piece of sheetmetal sandwiched between the carb base and intake manifold will prevent percolation in the float bowls, and also provide a denser air/fuel mixture. But why not have the heat shield perform double duty? This Camaro racer drilled holes in his carburetor heat plate and then inserted rubber grommets. Now he has a convenient place to store warm-up plugs and to take plug readings.

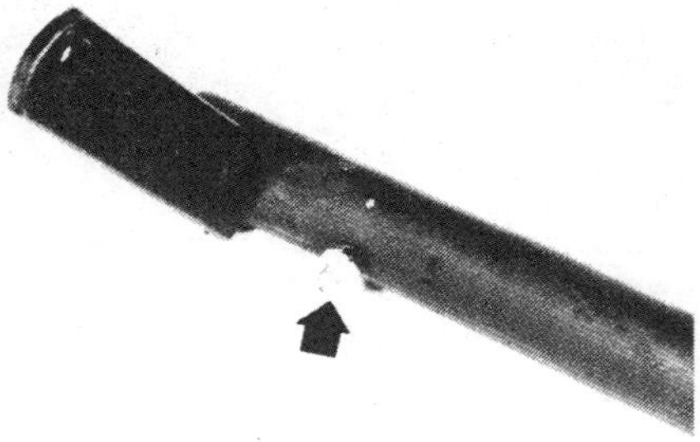

58. A worn clutch cross-shaft or pivot balls can make it difficult to shift a high-performance Camaro. Camaro cross-shafts are commonly drilled for lubrication fittings, but these fittings are almost never installed. All that's required to lubricate the clutch cross-shaft is to screw in a Zerk fitting, and then pump in some high-pressure chassis grease.

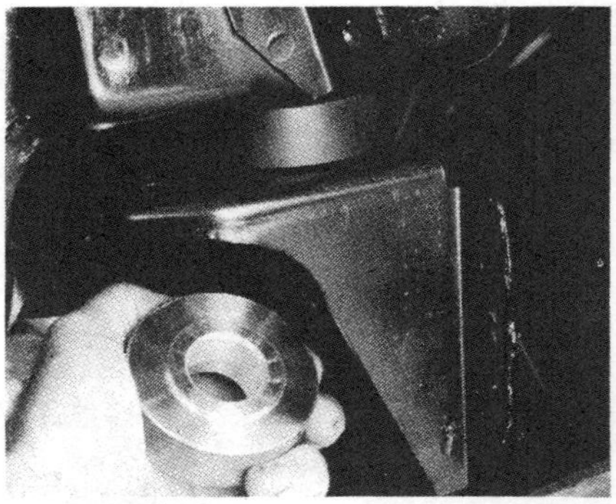

59. All Camaros built before 1982 have two distinct halves: a unit-body passenger compartment and a stamped-steel front subframe. The only connection between these parts is six rubber subframe mounts. As you might expect, a Camaro that's used for racing, autocrossing or simply hard driving can exhibit a startling amount of subframe flex. Solid aluminum subframe mounts are available to eliminate this deflection. You can expect a little more noise and harshness without the isolation offered by rubber body mounts, but the improvement in chassis rigidity is worthwhile for a performance-minded Camaro owner.

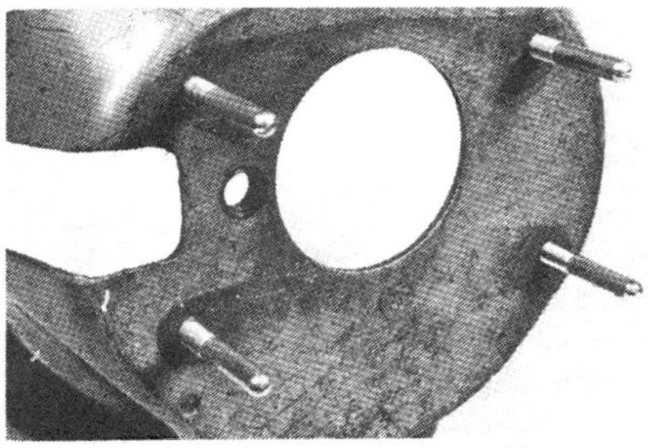

60. Whether you're installing a four-speed in your Camaro with a transmission jack or while lying flat on your back, a set of alignment studs will make the job easier. These round-headed studs from Mr. Gasket are screwed into the bellhousing to guide the transmission into place without damaging the clutch disc. Once the transmission is flush against the bellhousing, the studs can be removed one at a time with a screwdriver and replaced with hex-head bolts.

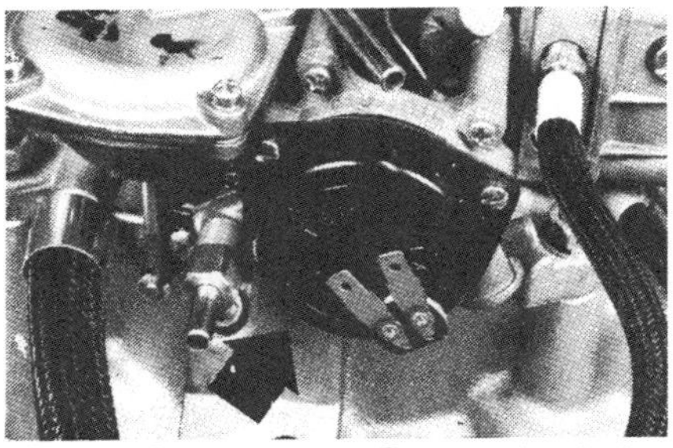

61. Coaxing your Camaro to start on a cold morning can be difficult, especially if you have already bolted on a high-performance carburetor and blocked off the intake manifold heat risers in search of more horsepower. An electric choke conversion is just what you need for chilly mornings. Holley offers easy-to-install electric choke conversions for all their popular performance carbs.

62. Camaro radiators are mounted with special shock-absorbing fasteners called "well nuts." These rubber bushings help isolate the radiator core from body and chassis vibrations, which prevents cracking the soldered radiator bracket joints. As these well nuts grow older, the rubber deteriorates; removing a used radiator usually destroys them. Few auto parts outlets carry these special fasteners, so you will probably have to find replacements at a dealership parts counter. Resist the temptation to bolt the radiator up solid; you'll soon be fixing broken brackets.

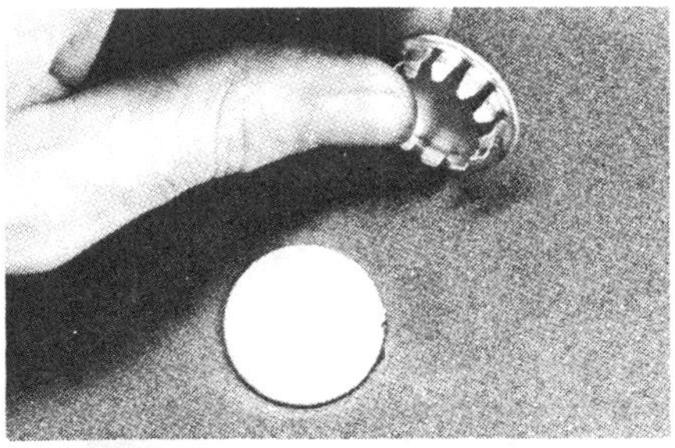

63. If you have removed the windshield wipers or radio antenna on your Camaro racer, you should finish the job by plugging the ugly holes left behind. Spring plugs used to cover unwanted openings in junction boxes are ideal; look for them in the electrical section of a well-stocked hardware store.

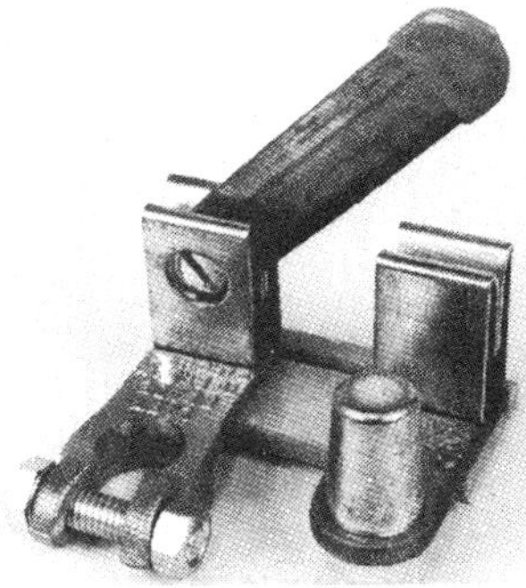

64. There are advantages to relocating your Camaro's battery to the trunk beyond the improvement in weight distribution and traction. If you install a battery-disconnect switch in the main cable, you'll have a very effective theft deterrent. Hot wiring or jumping the starter solenoid won't work as long as the current is disconnected. A blade-type switch like the one shown here will have enough capacity to handle the amperage requirements of a high-torque starter.

65. As the power output of high-tech ignition systems has climbed, the phenomenon of crossfire has become more prevalent. To realize the performance potential of these high-voltage systems, your Camaro's spark plug leads should be well separated from each other and from metal surfaces in the engine compartment. Although there are many wire separators on the market, a strip of stiff rubber weatherseal can be turned into an assortment of made-to-order looms. Slit the weatherstrip as illustrated to maintain at least a full inch of air between adjacent plug leads.

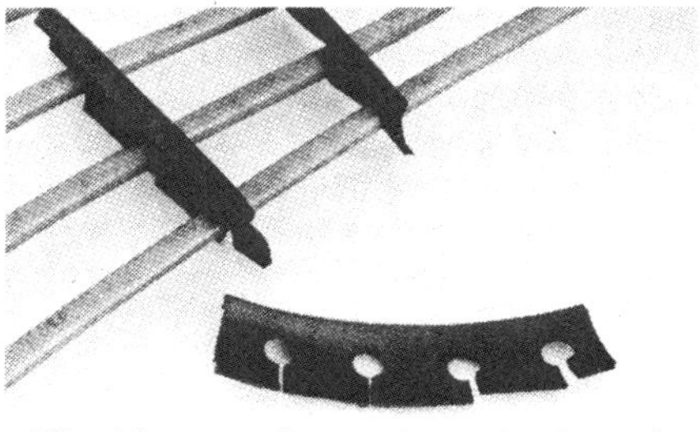

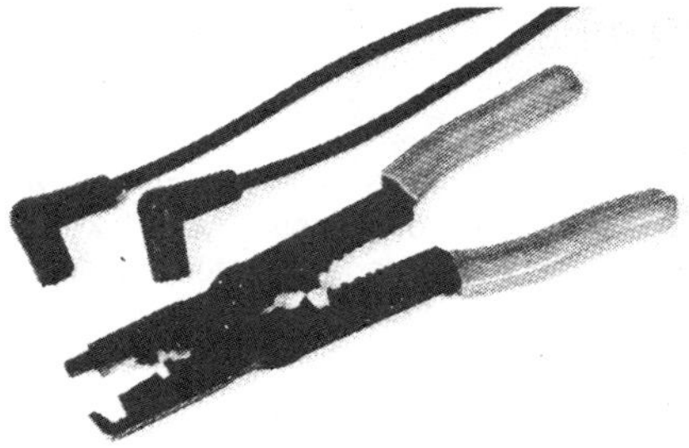

66. If you've ever tried to make up a set of custom spark plug wires for your Camaro with a pair of ordinary pliers, you were probably displeased with the results. Crimping the terminals takes the right tool, such as this all-in-one plier offered by Autotronic Controls and Moroso Performance. It works perfectly on both straight and 90-degree terminals, attaching the terminals neatly and securely.

67. Now that you've narrowed your Camaro's rear subframe and installed wall-to-wall rear rubber, what are you going to do for a gas tank? Fuel cells are expensive, and narrowing the stock tank is definitely no fun. Fear not—the answer is as close as the nearest derelict Vega (and there certainly isn't a shortage of junked Vegas). Once you fabricate a new pickup and toss the original in-the-tank fuel pump, the Vega tank will slip right between your Camaro's new framerails.

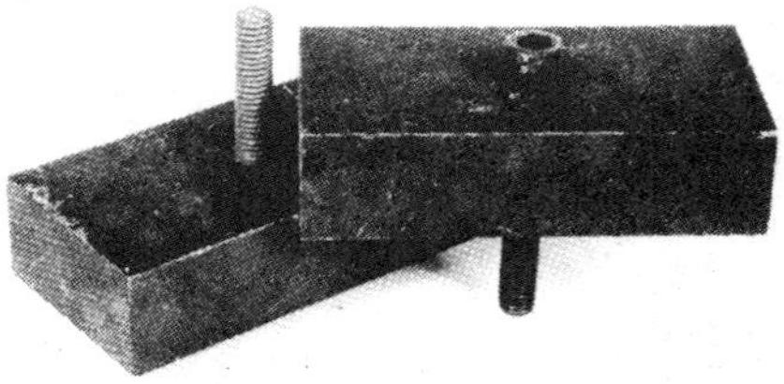

68. Thanks to the influence of both Pro Stock and road racing competition, the low-down look is much in demand for street Camaros. The easy way to bring your leaf spring-equipped Camaro down to earth is with lowering blocks. Combining perches designed for multi-leaf springs with monoleaf springs will restore the proper rear end height if you have added oversize rear tires. An aluminum spacer sandwiched between the spring and its perch makes up the difference in spring thickness.

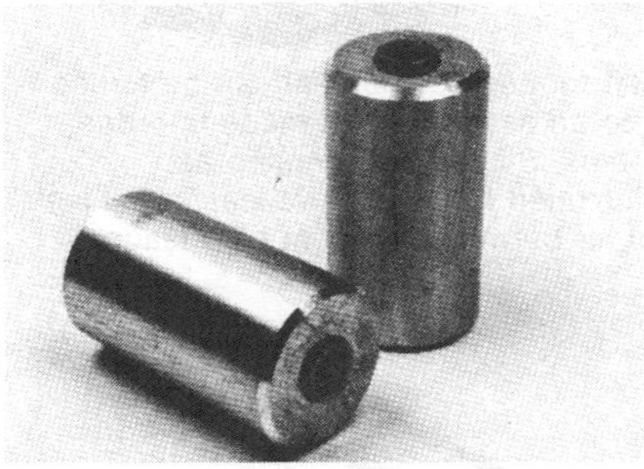

69. Part of the secret of making a Camaro chassis "work" with bolt-on traction bars is transferring the rearend torque into the body. Aluminum spring eye bushings eliminate the deflection of the stock soft rubber bushings installed in the front of Camaro leaf springs. The aluminum does not compress under a load, so the full force of the traction bars is harnessed to raise the body, which in turn presses the rear tires firmly against the pavement. Ride harshness and noise is increased with these rigid bushings, but many Camaro owners would consider that a small sacrifice for the traction improvment.

70. You don't need a polished aluminum motor plate to keep your Camaro's engine in place. In fact, for a street machine, a simple torque strap does a better job. You can retain the original rubber-cushioned motor mounts, which isolate the chassis for engine vibration. A simple limiter fashioned from bar stock, a turn buckle or a length of chain will prevent the engine torque from overwhelming the motor mounts. As the engine twists, it can bind the shifter, throttle and clutch linkages. Attach the torque strap from the driver's side cylinder head to the subframe.

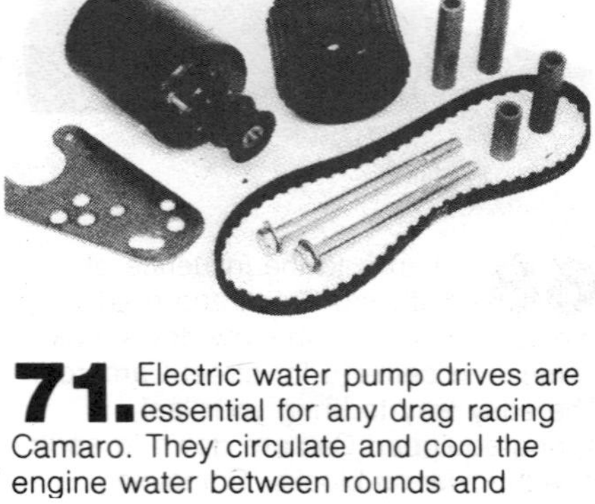

71. Electric water pump drives are essential for any drag racing Camaro. They circulate and cool the engine water between rounds and eliminate the power losses caused by water pump drag. These pump drives are not designed to keep up with the cooling requirements of a hot Camaro in rush-hour traffic, so using one on your commuter car is asking for a meltdown. When the tension of the toothed belt is adjusted properly, you should be able to twist it 90 degrees at a point halfway between the two pulleys.

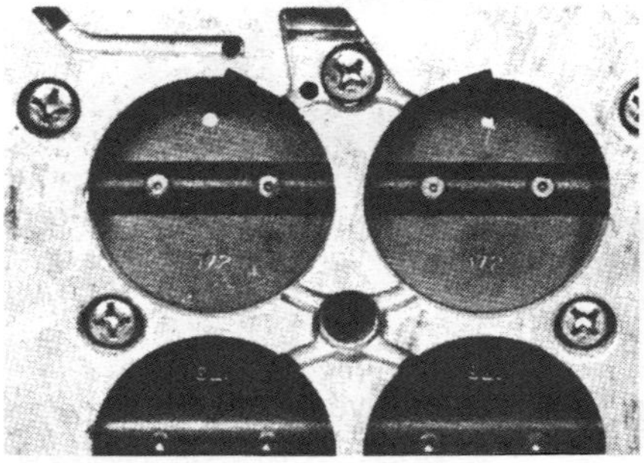

72. Camshaft profiles that used to be reserved for the racetrack are finding their way into Camaro street machines. A rump-rump idle may be impressive down at the drive-in, but it can hopelessly confuse the carburetor. If the engine will not idle properly with a long-duration cam profile, check the position of the carb's primary throttle blades; they should be over the idle mixture slots. If the blades have to be opened past the slots to produce a satisfactory idle, that's an indication that the engine needs more low-speed airflow. In these circumstances, a pair of .125-inch-diameter holes drilled in the primary throttle blades on the opposite side of the idle slots will clean up your carburetor's idle characteristics.

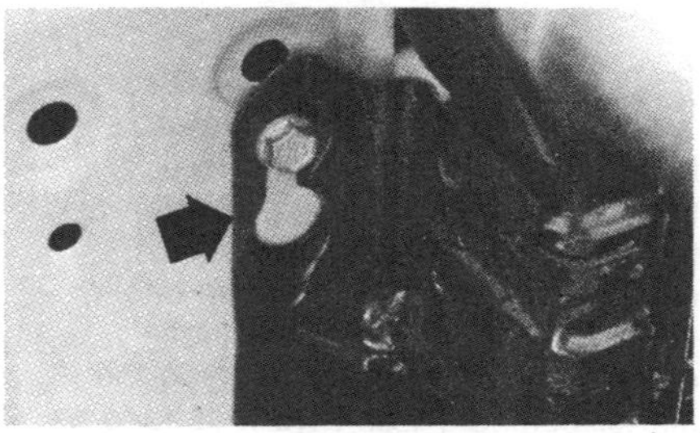

73. A Camaro radiator is in the way for almost any serious wrenching, from changing a camshaft to pulling the engine. You can make a quick-release release radiator by slotting its brackets as shown. Then all that's required to remove the radiator is to loosen its fasteners, pull it upward and slide it off the bolts; reinstalling the radiator is just as quick.

74. Camaro restorers have a trick for getting the right silver-gray finish on metal parts. First, paint the part with Rustoleum bright aluminum, then cover it with several coats of Krylon clear. The clear makes clean-ups easy, while providing an attractive metallic sheen. Next time, try this treatment for your Camaro's Rally wheels or intake manifold.

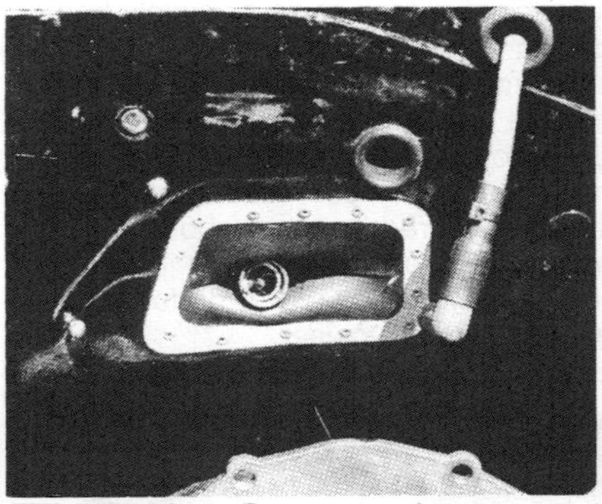

75. Many Camaro performance enthusiasts feel a mechanical tachometer is essential for reading exact engine rpm. Routing the tachometer cable to the passenger compartment can be a problem, however. A hole in the firewall that's large enough to allow the tach cable to move to accommodate changes in ignition timing is also large enough to allow noise and fumes into the passenger compartment as well. A shifter boot or rubber gasket around the cable opening will keep these nasties under the hood where they belong.

76. Keeping a Camaro's cast-iron exhaust manifolds looking good is a constant battle. It seems that rust always triumphs over header paint. Here's a tip for manifold preservation that some Camaro owners swear by: Paint the clean, sandblasted manifolds with graphite spray. This produces a handsome, dark gray finish that bakes on as the manifolds are heated by the exhaust gases.

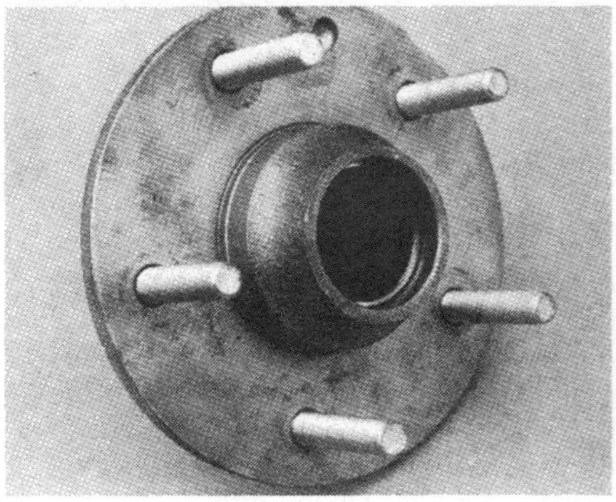

77. Stamped and spun-aluminum wheels are extremely popular with the Camaro street machine set because of their strong connections with Pro Stock. If you want to run your street legal rims at the weekend bracket races, however, you'll have to upgrade the front and rear wheel studs. The tech inspectors will want to see lugs that extend at least ½ inch past the wheels'

thick center hub. Fortunately, the aftermarket has come through with extra-long studs that press right into a Camaro's axle flanges and wheelhubs. Remember: Never use wheel spacers with aftermarket studs.

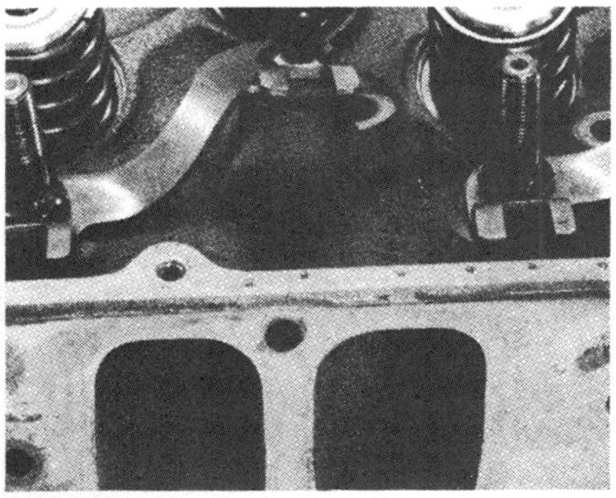

78. Is your big-block Camaro plagued with oil leaks? Does an oil slick on the garage floor greet you every morning? A few minutes' work with a centerpunch will eliminate one escape route for the engine lubricant. There's a long span between valve cover bolts on a big-block Chevy. If you dimple the gasket rail on the cylinder head with a punch, the raised metal will hold the gasket in place as you tighten down the valve covers.

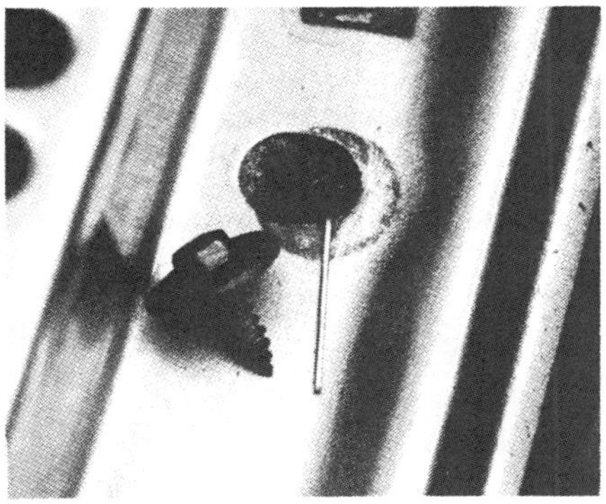

79. Every Camaro has at least one hole for a sheetmetal screw that is stripped after too many tightenings. Holes for armrests and interior trim are the most likely candidates. For a quick yet permanent fix, insert half a paper clip in the hole before starting the screw. The wire thickness will effectively reduce the diameter of the hole so that the screw threads can get a bite on the sheetmetal.

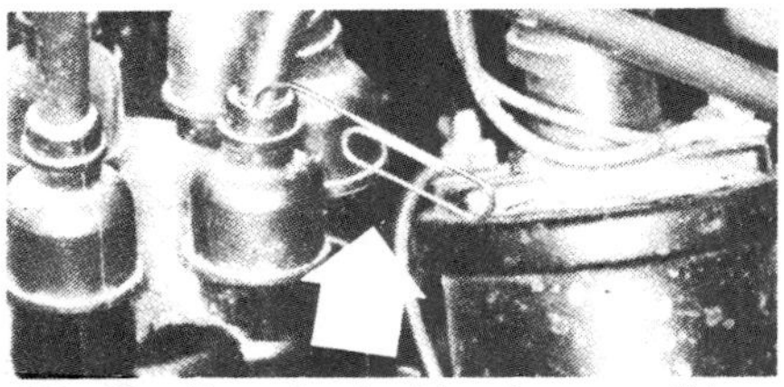

80. If you're looking for something useful to do with your leftover paper clips, stick one in your Camaro's distributor cap. If you slide it between the plug wire and boot for the number one cylinder, you'll have a convenient place to attach the high-voltage lead for your timing light.

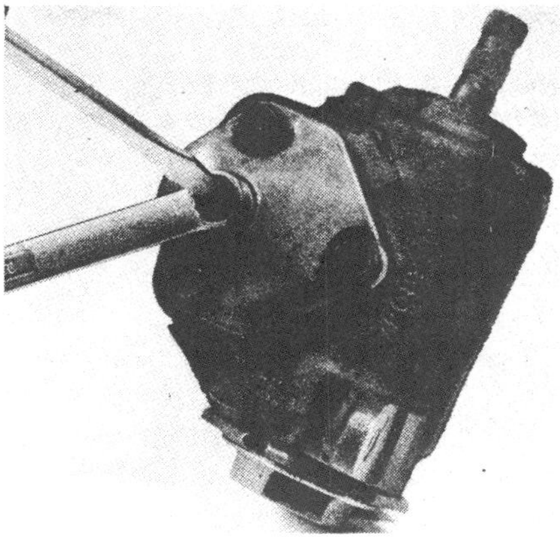

81. If your Camaro's handling is not as precise as it once was, the problem may be excessive lash in the steering box. Before investing time and money in rebuilding the entire front end, make sure the steering-lash adjustment is correct. The lash must be adjusted with the wheels pointed straight ahead and the gears at their high spot. Loosen the locknut on top of the steering box, tighten down the adjusting screw until it stops, then back it off a quarter turn. Cinch down the locknut and test the steering. If the steering becomes tight or binds as you turn the wheel, the gears were not at their high point when you adjusted the lash. The goal is to eliminate play in the steering box while still allowing the wheel to turn freely.

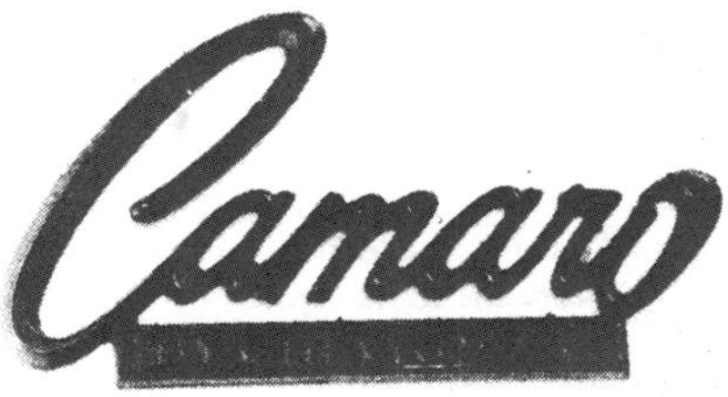

82. You can eliminate ugly pitted chrome trim and give your Camaro a European look by blacking out its medallions and emblems. Sandblast or glass bead the trim so paint will adhere, and use a good quality aerosol (Rustoleum Satin Black works well). Another alternative is to paint the trim a contrasting color, such as brown emblems for a beige or gold body color. Paint is definitely cheaper than buying new emblems or rechroming the old ones.

83. The hours spent working in a Camaro engine compartment will be more enjoyable if you install an

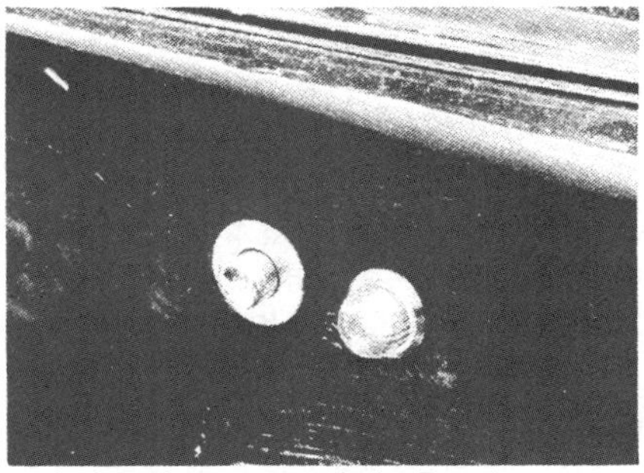

underhood starter switch. Routine chores such as setting valves and checking spark timing are much simpler if you don't have to use the ignition key to bump the starter motor. If you add a light alongside to indicate when the current is switched on in the ignition circuit, you won't be surprised by an engine that starts when you least expect it.

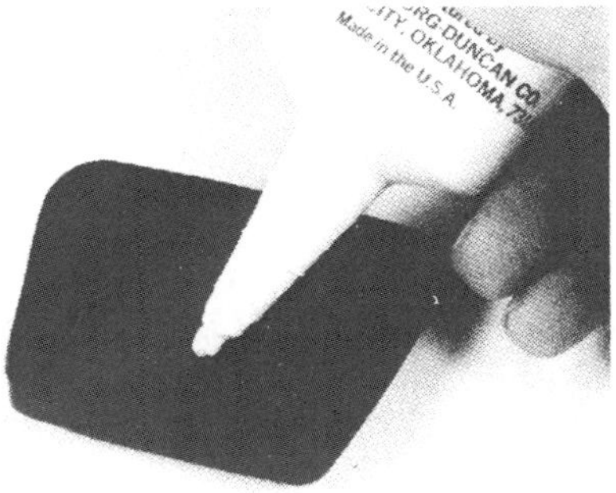

84. Worn pedal pads are an eyesore in any Camaro's passenger compartment. Worse yet, they can clip off the metal backing plates at the instant when you need to climb on the brakes or stab the clutch pedal. A generous dollop of silicone adhesive applied to the back of the rubber pads will help hold them in position even if their retaining grooves on the back side have cracked or split.

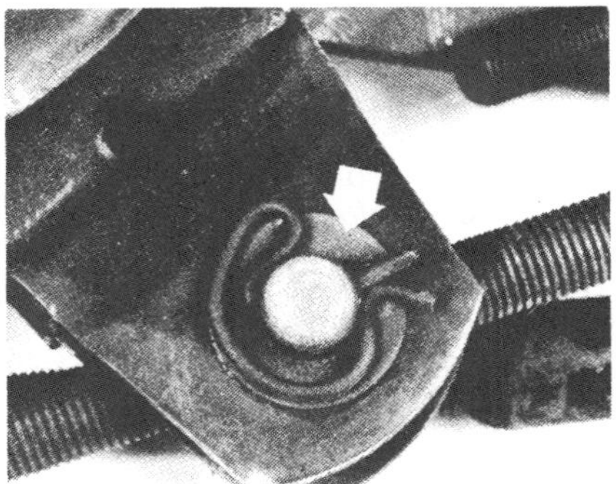

85. A sloppy four-speed shifter can endanger your reputation as a quick-shift artist. You can tighten up your Camaro's gear changes by adding shims between the transmission shifter arms and the shifter rods as shown. Select a clutch shim or thin washer that just fits between the shifter arm and the wire clip as shown.

86. No matter how careful you may be when overhauling your Camaro's engine, transmission or rearend, there are bound to be some tiny metal fragments that elude your cleaning efforts. If these chips are allowed to circulate, they can cause serious damage. Magnetic drain plugs installed in the oil pan, transmission filler plug and differential cover will snag these fragments before they can cause major mayhem. By inspecting the bits the magnets have collected at every oil change, you can detect developing problems before they turn into disasters.

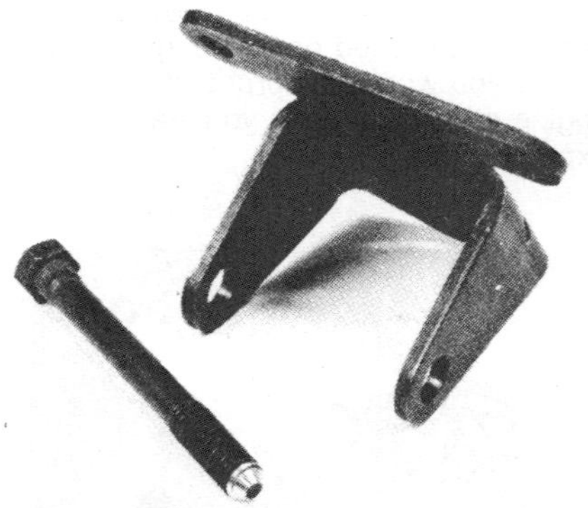

87. Inserting the bolts through a Camaro's engine mounts can be a real test of skill. Since it's virtually impossible to see what you're doing, you have to rely on feel to guide the bolts through their holes. You can improve your chances of getting everything aligned by tapering the ends of the bolts. A few moments with a bench grinder is all that is required; make sure that a nut will still start easily on the threads after you have finished grinding.

88. Now that many Camaro owners are on a restoration kick, factory wheel covers are enjoying the same popularity that was once reserved for mag wheels. Old hubcaps have their shortcomings, however. If your Rally wheel cap won't stay in place, bend the lip on the back so that it gets a better grip on the wheel, or build up the ridge on the wheel with several pieces of duct tape. Old Super Sport wheel covers that have developed a bad case of rattles can be quieted down by applying beads of clear silicone adhesive to the back side of the spinners and center medallion.

89. If your high-horsepower Camaro has a habit of leaving its driveshaft on the starting line every time you serve up a generous portion of horsepower, then it's probably time for a heavy-duty yoke. Chevy part No. 7815506 was originally used on big-block Chevelles; it fits both Powerglides and early Muncie four-speed transmissions and is much beefier than even aftermarket yokes.

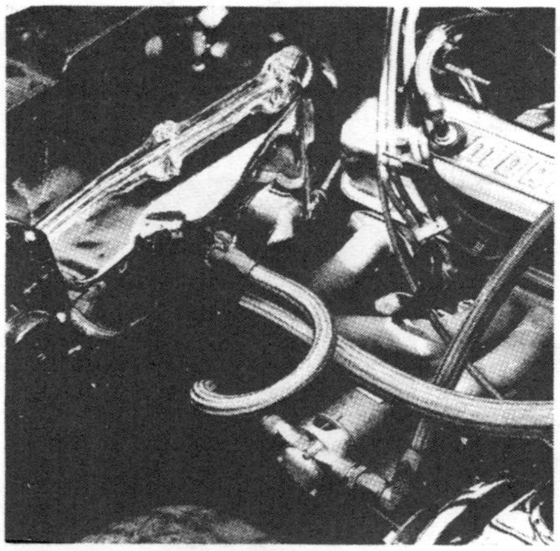

90. So you've swapped a big-block into your early-model street Camaro, and now you're wondering how to hook up the heater? The solution is to use a big-block heater core. The water outlets are relocated to the center of the heater housing, providing clearance for the passenger-side valve cover. This is a worthwhile modification even for small-block Camaros, since it relieves the congestion around the heater hoses that is frequently encountered with custom valve covers.

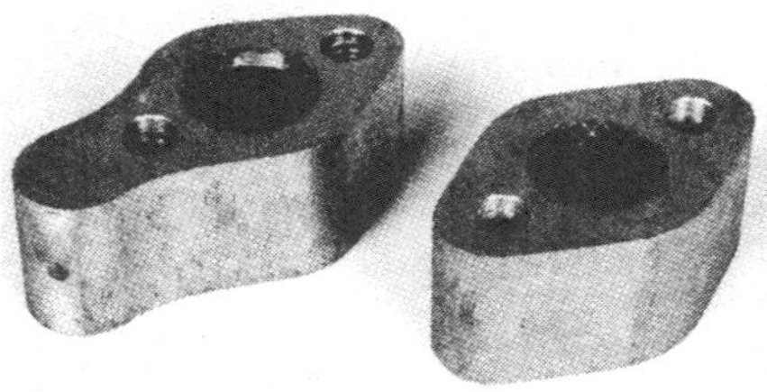

91. Aluminum water pumps are lighter than their cast-iron counterparts, and have larger bearings for improved durability. Unfortunately, lightweight pumps are only available in the short version used on '67-'68 Camaros. Don't despair, though, if you own a late-model Camaro. A pair of water pump spacers like these will allow you to use an alloy pump and still retain the stock pulleys and belts.

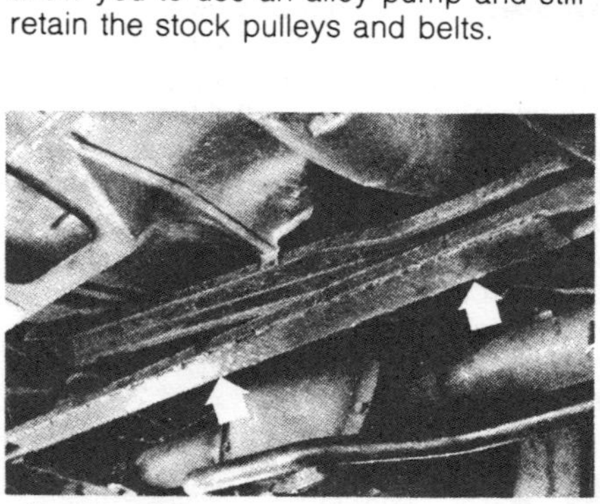

92. When the third generation of Camaros debuted in 1982, it introduced coil springs and a Panhard rod in place of the familiar rear leaf springs. Since the Panhard rod is responsible for keeping the rearend centered under the car, it has a considerable impact on handling. The factory-installed rods are a thin U-shaped stamping. If you don't want to pop for one of the heavy-duty tubular Panhard rods offered by aftermarket suspension specialists, you might consider boxing the stock rod by welding metal braces across the open part of the "U" as shown. The result? One stiff Panhard bar!

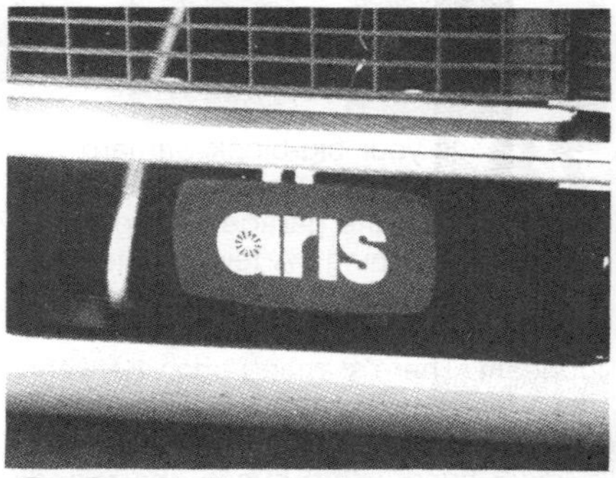

93. Auxiliary lights can help you find your way on dark country roads, but finding an inconspicuous place to mount them can be difficult. The owner of this cafe racer Camaro solved the problem by placing a pair of driving lights in the air intake beneath the grille. There's still adequate air for the radiator, and the lenses aren't as vulnerable to stone damage since they're tucked inside the bodywork.

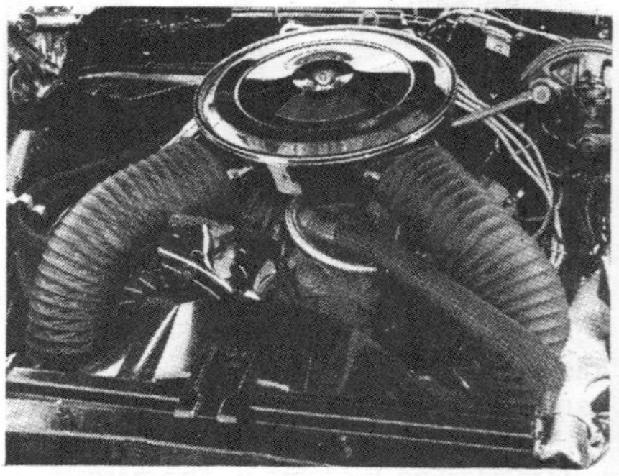

94. Here's a simple but effective way to supply cold air to a Camaro carburetor. Two lengths of flexible heater duct and a dual-inlet air cleaner are the only hardware required. The time and money you save on this easy ram air system can be put to good use detailing the engine compartment to rival this clean Camaro.

95. Chevrolet engines have always had a reputation for high-rpm reliability, but the same cannot be said for their fan belts. When the Chevy men introduced the short-stroke Z28 engine, they made sure the belts

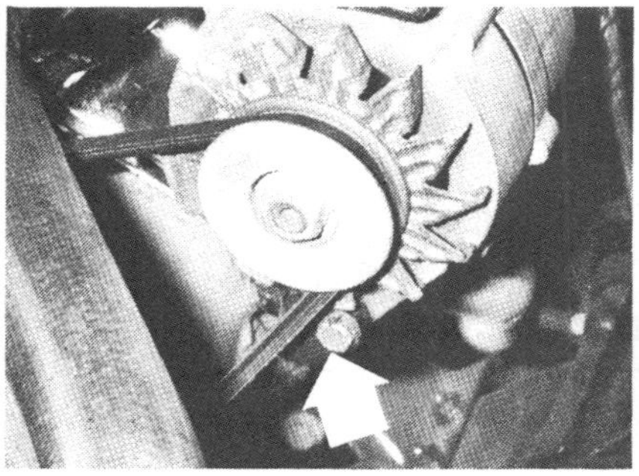

would stay in place by including deep-groove pulleys in the engine package. A large-diameter alternator pulley slows down the alternator armature to help keep the windings intact. Any quick-revving Chevy will benefit from the same equipment.

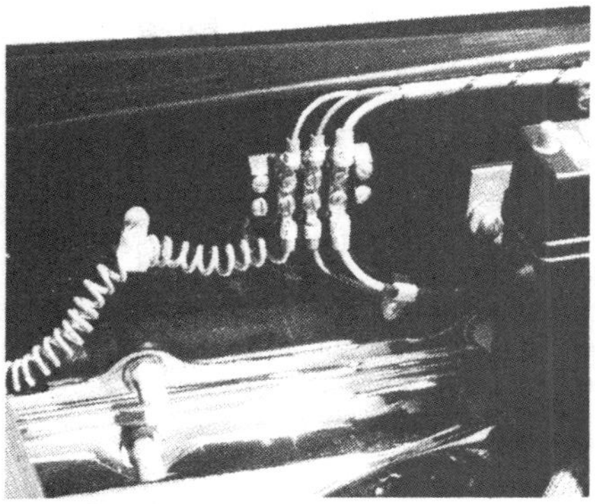

96. Usually, wiring is the least attractive element under a Camaro's hood. This engine compartment turns a liability into an asset, however, with careful attention to detail. A terminal block mounted on the firewall makes wiring changes easy and lends a race car look to the engine compartment. Thoughtfully routed wires, clamped in place and coiled, contribute to the sanitary appearance.

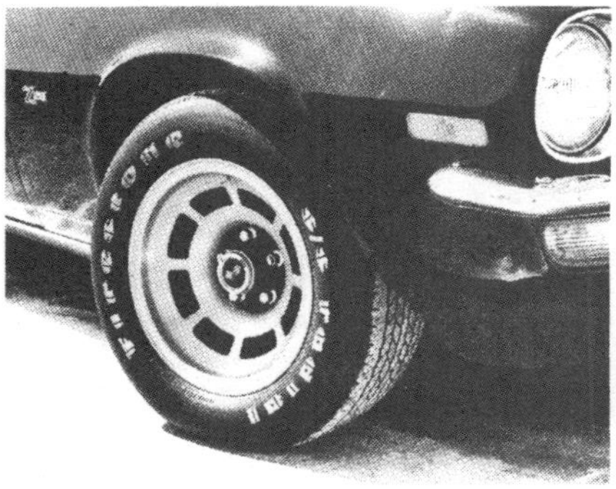

97. This Chevy enthusiast always dreamed of owning a Corvette, but practicality demanded the steel bodywork and lower insurance rates that come with owning a Camaro. Installing 15-inch slotted aluminum Corvette wheels on a classic Camaro brings together Chevrolet's two premier performance machines. This is an excellent example of customizing with factory parts.

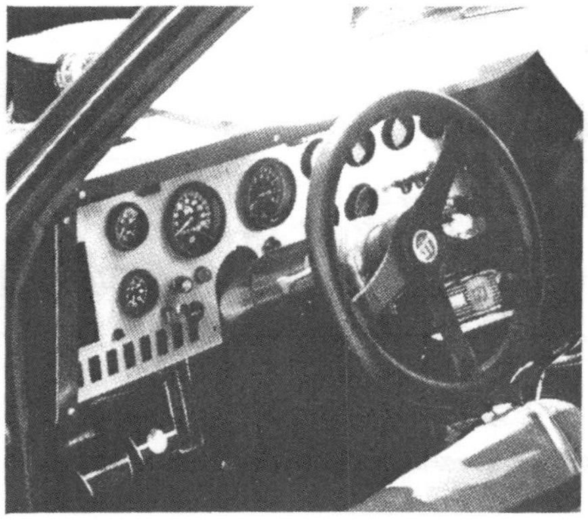

98. Even the factory didn't know where to put accessory gauges in an early-model Camaro. Chevy's solution was to mount them on the console, which is hardly in the driver's line of sight. A more radical answer is demonstrated by this well-instrumented Camaro dashboard. There's also considerable fabrication and rewiring required for such a conversion, but the result is at least as impressive as a jet fighter cockpit.

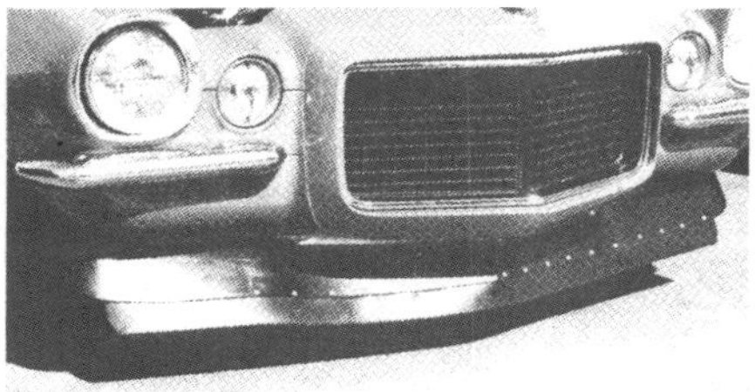

99. Front spoilers always require a compromise between ground clearance and aerodynamics. A rigid pavement-scraping chin spoiler surely won't survive very long on the streets if it encounters steep driveway entrances and parking lot stops. This Camaro has a flexible rubber skirt which extends the factory spoiler by two inches. The spoiler extension effectively directs air around the front end, yet it bounces back from confrontations with curbs.

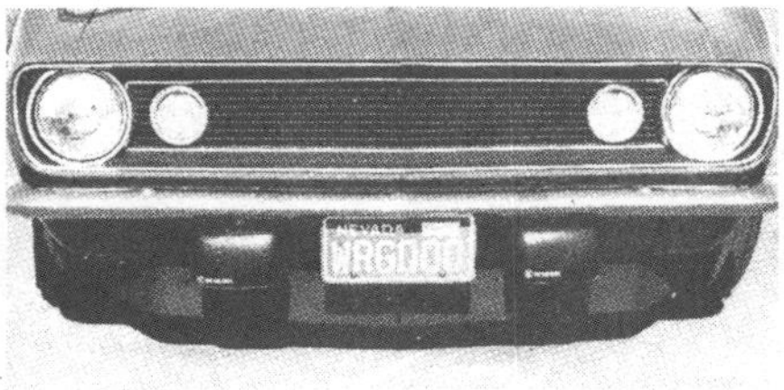

100. This early Camaro presents yet another aerodynamic alternative. This swept-back spoiler was originally installed on a new Chevy pickup; sectioning several inches out of its center tailored it to the Camaro's width. The owner reports that the truck spoiler lowered his engine's operating temperature, and successfully withstood the attack of a killer rabbit on the Nevada highways.

CALIFORNIA: THE LAND OF SURF, SUN AND PERSONALIZED CAMARO LICENSE PLATES

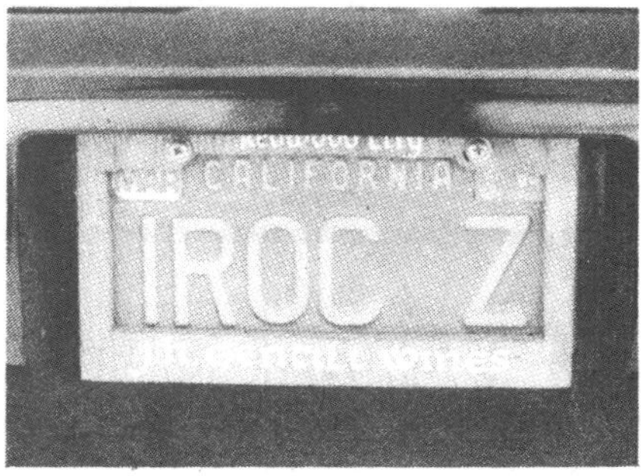

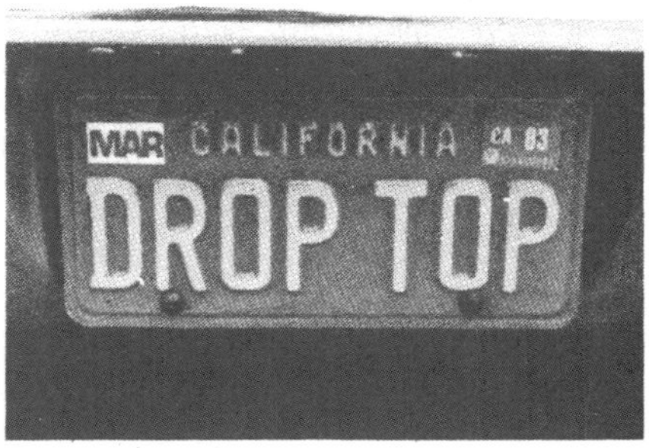

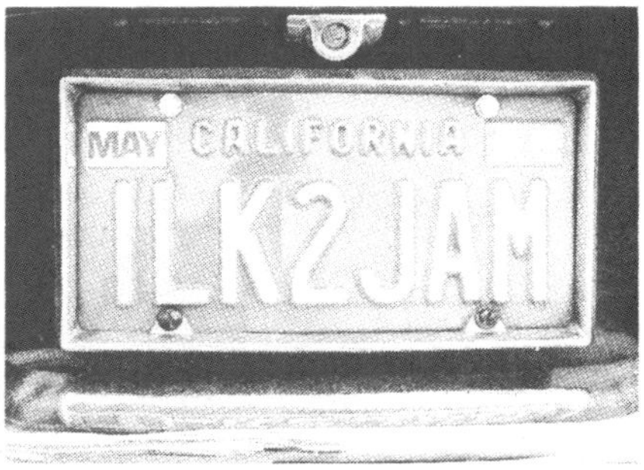

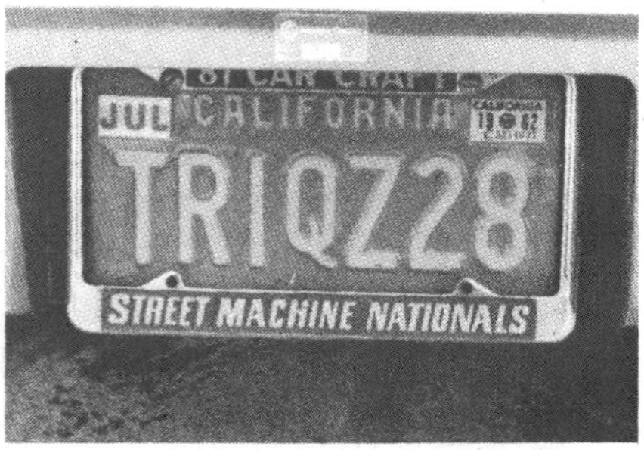

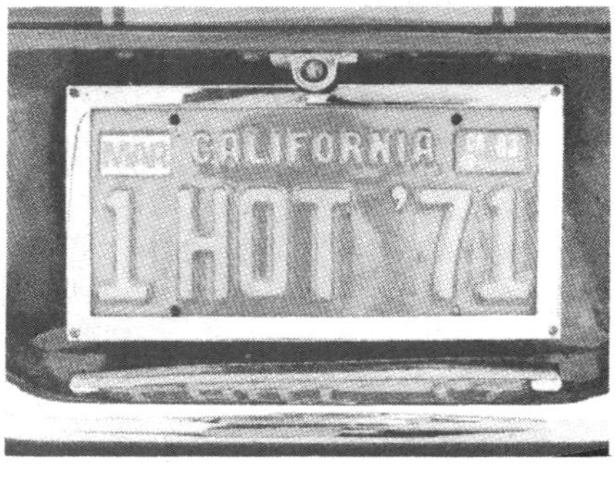

CAMARO FRONT END ALIGN

In the previous chapter some of the symptoms of poor front-end-alignment were discussed. Your front tires are perhaps the best indicators of the basic condition of your suspension components. Their wear pattern can indicate worn shock absorbers, improper alignment, even poor *driving* habits. The first two factors would also make themselves evident in the way the car *felt* on the road. Generally, alignment isn't a delicate arrangement and will last for many thousands of miles under normal conditions, but as suspension components wear, the precise alignment is bound to change, and such seemingly minor incidents as hitting a rut at moderate speed or banging the tires up against a curb when parking can set up enough force to change your alignment. It's best to have your alignment checked about every 20,000 miles, and it's also important to check it whenever you've installed a new set of tires, to protect your considerable investment in rubber.

There are basically three elements of suspension geometry that are adjustable to align the front wheels: caster, camber and toe-in. Caster is the rearward or forward inclination of the spindles, as measured by drawing a line through the center of the upper and lower ball joints to the ground. When looking at a dead side view of the suspension, a rearward tilt of this line is considered positive caster, and forward is negative caster. Most Chevrolets have a slight positive caster, as this is helpful in making the wheels track straight on the road (just think of the casters on shopping cart wheels).

The camber dimension is also concerned with the tilt of this ball-joint-centerline, but in the side-to-side plane rather than fore-and-aft. Looking at the car from a dead frontal view positive camber will make the tops of the tires further out than the bottom, and negative camber will make them appear to "lean in." Both of these suspension angles are measured in a very small amount of degrees though, so a car needs to be on an alignment rack with the proper

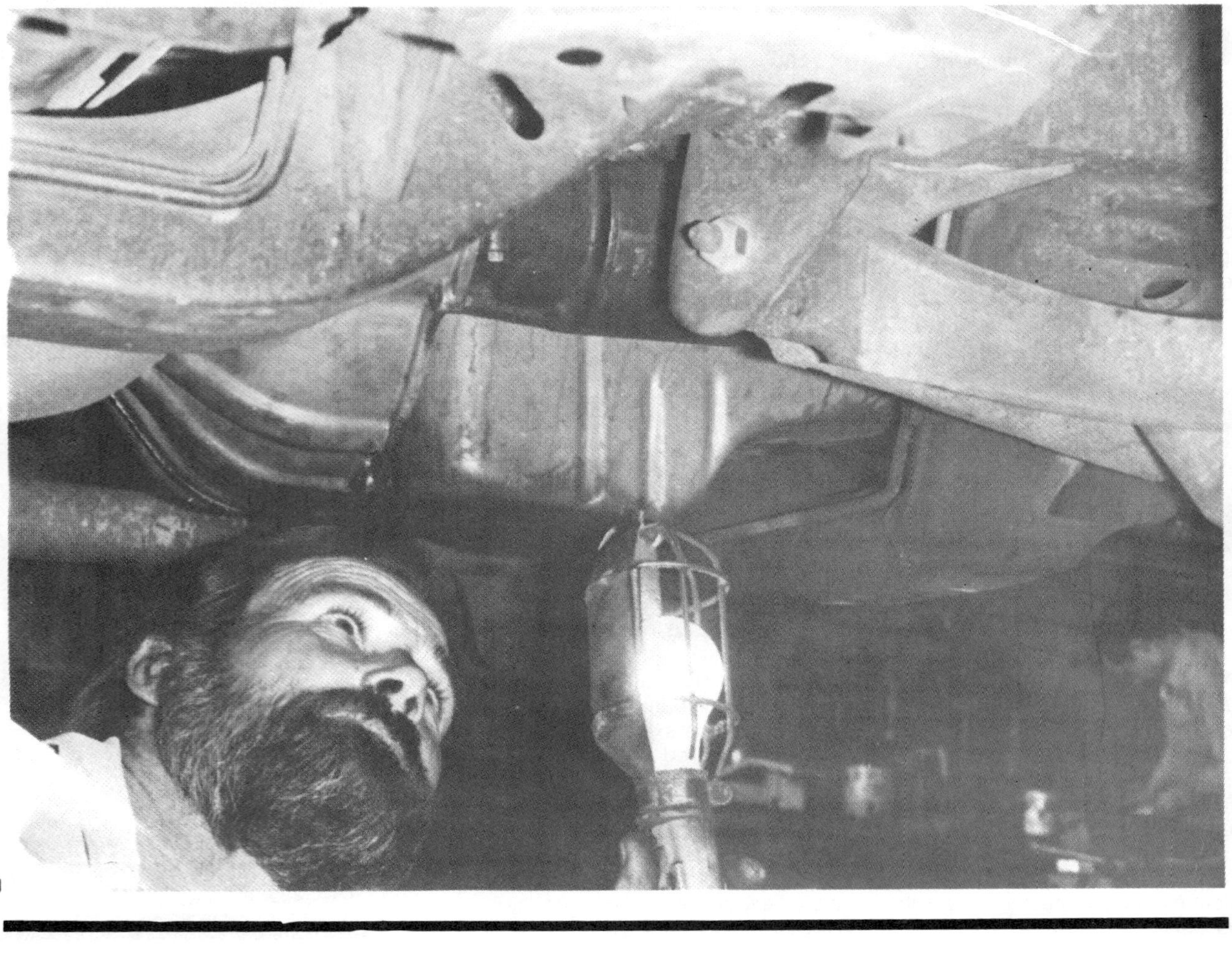

1

tools hooked to the wheels before the true status of wheel alignment can be determined.

The last factor in alignment is toe-in and toe-out. Looking at the front suspension from another plane, directly above it and looking *down*, the two tires may be either parallel, closer at the front or closer at the rear. When the fronts of the two tires are closer together than the rears of the tires, this is considered toe-*in*, and most cars have a slight amount of toe-in to keep the wheels tracking straight and to afford the proper steering geometry on turns.

Obviously, the equipment needed to precisely check out front-end alignment is beyond the home mechanic's budget, and you'll have to take your Chevrolet to a professional, but even if you can't do the job yourself, you ought to know what's involved in setting all of these factors correctly on your front end. Toe-in and toe-out are probably the easiest specs to set on your front end. All it takes is suitable measuring equipment and a few adjustments to the tie rod

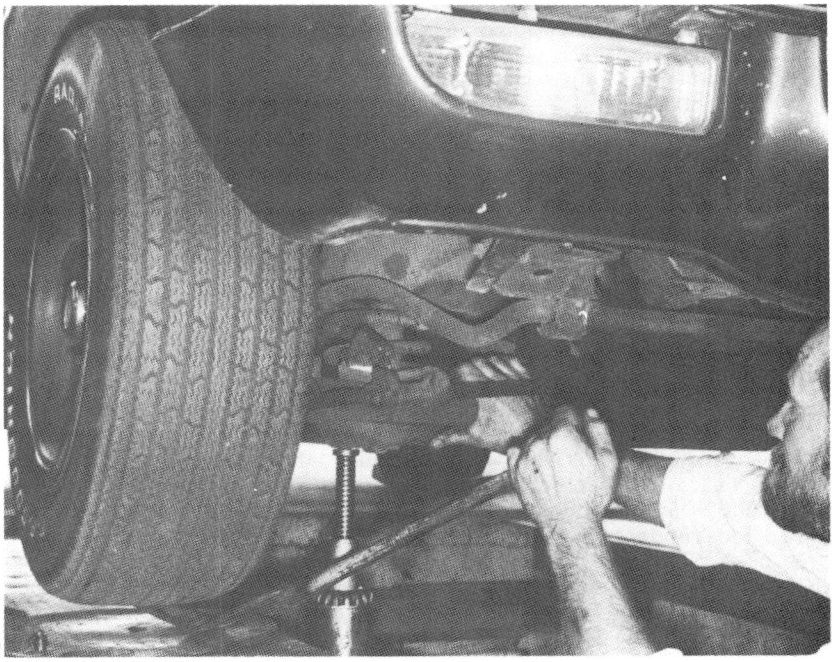
2

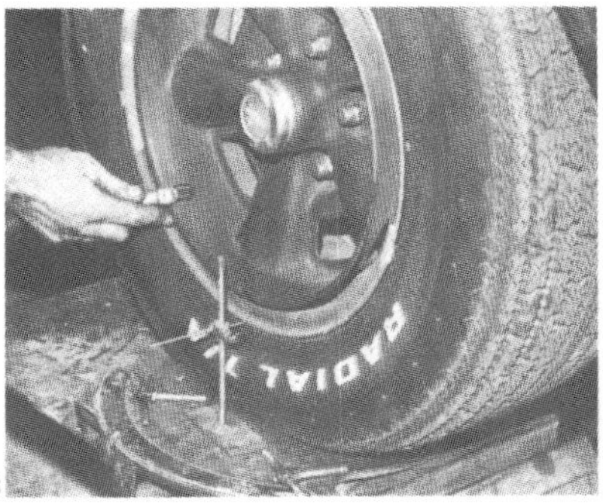
3

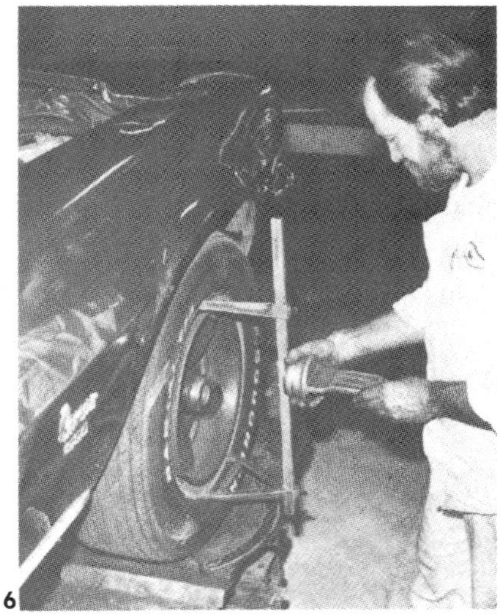
6

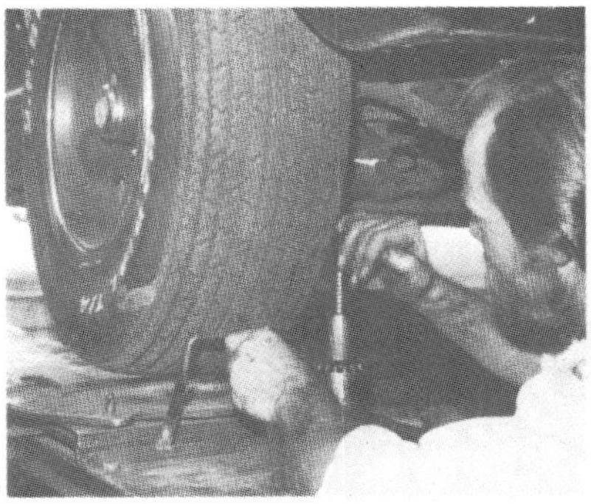
4

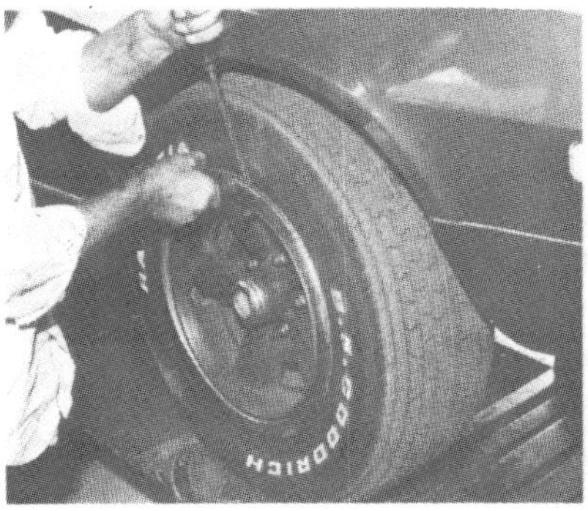
5

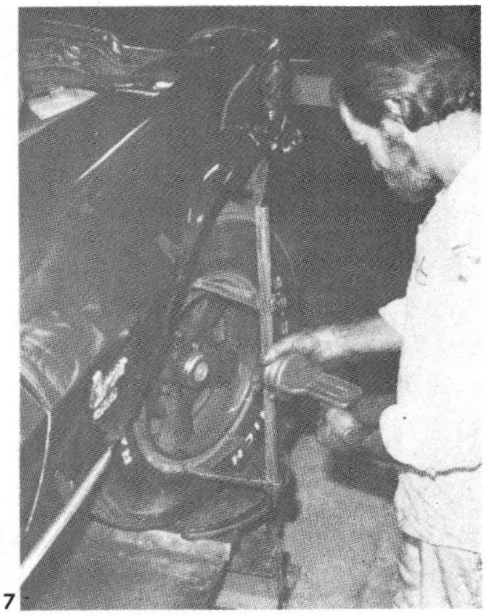
7

1. Before any front-end alignment, it's good practice to inspect all of the critical suspension components for wear, such as the upper and lower A-arm bushings, here being inspected by Jim Welch of Vanowen Brake & Wheel, our front-end experts in No. Hollywood.

2. Ball joints are another critical area to inspect. Jim demonstrates the proper way to check them. A jack is used to raise the tire off the ground with the suspension still taut. Now a prybar is used under the tire to see how much free play there is in the ball joints, if any.

3. It's important that both sides be equal in all respects before testing the alignment. This includes using a gauge to check for equal wheel height and adjustment of tire pressure.

4. Every shop has it's own methods that they've found best. One of Vanowen's steps is to hold a sharp scribe next to the tire center while the tire is rotated by hand. This scribes a perfect centerline that will be useful for toe-in checks.

5. In the case of our Rallye Sport Camaro, the wheel trim ring had to be removed so that the alignment tool could be fitted to the rim.

6. Jim attaches the gauge holder to the Camaro rim with the wheels in a straight-ahead position, and reads the bubble-levels on the gauge to find the camber of this wheel.

7. The caster reading isn't taken directly from the ball joint line; the wheel must be swung 40° to the rear and another reading taken.

length. As with any alignment problem, the condition and wearing of your tires will be your first clue that something is awry. When there's a feathered edge on the *inside* of the front tires' tread design, there's too much toe-in, meaning that the fronts of the tires are closer together than the rears. Almost all cars are supposed to have *some* toe-in, which essentially is designed to allow for the slack present in the steering gear and other front-end components.

Just as too much toe-in is wrong, toe-*out,* where tires are closer at the rear than at the front, is also wrong and will cause a feathered edge on the outside of the tread. Each tie rod on an independent front end has three basic parts; two male threaded ends and a female threaded centerpiece. By loosening the clamps, the centerpiece can be turned to effectively lengthen or shorten the tie rod as the ends thread in and out. Toe adjustments are simple. Making the tie rod shorter (where the tie rod is behind the spindles as it usually is) will give you toe-out, and making it longer will make the tires toe-in. But on an independent front end where there is a tie rod on each side of the car, you should make the same adjustments on each tie rod to keep the steering wheel in the correct center position.

Caster and camber are only slightly more complex to set than setting toe-in. On Chevrolet passenger cars, the upper A-arm is attached to the frame by a shaft that can be shimmed either closer to or away from its mounting bracket on the frame, thus tilting the wheel as desired. The camber setting is designed to reduce tire scrub as the wheels are turned and to spread wheel loading evenly over the wheel bearings and spindle. Some symptoms of improper camber in your front end would be rapid tire wear and a tendency for the car to pull to the side with the greatest camber. When there's more wear on the outside of the treads, then there's too much positive camber, while negative wears the inside treads. Most cars are designed to have some positive camber, or outward tilt, on the top of the wheels to allow for the change of angle that occurs when a normal load of cargo or passengers is added. As with other aspects of front-end alignment, camber is designed to make the wheels operate close to perpendicular to the road when driving, with maximum, even contact across the tire tread.

There are two adjustment points on each upper A-arm. While camber is adjusted by shimming the same amount at each point, the caster can also be changed merely by placing more or less shims at one point than the other, thereby tilting the spindle either forward or back.

Just as the methods of adjustment may vary from one vehicle type to another, the equipment used to align a front end also varies from shop to shop. In good working order, they all do the same job, which is to align your front wheels to factory specs. However, if you took your car to several different shops, you might get different readings at each shop! Find a good shop with a reputation and stick with it. Just remember that you shouldn't wait until your tires are almost bald before you think about having this service performed.

1
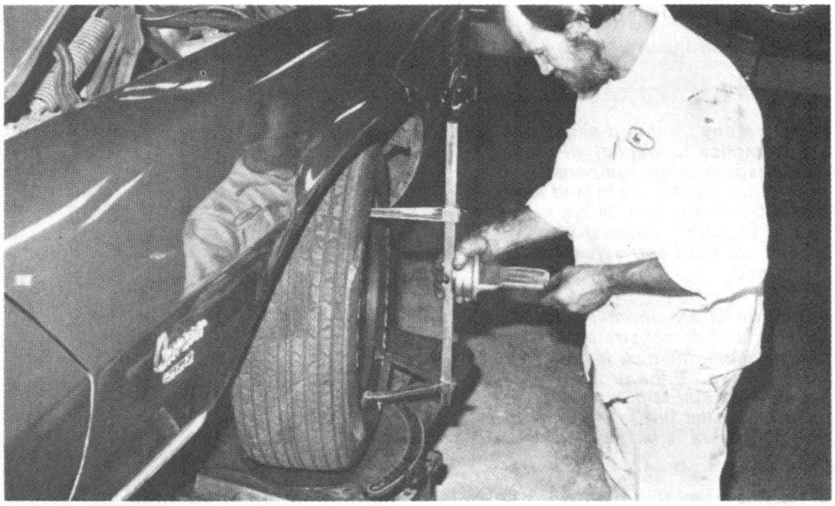

2
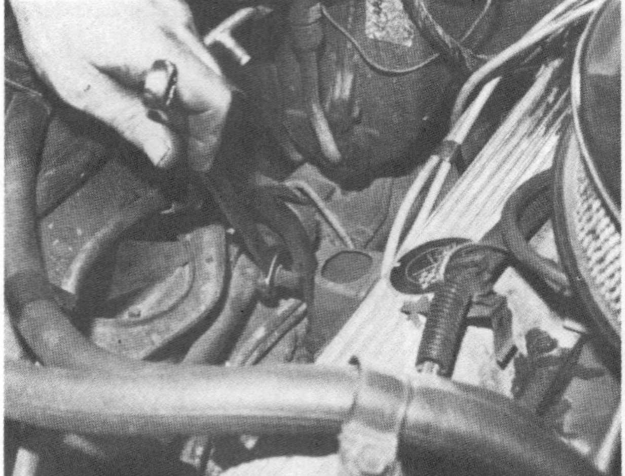

3
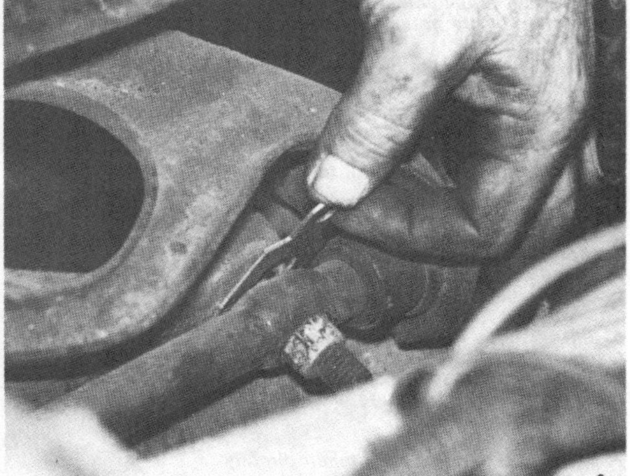

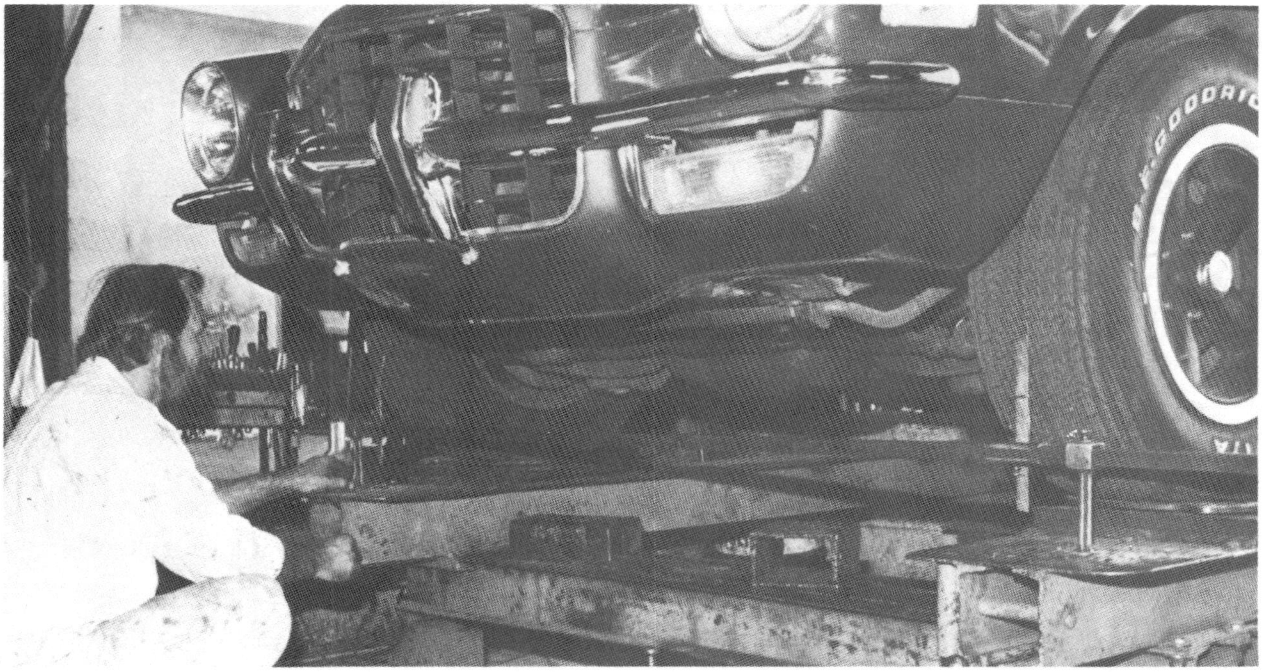

4

1. The same wheel is swung forward 40° this time, and the comparison of the readings taken indicates the rise and fall of the spindle during this turning, which can be translated to interpret the caster angle of the car.

2. In the case of this Camaro, the camber was within specifications, but the caster angle was slightly more than necessary, which may cause the steering effort to be higher. Jim loosens the nut at the rear of the A-arm support bar to change shims.

3. A thicker shim was inserted at the rear (which would tilt the ball joint centerline forward slightly) and the nut retightened. Sometimes a change of shim at one point means you must change the other end's shim, too, to maintain both readings, so you can see that alignment can get to be a time-consuming job.

4. Here's where those scribe lines come in. Jim sets the trammel bar in place in front of the tires, lining up the pointers with the centerlines of each tire and locking the setting.

5. Keeping the same setting, Jim moves the bar to the rear of the tires. Putting one pointer on the centerline of one tire, he can see how the other pointer lines up with the other tire, and whether there is toe-in or toe-out.

6. The tie rods must be loosened for toe-in adjustments, not easy if the bolts are rusty. Penetrating oil and an air wrench usually do the trick in loosening stubborn clamps.

7. Turning the tie rod center either lengthens or shortens the assembly, which changes the toe reading. The same amount of adjustment should be made at both left and right tie rods to keep the steering wheel alignment correct for straight-ahead driving.

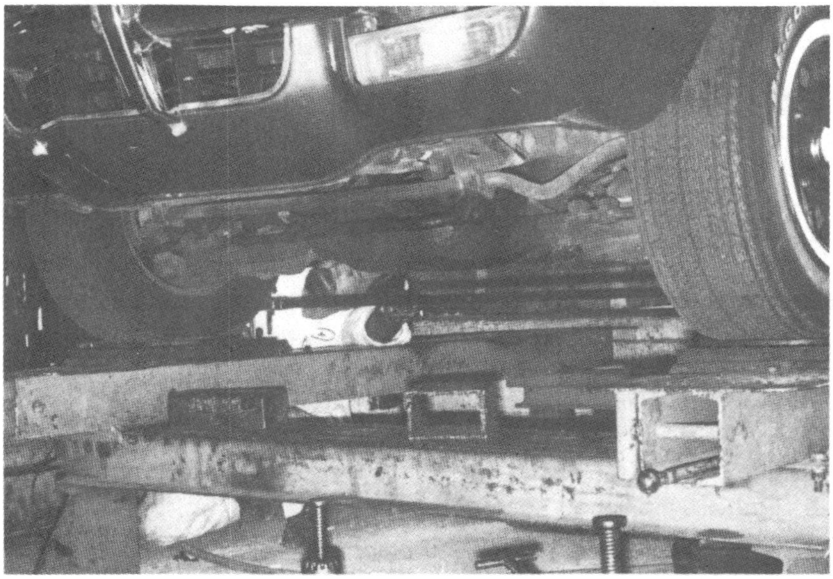

5

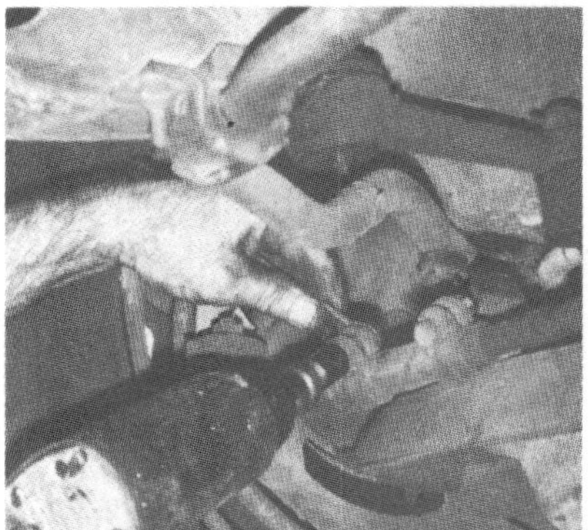

6

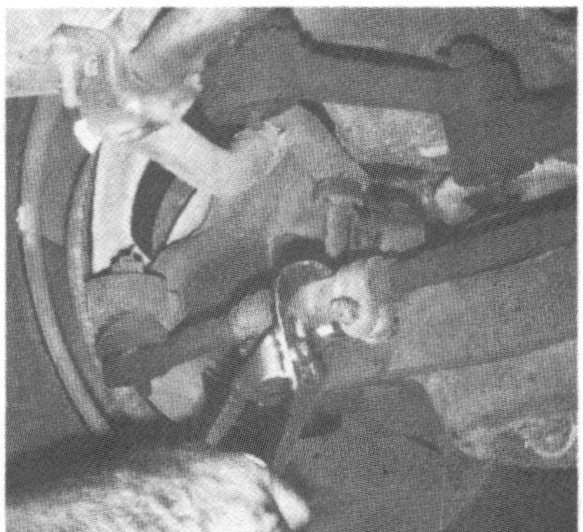

7

The Immaculate Deception

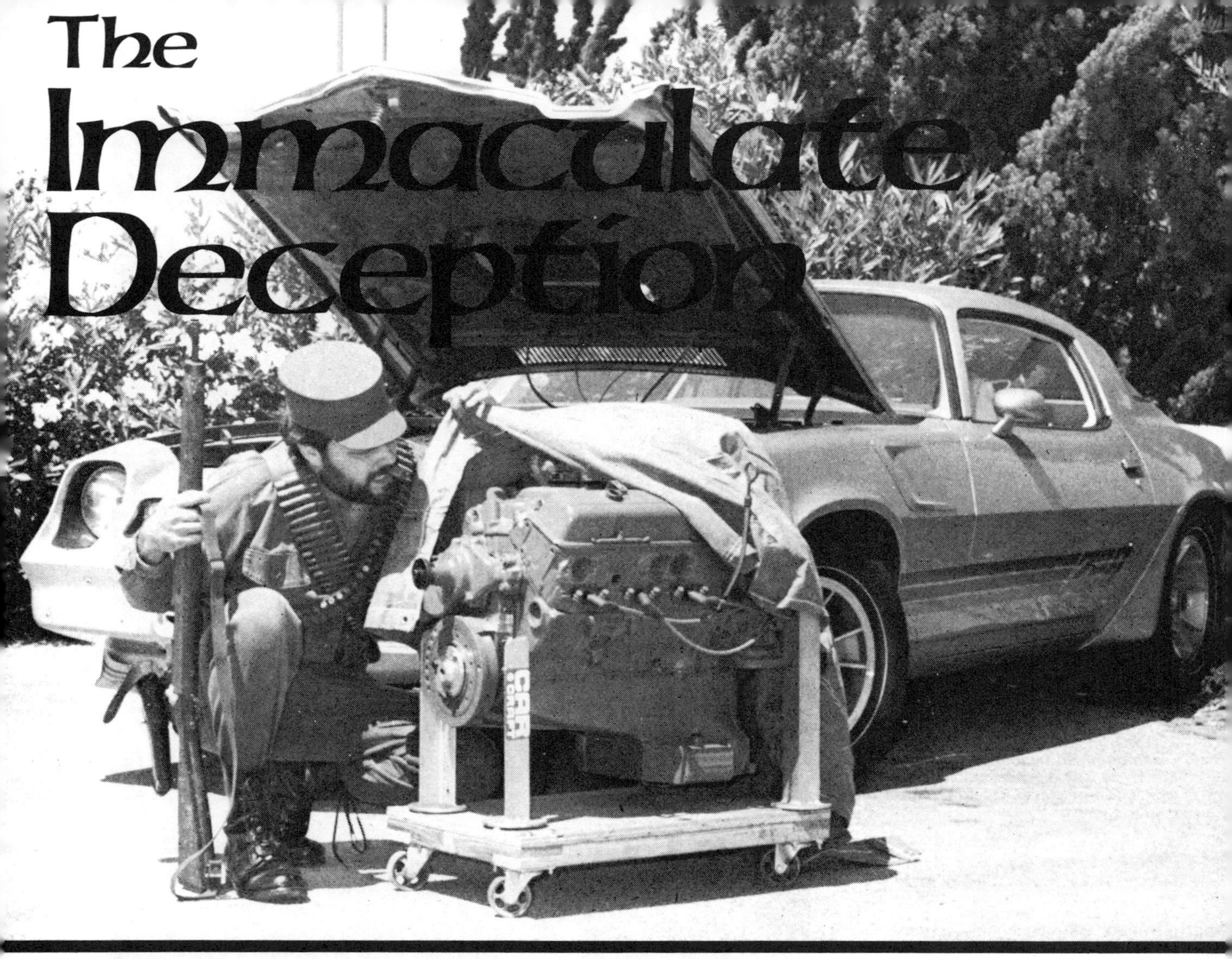

Cylinder Block

To build this engine you need a 4-inch bore large journal block in reasonably good condition. The standard 4.001-inch bore will yield 377 cubic inches with the 3.75-inch stroke of the 400 crankshaft. A 4.030-inch bore will yield a 383ci engine like ours, and a 4.060-inch bore will provide a whopping 388 cubic inches. The 400 crankshaft will fit into most 350 cylinder blocks after the large main journals have been turned down to the 350 journal size (see crank section). Some early two-bolt main blocks may interfere with the large counterweights, but minor grinding will cure the problem. The best block to use is a well-seasoned four-bolt main block, or lacking one of these, a new high performance four-bolt block from Chevrolet (PN 3970016). We used the special high tin alloy block (PN 366246), but its expense may be prohibitive for most street machiners. It is also a "green" casting that will make less power until subjected to considerable run-in. You must use a 350 or 327 block of the large journal variety because the 400 crank will not accept the large cut required to turn it down to small journal size. Hence you should avoid any blocks made prior to 1968.

Once you have settled on a final displacement, the block should be subjected to the standard preparation procedures. It should be deburred, thoroughly cleaned, bored, honed and align honed if necessary. The decks should be cut parallel to the crank centerline, but care should be taken that the cut is the absolute minimum required to achieve parallel decks. The more you cut here, the more you have to take off the top of the pistons later. Our new block required all of these operations, which were capably handled by the crew at Valley Head Service in Northridge, California. With all the special operations required for this engine, everyone in the shop was involved with it at one

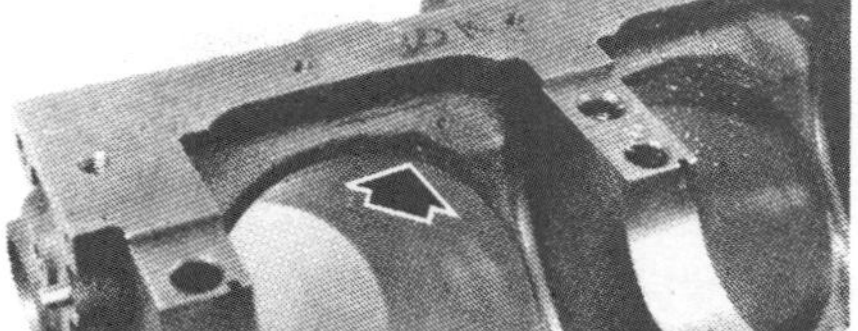

High performance cylinder blocks are already notched for connecting rod clearance. Once the main journals have been ground to the correct size, the 400 crankshaft drops in with no counterweight or connecting rod interference. The standard length connecting rods (5.7 inches) will hit the camshaft on some cylinders and minor grinding is required.

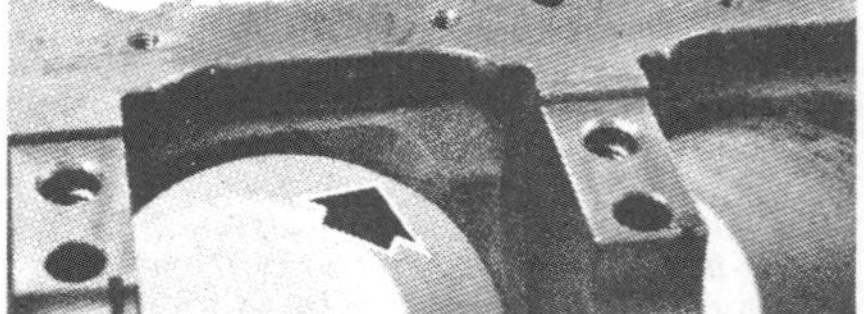

Standard two-bolt and four-bolt blocks are not always notched and may require grinding to provide clearance for the connecting rod bolts. A small notch is all that's necessary, but the final determination must be made by dummy-assembling the engine for clearance checking.

CC's Underground Camaro Gets a Strongarm Powerplant for Street Supremacy

By John Baechtel

In these overregulated times it's often difficult to identify a truly contemporary approach to hot street performance. Turbocharging probably gets the nod, but street supercharging continues to gain in popularity even as more and more street machiners look to modifications that stress fuel economy. The problem facing most street machiners is the continuing lack of an established middle ground. Those cars built with a generous application of economy pieces are reasonably snappy performers and save gas, but very few of them are capable of standing up to a well thought out turbo setup or even an early high-compression, 4-barrel supercar. Nitrous oxide is one of the best ways to put the snarl back into these economy-oriented street machines, but even generous doses of it aren't enough to help overcome a serious competitor. Late-model cars are worse off than most since they may have to contend with yearly inspections that require the presence of all OEM smog equipment in a functioning capacity. Headers, intake manifolds, carburetors and other equipment are all part of the high performance plan, but the sight of them is often grounds for unwarranted hassling by inspectors.

So what can we do to regain some of that old zip while maintaining some sort of quasi-legal facade that will keep the bad guys off our case? Furthermore, can we do it with an engine that could pass a smog test if necessary? We thought it might be interesting to find out, so we undertook the construction of a super street engine designed to fulfill these requirements. There are a number of ways to approach the problem, but we selected one that has become increasingly popular—a large displacement, long stroke small-block based on factory parts. Paramount in our consideration was a need for gobs of torque and a high reliability factor. And since our 1980 Z/28 test car was to be the recipient of this killer piece, it seemed reasonable to duplicate the original engine's external appearance as closely as possible. So return with us now to those glorious days of yesteryear when sleepers were in style and everyone had a 283 under the hood—if you know what we mean.

Real old-timers will tell you that the best path to performance is to "put an arm in it," or as we say, use the longest stroke you can get away with to ensure lots of torque. Throw in another tradition in the form of lots of cubic inches and you have the makings of a serious contender. Now, you can build a 400ci small-block with standard length rods and it will be a hard runner. You can build a 350-based motor with a 400 crank and the short 400 rods and it will also yield good performance. In fact, you can build about 10 different combinations involving these large displacement small-blocks just by jockeying blocks, stroke length, rod length and piston compression height. We happened on the 383ci engine by accident, but it turned us on and that's what we built. The combination is basically a 400 crank in a 350 block with standard length rods and a reduced piston compression height. We've seen them before and we know they run very well. And we have a sneaking suspicion that the ultimate street weapon may indeed be a long arm and a can of corporate blue engine paint.

---

point or another. Owner Larry Ofria was quick to stress the importance of checking the dimensional integrity of the new casting, and some of the problems we encountered reinforce this concern. The main saddles were off by a significant amount and precision realignment was called for before the decks could be trued. And even after the block was rendered dimensionally correct and the crank ground to the proper specs, it would not fit into the bearings and was not about to turn. A bit of examination was obviously in order.

It was determined that tolerance in the block casting was such that standard grinding and radiusing of the main bearing journals still did not provide enough clearance in some locations. The crank cheeks rubbed ever so slightly on the block bulkheads. The crank was put back in the grinder and the throw surfaces and radii were reground just enough to avoid the interference. Now the crank fit in the bearings and turned relatively easily, but a tight spot was felt and required further investigation. Finally the problem was traced to rear main cap distortion, evident when the cap was torqued to factory specs. When torque was backed off 10 ft.-lbs., the tight spot went away and the crank turned freely. But a very slight thump could still be heard as the crank was revolved. After much head scratching, frowning and mutual "I dunno's," we uncovered the problem—one that had not been encountered on previous

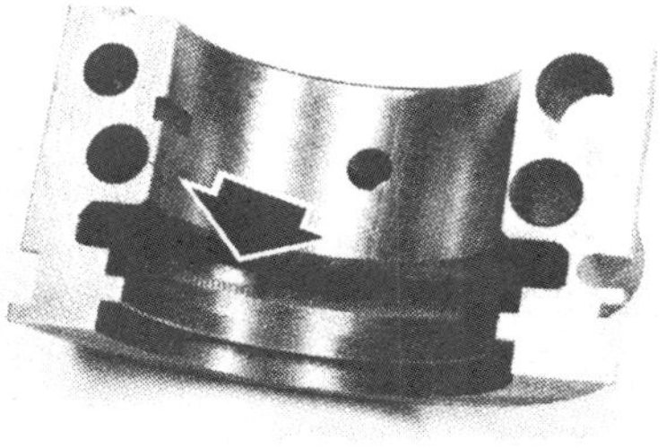

The arrow indicates the shiny spot where the outer surface of the thrust flange was contacting the inside of the rear maincap. Since the 350 maincap is smaller than the 400 maincap, the fit is relatively close and bears checking.

engines of this nature. A very small bump on the perimeter of the thrust flange was rubbing on the inside of the rear main cap. It was the last possible place you would expect a problem, but you have to remember that we were putting a 400 crank into a 350 block. The 400 block has larger main caps because of the larger 2.65-inch main journals. The thrust flange clears easily in a 400 main cap, but it gets very close in a 350 main cap. In most cases they clear, but the tolerance stack involving the tight cap and a slightly irregular thrust flange was enough to trip us up. Watch for it on your engine.

Crankshaft

The crankshaft is the heart of this powerplant. It is the standard cast crank from a 400ci small-block. There is nothing special about it other than the fact that it offers a 3.75-inch stroke which is highly attractive in terms of torque production and stout street

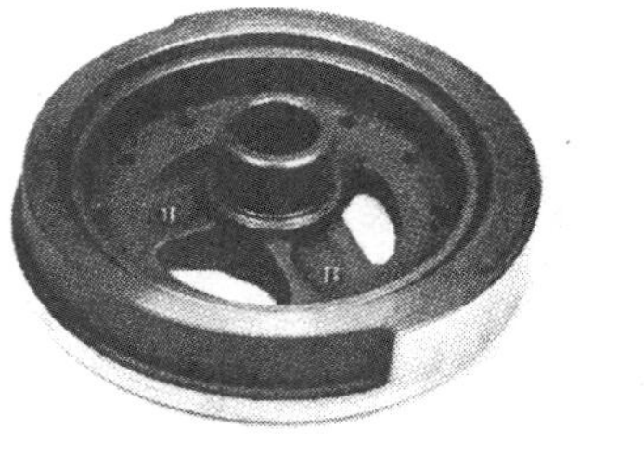

In the configuration we have assembled, the engine could be internally balanced, but it is easier to use the standard 400 balancer which provides external balancing. The 400 flexplate or flywheel should also be used.

performance. The crank is available over the counter under PN 3951527, and most engine shops have good ones in stock if your cash flow situation requires a used item. What you are looking for is a crank that is reasonably straight with minimal rod

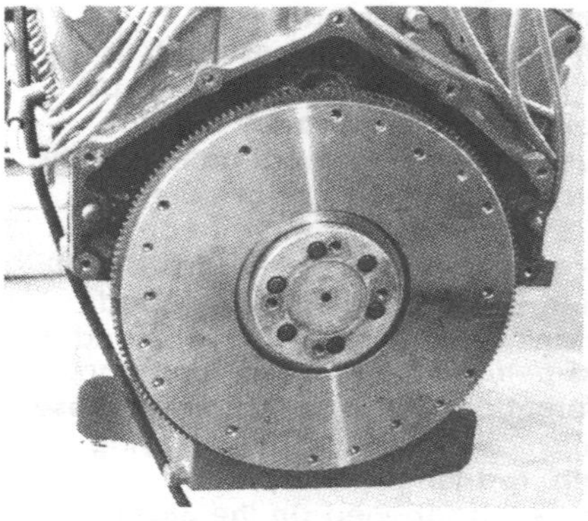

Automatic cars will make use of the standard GM flexplate for 400ci engines, but stock equipped cars should pass on the factory flywheel and use a high performance, explosion-resistant flywheel like this one from McLeod Industries Inc. The part number is 460231 and it is designed for an externally balanced application like the 454 or 400. Make certain you have it in time for your machinist to balance it properly with the rest of the internal engine parts; balancing is imperative with one of these engines.

journal wear. Main journal wear or damage isn't particularly significant in light of the large amount of material being removed. Once you uncover a likely prospect, take the time to have it Magnaflux inspected to ensure the absence of cracks. This is particularly important since you're going to be reducing the amount of overlap between the journals when the mains are reground. This doesn't appear to significantly weaken the crank, but a crack in the wrong spot may be enough to do you in if it isn't caught.

Other than main journal preparation, the crank is subjected to standard crank prep operations including deburring, oil hole chamfering, micro polishing and thrust clearance during mock assembly. It is also wise to have the flywheel flange checked to ensure that it is perpendicular to the crank centerline. The crank requires no other preparation other than balancing after the rest of the internal pieces have been modified and their weights determined.

Oiling System

The lubrication system is configured to match the car. Since the Z/28 is something of a handler, we anticipate more than a few curves, so we went with Moroso's road racing oil pan which is at home in either a left or right hand turn—or going straight. The screen windage tray was used in lieu of a factory unit since it tends to grab the oil and keep it from bouncing back into the crank assembly. A 1¾-inch extension and corresponding pump drive were used to lower the pump to the bottom of the sump. Their bottom cover pickup (PN 2490) was added to the stock Z/28 oil pump to eliminate a separate pickup tube.

Bearing clearances have been

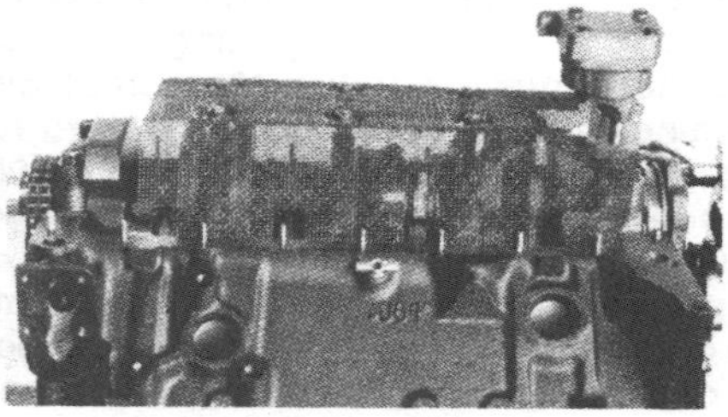

Moroso's screen windage tray required the use of two extra maincap bolts with extended studs. Note the 1¾-inch oil pump extension and the bottom cover pickup installed on the stock oil pump. The square cutouts in the screen windage tray are for connecting rod clearance. Here again, Specialty Fasteners studs are used to secure the oil pan.

kept relatively tight at .002-inch on the rods and .0028-inch on the mains. A .030-inch hole is drilled in the front center oil gallery plug to provide additional timing chain lubrication, and the oiling holes in the main bearings were drilled to match the size of the openings in the cylinder block.

Connecting Rods

The rods we selected for this engine are the standard Z/28, large journal rods (PN 3973386). You can use whatever rod you like as long as it is a large journal unit configured for a pressed pin application. The rods we used are those commonly referred to as "pink rods" because of the pink paint applied at the factory. They are the best choice because they receive additional preparation and Magnafluxing and are hand selected for the least amount of flash on the rod beams. Valley Head Service removed the stock bolts and installed high-strength bolts and nuts from Specialty Fasteners. Their new rod bolts have been tested to very high tensile strength and are recommended for this type of application. We did not polish the rod beams, but we did have to grind four of the rods where they hit the camshaft. This occurs on cylinders 1, 2, 5 and 6 and requires slight material removal from the rod near the bolt head. The only way to determine this accurately is during mock assembly of the engine. If you're using a small base circle cam like a mild roller profile, grinding may not be necessary, but stock cams and large lobe rollers are going to hit the rods. And don't take our word for it; check each and every rod and cam lobe by rotating the engine slowly with the cam installed and properly phased to the crankshaft. It all goes together very smoothly, but you have to check it. For final assembly we used TRW CL-77 bearings fitted at .002-inch.

Pistons

We've already said that the key to this engine is the long stroke and large displacement, but here is another item that helps bring it all together. Any 350 piston will bolt right into this motor if you use the short 400 connecting rods, which incidentally clear the camshaft and make the whole thing a bolt-together operation. We wanted to use the long rods (standard

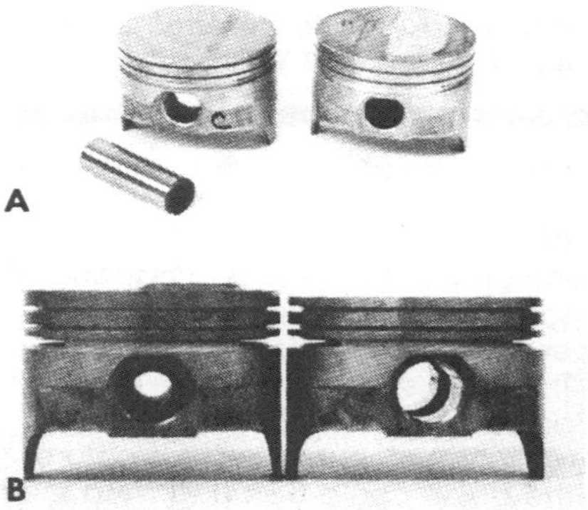

These photos show the amount of material removed from the top of the factory pistons to make them fit the engine. Photo ***(A)*** *is a comparison between the original domed piston and the flat top version prior to valve notch cutting. A side view,* ***(B),*** *shows how the dome and part of the deck surface are removed. There is still plenty of material left in the deck surface, and deep valve notches are still possible. Notice how this modification automatically moves the ring placement higher on the piston.*

5.7-inch) for several reasons. First of all, we wanted to avoid the severe rod angularity incurred with the short rods. We also wanted to spin the motor fairly fast if necessary, and the Z/28 rod is a significantly better piece than the 400 rod and better rod bolts are available.

But when you stack up the long stroke with a long rod and a standard 350 piston, you end up with a piston that sticks out the top of the block by about .100-inch. Now, normally a factory piston is about .030-inch down the bore at TDC, and when you add these figures together you come amazingly close to the difference in rod length between the standard 5.7-inch rod and the 5.565-inch 400 rod. Obviously we're going to have to whack off a bunch of deck material if we're ever going to put cylinder heads on the engine. The trick is in determining just how much material to remove and which pistons can safely handle the loss without collapsing under fire.

Ideally we want a minimum of .035-inch piston-to-cylinder head clearance, but we're also planning on using Chevrolet's recently released "O-ring" type head gasket (PN 14011041) which compresses to .040-inch. So in our case we can shoot for a zero deck height and minimize the amount of material removed from the piston deck. We used the factory LT-1 pistons because we wanted to maintain a factory parts approach as much as we could, but there are a lot of other pistons that will do the job. The factory piston and TRW's PN L2304F probably come up the lightest because they have more material removed. If you wanted to pick up some additional deck thickness for added reliability, you could use TRW's 3 9/16-inch stroker piston for the 350 (PN L2278F). It features a higher pin placement which lets you take less material off the deck surface. This piston and their Super Stock piston, PN L2327F, are both racing models and may require slightly more piston-to-wall clearance than some of the others. We haven't checked these other numbers, but some of them may provide a less expensive alternative since they are powerforged replacement pistons. If your dealer permits, try comparing the PN L2304F piston with the following numbers: PN L2417F, PN L2403F, PN L2369F and PN L2256F. In a Speed Pro piston you might investigate PN 2363P, PN 2363PA and PN 2244P.

In any case, a portion of the top of the piston has to come off and you only determine the amount by trial assembly and actual measurement of each piston as it protrudes from the block. Write it on the top of the piston and make certain you have them numbered correctly and that you match the pistons to the rods each time.

After the pistons are cut, you'll have to go back and determine piston-to-valve clearance. We determined that a .100-inch deeper valve notch was not unreasonable with the factory piston and Valley Head Service cut and blended the notches for us. When you do this it is important to measure the angle of the valve in the cylinder head and transfer this angle to the valve notches in the piston. By matching these angles, the piston will contact the valve squarely should valve float occur, and you may escape with little or no damage, depending on the severity of the valve float. Make certain you blend and smooth all the sharp edges on the piston top and then have the decks glass-beaded to obtain a uniform surface. Install them in the motor with .0035-.004-inch piston-to-wall clearance.

Use TRW's single moly ring set (PN T7945M) with the top ring gap set at .018-inch and the second ring set at .012-inch. This ring set has standard tension oil rings which require no special fitting. Be certain the rings are installed with the small dots facing up. Spray the cylinder walls with 3-In-1 oil, install the rings dry and lightly coat the piston skirts with engine oil.

Camshaft & Valvetrain

Chevrolet engineers felt that we should try one of their factory grinds in this engine. We weren't opposed to doing so based on the stout performance turned in by a factory cam in our 327ci bracket motor earlier this year. They suggested the first-design off-road cam (PN 3927140) which we have used successfully in 327ci and 350ci engines in the past. They felt that the 140 cam would be very smooth and tractable in the large displacement, torque-efficient engine. The corresponding valve lifters are the piddle valve variety (PN 5232695). Z/28 pushrods (PN 3796243), and Chevrolet's dual valvespring and hardened retainer, PN 330585 and PN 330586 respectively, are used to control the valves. Aluminum retainers are generally regarded as something of a no-no for street performance applications, but you can get away with it with the Chevy pieces because the retainers are hardened and there is no valvespring dampener to chew up the aluminum. When installing factory guideplates, it is a good idea to lightly dress and round the inside of the notches on a small belt sander. They are stamped pieces that sometimes have rough surfaces that can damage even the hardened pushrods. Moreover, you want to avoid any binding they might cause Make certain each pushrod spins freely in the guideplate.

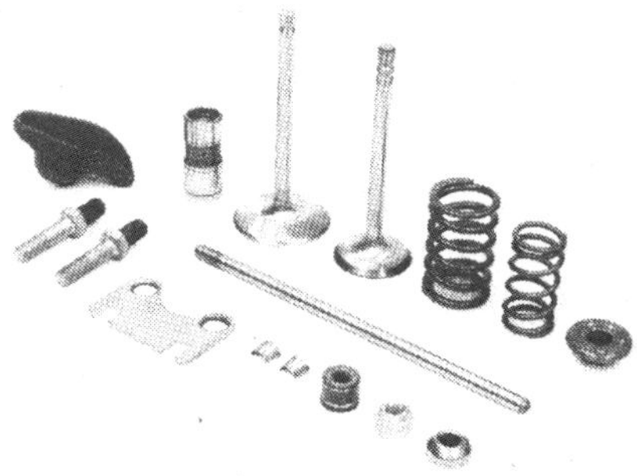

Chevrolet's high performance valvetrain hardware is more than adequate for an all-out street assault. Swirl polished valves, dual valvesprings, hardened retainers and pushrods, screw-in studs and guideplates make up a reliable yet formidable package.

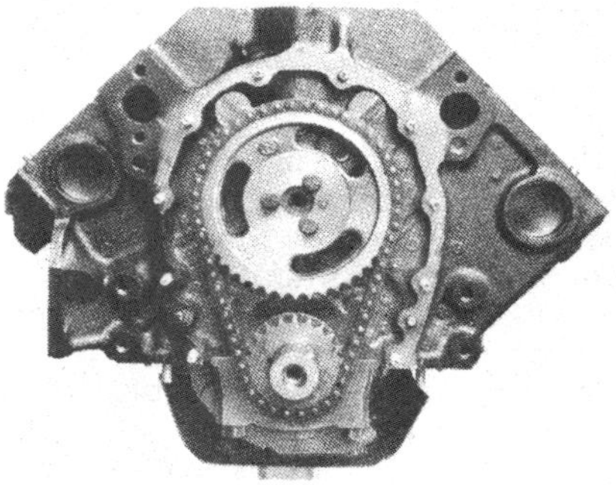

The factory camshaft was installed four degrees advanced with the aid of a Cloyes true roller timing set that features three separate keyways for timing adjustments. Note the front cover studs from Specialty Fasteners.

Cylinder Heads

Although a reasonably strong engine could be built with the early 1.94-inch valve, straight plug cylinder heads, we opted for the large valve 292 turbo castings because of their improved port configuration. For our purposes, Valley Head Service suggested that their "Streetmaster Porting" would provide the best overall performance. In what they refer to as their "Power Play" approach to engine building and head porting, they felt that the cylinder heads would provide most of the significant gains for our engine.

On the intake side **(A)** they work into the port approximately 1½ inches, blending and opening the port to match the intake gasket (Mr. Gasket PN R-102). The valve pocket is reworked slightly from the base of the valve guide to the valve seat, but the port is not ported all the way through because they want to maintain a rough surface for improved atomization. Furthermore, there is no need to drastically enlarge the port since this would kill the velocity required for good street performance.

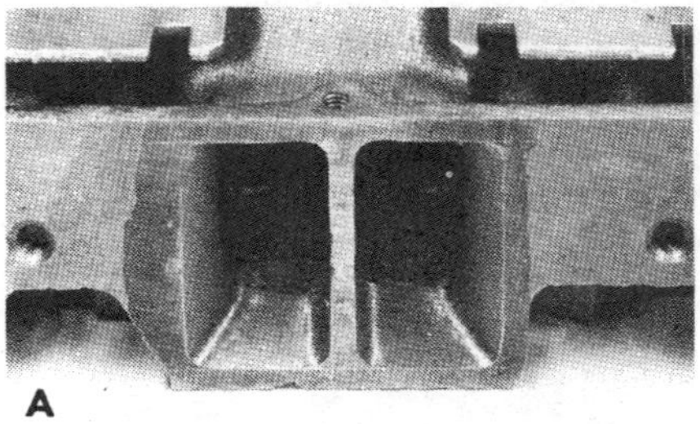

A

The exhaust port **(B)**, on the other hand, is ported all the way through because it needs all the help it can get. It is given a smooth surface, with the short side radius being very critical. The port floor is made as wide, flat and high as possible to optimize exhaust flow. The combustion chamber is polished and the area on the intake side of the spark is laid back slightly.

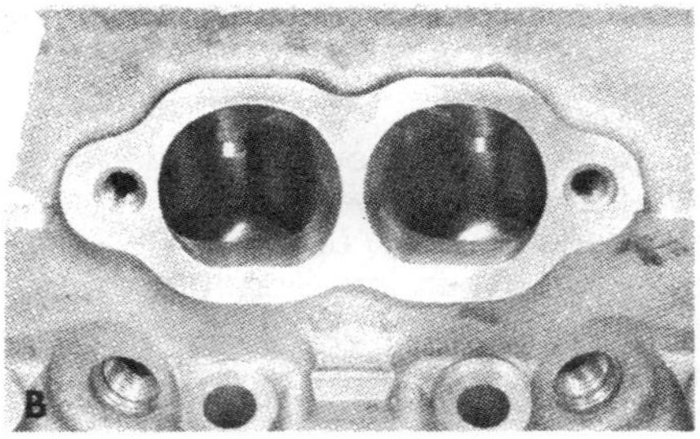

Their "Marathon" valve job **(C)** is a three-angle approach with .060-inch intake seats and .090-inch exhaust seats, both beginning .015-inch from the edge of the valve. Final cc'ing and milling brings the chambers to 64cc's. The Chevrolet dual valvesprings were assembled to within ±.005-inch of the installed height required for proper spring pressure (135 pounds at 1.720 inches).

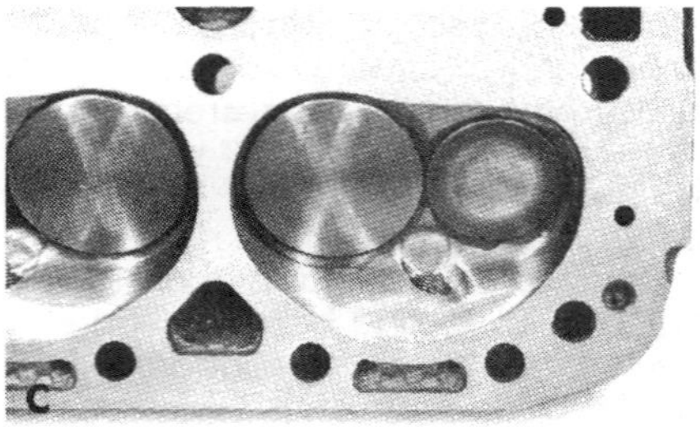

Compression Ratio

Since flat top pistons give the most desirable flame travel, we weren't particularly distraught about chopping off the piston dome and part of the deck, but we had to juggle head gasket thickness, deck height and combustion chamber volume to arrive at a desired compression ratio. With premium gas and fuel additives, 11.0:1 doesn't seem entirely unreasonable, and our CR works out to slightly more than this with the zero deck height, .040-inch thick compressed gasket thickness and a 64cc combustion chamber. This also takes into account the deepened valve

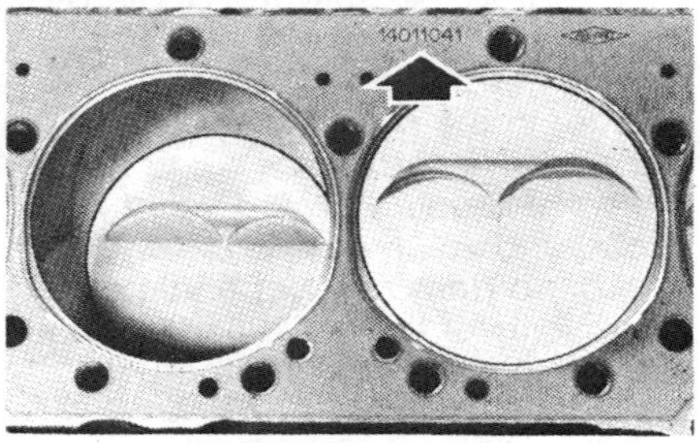

Chevrolet's special O-ring head gasket is designed for use with 4 and 4⅛-inch bores. It compresses to .040-inch thickness and the extra area helps reduce compression on our 4-inch bore engine. The 2.05-inch intake valves and 1.6-inch exhaust valves required larger valve reliefs than the original valve clearance slot in the factory pistons.

notches. The combination of overbore, reduced deck height and small chamber volume has placed us back in the vicinity of the 11.0:1 CR originally obtainable with the factory piston. If you want less compression, use a larger combustion chamber. You can also add a little deck height, but don't get crazy here, because you've already taken a healthy slice off the top of the piston. The turbo heads we used normally come with approximately 71cc chambers that would get your CR down to a reasonable figure if necessary. Compute it on paper first and you'll be able to determine the exact procedure to follow to achieve the CR you want.

Intake Manifold

We all felt that an aftermarket manifold would be necessary to feed this monster, but it didn't really fit into our deceptive plan. That's why we're using Chevrolet's aluminum Quadrajet manifold with a few subtle modifications. When this machine rolls into an inspection station resplendent in corporate blue engine paint, with a stock looking Quadrajet carb and an EGR valve on the manifold, it's going to take a well-trained eye to spot the ruse. There are still some obvious clues, but we're not talking. Marvin Miller plumbed the hidden nitrous system under our manifold so there's no plate to expose us. The fittings go in through the back of the block where they are easily hidden beneath the ignition system. Gene Kuklovski at Marvin Miller informed us that the modification is much easier to perform on the aluminum manifold and that owners of cast iron manifolds can expect to pay a good deal more for the same plumbing. We elected not to port-match the manifold at first because we wanted to see how well it worked with the heads in a stock condition. The ports are quite a bit smaller than the ports in the heads,

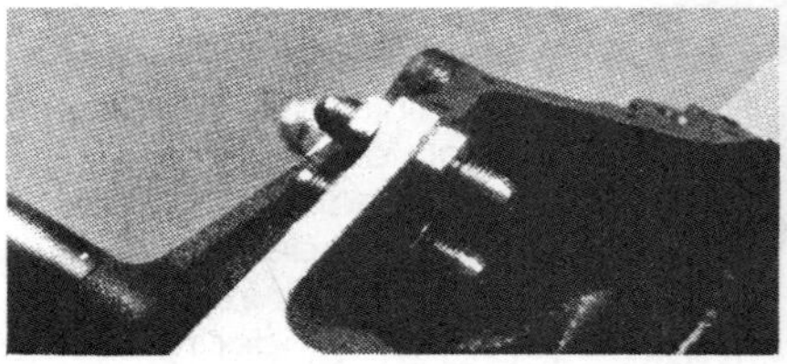

Marvin Miller provided these special bulkhead fittings for their nitrous oxide injection system. The block was extra thick in the area where we chose to install them. We thinned it with a hand grinder, drilled two ⅜-inch pilot holes, and spot-faced the surface with a screw-in stud cutter. The holes were then drilled to ½-inch and the fittings installed with silicon sealer. The lines clear the number eight pushrods and the distributor housing with room to spare.

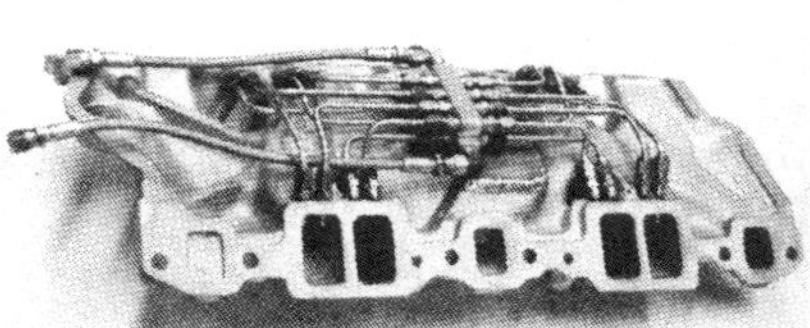

A stock aluminum Quadrajet intake manifold was delivered to Marvin Miller for transformation into a direct port nitrous injection setup. They found it necessary to cut away part of the exhaust heat crossover passage and weld in a flat plate to seal it. The manifold was then plumbed with nitrous and fuel lines.

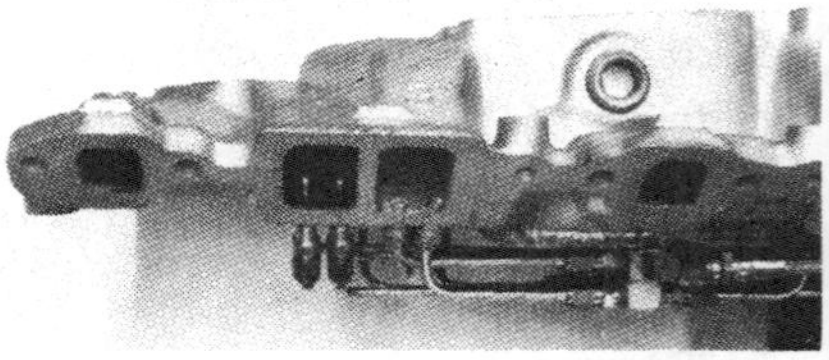

A nitrous line and a fuel line are run to each port, but the ports get different nozzles according to their entry angle. Runners leading from the lower chamber of the intake plenum enter the port at a flatter angle; thus they receive extended nozzles that are drilled to direct the extra fuel and nitrous oxide downstream. The steep runners from the upper chamber of the plenum have nozzles that are already aimed slightly downstream via their entry angle into the port.

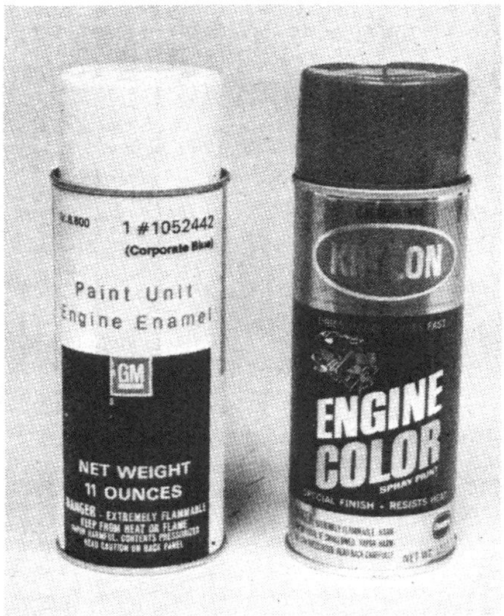

Corporate blue engine paint (PN 1052442) is an important part of our deceptive plan, but it is often difficult to find. Many GM dealers don't care to stock it because of government shipping regulations regarding paint and other materials. A very close substitute is Krylon's GM Blue (PN 1930). Be certain to paint the entire engine, and if you really want to get sneaky, try installing a set of Targetmaster decals on the valve covers.

and the mismatch is pretty substantial. Still, we're actually dumping into a larger runner after the manifold, and the mismatch may help slightly in the reversion department, which we know to be a significant problem with this manifold. If dyno time is available, we may port-match the manifold and try a high-velocity aftermarket manifold to observe the difference.

For What It's Worth...

If you study the parameters by which our underground powerplant was built, it quickly becomes evident that the entire effort is a massive compromise. No one in their right mind would go with a stock intake manifold and carburetor on top of an engine like this one. And even more objectionable is the fact that we're going to run through the stock exhaust manifolds and the stock exhaust system, catalytic converter and all. The compression ratio is way too high for unleaded fuel, but we only have to worry about that when we baby the car in for a landing at the emissions test center where everything has to look just right. We can't really recommend the use of dummy converters with straight-through pipes, or a dual converter setup, but we're willing to concede that a set of headers and a similar exhaust system arrangement would be highly beneficial.

Power output is certainly acceptable with the configuration we're using, and the addition of an Edelbrock Streetmaster manifold provided an increase indicative of the engine's real capability if fitted with serious induction and exhaust system hardware. But there's really no need for it. We're already pushing 400 horsepower with the stock intake manifold and well over 400 ft.-lbs. of torque in the mid-range. The nitrous oxide injection system will easily put the engine into the grey area where parts breakage becomes frequent. Once the engine is in the car, we have a feeling it's going to show us that the cylinder heads are actually a little too big and the cam is a little too wild. They work well with a good manifold and headers, but we think the car is going to want a smaller cam and smaller heads and probably a torque converter to help overcome the 3.42:1 gear ratio. In fact, one of Kenne-Bell's "Switch Pitch" torque converters seems like just the ticket for hot street performance with this combination. We're getting into a lot of maybes here, but it's really the only way to pin this thing down. At any rate it's certainly better for us to do all the legwork and make all the mistakes so that you only have to do it once—the right way. It's all coming up in following installments on CC's Underground Camaro.

383ci Chevy

Factory Parts Guide

PART NUMBER	ITEM AND/OR ASSEMBLY
366246	Cylinder block, 4-bolt mains, tin alloy
3951527	Crankshaft, 400ci engine, 3.750-inch stroke
3973386	Connecting rods, Z/28, large journal
3942543	LT-1 piston, .030-inch oversize
3965784	Cylinder heads, bare 292 turbo castings
14011041	Head gasket, integral O-ring, .040-inch compressed thickness
3796243	Pushrods, Z/28 type with hardened tip
3973418	Guideplates
3973416	Rocker arm stud, screw-in type
3764389	Valve guide seal
330585	Dual valvespring for HD camshafts
330586	Aluminum retainer (hardened)
3947770	Keeper, purple
361976	Intake valve, 2.05-inch diameter
3849818	Exhaust valve, 1.6-inch diameter
6272225	Balancer for 400ci engine
340298	Flexplate for 400ci engine, auto trans
1103291	Corvette distributor, HEI unit
14007377	Aluminum Q-jet intake manifold
8629423	M-40 Turbo 400 trans, Z/28 type
8626648	Torque converter, high performance

Dyno Results

Stock Intake Manifold

RPM	HP	Torque
3000	254	444
3500	283	424
4000	320	421
4500	372	434
5000	389	409
5500	387	370
6000	385	337

Edelbrock Streetmaster Manifold

RPM	HP	Torque
3000	246	431
3500	284	426
4000	326	428
4500	382	446
5000	401	421
5500	408	389
6000	403	353

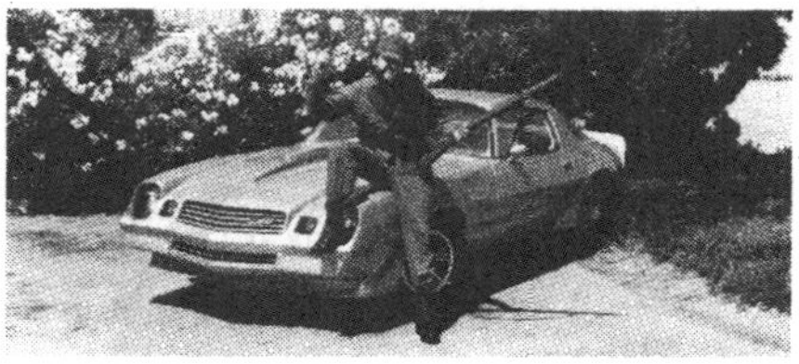

Specialty Equipment Sources

Autotronic Controls Corp.
6908 Commerce
El Paso, TX 79915

Cloyes Gear and Products
4520 Beidler Rd.
Willoughby, OH 44094

Marvin Miller Manufacturing
7745 S. Greenleaf Ave.
Whittier, CA 90602

McLeod Industries
1125 N. Armando St.
Anaheim, CA 92806

Moroso Performance Products Inc.
Carter Drive
Guilford, CT 06437

Specialty Fasteners
2601 Redlands Dr.
Costa Mesa, CA 92627

Valley Head Service
19340 Londelius St.
Northridge, CA 91324

CAMARO ENGINE EXCHANGE

HOW TO KEEP YOUR CAMARO ALIVE

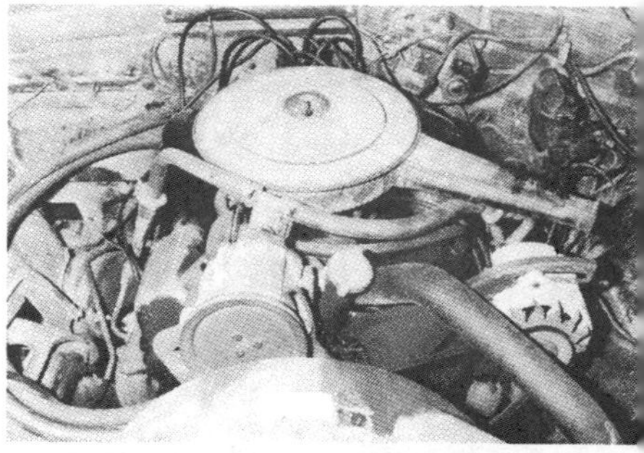

After 150,000 miles of loyal service, this '68 Camaro's 327 has finally earned a thorough overhaul. Yes, it isn't a pretty sight—but that will soon change.

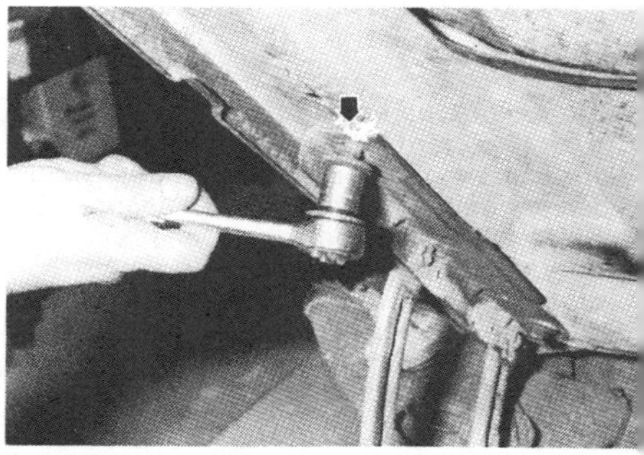

Engine exchange begins by scribing around edges of hood hinges to simplify realigning hood. Then, four hood fasteners are removed.

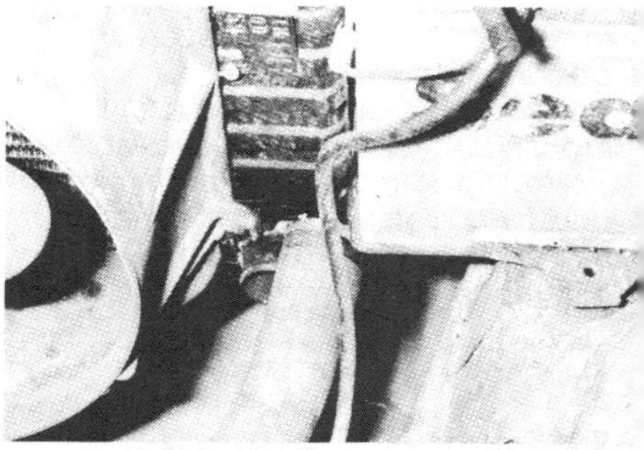

. . . and then battery hold-down clamp is removed.

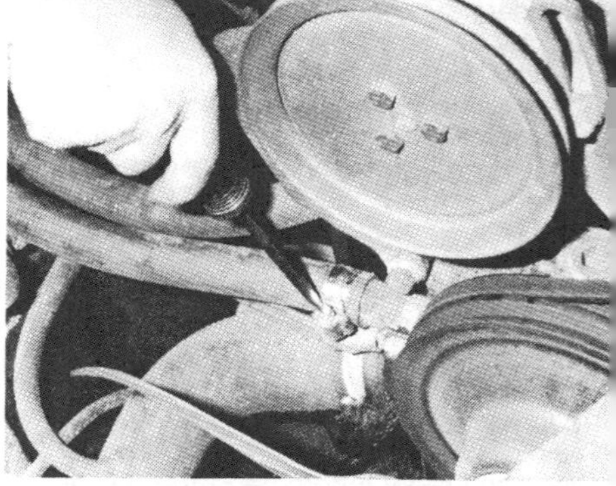

Two heater hoses are removed, one at water pump inlet and second at front of intake manifold. Disconnect and plug gas line at fuel pump.

Even Chevrolet engines wear out. Most early-model Camaros are pushing 150,000 miles by now; it's rare to find a model from the Seventies that isn't edging up on the 100,000 mile mark. There are always those Camaros that are driven only on sunny weekends, and then never more than five miles from the garage. And while you may hear about such cars, very few Camaro enthusiasts ever get to own one. Most Camaros lead ordinary lives. They're used for everyday commuting, weekend cruising and general transportation. And that means that a powerplant transplant is probably due for any Camaro that's seen its odometer come back to six zeroes.

There are a number of excellent publications detailing the steps in rebuilding a Chevrolet engine; we suggest you consult one of these if you're planning a do-it-yourself overhaul. We'll concentrate here on what it takes to remove and replace a Camaro motor, with special emphasis on the details that make the difference between a successful engine exchange and one that creates more problems than it solves. And since there's no better time to clean up an engine compartment than when the motor is elsewhere, we'll also outline how to bring the underhood area up to standards. Although these efforts will be far short of a "concours" restoration, they will restore some of the lost sheen to any Camaro engine room.

The one piece of equipment that's indispensable for this operation is an engine hoist. While throwing a chain over a convenient limb may have been standard procedure for the great shade-tree mechanics of the past, a rolling engine hoist (begged, borrowed or rented) will make the entire job much easier and safer. An A-frame with a ratcheting "come along" is a workable alternative, although you'll have to roll the car to the engine. And before you start eyeing the garage rafters as candidates for an engine hoist, remember that a fully outfitted small-block V8 weighs more than 600 pounds and a big-block will add another 200. Don't suffer the embarrassment of collapsing the garage on top of your Camaro.

There are two strategies to a Camaro revival: pull the entire drivetrain from fan belt to transmission tailshaft as a unit, or hoist the block alone. The first approach requires removing the shifter, driveshaft, and transmission crossmember, while the second commits you to removing the bellhousing bolts (which are none too easy to reach in a Camaro chassis). Since the transmission in this particular car had recently been overhauled, it was decided to leave it in place.

Remembering where all the bolts and brackets go can be one of the most frustrating parts of an engine exchange. In the days and weeks it takes to overhaul a motor, the memory of where everything goes can fade away. To prevent being left with a pile of leftover fasteners, make a photographic record of your work as the engine is removed. An inexpensive snapshot camera (or better still, a discount store Polaroid) will provide the pictures to jog your recollection of just where the vacuum lines are attached and how the wires are routed. If possible, reinstall fasteners in their original holes. Paper bags, coffee cans and plastic margarine tubs are all good storage containers for bolts and brackets. Try to keep the parts in logical groups, like "bellhousing bolts" and "radiator fasteners."

Accomplished racers can change a Camaro engine in less than an hour if they set their minds to it. First-time engine exchangers shouldn't try for such a quick turnover; reserve a full weekend for the job if you've got a fresh powerplant already on the stand ready for installation. Anticipate the parts you'll need to prevent making six runs to the auto supply on Sunday afternoon. Have you got exhaust packing, a new throwout bearing, fresh belts and water hoses, a new oil filter, thermostat gaskets and all the rest? If so, you're ready to become an engine exchange student.

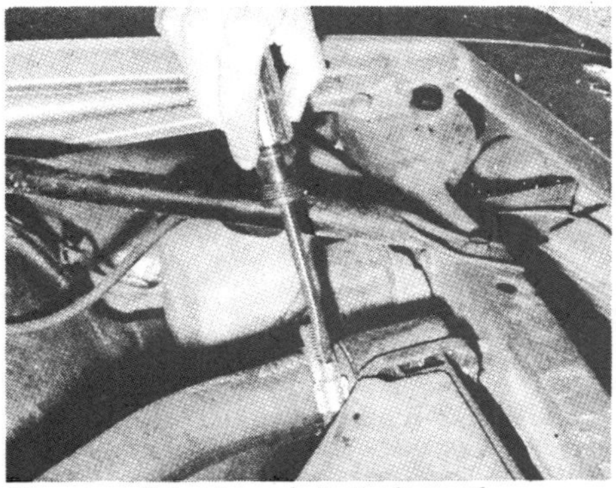

Upper radiator hose is last piece of plumbing to be disconnected.

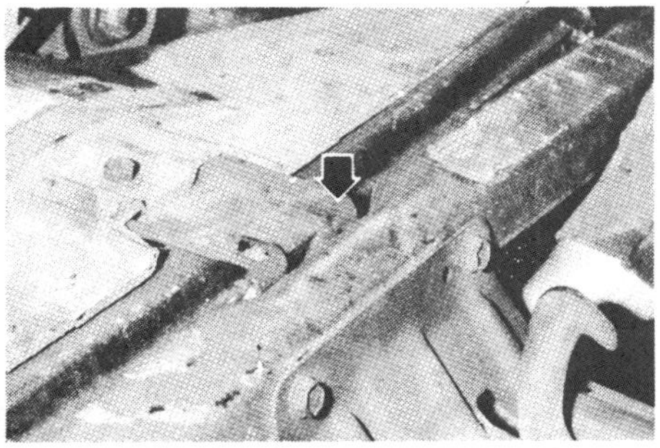

Two bolts for fan shroud brace are well hidden under radiator mount. If they are too difficult to remove, wait until radiator has been removed before working on them.

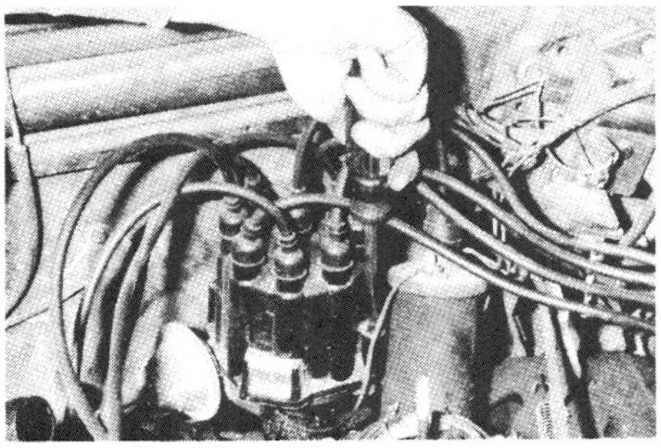

Remove air cleaner, disconnect throttle linkage, and unfasten distributor cap. Remove ground straps from firewall to valve cover hold-downs.

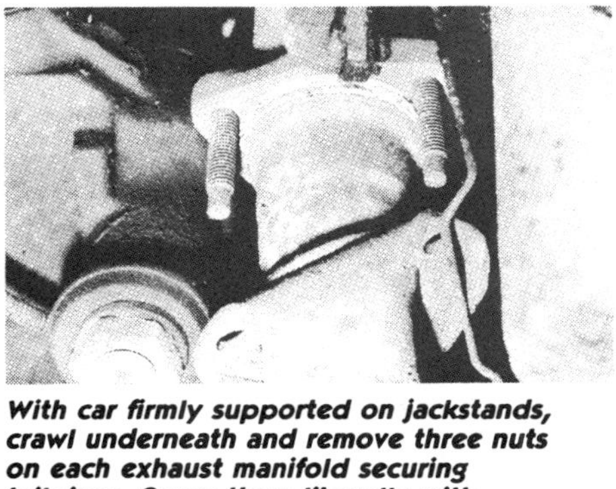

With car firmly supported on jackstands, crawl underneath and remove three nuts on each exhaust manifold securing tailpipes. Spray them liberally with penetrating oil before removing. And don't be surprised if one or more studs breaks off before the nuts loosen. Remove the clutch linkage and bellhousing dust cover (for automatics, remove dust cover and three flexplate-to-converter bolts).

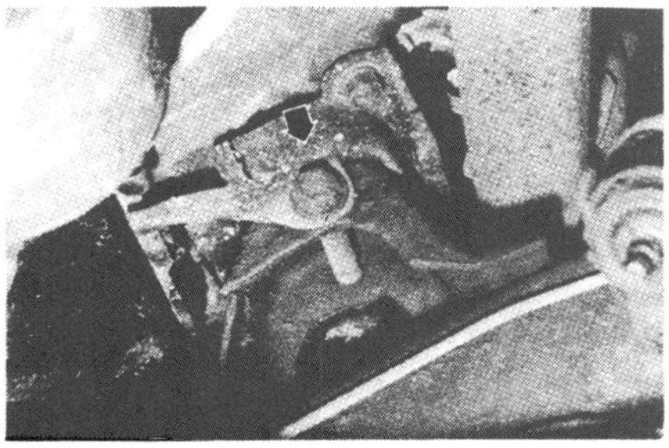

Remove nuts on engine mount through-bolts and disconnect wires to starter solenoid. Now is also the time to remove all the bellhousing-to-block bolts you can reach from underneath the car. A long extension with a universal 9/16-inch socket is an invaluable tool for this operation.

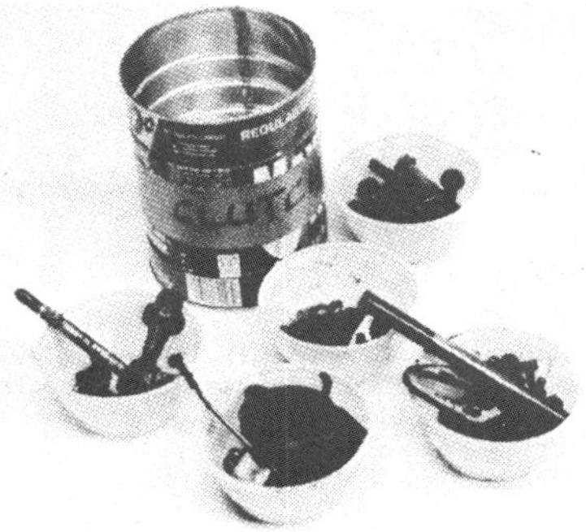

If you've been conscientious so far, your margarine cups and coffee cans should contain all the fasteners for each component.

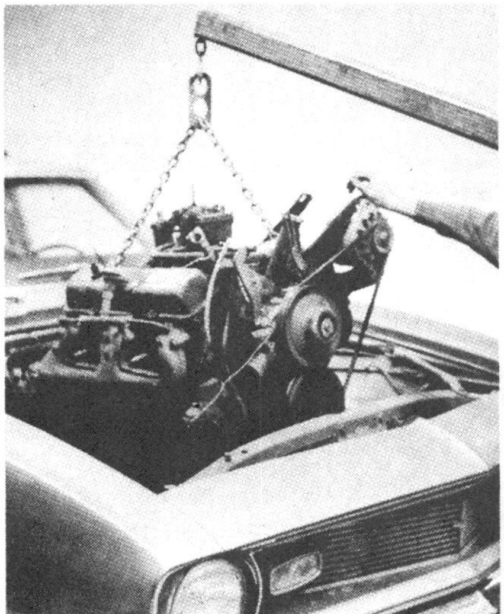

Upper two bellhousing bolts will have to be removed from above. Attach engine hoist to block and support transmission with jack. Raise hoist boom until weight comes off engine mounts, then pull out two through-bolts. Make a final check that everything is disconnected and hoist away, simultaneously lifting and pulling forward on block to disengage transmission input shaft.

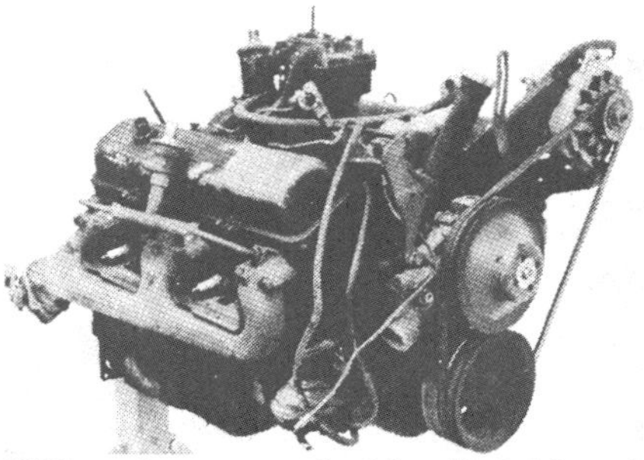

With engine removed, strip off clutch and flywheel. Now is not the time to start madly removing parts and brackets. Take photos or make notes to serve as guides when it comes time to reinstall all the hardware.

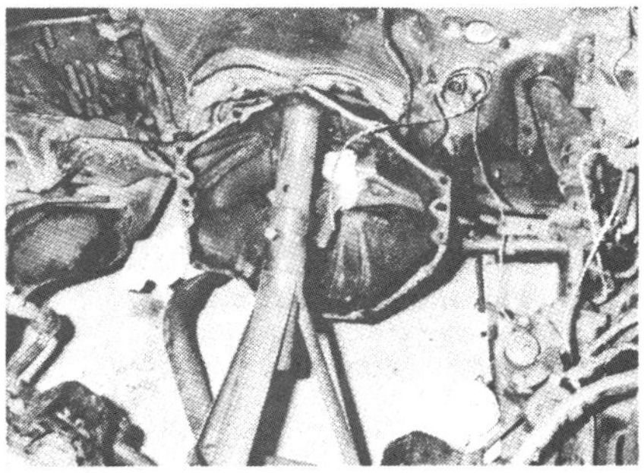

Bellhousing can be supported by wedging jackstand between bellhousing and front crossmember. With automatic transmission, torque converter should be secured to prevent it from falling off input shaft.

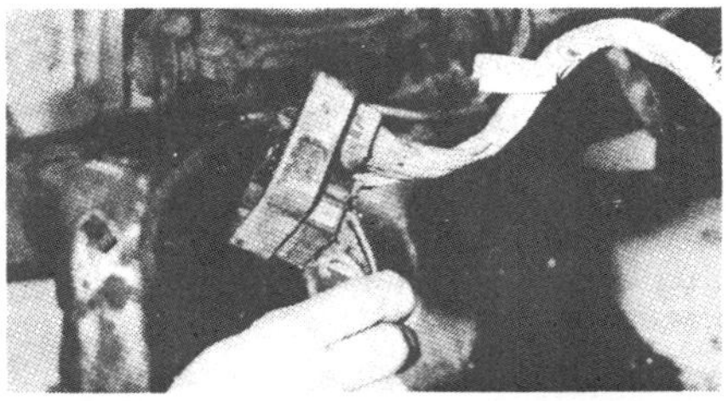

Main wiring harness unplugs from driver's side firewall. Mask off junction box before painting engine compartment.

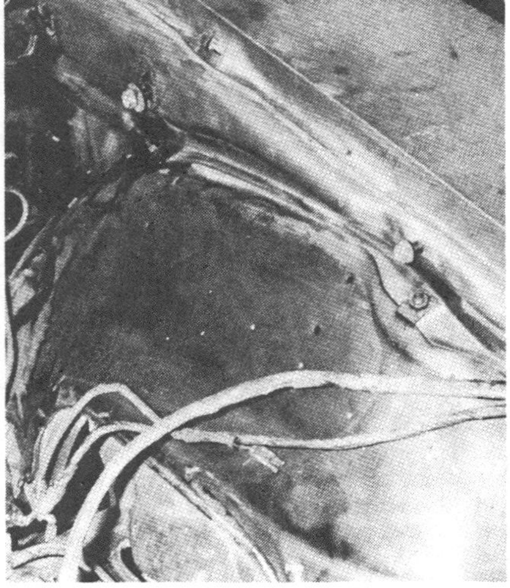

After final degreasing, all surfaces are wiped down with acetone and a tack rag.

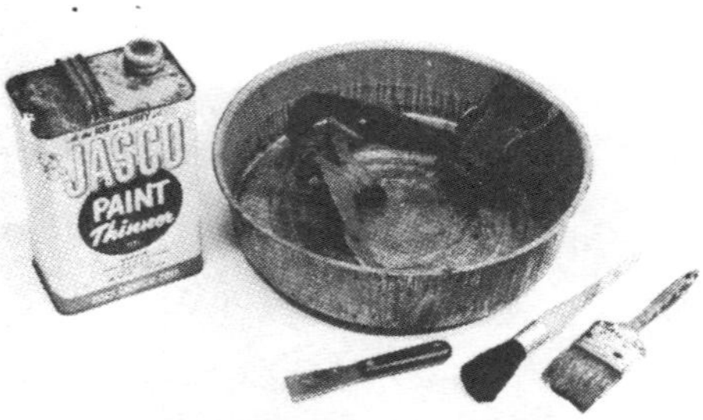

Brackets and small hardware are soaked in solvent. Putty knife and wire brush are good for degreasing chores, while an old paint brush finishes the cleaning job.

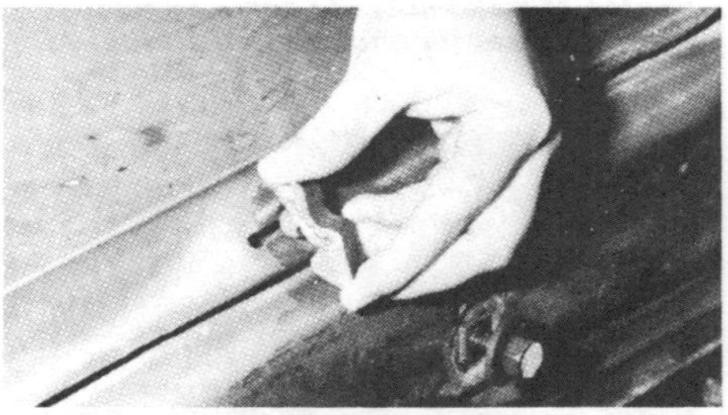

Rubber parts such as hood bumpers should be cleaned with household detergent and treated with Armor-all.

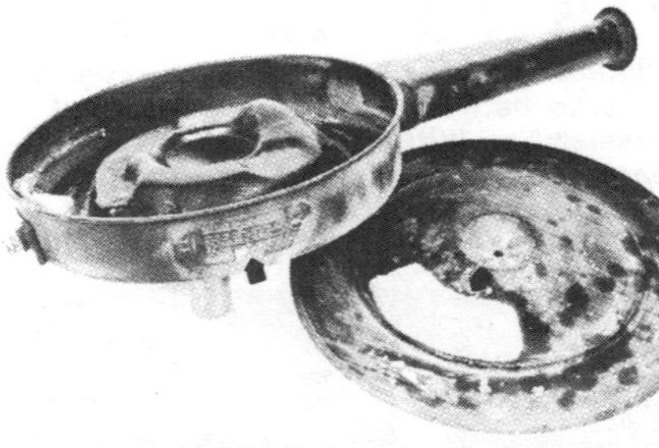

Air cleaner has service instructions and engine designation stickers which should be replaced for "factory original" appearance.

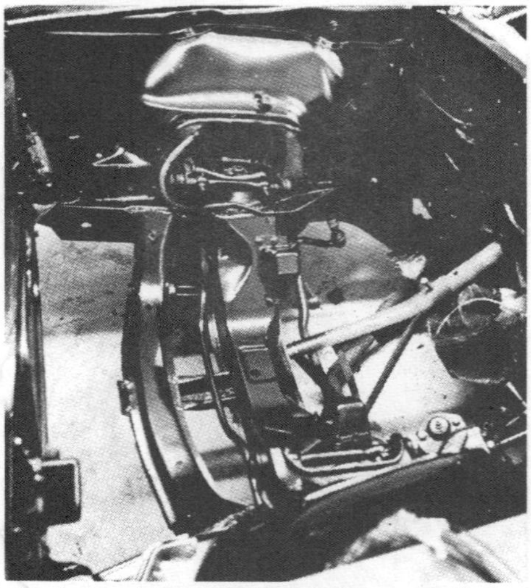

Semi-gloss black acrylic paint is sprayed with 35psi pressure. Small touch-up gun is best for this application.

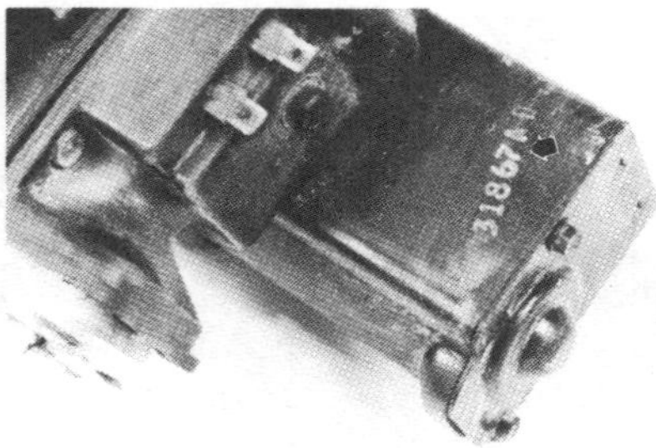

Thoughtful Camaro restorer is careful not to paint over original part numbers, such as these yellow numerals on wiper motor.

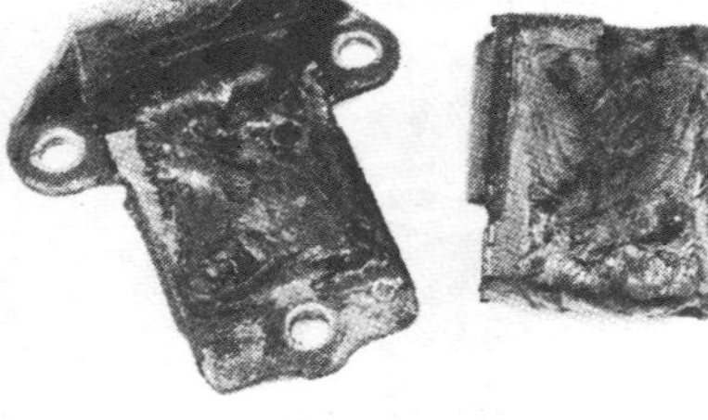

Close inspection of motor mounts revealed cracks between rubber insulator and steel bracket. This mount is overdue for replacement.

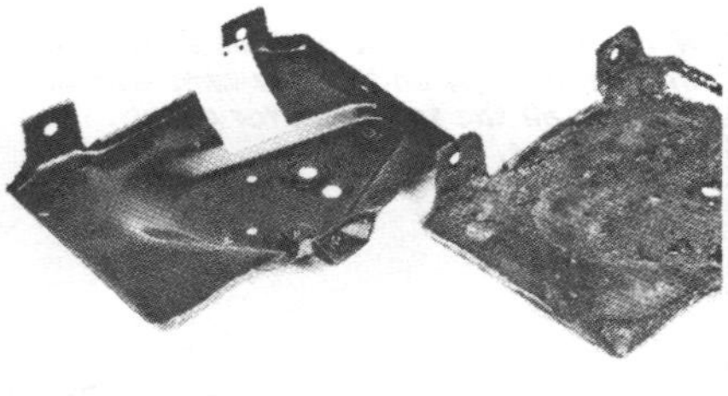

Original battery trays and hold-downs are still available from Chevrolet parts division for early Camaros.

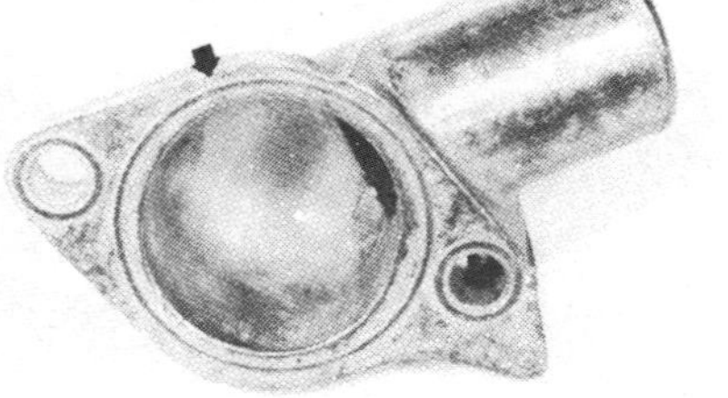

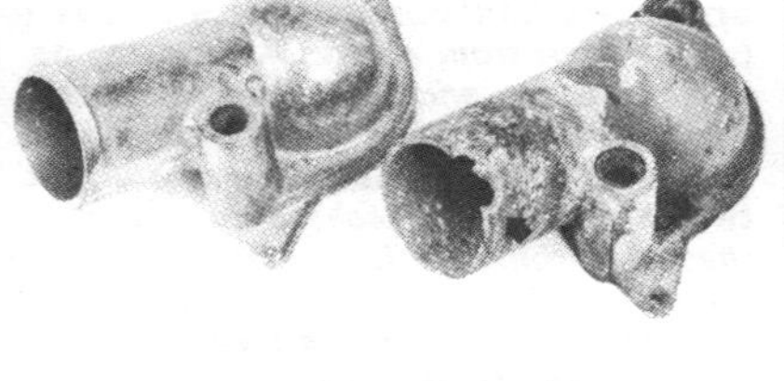

Electroylsis corrodes original aluminum water neck, causing seepage around radiator hose. New water neck has well-defined ridge around base to prevent leaks at thermostat.

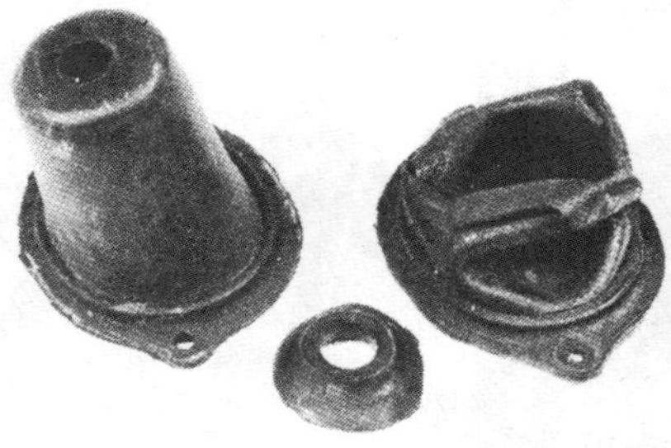

Deteriorated rubber boot for clutch linkage (right) lets engine noise and fumes get into passenger compartment. A replacement is inexpensive.

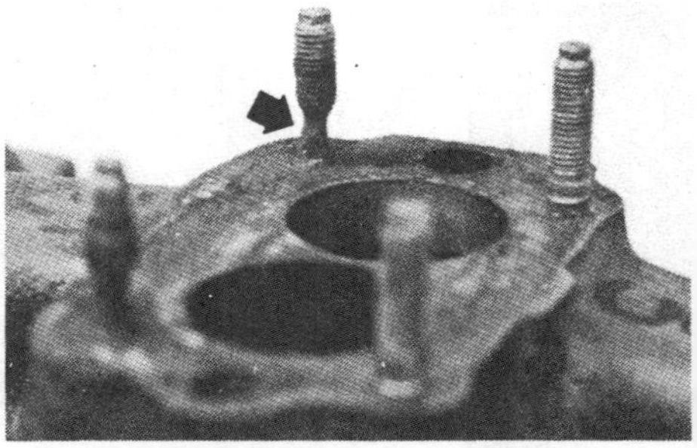

Exhaust gases from heat riser passage corrode front carburetor studs. Old studs usually must be drilled out and holes retapped.

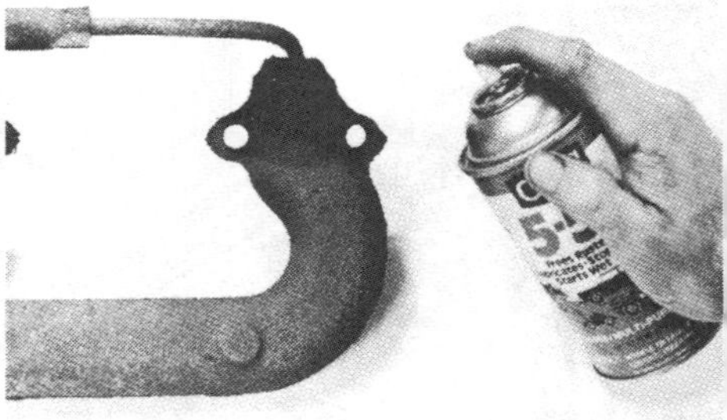

Frozen air injector pipes can often be loosened if threads are soaked with penetrating oil before sandblasting exhaust manifolds.

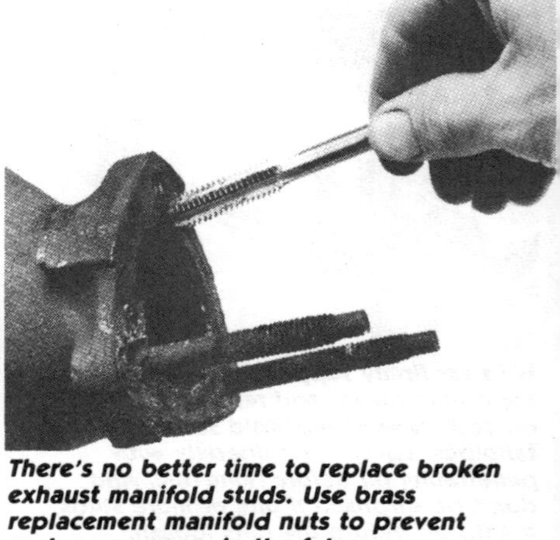

There's no better time to replace broken exhaust manifold studs. Use brass replacement manifold nuts to prevent such occurrences in the future.

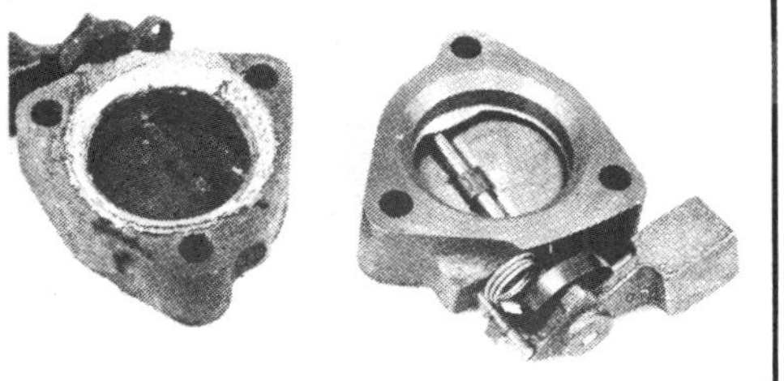

Old worn heat riser valve is often the source of exhaust leaks and long warm-ups. New valve is easy to install during engine exchange.

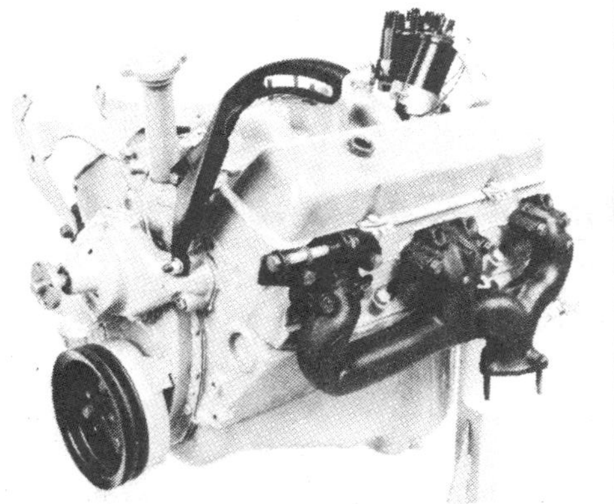

Replacement engine is outfitted with original accessories. Block is spray-painted Chevy orange, exhaust manifolds black.

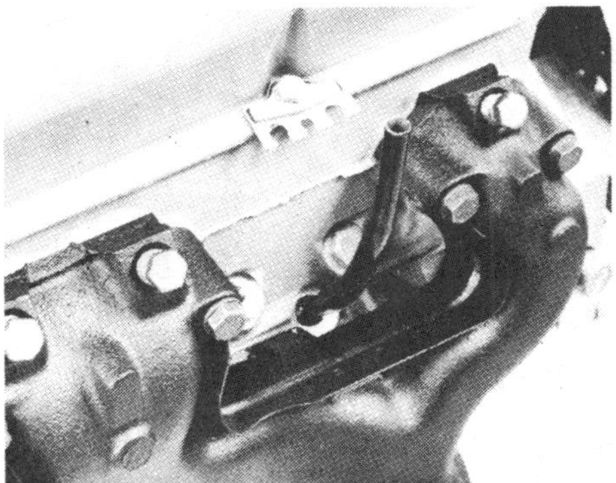

Oil dipstick tube presses into new block.

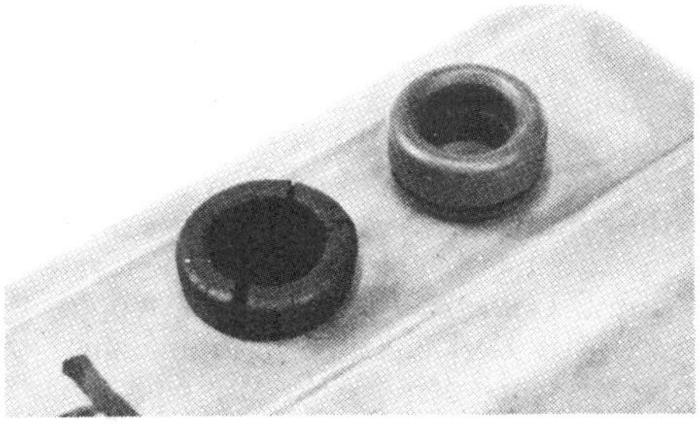

Dried and cracked valve cover grommets are replaced with new Chevy pieces to forestall oil leaks on clean engine.

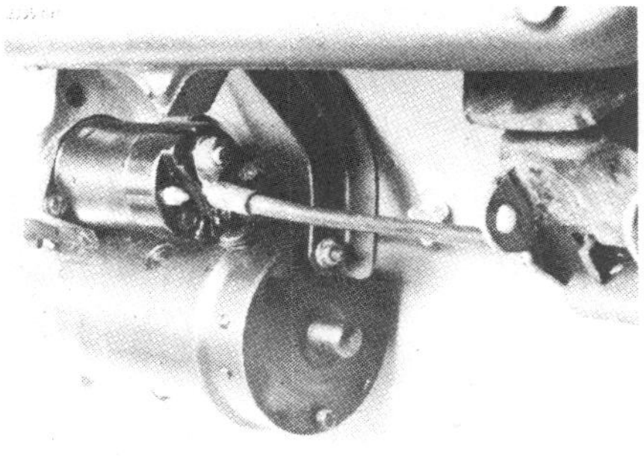

Starter support strap is necessary to prevent cracking starter noses. New battery cable is retained with clips under oil pan bolts.

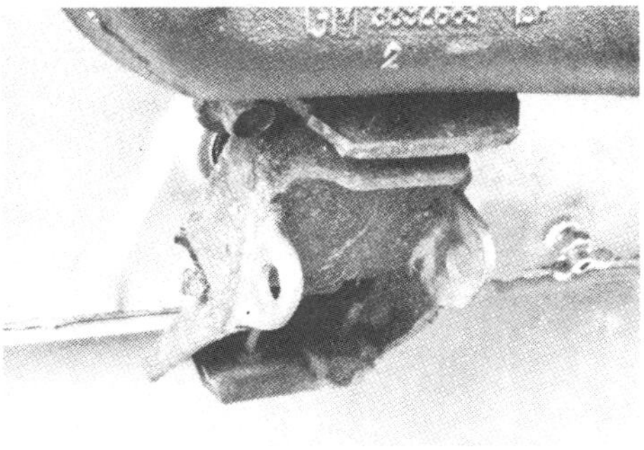

New "safety-type" engine mounts have positive interlock to prevent breakage. Tighten bolts to 30 ft.-lbs.

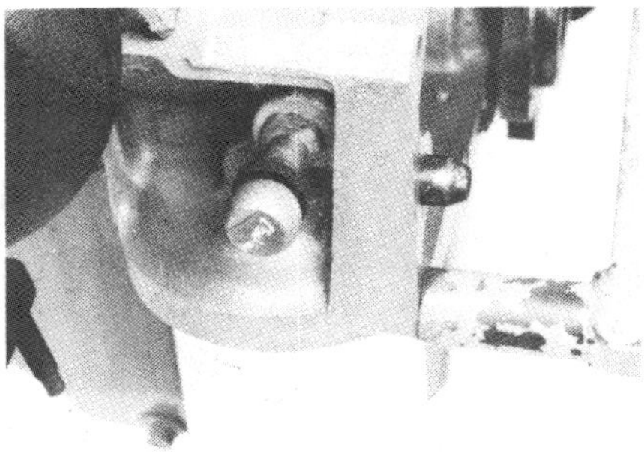

Clutch cross-shaft pivot screws into block above oil filter. Ball should be well greased.

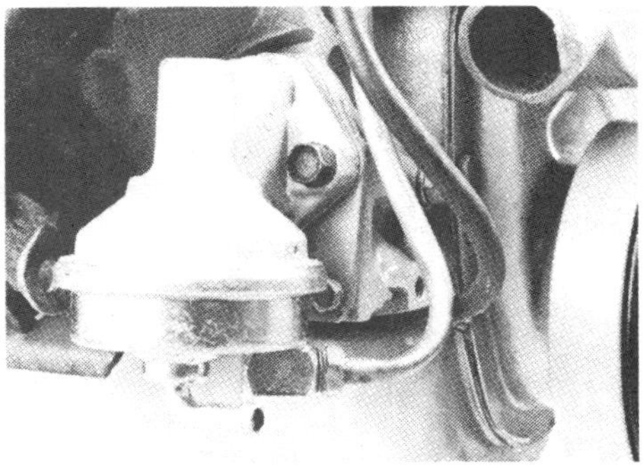

Steel fuel line is cleaned with steel wool, painted with clear acrylic to prevent rust.

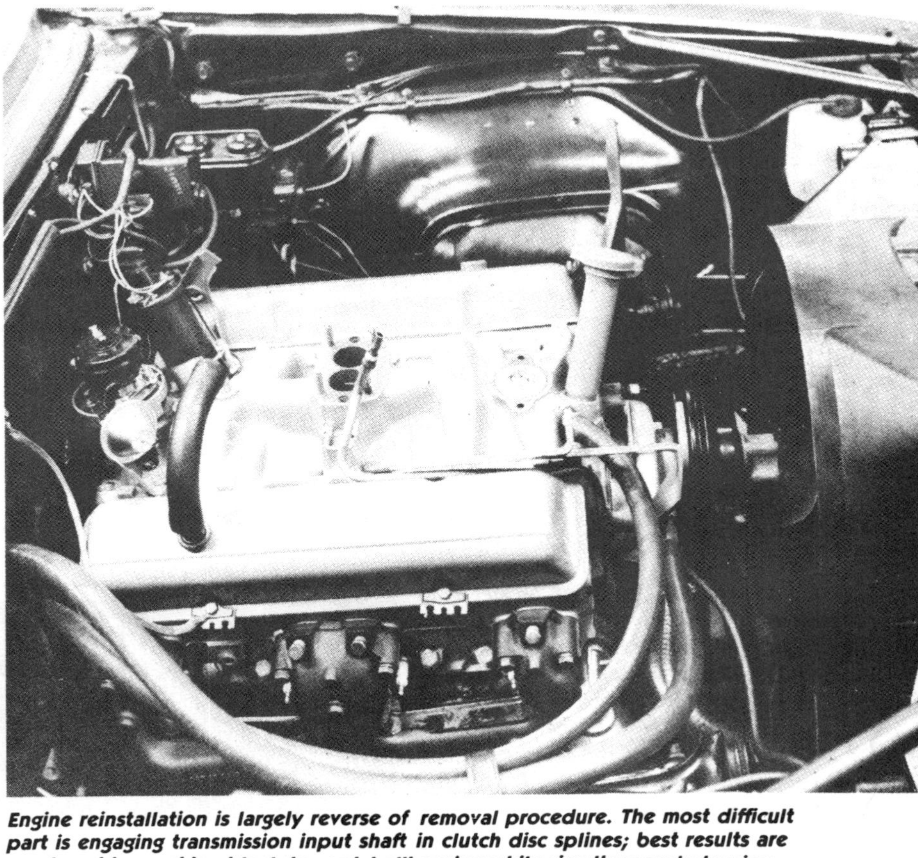

Engine reinstallation is largely reverse of removal procedure. The most difficult part is engaging transmission input shaft in clutch disc splines; best results are produced by pushing block toward bellhousing while simultaneously turning crankshaft. (Engine installation with automatics is much easier—make sure that torque converter fully engages transmission pump, and guide hub on converter into crank flange.) Motor mount through-bolts are tightened to 55 ft.-lbs., bellhousing bolts to 30 ft.-lbs. With new hoses and reconditioned radiator, this Camaro is ready for another 100,000 miles!

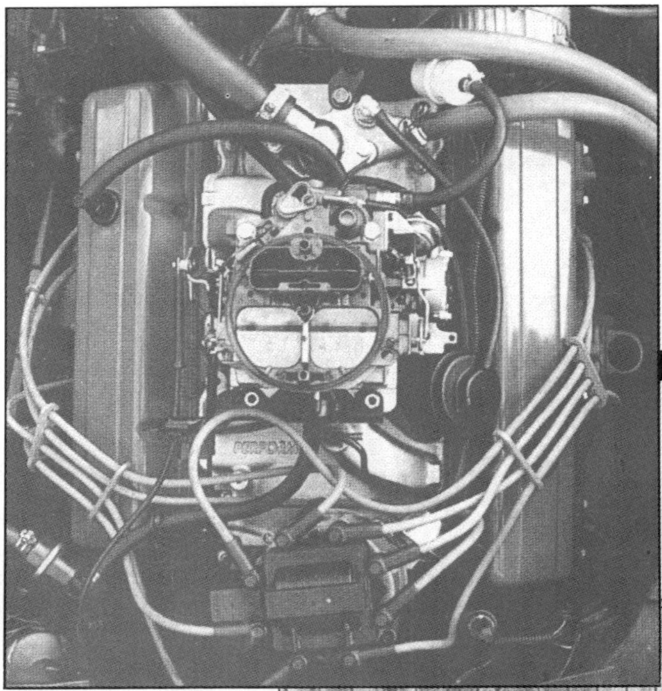

CLEAN AIR CAMARO

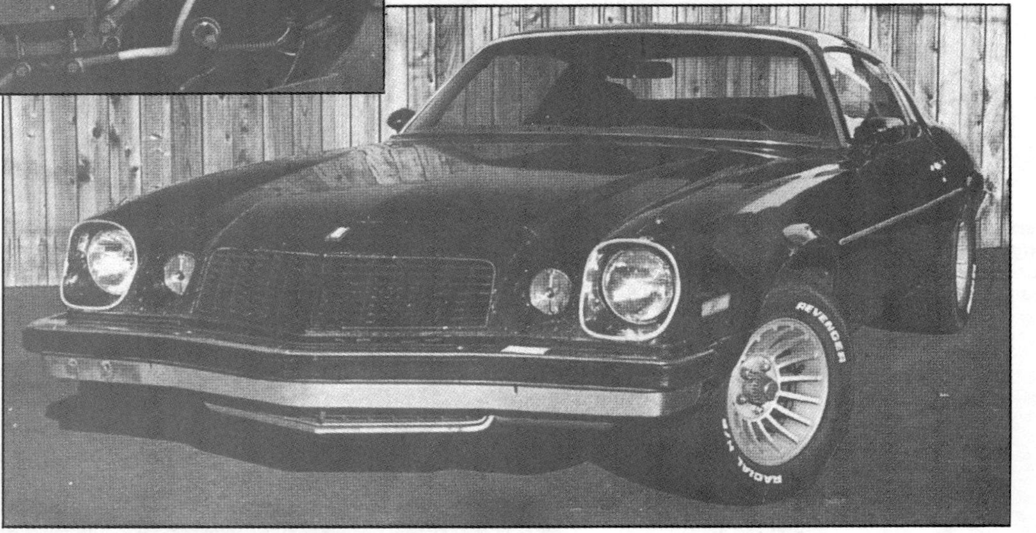

A HOT ROD THAT WILL LET YOU BREATHE EASY

By Cole Quinnell

> "Signs, signs, everywhere a sign
> Blocking out the scenery,
> Breakin' my mind
> Signs say: Do this, don't do that
> Can't you read the signs?"
>
> The Five-Man Electrical Band

The signs of our times say that hot rodding is moving toward heavy regulations and increased emissions requirements. There are those who believe that Big Brother has too heavy a hand in performance automobiles and that rodders have been singled out as a special interest group and penalized as such. Despite this feeling, we're going to see what we can do to legally improve performance on an emissions-controlled vehicle. Believe it or not, there are numerous things we can do! Even with mid-'70s cars—perhaps the most heavily smog-laden cars—respectable performance is obtainable while staying within the letter of the law.

Our plan was to take a slow, dirty, and illegal road urchin and transform it through the miracle of smog-legal hi-po parts into a fast, clean, and oh-so-legal street stormer! We started with a '76 Camaro that had been victimized to the point that every piece of emissions equipment had been removed and replaced with what most people would consider typical hot rod parts such as headers, a non-EGR Performer manifold, and a universal Holley carburetor. The plan was to baseline the car in its illegal condition, testing for emissions and quarter-mile performance. We would then return the car to stock, retest it, add a few certified emissions-legal performance components, and test it a third time.

Unfortunately, the Camaro had a very tired 307 that didn't belong between the fenderwells. A phone call to California Discount Warehouse (CDW) landed a 350 replacement engine in the engine bay. We opted for forged replacement pistons in our engine kit and used Dorman's HPX-series fasteners for durability, since we knew that we would be flogging the car on the strip, trying to squeeze every last hundredth out of this emissions Mouse.

The tailpipe emissions changes varied according to the test conditions between the 307 and the new 350, but they didn't increase greatly. It's only fair to admit that we added a smog-legal cam (a Crane 266) in order to save time when we later changed the remaining performance parts. We also rebuilt the distributor during the motor change when we discovered that the mechanical and vacuum advance were nonoperational!

We were still running from the smog cops, but with the newer 350, we improved our dragstrip e.t. by 1.22 seconds and did not significantly raise the emissions. The next step was to meet

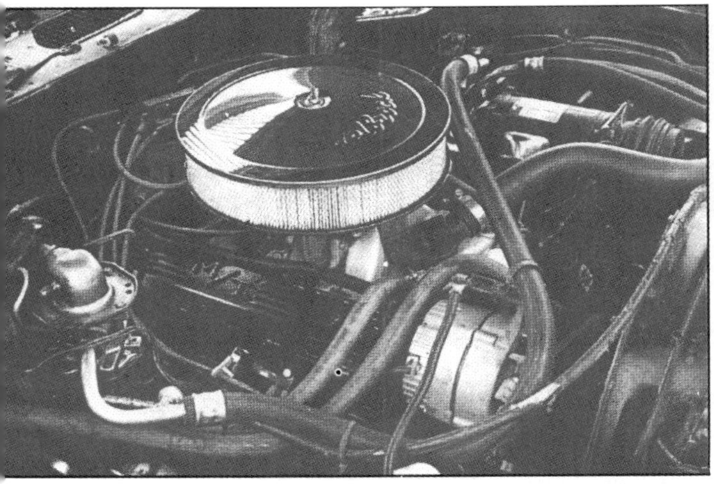

The setup under the hood was aesthetically appalling. The Camaro came with a very tired '68 307 that was outfitted with headers, an Edelbrock Performer, and a universal replacement Holley, which had seen better days. This combination violates both federal and California emissions regulations, so out it came!

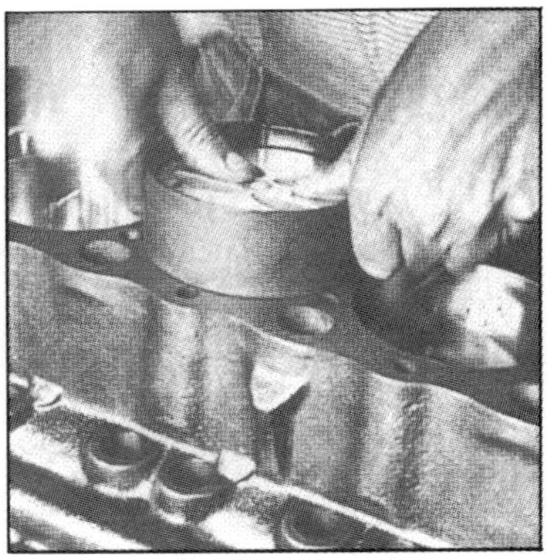

The first step was to install a replacement '76 350. To re-establish our baseline we reused all of the original 307 bolt-ons. We decided on a short-block kit from CDW that came complete with the block, crank, rods, forged replacement pistons, rings, bearings, and Fel-Pro gaskets. We also used a pair of CDW's assembled stock production '76 replacement heads.

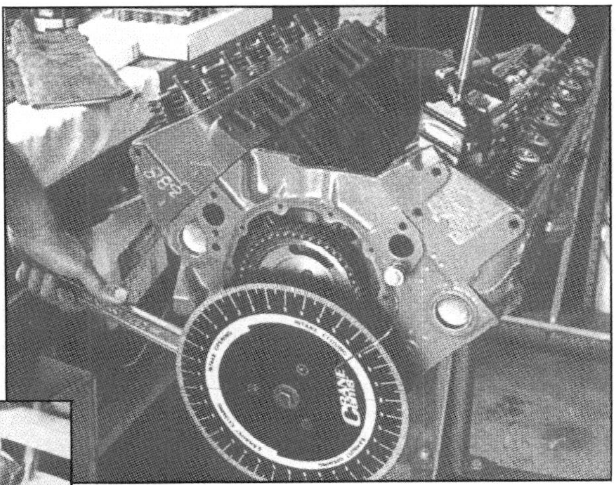

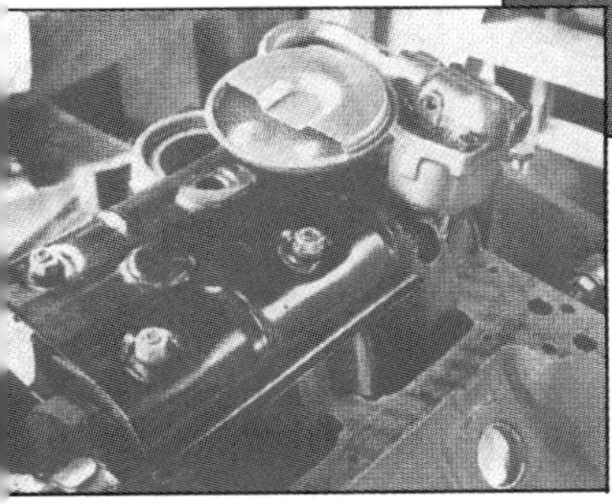

OK, we admit it. We cheated a little bit as far as re-establishing our baseline. Due to the amount of effort and time required to change a cam, we installed the Crane 266 cam before installing the 350. We also took the time to degree the cam to ensure timing accuracy.

On the bottom end, we added a TRW high-volume oil pump and a GM windage tray. Cone Chevrolet in Fullerton, California, usually has these in stock, as well as Z/28 oil pans that are designed to work with the windage tray. The windage tray will keep oil from splashing onto the crank, robbing horsepower at higher rpms.

AT THE TRACK

Engine Configuration	e.t.	mph
307 Baseline	16.34	85.95
350 Baseline (with headers, Holley, 266 cam, and no emissions equipment)	15.12	95.63

. Test performed at LACR, 3000-foot altitude; e.t. and mph corrected to sea-level performance.

ALL THE RIGHT PARTS

Description	GM part No.
EGR Valve	17052669
EGR-TVS	364788
EEF Canister	17065031
Z/28 Oil Pan	465220
Windage Tray	3927136
Studs for Windage Tray (for '68-and-later 327 and all 350 engines)	14087508

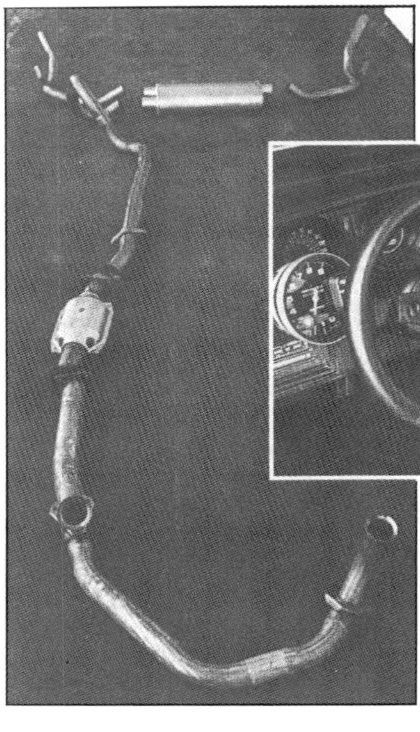

A Walker factory-replacement exhaust system was used, including a high-flow, replacement catalytic converter and a DynoMax muffler.

THE INSIDE STORY

Since building a car is a package deal, we couldn't ignore the Camaro's interior while we were tuning up the performance and reducing emissions. Obviously, the emphasis of this article is on the engine, but the interior needed some help to make this car a true pleasure to drive.

Thanks to the great restoration products that Classic Camaro carries, the interior retains a factory look. The seat covers, headliner, door panels, and all the small detail pieces came from Classic Camaro. The New-Dash Group reskinned our dash, and an Auto Custom Carpet set completed the restored look. We also added a couple of functional performance statements such as the Auto Meter Monster tach, a Grant G.T. steering wheel, and the Hurst Quarter Stick shifter. Between the new interior and a much quieter exhaust that doesn't set off car alarms anymore, the Clean Air Camaro has already become a blast to drive!

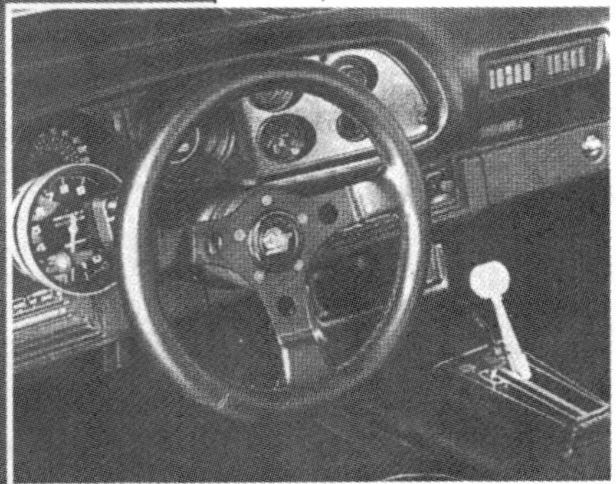

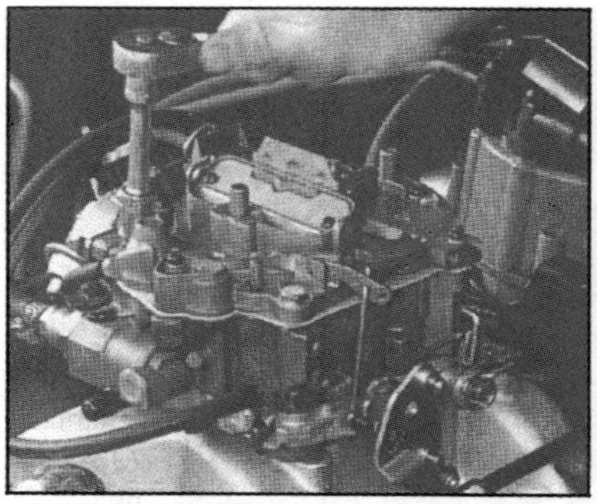

To top off this emissions-legal 350, Brad Urban of The Carburetor Shop assembled a stock-calibrated replacement Q-jet that has all of the correct vacuum ports.

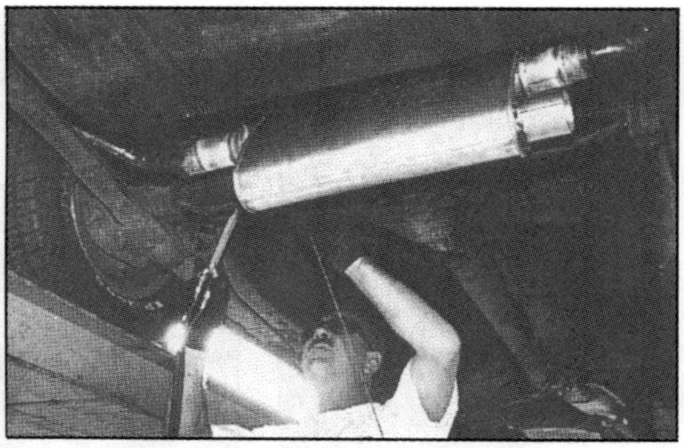

When using Walker's complete factory-replacement components, the system is easy to install at home. We incorporated a DynoMax muffler into the system that required a little adapting and welding, so the crew at Performance Specialties helped us out. They specialize in exhaust systems that are designed to optimize an engine's combination.

the sniffer requirement for a catalytic-equipped car.

We began the road back to legal condition with the exhaust. Using stock manifolds, we traded our headers and very loud header mufflers for a complete Walker exhaust system, using a replacement high-flow catalytic converter and a hi-po DynoMax muffler. The addition of a catalytic converter alone will drastically improve emissions, but at the same time, the manifolds and converter will undoubtedly hurt performance.

We've already mentioned that we installed a Crane 266 cam that is legal for all passenger-car Chevy small-blocks (350 cid and smaller) in pre-'80, pre-feedback carbureted cars. The cam offers 210/216 intake/exhaust degrees duration at .050- and .440-inch intake and a .454-inch exhaust net lift when using 1.5-inch rockers. Crane's 1.6-inch roller rockers, which are also legal for this application, would increase net lift to .469-inch intake and .484 of an inch on the exhaust.

From here, we turned our attention to the induction side of the engine. We swapped the early Edelbrock Performer for an Exhaust Gas Recirculation (EGR)-equipped Performer, which is considered a replacement part on engines that originally came with EGR valves. Next, we added a stock replacement '76 Q-jet from The Carburetor Shop, complete with all the necessary vacuum ports.

A fact-finding effort revealed that this particular Camaro, a 49-state, federally equipped car, didn't come with an AIR pump, but it did need a new EGR valve, an EGR thermal vacuum switch, and an early evaporation fuel (EEF) system. These items are still available from Chevrolet; we obtained ours from Cone Chevrolet in Fullerton, California. We also needed a vacuum-hose-routing diagram that our pal Mark Sanchez provided to help us find our way through the vacuum hose maze.

With all these components and the factory air cleaner in place, we were completely legal. The question was whether the car was still faster with all the smog eq,uipment in place than with headers, a Holley, and an open-element air cleaner. Unfortunately, the story deadline didn't allow time for track and emissions testing. In the next issue of HRM, we'll have the results from this emissions-legal configuration. We'll also see if we can improve performance and find out how quick this Clean Air Camaro can go! **HR**

SNIFFER RESULTS

Our testing for this story took place at Edelbrock's emissions lab that (unlike most local smog stations) has the capability to measure NO_X in a loaded, dynamic test. We expected tailpipe emissions to be worse with the 350 compared with the 307, but that wasn't necessarily true. By adjusting the idle-mixture

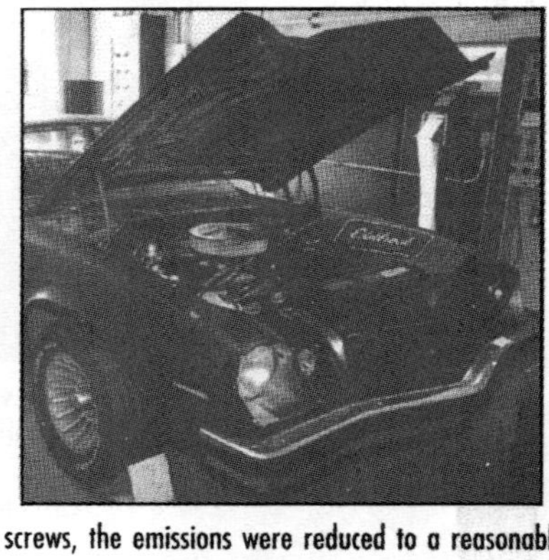

screws, the emissions were reduced to a reasonable level for a noncatalytic-equipped car. Keep in mind, even though it was cleaner, the car was still illegal without the catalytic converter and other smog pieces.

SNIFFERS AT 10 PACES

engine	rpm	loaded	O_2%	CO_2%	CO%	NOx ppm	HC ppm
307	idle	no	14.0	10.5	5.6	128	804
307	2500	no	14.0	13.6	1.6	226	231
307	2500	yes	13.7	11.3	4.98	564	404
350/dirty	idle	no	6.3	9.0	2.3	90	211
350/dirty	2500	no	5.2	11.1	0.4	326	153
350/dirty	2500	yes	1.7	11.8	3.0	989	269

SOURCES

Auto Custom Carpets, Inc.
Dept. HR
P.O. Box 1167
1429 Noble St.
Anniston, AL 36202
(205) 236-1118

Auto Meter Products, Inc.
Dept. HR
413 W. Elm St.
Sycamore, IL 60178
(815) 895-8141

California Discount Warehouse (CDW)
Dept. HR
2320 E. Artesia Blvd.
Long Beach, CA 90805
(310) 423-4346

The Carburetor Shop
Dept. HR
1457 Philadelphia Unit 24
Ontario, CA 91761
(714) 947-7744

Classic Camaro Parts & Accessories
Dept. HR
17832 Gothard St.
Huntington Beach, CA 92647
(714) 894-0651

Cone Chevrolet
Dept. HR
600 W. Commonwealth
Fullerton, CA 92632
(714) 871-9110

Crane Cams
Dept. HR
530 Fentress Blvd.
Daytona Beach, FL 32014
(904) 252-1151

Dorman Products
Dept. HR
1 Dorman Dr.
Warsaw, KY 41095
(606) 567-7000

Edelbrock Corporation
Dept. HR
2700 California St.
Torrance, CA 90503
(213) 781-2222

Fel-Pro
Dept. HR
P.O. Box 1103
7450 N. McCormick Blvd.
Skokie, IL 60076-8103
(708) 674-7700

Grant Products
Dept. HR
700 Allen Ave.
Glendale, CA 91201
(213) 849-3171

Hurst Peformance
Dept. HR
8700 Brookpark Rd.
Cleveland, OH 44129
(216) 398-3800

NewDash Group
Dept. HR
12221 S. Appaloosa Dr.
Phoenix, AZ 85044
(602) 893-3030

Performance Specialties
Dept. HR
397 W. Baseline
San Bernardino, CA 92405
(714) 889-2329

Walker Manufacturing
Dept. HR
1201 Michigan Blvd.
Racine, WI 53402
(800) 767-DYNO

CLEAN AIR CAMARO PART 2

LEGAL PERFORMANCE

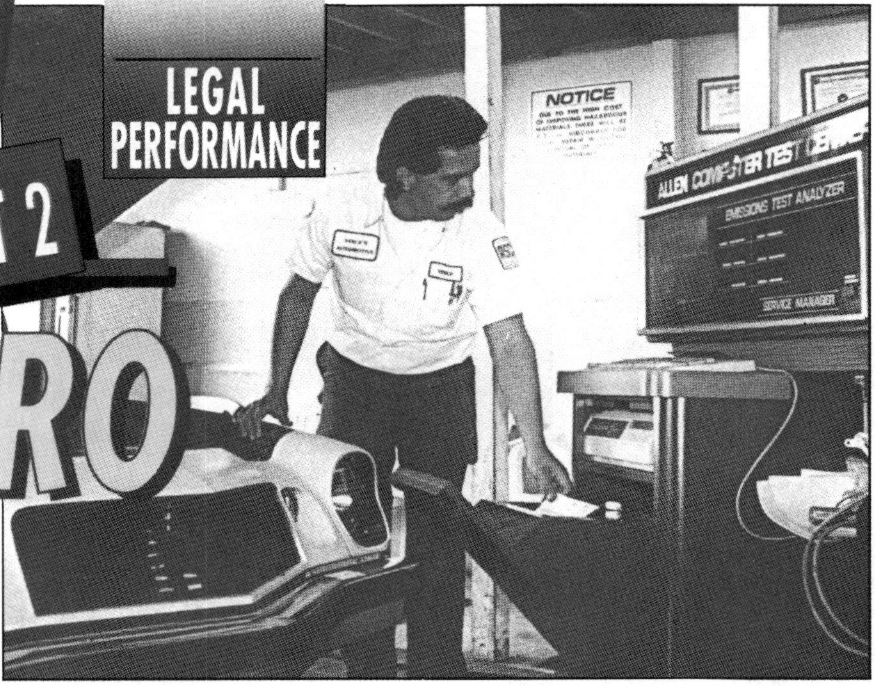

CONVERTING A DIRTY SLUG INTO A CLEAN PERFORMER

By Cole Quinnell

When the tire smoke had cleared and the numbers were crunched, we had reached our goal. We started last month with a terribly illegal, slow '76 Camaro and transformed it into an emissions-legal performer. On the surface that appears easy, right? After all, those cars were slugs to begin with. The '76 model was equipped with a four- barrel and a TH350 that ran a yawn-inspiring 16.10 e.t., according to the HOT ROD archives. We figured we could do better than that, even with one plug wire tied behind our back! And, in a matter of speaking, that's what we did.

The handicap: We could only use emissions-legal parts, which presented an interesting challenge. Last month we installed a replacement engine from California Discount Warehouse, slid in a Crane cam, added an Edelbrock Performer intake, a K&N air filter, a Walker factory exhaust, and all the necessary emissions equipment.

In this configuration, the car ran a slower quarter-mile compared with the 350 equipped with headers, header mufflers, a Holley carburetor, a non-EGR Performer, and an open-element air cleaner. It ran slower due to the restrictive cast-iron exhaust manifolds and catalytic converter. Power was also lost due to the single-snorkel air cleaner versus the open element used in the previous test. We're still a little over .60-second quicker than a stocker, but that wasn't good enough.

The most significant restriction of power was the exhaust system. A set of headers was exactly what we needed to boost bottom-end torque and allow the engine to breathe better throughout the powerband. The solution came in the form of Doug Thorley headers that are emissions-legal for our '76 Camaro application. A custom two-into-one pipe must be constructed to direct the exhaust gases through the catalytic converter, while maintaining the converter *in the stock location.* Ron Scranton's Performance Specialties was able to construct this pipe for us and install the exhaust system in our Camaro. We also replaced the Walker high-flow cat used in the first installment with Walker's new Super Cat. Although Walker offers the Super Cat with inlet/outlet sizes as large as three inches, we selected a 2½-inch inlet version. Walker's new cat offers greater flow capabilities because of a larger frontal area in conjuction with a larger inlet/outlet. Legally, an exhaust shop cannot replace your catalytic converter unless it is damaged, not functional, or missing; we made this change solely for test purposes.

Then it was time to test our legal combination and tune the engine for maximum performance. A couple of quarter-mile passes were made with the factory air cleaner in place and General XP2000 street tires. Then we got serious. A pair of Mickey Thompson 26x10.5 Sportsman Pros came out of the trunk (actually, one was in the back seat because the trunk is too small). All of our tires were mounted on Cragar Racing Superlite wheels to keep the rotating mass as low as possible and to improve the appearance of the car.

We also installed a K&N 14-inch open-

THE NUMBERS GAME

The parts involved in this buildup are very important because they are emissions-legal components. Some have a California exemption Executive Order (E.O.) while others are marketed as replacement parts and don't require an E.O.

COMPANY	PART	PART NO.
Crane	camshaft	113931
Crane	1.6 roller rocker arms	3005
Edelbrock	EGR Performer	3701
Doug Thorley	headers (without AIR fittings)	375Y
Doug Thorley	hot air pick-up	HA112
Jacobs Electronics	Energy Team	370412
Jacobs Electronics	HEI distributor	390402
K&N Engineering	air filter	E-1500
Walker Manufacturing	Dynomax crossflow muffler	21598
Walker Manufacturing	Super Cat	15142

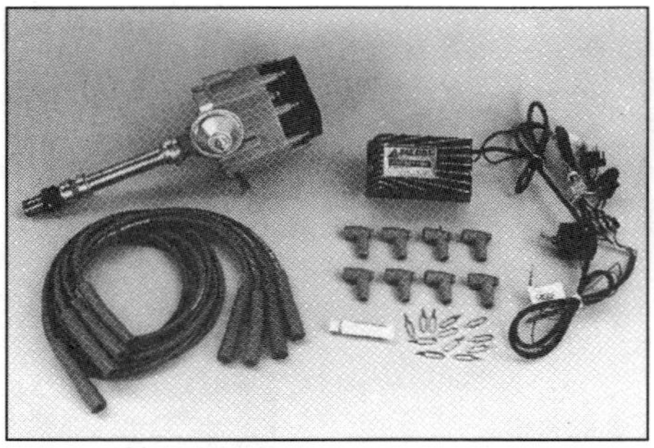

By installing a complete Jacobs ignition, we eliminated the ignition system as a possible source of inconsistency. In addition to its Energy Team, which consists of an ignition computer and Energy Core spark plug wires, we added a Jacobs blueprinted HEI distributor. We also used SplitFire spark plugs to ensure a quality spark.

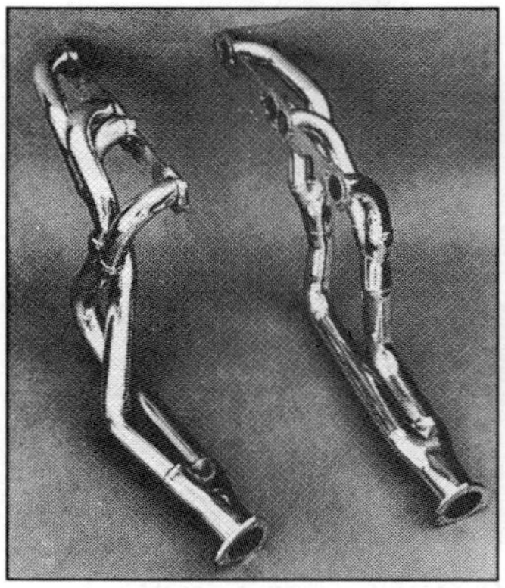

We felt that the exhaust was definitely a limiting factor, starting with the exhaust manifolds. Doug Thorley offers Tri-Y headers that are legal for 350-and-smaller V8 '75-'79 Chevrolet passenger cars with or without an AIR pump. The Tri-Y design helped the bottom-end torque and enabled the headers to fit perfectly.

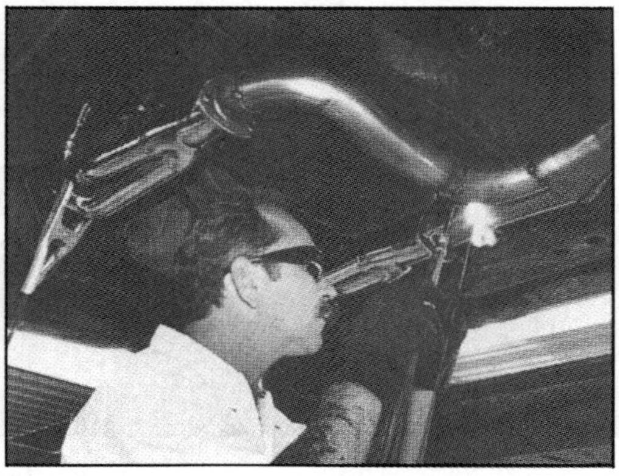

Ron Scranton at Performance Specialties constructed the two-into-one pipe that is necessary to route the exhaust gases into the catalytic converter. To be technically legal, the converter must be located in the stock location. Scranton mocked up the tubing necessary to properly locate the converter, tack welded the pipes, then removed the converter before completely welding the custom pipe.

element air cleaner that can produce as much as 14 hp over the factory air cleaner, according to Kevin McClelland at McFarland. Adjusting the timing with a Jacobs timing light and making fuel changes with secondary metering rods from The Carb Shop, we were able to squeeze out a corrected 15.15-second quarter-mile at 92.97 mph! Even this is not the best the car can do. The current limitation is fuel starvation that prevents us from running high 14s at around 94 mph.

So far we've obtained two-thirds of our goal: We've improved the quarter-mile times and equipped the Camaro so that it is technically emissions legal. The last of our three goals is to hook the sniffer to the tailpipe and test this combination for exhaust cleanliness. According to Mark Sanchez at Advanced Engineering West (AEW), the California standards that our '76 Camaro must meet are 1.20 percent carbon monoxide (CO) and 150 parts per million (ppm) hydrocarbons (HC), or unburned fuel at idle. In the car's initial illegal state with an open-element air cleaner, a non-EGR intake, a Holley carb, and without a catalytic converter, the tailpipe emissions were almost double California emissions standards. However, with the emissions equipment and various legal performance components in place, emissions were considerably cleaner than the standard. Edelbrock's sniffer showed 0.0 percent CO at idle and approximately 83 ppm HC!

The proof is in the numbers. We didn't create a 12-second quarter-pounder, but we did put together a strong, easy-to-drive street car that is above and beyond the legal emissions requirements. We improved the bottom end of the powerband significantly through the addition of legal headers and overall power with the Crane 266 camshaft. Everything we've done is technically legal, and there are more improvements that can be realized. For instance, our Camaro is equipped with an open carrier 2.80 rear gear, which partially explains the rather slow 60-foot times in the

CLEAN IN SIGHT, SOUND, AND BODY

In our search for environmentally conscious performance products, we discovered some interesting information. Any body and paint person will tell you that this industry is undergoing at least as much change as the hot rod industry for the sake of the environment. The concept is the same—to reduce emissions. In the automotive refinishing industry, this means changes in the delivery of the paint and the chemical structure of the products. The main focus is on the amount of volatile organic compounds (VOCs) emitted into the air. These VOC reductions are accomplished by using reduced VOC paint products and by cutting down on emitted vapors produced by spray gun equipment.

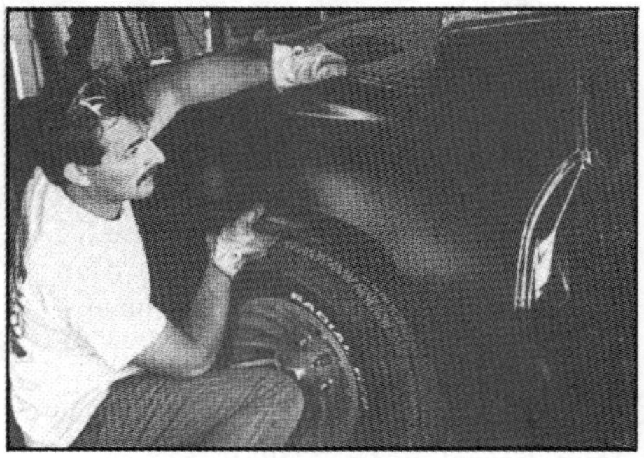

The Clean Air Camaro required metal surgery on the rear quarters. Donny Ekstedt at Santini's Paint & Body Werkes performed the replacement by using quality Classic Camaro quarter panels.

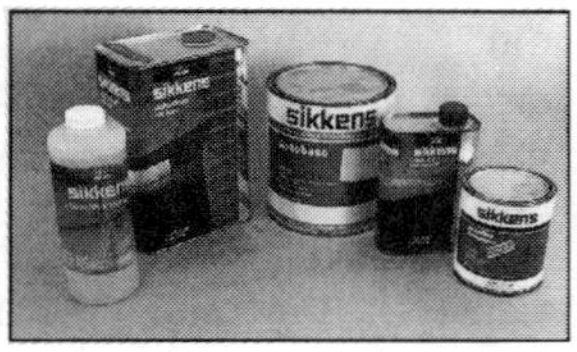

Akzo Coating is one of the refinishing companies that has developed environment friendly products, such as its Sikkens brand paints. It also has a full line of body filler products that are extremely easy to work with.

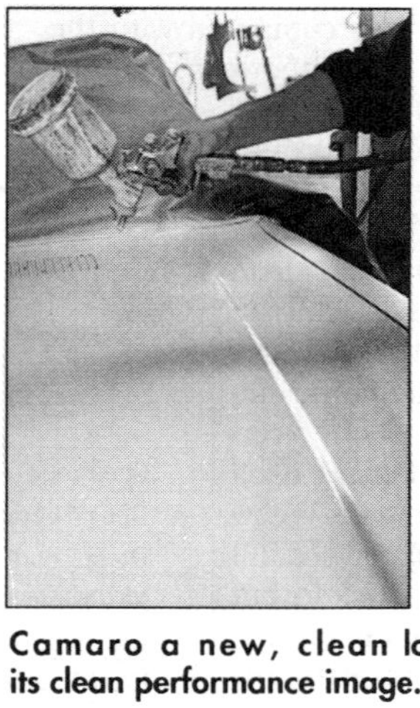

The crew at Western Collision finished the bodywork and sprayed on the color. The company also used Sikkens etching primer and base/clearcoat system, as well as Sikkens-brand body filler. This system gave the Camaro a new, clean look that matches its clean performance image.

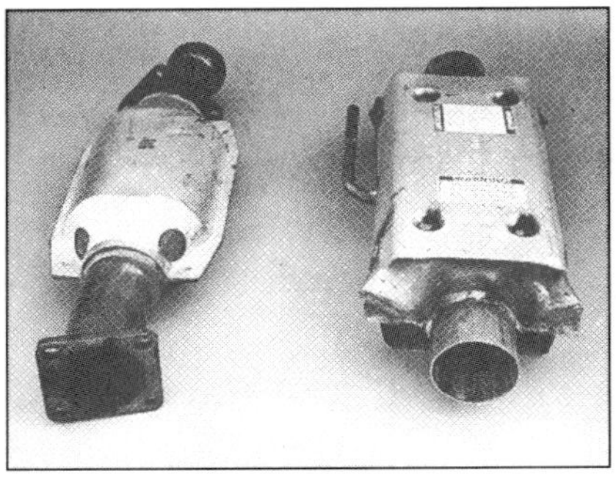

With catalytic converters, the "bigger is better" rule almost always applies. The Walker Super Cat (on the right) is available in up to a three-inch inlet/outlet diameter and has a larger frontal area than a stock, high-flow unit.

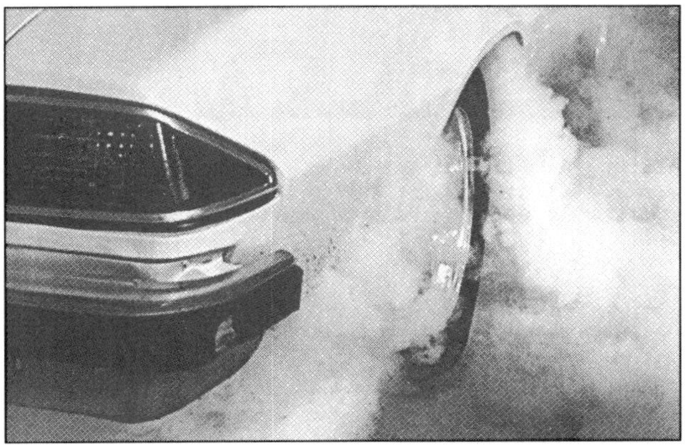

The performance gains were more noticeable during the seat-of-the-pants evaluation than evidenced by the e.t.'s, but the track times did improve. Most of the power improvements were made in the bottom portion of the torque curve, which really makes the car feel strong stoplight to stoplight.

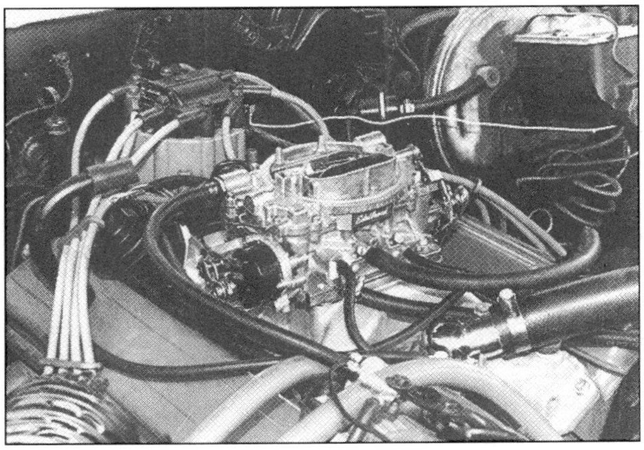

For curiosity's sake, we bolted on an Edelbrock Performer carburetor to evaluate its emissions and performance virtues. The carburetor ran clean on the sniffer and yielded performance figures comparable to the Q-jet at the track! Unfortunately, it is not a replacement carburetor (primarily because a stock air cleaner will not fit properly), making it illegal for our application.

TRACK TALES

Configuration	e.t.	mph
350 with Holley carb and header mufflers	15.12	95.63
350 with exhaust manifolds, Q-jet, and emissions equipment	15.32	91.05
350 with Tri-Y headers, larger catalytic converter, and Jacobs ignition	15.14	93.97

Test performed at LACR (3000 foot altitude); e.t. and mph corrected to sea-level performance

CLEAN THING

engine	rpm	loaded	O_2 (%)	CO_2 (%)	CO (%)	NO_x (ppm)	HC(ppm)
baseline with illegal headers, Holley, 266 cam, and no emissions equipment							
350	idle	no	6.3	9.0	2.3	90	211
350	2500	no	5.2	11.1	0.4	326	153
350	2500	yes	1.7	11.8	3.0	989	269
Crane 266 cam, Performer EGR manifold, and emissions equipment							
350	idle	no	5.3	12.6	0.0	50	99
350	2500	no	0.5	12.8	2.1	216	202
350	2500	yes	0.5	12.8	2.7	1004	289
stock with Edelbrock carb							
350	idle	no	4.2	12.1	0.0	52	81
350	2500	no	0.2	14.6	0.4	31	44
350	2500	yes	0.2	14.5	1.0	640	85
legal headers and ignition							
350	idle	no	3.1	12.8	0.0	52	83
350	2500	no	0.3	13.4	2.3	218	217
350	2500	yes	0.2	13.0	3.0	961	258
California standard for '76 Camaro with 350 engine							

2.5-second range. The launch could also be improved with a steeper first gear in the TH350 transmission. Overall, we improved the performance of the Camaro while cleaning up the emissions, making it a completely emissions-legal car! What more could a hot rodder want? **HR**

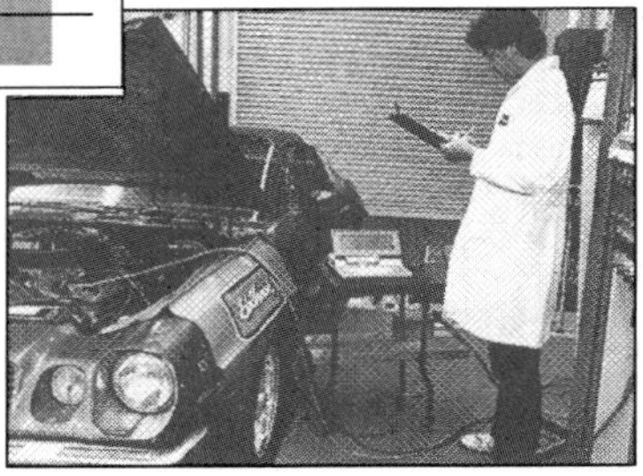

With the engine in its final form, it was time to try the emissions sniffer again. The folks at Edelbrock let us roll our Camaro back onto their emissions dyno to determine if it was legal. The results were incredibly clean for a mid-'70s car, registering 0.0 percent CO and only 83 ppm HC at idle.

SOURCES

Akzo Coatings, Inc.
3295 River Exchange Dr., Ste. 500
Norcross, GA 30092
(404) 662-8464

Classic Camaro Parts & Accessories
17832 Gothard St.
Huntington Beach, CA 92647
(714) 894-0651

Cragar Industries
2401 W. First St.
Tempe, AZ 85281
(602) 894-8300

Crane Cams
530 Fentress Blvd.
Daytona Beach, FL 32014
(904) 252-1151

Edelbrock Corporation
2700 California St.
Torrance, CA 90503
(310) 781-2222

Jacobs Electronics
500 N. Baird St.
Midland, TX 79701
(800) 375-5226

K&N Engineering
P.O. Box 1329
Riverside, CA 92502
(714) 684-9762

Performance Specialties
397 W. Baseline
San Bernardino, CA 92405
(714) 889-2329

Santini's Paint & Body Werkes
7416 Westminster Blvd.
Westminster, CA 92683
(714) 891-8895

SplitFire, Inc.
4065 Commercial Ave.
Northbrook, IL 60062
(708) 559-2000

Mickey Thompson Performance Tires
P.O. Box 227
Cuyahoga Falls, OH 44222
(800) 222-9092

Doug Thorley Headers
1561 Commerce St.
Corona, CA 91720
(714) 735-7280

Walker Manufacturing
1201 Michigan Blvd.
Racine, WI 53402
(800) 767-DYNO

Western Collision
811 N. Western Ave.
Hollywood, CA 90029
(213) 465-5191

CAMARO BUYER'S GUIDE

What To Look For When Buying The Camaro Of Your Dreams

By Bruce Caldwell

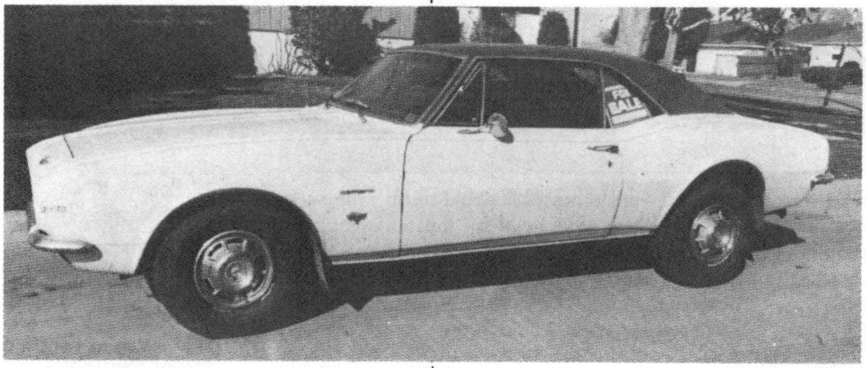

1. Every Camaro buyer's dream—an original, one-owner Rally Sport parked in a quiet neighborhood with a for sale sign—offered for under $1000. These deals are getting rare, but they are still out there.

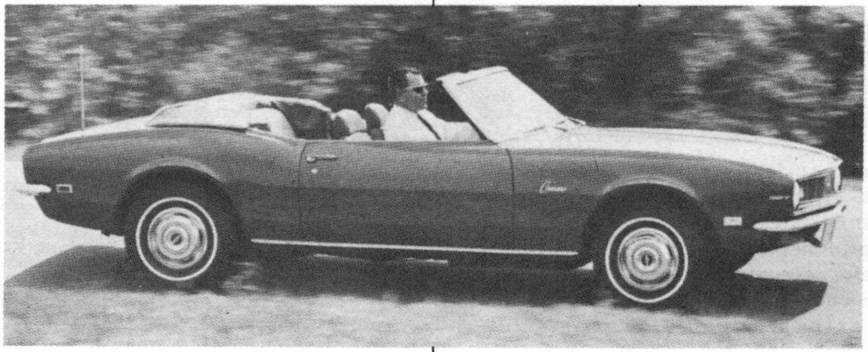

2. Camaro convertibles are excellent collector's cars. You can still drive one every day while its value continues to rise. There are several more desirable convertibles than this '68 327-powered one, but they are all worth saving.

The Chevrolet Camaro is one of those special cars that has maintained a tremendous popularity ever since the day it was formally introduced to the buying public. In the 15 years of its production, the Camaro has sold very well both in terms of new car and used car sales. The only exception was during the early Seventies when there was corporate talk of dropping all F-body cars (Camaro and Firebird) because of a long strike, rising gas costs and increased insurance premiums. By the end of the decade, sales quickly rebounded in spite of the coming small car boom, showing that there were hundreds of thousands of car buyers still seeking style, comfort and performance for their driving needs.

You might think that with about two million Camaros sold by General Motors that the used Camaro market would be flooded with cheap Camaros. Unfortunately, that is not the case. Demand always seems to exceed supply (especially in certain special models) so prices remain relatively high. Many years and models of Camaros have long ago reached the bottom of their price curve and have started going back up in price. In other cases, the price structure is closely related to the condition of the cars. Premium condition Camaros have always brought premium prices.

Camaros are not like vintage Corvettes or T-Birds where even old hulks are super valuable. There are still plenty of good deals left in the Camaro market. Even the cars that have gone up in price can be great deals because their prices are almost certain to continue rising. The key to finding a good deal is to know exactly what to look for.

INVESTMENT OR TRANSPORTATION?

The Camaro is such a new car that most of them are still being used for everyday transportation. However, there are a growing number of Camaros that are now considered as investments. The investment Camaros are the rare or unique ones like certain Z-28s, many of the convertibles, extremely low-mileage cars and certain other high-performance models. As prices continue to rise, the extremely collectible Camaros are being treated like other valuable collector's cars, which means they are seldom driven. That is a shame because the Camaro is such an enjoyable car to drive.

If you are looking for an investment Camaro, your best bet is to find the rarest, most original and lowest-mileage Camaro you can find. Rarity, originality and newness are all factors that are vital to a good automotive investment. To determine rarity, check production figures. Certain option combinations might be very rare but unless the options are super-desirable (like four-wheel disc brakes or a cross-ram induction system) options alone don't make a particular car desirable, although they can enhance the value of a car somewhat.

Any convertible is currently collectible and that goes for Camaros, too. Since Chevrolet only made factory convertibles (the late-model aftermarket conversions are a special case by themselves) for the years 1967, 1968 and 1969, any Camaro convertible is worth preserving. Of the convertibles, the most desirable ones are the high-performance models and the Indianapolis pace car replicas. The best-known Indy pace car replicas are those made in 1969 when 3675 units were produced. Approximately 100 Indy pace car replicas were also made in 1967, making those cars extremely rare and valuable. Total convertible production for the three years breaks down this way:

1967	25,141
1968	20,440
1969	17,573

The other best bet for a collectible Camaro is an early-model Z-28. Only 602 Z-28s were made in 1967, making them highly sought after. A genuine '67 Z-28 with all its original equipment is worth buying at almost any price. Z-28s from the next three years are also worth collecting, but their value depends as much on condition as on the fact that it is a Z-28. Besides an original '67 Z-28, a top investment would be a fully optioned (meaning four-wheel disc brakes and cross-ram cowl induction) '69 Z-28. Of the later Z-28s, the only ones attracting much collector's attention are the premium '70½ Z-28s. These cars were the last ones with really strong engines. In 1971, the horsepower ratings started on their downward spiral, a trend that leaves current Z-28s powered by an anemic 305-cubic-inch engine.

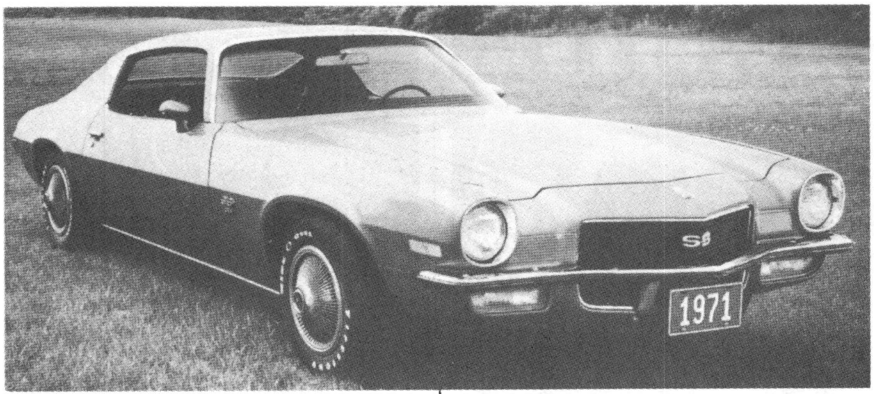

3. Big-block-equipped Camaros like this '71 SS 396 are becoming quite scarce. These power-packed rockets are prime collector's cars.

When considering a good transportation car, the Camaro is an excellent choice. By picking a Camaro in good condition you can enjoy comfortable, low-maintenance driving for years and will be likely to sell the car for more than you paid for it. The components of the Camaro are so familiar and common that getting parts and repairs is as easy as for any car ever made. Its small-block Chevy engine is the most versatile source of performance and/or economy ever produced by Detroit.

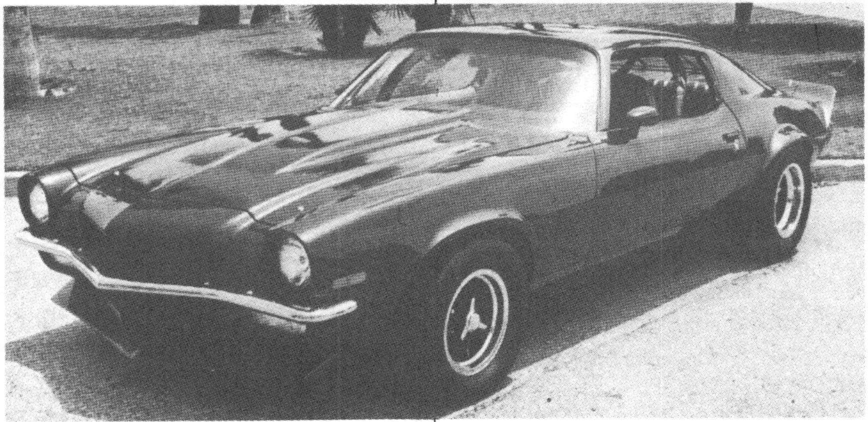

4. Camaros make great street machines. Modified Camaros are very popular and there is a good market for them. If you are buying a Camaro for just an investment, though, stick with stock models.

The key to finding the best Camaro transportation value is to either find one in mint shape or one needing repairs that are financially lucrative. If by doing the repairs yourself you have invested less money than the car is worth, then it was a good deal. If the cost of the repairs plus the cost of the car is more than the cost of a better Camaro, save yourself the trouble and buy the better machine.

Camaros are very popular modified vehicles. It would be tough to find a better street machine or basis for a race car. If you are looking for an investment Camaro, though, modified examples are seldom a good choice. Buy a modified Camaro if they please you but remember that modified cars are tougher to sell than stock ones. The way many performance enthusiasts get around the modification problem is to keep the original equipment and only make bolt-on changes that can be easily removed.

DRIVETRAIN COMBINATIONS

Camaros have been offered with a variety of engines and transmissions over the years. A look at production statistics will quickly reveal one overwhelming favorite: the small-block V8 backed by an automatic transmission. The V8 engine has always outsold the 6-cylinder, sometimes by as much as 25 to 1. Manual transmissions have been more popular than 6-cylinder engines, but in recent years the automatic has begun to really out-distance manual transmission sales.

Whether buying a Camaro for an investment or transportation, the best engine choice is the small-block V8. Unless you have a particular fondness for either the 230-cubic-inch or 250-cubic-inch six, stay away from cars with those engines. The V8-equipped models are much easier to resell. Depending on the year of Camaro selected, the best engine is either a 327 or 350 with a small 4-barrel carburetor. Such an engine will give snappy performance by current standards and still return respectable gas mileage on regular fuel. The higher horsepower models are great if you want a little more excitement in your life.

If you really want to inject a little adrenalin into your driving, consider a Camaro with either a big-block engine or one of the high-performance small-blocks. The big-blocks are relatively rare since they were only offered through 1972. An early Camaro with its original big-block engine is a good investment. Probably the best investment, engine-wise, is either a 302 '69 Z-28 engine or a '70½ 350 as found in the Z-28s. These high-compression engines are real screamers but can be difficult to run on the garden variety junk gasoline found at most service stations today.

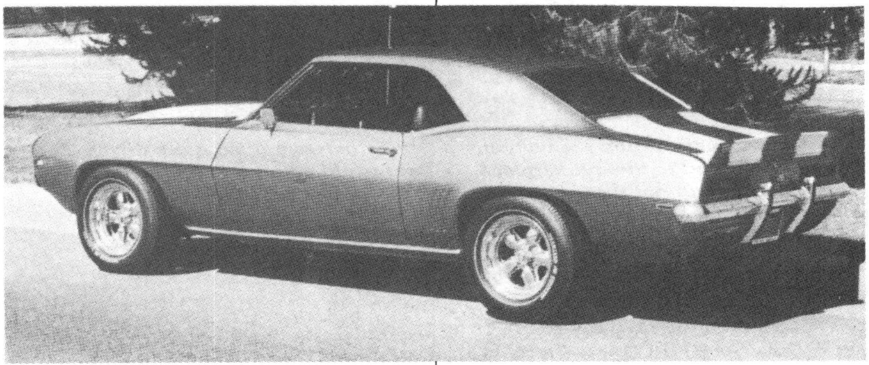

5. Early Z-28's like this pristine '69 are among the most desirable Camaros. This example features aftermarket wheels and tires, but it could easily be returned to 100% stock.

When it comes to transmissions, the 4-speed is the most popular from a collector's and enthusiast's standpoint. For everyday driving either the Turbo 350 or the Powerglide are probably better choices than a manual transmission. The M-22 4-speed is preferred over the M-21 model, but you are most likely to encounter the M-21 Muncie, which is a fine transmission in its own right.

DESIRABLE OPTIONS

Serious collectors are always concerned about options or in some rare cases (like ultra high-performance Corvettes) the lack of options. A fully equipped car is more desirable and valuable than a stripped one, unless the stripped one came with a killer engine. Even so, the well-optioned model is usually preferred over the plain-Jane version.

BUYER'S GUIDE

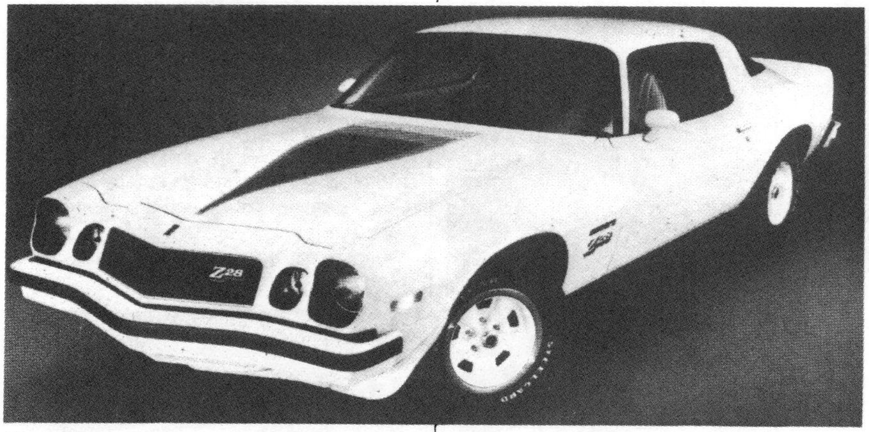

6. After two years of non-production, the Z-28 returned to the marketplace in 1977. Although not the high-performance terror of the '60s and early '70s, the '77 Z-28 is nonetheless a great daily driver. The car is old enough to be reasonably priced, but not so old that it is a collector's item.

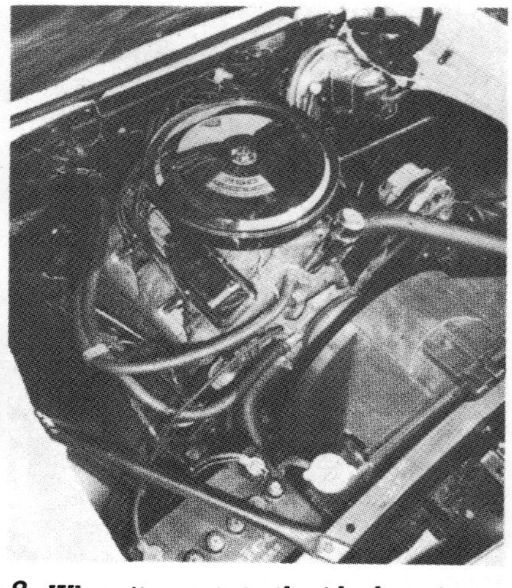

8. When it comes to the ideal engine for a Camaro, it would be tough to beat the 350-cubic-inch small-block equipped with a small 4-barrel carburetor. The engine is tops for performance, economy and reliability—a proven winner.

When considering options as a basis for purchasing a Camaro the best options to look for are ones such as the Z-28 option, the RS (Rally Sport) package or the SS (Super Sport) package. High-performance engines are desirable options. The legendary and super-rare four-wheel disc brakes offered in late '68 and '69 are highly sought options. Other performance items like the two 4-barrel induction system and cowl induction hoods are great finds.

Interior options that are rare include a front bench seat, fold-down rear seat, power windows, factory stereo systems and tilt steering wheels. Air conditioning is considered a rare option on a convertible.

Other worthwhile options include body-color Endura front bumpers and the various special wheels. The Endura bumper coating was found on '69 Camaros and as a part of the grille on '70½-'73 Rally Sport models. The wheels that are worth seeking are the slotted Rally wheels found on early cars and the Z-28 five-spoke mag-style wheels found on later models. The later models also could be ordered with the six-slot ZJ-7 Rally wheel but this version isn't as popular as the earlier five-slot model. The '67 Rally wheel with the flat style hub cap as found on disc brake-equipped cars is the most difficult wheel to find.

CAMARO PROBLEM AREAS

Overall, Camaros are sturdy, well-built cars, but they are not without their faults. Knowing how to spot these flaws can save you considerable money and agony. As with any used car, it pays to give the prospective purchase as thorough an inspection as possible. It never hurts to get an unbiased outsider's opinion either. The old advice about taking the car to a qualified mechanic for his inspection is still a good idea. If you don't want to go to a mechanic, take along a knowledgeable friend who isn't as emotionally involved with the car as you.

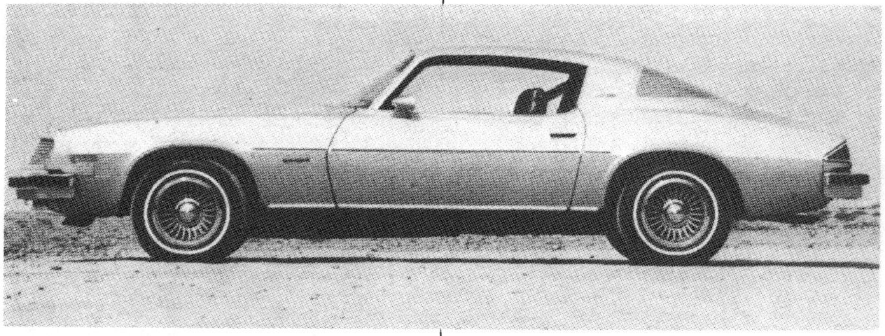

7. For transportation value, the mid-'70s Camaros like this '75 are great buys. The cars are nearing the bottom of their price curve while still being new enough that major repairs aren't needed.

Rust is a problem with any car especially those from states that use salt on the roads during the winter. Camaros have a few special rust areas to check. Most trunks seem to be poorly sealed. Rust is quite common in trunks, so check under the trunk mat. Water can also leak into the area under the rear window; this is especially true of cars with vinyl roofs. A Camaro with a blistered vinyl roof is one to avoid. The blisters mean there is bad rust underneath the vinyl top. Any vinyl top equipped Camaro is one to be careful with. In the '70 and later models, water also can leak around the windshield posts.

On the early model Camaros, check the condition of the headlight doors on Rally Sport models. These headlight doors often work poorly or are stuck in the open position. Repairs and parts are expensive.

On the newer Camaros there are a couple of problems associated with the doors. The doors are wide (to allow for rear seat access) and heavy, thanks to side impact beams. The weight makes sagging hinges a distinct possibility. The large expanse of curved side glass often rattles and gets loose in the window channel tracks. The windshields of the newer models are more expensive to replace because of the hidden antennas imbedded in them. Consider the extra cost if you buy a Camaro with a damaged windshield.

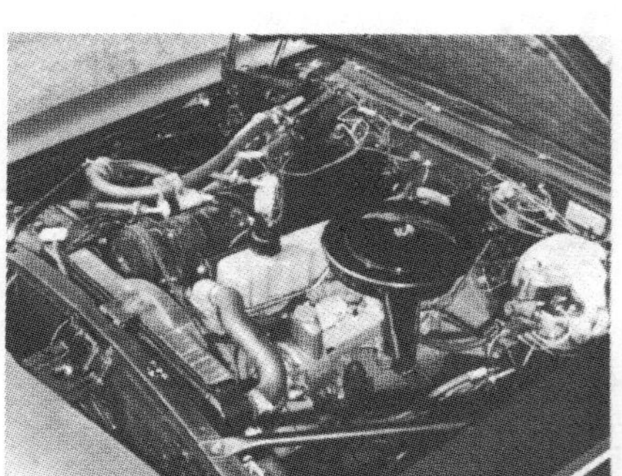

9. Six-cylinder engines are definitely not the popular setup for Camaros. It's not that it is a poor engine, it's just that the V8s are considered the definitive Camaro engine.

Another late-model problem is warped dash pads. This problem is very common. Although replacement dash pads are priced in the $100 range, installing a new

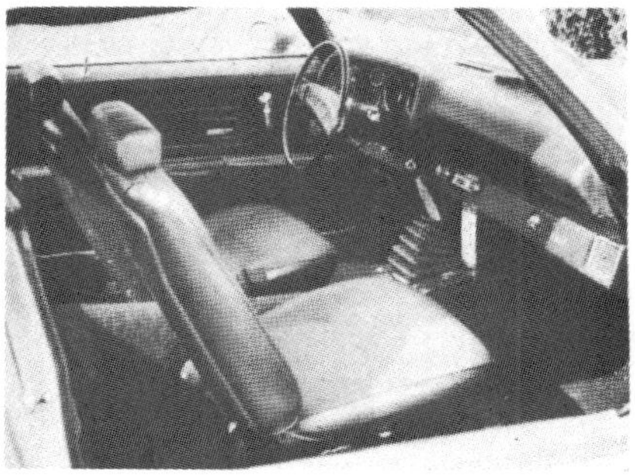

10. The low-back seats with the separate headrests are generally considered superior seats to the high-back seats. The low-back seats were only found on the '70½ Camaros.

pad is a pain. The large hoods of the '70 and newer Camaros are heavy items which makes loose hood bolts an unfortunate possibility.

The Rally Sport option found on the '70½ through '73 Camaros is a desirable option, but damaged cars with this option are more expensive to repair than standard bumper models. The Endura surround for the grille is prone to dings as well as cracking.

On big-block-engine-equipped Camaros, check the condition of the front suspension. The extra weight of the bigger engine can cause the springs to wear faster than on a small-block equipped car. The big-block and high-performance Camaros are also more often subjected to rearend and transmission abuse.

WHERE TO FIND USED CAMAROS

Used Camaros are found just about everywhere, but the more desirable models are getting harder and harder to find. There are always the lucky few Camaro enthusiasts who run across an original owner cream puff for a bargain price, but with the increasing number of Camaro enthusiasts out there looking, those dream deals are pretty rare. The best cars and the most collectible ones are already in the hands of knowledgable people. When those cars

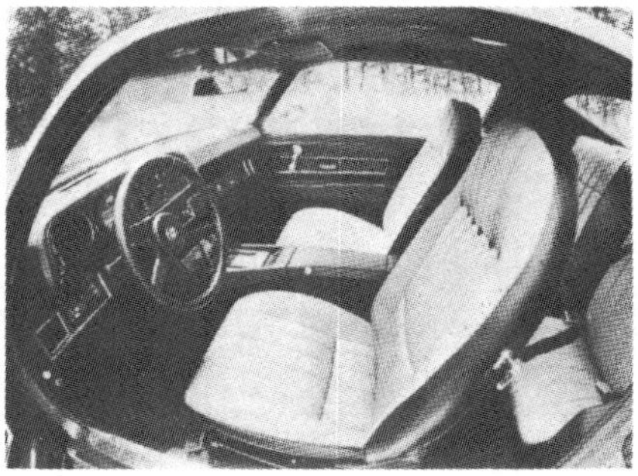

11. In 1971 a modified Vega seat was introduced. The deluxe steering wheel and console shown on this '71 Camaro are desirable options.

are offered for sale it is usually in specialty publications like Hemmings Motor News, Old Cars and the ads in HOT ROD Magazine. These cars also show up at old swap meets and in the showrooms of specialty used car dealers. The rare models appear less frequently in the "Classic Cars" section of local newspapers. When such cars appear locally, that is your best chance of getting the car without as many competing buyers as you would encounter with a nationally advertised car.

Everyday Camaros are mostly found in the local classified sections but good deals can also be found in the smaller papers like the "throwaways" found in neighborhood stores or delivered to your door. Many cities have bargain rate classified-only weekly papers. These papers are good sources of transporation Camaros. There are also many cities that have a weekly publication that features photos of every vehicle offered for sale. Lesser sources are bulletin boards in auto parts stores or grocery stores, cars parked with for sale signs at gas stations and cars offered for sale on vacant lots.

Wherever you find your used Camaro or whatever model appeals to you, it is hard to go wrong with a good used Camaro. Camaros are fun to drive, easy to maintain and they have a lasting appeal. As an investment, as transportation, or both, the Camaro is tough to beat.

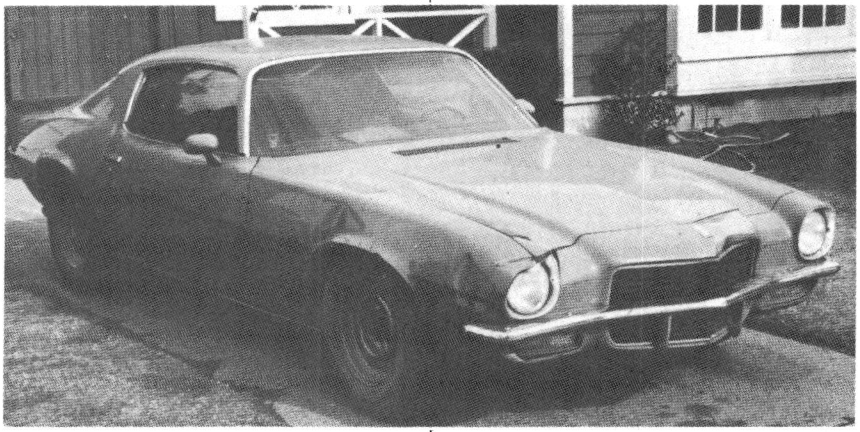

12. Damaged Camaros can be a good deal but only if the cost of the car plus the repairs is less than a non-damaged version of the same car. Rare cars like this '70 SS 396 are definitely worth fixing.

BROOKLANDS *ROAD TEST* BOOKS

MOTORING

BROOKLANDS ROAD TEST SERIES

Abarth Gold Portfolio 1950-1971
AC Ace & Aceca 1953-1983
Alfa Romeo Giulietta Gold Portfolio 1954-1965
Alfa Romeo Giulia Coupés 1963-1976
Alfa Romeo Giulia Coupés Gold Port. 1963-1976
Alfa Romeo Spider 1966-1990
Alfa Romeo Spider Gold Portfolio 1966-1991
Alfa Romeo Alfasud 1972-1984
Alfa Romeo Alfetta Gold Portfolio 1972-1987
Alfa Romeo Alfetta GTV6 1980-1986
Allard Gold Portfolio 1937-1959
Alvis Gold Portfolio 1919-1967
AMX & Javelin Muscle Portfolio 1968-1974
Armstrong Siddeley Gold Portfolio 1945-1960
Aston Martin Gold Portfolio 1948-1971
Aston Martin Gold Portfolio 1972-1985
Aston Martin Gold Portfolio 1985-1995
Audi Quattro Gold Portfolio 1980-1991
Austin A30 & A35 1951-1962
Austin Healey 100 & 100/6 Gold Port. 1952-1959
Austin Healey 3000 Gold Portfolio 1959-1967
Austin Healey Sprite Gold Portfolio 1958-1971
Barracuda Muscle Portfolio 1964-1974
BMW 1600 Collection No.1 1966-1981
BMW 2002 Gold Portfolio 1968-1976
BMW 316, 318, 320 (4 cyl.) Gold Port. 1975-1990
BMW 320, 323, 325 (6 cyl.) Gold Port. 1977-1990
BMW M Series Performance Portfolio 1976-1993
BMW 5 Series Gold Portfolio 1981-1987
BMW 6 Series Gold Portfolio 1976-1989
Bricklin Gold Portfolio 1974-1975
Bristol Cars Gold Portfolio 1946-1992
Buick Automobiles 1947-1960
Buick Muscle Cars 1965-1970
Cadillac Allanté 1986-1993
Cadillac Automobiles 1949-1959
Cadillac Automobiles 1960-1969
Caprice 1965-1976 ☆ Limited Edition
Charger Muscle Portfolio 1966-1974
Checker ☆ Limited Edition
Chevrolet 1955-1957
Impala & SS Muscle Portfolio 1958-1972
Chevrolet Corvair 1959-1969
Chevy II & Nova SS Muscle Portfolio 1962-1974
Chevy El Camino & SS 1959-1987
Chevelle & SS Muscle Portfolio 1964-1972
Chevrolet Muscle Cars 1966-1971
Chevy Blazer 1969-1981
Chevrolet Corvette Gold Portfolio 1953-1962
Chevrolet Corvette Sting Ray Gold Port. 1963-1967
Chevrolet Corvette Gold Portfolio 1968-1977
High Performance Corvettes 1983-1989
Camaro Muscle Portfolio 1967-1973
Chevrolet Camaro & Z28 1973-1981
High Performance Camaros 1982-1988
Chrysler 300 Gold Portfolio 1955-1970
Chrysler Valiant 1960-1962
Citroen Traction Avant Gold Portfolio 1934-1957
Citroen 2CV Gold Portfolio 1948-1989
Citroen DS & ID 1955-1975
Citroen DS & ID Gold Portfolio 1955-1975
Citroen SM 1970-1975
Cobras & Replicas 1962-1983
Shelby Cobra Gold Portfolio 1962-1969
Cobras & Cobra Replicas Gold Portfolio 1962-1989
Cunningham Automobiles 1951-1955
Daimler SP250 Sports & V-8 250 Saloon Gold P. 1959-1969
Datsun Roadsters 1962-1971
Datsun 240Z 1970-1973
Datsun 280Z & ZX 1975-1983
DeLorean Gold Portfolio 1977-1995
Dodge Muscle Cars 1967-1970
Dodge Viper on the Road
Edsel 1957-1960 ☆ Limited Edition
ERA Gold Portfolio 1934-1994
Excalibur Collection No.1 1952-1981
Facel Vega 1954-1964
Ferrari 1947-1957 ☆ Limited Edition
Ferrari 1958-1963 ☆ Limited Edition
Ferrari Dino 1965-1974
Ferrari Dino 308 & Mondial Gold Portfolio1974-1985
Ferrari 328 • 348 • Mondial Gold Portfolio 1986-1994
Fiat 500 Gold Portfolio 1936-1972
Fiat 600 & 850 Gold Portfolio 1955-1972
Fiat Pininfarina 124 & 2000 Spider 1968-1985
Fiat X1/9 Gold Portfolio1973-1989
Fiat Abarth Performance Portfolio 1972-1987
Ford Consul, Zephyr, Zodiac Mk.I & II 1950-1962
Ford Zephyr, Zodiac, Executive, Mk.III & Mk.IV 1962-1971
Ford Cortina 1600E & GT 1967-1970
High Performance Capris Gold Portfolio 1969-1987
Capri Muscle Portfolio 1974-1987
High Performance Fiestas 1979-1991
High Performance Escorts Mk.I 1968-1974
High Performance Escorts Mk.II 1975-1980
High Performance Escorts 1980-1985
High Performance Escorts 1985-1990
High Performance Sierras & Merkurs
Gold Portfolio 1983-1990
Ford Automobiles 1949-1959
Ford Fairlane 1955-1970
Ford Ranchero 1957-1959
Ford Thunderbird 1955-1957
Ford Thunderbird 1958-1963
Ford GT40 Gold Portfolio 1964-1987
Ford Bronco 1966-1977
Ford Bronco 1978-1988
Goggomobil ☆ Limited Edition
Holden 1948-1962
Honda CRX 1983-1987
Imperial 1955-1970 ☆ Limited Edition
International Scout Gold Portfolio 1961-1980
Isetta Gold Portfolio 1953-1964
Iso & Bizzarrini Gold Portfolio 1962-1974
Kaiser • Frazer 1946-1955 ☆ Limited Edition
Jaguar and SS Gold Portfolio 1931-1951
Jaguar XK120, 140, 150 Gold Port. 1948-1960
Jaguar Mk.VII, VIII, IX, X, 420 Gold Port. 1950-1970
Jaguar Mk.1 & Mk.2 Gold Portfolio 1959-1969
Jaguar E-Type Gold Portfolio 1961-1971
Jaguar E-Type V-12 1971-1975
Jaguar S-Type & 420 ☆ Limited Edition
Jaguar XJ12, XJ5.3, V12 Gold Portfolio 1972-1990
Jaguar XJ6 Series I & II Gold Portfolio 1968-1979
Jaguar XJ6 Series III Perf. Portfolio 1979-1986
Jaguar XJ6 Gold Portfolio 1986-1994
Jaguar XJS Gold Portfolio 1975-1988
Jaguar XJS Gold Portfolio 1988-1995
Jeep CJ5 & CJ6 1960-1976
Jeep CJ5 & CJ7 1976-1986
Jensen Interceptor Gold Portfolio 1966-1986
Jensen Healey 1972-1976
Lagonda Gold Portfolio 1919-1964
Lancia Aurelia & Flaminia Gold Portfolio 1950-1970
Lancia Fulvia Gold Portfolio 1963-1976
Lancia Beta Gold Portfolio 1972-1984
Lancia Delta Gold Portfolio 1979-1994
Lancia Stratos 1972-1985
Land Rover Series I 1948-1958
Land Rover Series II & IIa 1958-1971
Land Rover Series III 1971-1985
Land Rover 90 110 Defender Gold Portfolio 1983-1994
Land Rover Discovery 1989-1994
Land Rover Story Part One 1948-1971
Lincoln Gold Portfolio 1949-1960
Lincoln Continental 1961-1969
Lincoln Continental 1969-1976
Lotus Sports Racers Gold Portfolio 1953-1965
Lotus Seven Gold Portfolio 1957-1974
Lotus Caterham Seven Gold Portfolio 1974-1995
Lotus Elan Gold Portfolio 1962-1974
Lotus Elan Collection No. 2 1963-1972
Lotus Elan & SE 1989-1992
Lotus Europa Gold Portfolio 1966-1975
Lotus Elite & Eclat 1974-1982
Lotus Turbo Esprit 1980-1986
Maserati 1965-1970
Matra 1965-1983 ☆ Limited Edition
Mazda Miata MX-5 Performance Portfolio 1989-1996
Mazda RX-7 Gold Portfolio 1978-1991
Mercedes 190 & 300 SL 1954-1963
Mercedes G Wagen 1981-1994
Mercedes S & 600 1965-1972
Mercedes S Class 1972-1979
Mercedes 230 • 250 • 280SL Gold Portfolio 1963-1971
Mercedes SLs & SLCs Gold Portfolio 1971-1989
Mercedes SLs Performance Portfolio 1989-1994
Mercury Muscle Cars 1966-1971
Messerschmitt Gold Portfolio 1954-1964
MG Gold Portfolio 1929-1939
MG TA & TC Gold Portfolio 1936-1949
MG TD & TF Gold Portfolio 1949-1955
MGA & Twin Cam Gold Portfolio 1955-1962
MG Midget Gold Portfolio 1961-1979
MGB Roadsters 1962-1980
MGB MGC & V8 Gold Portfolio 1962-1980
MGB GT 1965-1980
MGC & MGB GT V8 ☆ Limited Edition
MG Y-Type & Magnette ZA/ZB ☆ Limited Edition
Mini Gold Portfolio 1959-1969
Mini Gold Portfolio 1969-1980
High Performance Minis Gold Portfolio 1960-1973
Mini Cooper Gold Portfolio 1961-1971
Mini Moke Gold Portfolio 1964-1994
Morgan Three-Wheeler Gold Portfolio 1910-1952
Morgan Plus 4 & Four 4 Gold Portfolio 1936-1967
Morgan Cars 1960-1970
Morgan Cars Gold Portfolio 1968-1989
Morris Minor Collection No. 1 1948-1980
Shelby Mustang Muscle Portfolio 1965-1970
High Performance Mustang IIs 1974-1978
High Performance Mustangs 1982-1988
Nash & Nash-Healey 1949-1957 ☆ Limited Edition
Nash-Austin Metropolitan Gold Portfolio 1954-1962
Oldsmobile Automobiles 1955-1963
Oldsmobile Toronado 1966-1978
Opel GT Gold Portfolio 1968-1973
Opel Manta 1970-1975 ☆ Limited Edition
Packard Gold Portfolio 1946-1958
Pantera Gold Portfolio 1970-1989
Panther Gold Portfolio 1972-1990
Pontiac Tempest & GTO 1961-1965
Firebird & Trans-Am Muscle Portfolio 1973-1981
High Performance Firebirds 1982-1988
Pontiac Fiero 1984-1988
Porsche 356 Gold Portfolio 1953-1965
Porsche 911 1965-1969
Porsche 911 1970-1972
Porsche 911 1973-1977
Porsche 911 SC & Turbo Gold Portfolio 1978-1983
Porsche 911 Carrera & Turbo Gold Port. 1984-1989
Porsche 924 Gold Portfolio 1975-1988
Porsche 928 Performance Portfolio 1977-1994
Porsche 944 Gold Portfolio 1981-1991
Range Rover Gold Portfolio 1970-1985
Range Rover Gold Portfolio 1986-1995
Reliant Scimitar 1964-1986
Renault Alpine Gold Portfolio 1958-1994
Riley Gold Portfolio 1924-1939
Rolls Royce Silver Cloud & Bentley 'S' Series
Gold Portfolio 1955-1965
Rolls Royce Silver Shadow Gold Port. 1965-1980
Rolls Royce & Bentley Gold Port. 1980-1989
Rover P4 1949-1959
Rover P4 1955-1964
Rover 3 & 3.5 Litre Gold Portfolio 1958-1973
Rover 2000 & 2200 1963-1977
Rover 3500 & Vitesse 1976-1986
Saab Sonett Collection No.1 1966-1974
Saab Turbo 1976-1983
Studebaker Gold Portfolio 1947-1966
Studebaker Hawks & Larks 1956-1963
Avanti 1962-1990
Sunbeam Tiger & Alpine Gold Portfolio 1959-1967
Triumph Dolomite Sprint ☆ Limited Edition
Triumph TR2 & TR3 Gold Portfolio 1952-1961
Triumph TR4, TR5, TR250 1961-1968
Triumph TR6 Gold Portfolio 1969-1976
Triumph TR7 & TR8 Gold Portfolio 1975-1982
Triumph Herald 1959-1971
Triumph Vitesse 1962-1971
Triumph Spitfire Gold Portfolio 1962-1980
Triumph 2000, 2.5, 2500 1963-1977
Triumph GT6 Gold Portfolio 1966-1974
Triumph Stag Gold Portfolio 1970-1977
TVR Gold Portfolio 1959-1986
TVR Performance Portfolio 1986-1994
VW Beetle Gold Portfolio 1935-1967
VW Beetle Gold Portfolio 1968-1991
VW Beetle Collection No.1 1970-1982
VW Karmann Ghia 1955-1982
VW Bus, Camper, Van 1954-1967
VW Bus, Camper, Van 1968-1979
VW Bus, Camper, Van 1979-1989
VW Scirocco 1974-1981
VW Golf GTI 1976-1986
Volvo PV444 & PV544 1945-1965
Volvo Amazon-120 Gold Portfolio 1956-1970
Volvo 1800 Gold Portfolio 1960-1973
Volvo 140 & 160 Series Gold Portfolio 1966-1975
Westfield ☆ Limited Edition

Forty Years of Selling Volvo

BROOKLANDS ROAD & TRACK SERIES

Road & Track on Alfa Romeo 1964-1970
Road & Track on Alfa Romeo 1971-1976
Road & Track on Aston Martin 1962-1990
R & T on Auburn Cord and Duesenburg 1952-84
Road & Track on Audi & Auto Union 1952-1980
Road & Track on Audi & Auto Union 1980-1986
Road & Track on Austin Healey 1953-1970
Road & Track on BMW Cars 1966-1974
Road & Track on BMW Cars 1975-1978
Road & Track on BMW Cars 1979-1983
R & T on Cobra, Shelby & Ford GT40 1962-1992
Road & Track on Corvette 1953-1967
Road & Track on Corvette 1968-1982
Road & Track on Corvette 1982-1986
Road & Track on Corvette 1986-1990
Road & Track on Ferrari 1975-1981
Road & Track on Ferrari 1981-1984
Road & Track on Ferrari 1984-1988
Road & Track on Fiat Sports Cars 1968-1987
Road & Track on Jaguar 1950-1960
Road & Track on Jaguar 1961-1968
Road & Track on Jaguar 1968-1974
Road & Track on Jaguar 1974-1982
Road & Track on Jaguar 1983-1989
Road & Track on Lamborghini 1964-1985
Road & Track on Lotus 1972-1981
Road & Track on Maserati 1975-1983
R & T on Mazda RX-7 & MX-5 Miata 1986-1991
Road & Track on Mercedes 1952-1962
Road & Track on Mercedes 1963-1970
Road & Track on Mercedes 1971-1979
Road & Track on Mercedes 1980-1987
Road & Track on MG Sports Cars 1949-1961
Road & Track on MG Sports Cars 1962-1980
Road & Track on Mustang 1964-1977
R & T on Nissan 300-ZX & Turbo 1984-1989
Road & Track on Pontiac 1960-1983
Road & Track on Porsche 1951-1967
Road & Track on Porsche 1968-1971
Road & Track on Porsche 1972-1975
Road & Track on Porsche 1975-1978
Road & Track on Porsche 1985-1988
R & T on Rolls Royce & Bentley 1950-1965
R & T on Rolls Royce & Bentley 1966-1984
Road & Track on Saab 1972-1992
R & T on Toyota Sports & GT Cars 1966-1984
R & T on Triumph Sports Cars 1953-1967
R & T on Triumph Sports Cars 1967-1974
R & T on Triumph Sports Cars 1974-1982
Road & Track on Volkswagen 1951-1968
Road & Track on Volkswagen 1968-1978
Road & Track on Volkswagen 1978-1985
Road & Track on Volvo 1957-1974
Road & Track on Volvo 1977-1994
R & T - Henry Manney at Large & Abroad
R & T - Peter Egan's "Side Glances"
R & T - Peter Egan "At Large"

BROOKLANDS CAR AND DRIVER SERIES

Car and Driver on BMW 1955-1977
Car and Driver on BMW 1977-1985
C and D on Cobra, Shelby & Ford GT40 1963-84
Car and Driver on Corvette 1978-1982
Car and Driver on Corvette 1983-1988
C and D on Datsun Z 1600 & 2000 1966-1984
Car and Driver on Ferrari 1955-1962
Car and Driver on Ferrari 1963-1975
Car and Driver on Ferrari 1976-1983
Car and Driver on Mopar 1956-1967
Car and Driver on Mopar 1968-1975
Car and Driver on Mustang 1964-1972
Car and Driver on Pontiac 1961-1975
Car and Driver on Porsche 1955-1962
Car and Driver on Porsche 1963-1970
Car and Driver on Porsche 1970-1976
Car and Driver on Porsche 1977-1981
Car and Driver on Porsche 1982-1986
Car and Driver on Saab 1956-1985
Car and Driver on Volvo 1955-1986

BROOKLANDS PRACTICAL CLASSICS SERIES

PC on Austin A40 Restoration
PC on Land Rover Restoration
PC on Metalworking in Restoration
PC on Midget/Sprite Restoration
PC on MGB Restoration
PC on Sunbeam Rapier Restoration
PC on Triumph Herald/Vitesse
PC on Spitfire Restoration
PC on 1930s Car Restoration

BROOKLANDS HOT ROD 'MUSCLECAR & HI-PO ENGINES' SERIES

Chevy 265 & 283
Chevy 302 & 327
Chevy 348 & 409
Chevy 350 & 400
Chevy 396 & 427
Chevy 454 thru 512
Chrysler Hemi
Chrysler 273, 318, 340 & 360
Chrysler 361, 383, 400, 413, 426, 440
Ford 289, 302, Boss 302 & 351W
Ford 351C & Boss 351
Ford Big Block

BROOKLANDS RESTORATION SERIES

Auto Restoration Tips & Techniques
Basic Bodywork Tips & Techniques
Camaro Restoration Tips & Techniques
Chevrolet High Performance Tips & Techniques
Chevy Engine Swapping Tips & Techniques
Chevy-GMC Pickup Repair
Chrysler Engine Swapping Tips & Techniques
Engine Swapping Tips & Techniques
Ford Pickup Repair
Land Rover Restoration Tips & Techniques
MG 'T' Series Restoration Guide
MGA Restoration Guide
Mustang Restoration Tips & Techniques

MOTORCYCLING

BROOKLANDS ROAD TEST SERIES

AJS & Matchless Gold Portfolio 1945-1966
BSA Twins A7 & A10 Gold Portfolio 1946-1962
BSA Twins A50 & A65 Gold Portfolio 1962-1973
BMW Motorcycles Gold Portfolio 1950-1971
BMW Motorcycles Gold Portfolio 1971-1976
Ducati Gold Portfolio 1960-1974
Ducati Gold Portfolio 1974-1978
Ducati Gold Portfolio 1978-1982
Laverda Gold Portfolio 1967-1977
Moto Guzzi Gold Portfolio 1949-1973
Norton Commando Gold Portfolio 1968-1977
Triumph Bonneville Gold Portfolio 1959-1983

BROOKLANDS CYCLE WORLD SERIES

Cycle World on BMW 1974-1980
Cycle World on BMW 1981-1986
Cycle World on Ducati 1982-1991
Cycle World on Harley-Davidson 1962-1968
Cycle World on Harley-Davidson 1978-1983
Cycle World on Harley-Davidson 1983-1987
Cycle World on Harley-Davidson 1987-1990
Cycle World on Harley-Davidson 1990-1992
Cycle World on Honda 1962-1967
Cycle World on Honda 1968-1971
Cycle World on Honda 1971-1974
Cycle World on Husqvarna 1966-1976
Cycle World on Husqvarna 1977-1984
Cycle World on Kawasaki 1966-1971
Cycle World on Kawasaki Off-Road Bikes 1972-1979
Cycle World on Kawasaki Street Bikes 1972-1976
Cycle World on Norton 1962-1971
Cycle World on Suzuki 1962-1970
Cycle World on Suzuki Off-Road Bikes 1971-1976
Cycle World on Suzuki Street Bikes 1971-1976
Cycle World on Triumph 1967-1972
Cycle World on Yamaha 1962-1969
Cycle World on Yamaha Off-Road Bikes 1970-1974
Cycle World on Yamaha Street Bikes 1970-1974

MILITARY

BROOKLANDS MILITARY VEHICLES SERIES

Allied Military Vehicles No.2 1941-1946
Complete WW2 Military Jeep Manual
Dodge Military Vehicles No.1 1940-1945
Hail To The Jeep
Military & Civilian Amphibians 1940-1990
Off Road Jeeps: Civ. & Mil. 1944-1971
US Military Vehicles 1941-1945
US Army Military Vehicles WW2-TM9-2800
VW Kubelwagen Military Portfolio 1940-1990
WW 2 Jeep Military Portfolio 1941-1945

RACING

Le Mans - The Jaguar Years - 1949-1957
Le Mans - The Ferrari Years - 1958-1965
Le Mans - The Ford & Matra Years - 1966-1974